T0324118

300 PROBLEMS IN SPECIAL AND GENERAL RELATIVITY

Einstein's theories of special relativity and general relativity form a core part of today's undergraduate (or master's-level) physics curriculum. This is a supplementary problem book or student's manual, consisting of 150 problems in each of special and general relativity i.e., in total 300 problems. The problems have been collected, developed, tested, and refined by the authors over the past two decades from homework and exams given at KTH Royal Institute of Technology, Stockholm, Sweden, starting in the late 1990s. They are a mixture of short-form and multipart extended problems, with hints provided where appropriate. Complete solutions are elaborated for every problem, in a different section of the book; some solutions include brief discussions on their physical or historical significance. The extensive and fully worked out solutions are the main feature of the book and have been revised several times by the authors. Designed as a companion text to complement a main relativity textbook, it does not assume access to any specific textbook. This is a helpful resource for advanced students, for self-study, as a source of problems for university teaching assistants, or as an inspiration for instructors and examiners constructing problems for their lectures, homework, or exams.

MATTIAS BLENNOW is Associate Professor in Theoretical Astroparticle Physics at KTH Royal Institute of Technology, Stockholm, Sweden. His research is mainly directed toward the physics of neutrinos and dark matter and physics beyond the Standard Model. He is the author of the textbook *Mathematical Methods for Physics and Engineering* (CRC Press, 2018). He has more than 15 years of experience in teaching and has taught special and general relativity both as a lecturer and as a teaching assistant.

TOMMY OHLSSON is Professor of Theoretical Physics at KTH Royal Institute of Technology, Stockholm, Sweden. He has also been a visiting professor at the University of Iceland, Reykjavik, Iceland, over several years. His main research field is theoretical particle physics, especially neutrino physics and physics beyond the Standard Model. He has written the textbook *Relativistic Quantum Physics: From Advanced Quantum Mechanics to Introductory Quantum Field Theory* (Cambridge University Press, 2011). He has more than 25 years of university-level teaching experience in relativity theory and physics in general.

300 PROBLEMS IN SPECIAL AND GENERAL RELATIVITY

With Complete Solutions

MATTIAS BLENNOW
KTH Royal Institute of Technology

TOMMY OHLSSON
KTH Royal Institute of Technology

CAMBRIDGE
UNIVERSITY PRESS

University Printing House, Cambridge CB2 8BS, United Kingdom

One Liberty Plaza, 20th Floor, New York, NY 10006, USA

477 Williamstown Road, Port Melbourne, VIC 3207, Australia

314–321, 3rd Floor, Plot 3, Splendor Forum, Jasola District Centre, New Delhi – 110025, India

103 Penang Road, #05–06/07, Visioncrest Commercial, Singapore 238467

Cambridge University Press is part of the University of Cambridge.

It furthers the University's mission by disseminating knowledge in the pursuit of education, learning, and research at the highest international levels of excellence.

www.cambridge.org
Information on this title: www.cambridge.org/9781316510674
DOI: 10.1017/9781009039345

First published 2022

A catalogue record for this publication is available from the British Library.

Library of Congress Cataloging-in-Publication Data
Names: Blennow, Mattias, 1980– author. | Ohlsson, Tommy, 1973– author.
Title: 300 problems in special and general relativity : with complete solutions / M. Blennow & T. Ohlsson.
Other titles: Three hundred problems in special and general relativity
Description: New York : Cambridge University Press, [2021] | Includes bibliographical references and index.
Identifiers: LCCN 2021017342 (print) | LCCN 2021017343 (ebook) | ISBN 9781316510674 (hardback) | ISBN 9781009017732 (paperback) | ISBN 9781009039345 (epub)
Subjects: LCSH: Special relativity (Physics)–Problems, exercises, etc. | General relativity (Physics)–Problems, exercises, etc. | LCGFT: Problems and exercises.
Classification: LCC QC173.65 .B44 2021 (print) | LCC QC173.65 (ebook) | DDC 530.11–dc23
LC record available at https://lccn.loc.gov/2021017342
LC ebook record available at https://lccn.loc.gov/2021017343

ISBN 978-1-316-51067-4 Hardback
ISBN 978-1-009-01773-2 Paperback

To our wives, Ana and Linda

Contents

Preface

This book is a supplementary book in the form of a "problem book" or "student's manual" in special and general relativity consisting of a total of 300 problems (150 problems each in special and general relativity) with complete and elaborate solutions. It is intended as a companion text to a main textbook, but does not assume any particular textbook. It may be used for self-study act as a source of problems for classes, or as inspiration for teachers and examiners looking to construct new problems for lectures, homework, or exams.

The problems have been collected over the course of about two decades from homework and exams given at KTH Royal Institute of Technology, Stockholm, Sweden, starting in the late 1990s. The extensive and fully worked-out solutions are the main feature of the book and have been revised several times by the authors.

The book is divided into the following chapters:

"Notation, Concepts, and Conventions in Relativity Theory";

1. Problems in "Special Relativity Theory";
2. Problems in "General Relativity Theory"; and
3. "Solutions to Problems" in both special and general relativity,

where the first, unnumbered chapter introduces and sets the stage for both special and general relativity and is intended to be a brief review. The structure of the book is to first present the problems belonging to each main chapter (i.e., Chapters 1 and 2), which are further split into sections in order to obtain a better overview. The solutions are then presented in Chapter 3 (i.e., they do not follow immediately after the problem formulations). The main purpose of this is to suppress the urge for the reader to look at the solution to a problem before making a proper attempt. Some of the problems and solutions are illustrated by figures.

The target audience of the book is students and teachers of special and/or general relativity courses at the master's level that may benefit from it in the way described above. It will generally be too advanced for the relativity covered by the typical introductory modern physics courses at the bachelor's level, and most likely not advanced enough for an in-depth study at the PhD level.

Finally, we would like to acknowledge our colleagues Jouko Mickelsson, Håkan Snellman, Edwin Langmann, and Teresia Månsson, who have given important contributions to some of the problem statements included in this book. We would also like to thank our editor, Vince Higgs, at Cambridge University Press for a smooth and constructive process with the publication of this book, Torbjörn Bäck for supporting us in developing this book, and Marcus Pernow for proofreading earlier versions of the problem statements and solutions in special relativity. In addition, the KTH Royal Institute of Technology in Stockholm, Sweden, and the University of Iceland in Reykjavik, Iceland, are acknowledged for their hospitality and financial support.

Notation, Concepts, and Conventions in Relativity Theory

This chapter serves to briefly review the concepts relevant to the problems presented in this book. Its purpose is to remind the reader of the basic concepts as well as to introduce the notations and conventions that will be used. In particular, some notations and conventions will vary throughout the different textbooks available on the subject. Some of the different notations have been deliberately used in a number of problems in order to familiarize the reader with the fact that different notations occur in the literature.

General Notation

The components of a *vector* V will be written as V^μ in *contravariant* form and V_μ in *covariant* form with the index μ running over all the spacetime coordinates. When referring to 3-vectors, Latin letters will be used for the spatial indices rather than Greek ones, which we use for spacetime coordinates. In the case when an explicit basis for a given vector space is needed, we will use the *partial derivatives* ∂_μ to denote such a basis, i.e.,

$$V = V^\mu \partial_\mu. \tag{0.1}$$

Similarly, *tensor* components will be denoted with superscripts for contravariant indices and subscripts for covariant indices. Tensors with n indices, all down, are called *covariant tensors of rank n* and tensors with n indices, all up, are called *contravariant tensors of rank n*. Tensors with indices both up and down are so-called *mixed tensors*. Thus, a vector is a tensor of rank 1 (one index) and a scalar is a tensor of rank zero (no indices). The *Einstein summation convention* is used throughout the book, implying that indices that are repeated are to be summed over the relevant range. For example, in a four-dimensional spacetime, we have

$$V^\mu U_\mu \equiv \sum_{\mu=0}^{3} V^\mu U_\mu = V^0 U_0 + V^1 U_1 + V^2 U_2 + V^3 U_3, \tag{0.2}$$

where U and V are vectors. Contravariant and covariant components are related by lowering and raising with the metric tensor $g = (g_{\mu\nu})$ and its inverse $g^{-1} = (g^{\mu\nu})$, respectively, i.e.,

$$V_\mu = g_{\mu\nu} V^\nu \qquad \text{and} \qquad V^\mu = g^{\mu\nu} V_\nu. \tag{0.3}$$

Partial derivatives of a given function f may be denoted in several ways, e.g.,

$$\frac{\partial f}{\partial x^\mu} = \partial_\mu f = f_{,\mu}. \tag{0.4}$$

Several indices after the comma in the latter notation represent higher-order derivatives and the notation may also be used for vector components, for which indices belonging to the vector component are written before the comma and indices denoting derivatives after the comma, i.e.,

$$\partial_\mu \partial_\nu f = f_{,\mu\nu} \qquad \text{and} \qquad \partial_\mu V_\nu = V_{\nu,\mu}. \tag{0.5}$$

Objects with two indices may be represented in matrix form. We will indicate this by putting parentheses around the considered objects. For example, we can write the object A with two indices as

$$A = (A_{\mu\nu}) = \begin{pmatrix} A_{00} & A_{01} & A_{02} & A_{03} \\ A_{10} & A_{11} & A_{12} & A_{13} \\ A_{20} & A_{21} & A_{22} & A_{23} \\ A_{30} & A_{31} & A_{32} & A_{33} \end{pmatrix}. \tag{0.6}$$

By convention, the first index represents the row of the matrix and the second index represents the column. When this is used for one covariant index and one contravariant index, the contravariant index is taken as the row index and the covariant index as the column index.

For objects with more than two indices, we may use matrix notation to represent parts of such objects by inserting a bullet ('•') in place of the indices being considered. For example, the components $A^\mu_{1\nu}$ of the three-index object $A^\mu_{\lambda\nu}$ would be represented as the matrix

$$A^\bullet_{1\bullet} = (A^\mu_{1\nu}) = \begin{pmatrix} A^0_{10} & A^0_{11} & A^0_{12} & A^0_{13} \\ A^1_{10} & A^1_{11} & A^1_{12} & A^1_{13} \\ A^2_{10} & A^2_{11} & A^2_{12} & A^2_{13} \\ A^3_{10} & A^3_{11} & A^3_{12} & A^3_{13} \end{pmatrix}. \tag{0.7}$$

Special Relativity

In a flat 1 + 3-dimensional spacetime and in Cartesian coordinates, the *Minkowski metric* is given by

$$ds^2 = \eta_{\mu\nu} dx^\mu dx^\nu = c^2 dt^2 - dx^2 - dy^2 - dz^2, \tag{0.8}$$

where c is the speed of light in vacuum and $x^0 = ct$. In units of $c = 1$, so-called *natural units*, it holds that $x^0 = t$. The *metric tensor* and its inverse, i.e., the *inverse metric tensor*, can be written as

$$\eta = (\eta_{\mu\nu}) = \begin{pmatrix} 1 & 0 & 0 & 0 \\ 0 & -1 & 0 & 0 \\ 0 & 0 & -1 & 0 \\ 0 & 0 & 0 & -1 \end{pmatrix} \quad \Leftrightarrow \quad \eta^{-1} = (\eta^{\mu\nu}) = \begin{pmatrix} 1 & 0 & 0 & 0 \\ 0 & -1 & 0 & 0 \\ 0 & 0 & -1 & 0 \\ 0 & 0 & 0 & -1 \end{pmatrix}. \tag{0.9}$$

For any two vectors $x = (x^\mu) = (x^0, x^1, x^2, x^3)$ and $y = (y^\nu) = (y^0, y^1, y^2, y^3)$ in *Minkowski space* described by their contravariant components expressed in Cartesian coordinates, the *Minkowski inner product* is introduced as

$$x \cdot y \equiv x^0 y^0 - x^1 y^1 - x^2 y^2 - x^3 y^3, \tag{0.10}$$

which is obviously commutative, *i.e.*, $x \cdot y = y \cdot x$. We also define the notation $x^2 \equiv x \cdot x = (x^0)^2 - (x^1)^2 - (x^2)^2 - (x^3)^2$ for the squared norm ('length') of the vector x, which is indefinite, since it can be either positive or negative.[1] The Minkowski metric η and its inverse η^{-1} fulfill the relation

$$\eta_{\mu\lambda}\eta^{\lambda\nu} = \eta_\mu{}^\nu = \eta^\nu{}_\mu = \delta^\nu_\mu, \tag{0.11}$$

where δ^ν_μ is the *Kronecker delta* such that $\delta^\nu_\mu = 1$ if $\mu = \nu$ and $\delta^\nu_\mu = 0$ if $\mu \neq \nu$. We can write the Minkowski inner product in multiple ways as

$$x \cdot y = x^\mu \eta_{\mu\nu} y^\nu = \eta_{\mu\nu} x^\mu y^\nu = x^\mu y_\mu = x^0 y_0 + x^i y_i = x^0 y^0 - x^i y^i, \tag{0.12}$$

where, e.g., x^μ can be considered as the contravariant components of the vector x and y_μ the covariant components of the vector y, i.e., $y_0 = y^0$ and $y_i = -y^i$, and it also holds that $x^\mu y_\mu = x_\nu y^\nu$. Furthermore, we say that the vector x is *timelike* if $x^2 > 0$, *lightlike* if $x^2 = 0$, and *spacelike* if $x^2 < 0$. Note that lightlike vectors x form a cone $(x^0)^2 = (x^1)^2 + (x^2)^2 + (x^3)^2$ and a nonspacelike vector x is future pointing if $x^0 > 0$ and past pointing if $x^0 < 0$.

[1] Note the abuse of notation – the symbol x^2 denotes both the 'length' of the vector x and the second spatial contravariant component of the vector x. Unfortunately, this type of abuse of notation is difficult to avoid in relativity theory, since the notation would otherwise be too cumbersome.

In general, a *Lorentz transformation* Λ between two coordinate systems S and S' described by coordinates x and x', respectively, is given by

$$x' = \Lambda x \quad \Leftrightarrow \quad x'^{\mu} = \Lambda^{\mu}{}_{\nu} x^{\nu}. \tag{0.13}$$

In particular, if the Lorentz transformation is a boost in the x^1-direction, we can write

$$\Lambda^{(01)} = \begin{pmatrix} \cosh\theta & -\sinh\theta & 0 & 0 \\ -\sinh\theta & \cosh\theta & 0 & 0 \\ 0 & 0 & 1 & 0 \\ 0 & 0 & 0 & 1 \end{pmatrix} = \begin{pmatrix} \gamma & -\beta\gamma & 0 & 0 \\ -\beta\gamma & \gamma & 0 & 0 \\ 0 & 0 & 1 & 0 \\ 0 & 0 & 0 & 1 \end{pmatrix}, \tag{0.14}$$

where θ is the *rapidity*, $\beta \equiv v/c$, and γ is the so-called *gamma factor*, i.e., $\gamma \equiv \gamma(v) \equiv (1-v^2/c^2)^{-1/2}$, with v being the relative speed between the coordinate systems S and S'. Furthermore, it holds that

$$\cosh\theta = \gamma = \frac{1}{\sqrt{1-v^2/c^2}}, \quad \sinh\theta = \beta\gamma = \frac{v}{c}\frac{1}{\sqrt{1-v^2/c^2}}, \quad \tanh\theta = \beta = \frac{v}{c}. \tag{0.15}$$

The formulas for *(Lorentz) length contraction* and *time dilation* are given by

$$\ell' = \frac{\ell}{\gamma(v)} = \ell\sqrt{1-v^2/c^2}, \quad t = t'\gamma(v) = \frac{t'}{\sqrt{1-v^2/c^2}}, \tag{0.16}$$

respectively.

The *relativistic energy–momentum dispersion* relation is given by

$$E' = m\gamma(v)c^2 = \frac{mc^2}{\sqrt{1-v^2/c^2}}, \tag{0.17}$$

where m is the mass of an object, v its speed, and E its energy. In the rest frame of the object, it leads to *Einstein's famous formula*

$$E = mc^2. \tag{0.18}$$

The *relativistic addition of velocities* for colinear velocities v and v' is given by

$$v'' = \frac{v+v'}{1+vv'/c^2}. \tag{0.19}$$

In the nonrelativistic limit, i.e., $v, v' \ll c$, the classical formula $v'' \simeq v + v'$ is recovered.

Consider radiation of light in a specific coordinate direction of the coordinate system S. One should think of the radiation as coming from a fixed source in this coordinate system, where the radiation has frequency ν. For an observer in a coordinate system S' moving along the same coordinate direction with the relative velocity v, a frequency ν' is observed that is given by the *relativistic Doppler formula*, i.e.,

$$\nu' = \nu\sqrt{\frac{c-v}{c+v}}. \tag{0.20}$$

If the observer is moving away from the source, there is a *redshift* in the frequency of light, whereas if the observer is moving toward the source, there is a corresponding *blueshift*.

For the (binary) reaction $A + B \longrightarrow a + b + \cdots$, where two particles with 4-momenta p_A and p_B collide, using *conservation of 4-momentum*, we have

$$p_A + p_B = p_a + p_b + \cdots, \tag{0.21}$$

whereas for the decay $A \longrightarrow a + b + \cdots$, we have the simpler relation

$$p_A = p_a + p_b + p_c + \cdots. \tag{0.22}$$

This can be generalized to any number of particles with corresponding 4-momenta before and after a reaction, i.e.,

$$P_{in} = \sum_{i=A,B,\ldots} p_i = \sum_{j=a,b,\ldots} p_j = P_{out}. \tag{0.23}$$

Note that P^2 is invariant for any $P = \sum_{k=1}^{N} p_k$, where N is the number of particles, and actually, for any two 4-vectors A and B, the Minkowski inner product $A \cdot B = \eta_{\mu\nu} A^\mu B^\nu = A^\mu B_\mu$ is invariant under Lorentz transformations. Especially, $A^2 = A^\mu A_\mu$ is invariant. This is useful in many applications.

In electromagnetism, the *electromagnetic field strength tensor* F is defined as

$$F^{\mu\nu} = \partial^\mu A^\nu - \partial^\nu A^\mu, \tag{0.24}$$

where $A = (A^\mu) = (\phi, c\boldsymbol{A})$ is the 4-vector potential with ϕ and $\boldsymbol{A} = \boldsymbol{A}(x)$ being the electric scalar potential and the magnetic 3-vector potential, respectively, and can be written as

$$F = (F^{\mu\nu}) = \begin{pmatrix} 0 & -E^1 & -E^2 & -E^3 \\ E^1 & 0 & -cB^3 & cB^2 \\ E^2 & cB^3 & 0 & -cB^1 \\ E^3 & -cB^2 & cB^1 & 0 \end{pmatrix}, \tag{0.25}$$

which is a real antisymmetric matrix, i.e., $F^{\mu\nu} = -F^{\nu\mu}$, that combines both the electric and magnetic field strengths, i.e., $\boldsymbol{E} = (E^1, E^2, E^3)$ and $\boldsymbol{B} = (B^1, B^2, B^3)$. Using this tensor, *Maxwell's equations* can be written as

$$\partial_\mu F^{\mu\nu} = j^\nu, \quad \partial^\mu F^{\nu\lambda} + \partial^\nu F^{\lambda\mu} + \partial^\lambda F^{\mu\nu} = 0, \tag{0.26}$$

where $j = (j^\mu) = (\rho, \boldsymbol{j})$ is the 4-current density with $\rho = \rho(x)$ and $\boldsymbol{j} = \boldsymbol{j}(x)$ being the charge density and the electric 3-current density, respectively. In addition, we have the two Lorentz invariants

$$F^{\mu\nu} F_{\mu\nu} = 2\left(c^2 \boldsymbol{B}^2 - \boldsymbol{E}^2\right), \quad \epsilon_{\mu\nu\lambda\omega} F^{\mu\nu} F^{\lambda\omega} = -8c\boldsymbol{B} \cdot \boldsymbol{E}, \tag{0.27}$$

where $\epsilon_{\mu\nu\lambda\omega}$ is the *Levi-Civita tensor* with $\epsilon^{0123} = -\epsilon_{0123} = 1$. Maxwell's equations describe how sources (charges and currents) give rise to electric and magnetic fields. Assuming a moving test charge q with rest mass m and parametrizing the trajectory of the test charge as $x = x(s)$, where s is the proper time parameter, the *Lorentz force law* describes how the field strengths determine the trajectory of the test charge and is given by

$$mc^2 \ddot{x}^\mu(s) = q\dot{x}_\nu(s) F^{\mu\nu}(x(s)), \tag{0.28}$$

which is covariant under Lorentz transformations. The *energy–momentum tensor* T of the electromagnetic field is defined as

$$T^{\mu\nu} = \epsilon_0 F^\mu{}_\lambda F^{\lambda\nu} + \frac{\epsilon_0}{4} \eta^{\mu\nu} F_{\lambda\omega} F^{\lambda\omega}, \tag{0.29}$$

where ϵ_0 is the electric constant (or permittivity of free space). It holds that T is symmetric, i.e., $T^{\mu\nu} = T^{\nu\mu}$, and $T^\mu{}_\mu = \eta_{\mu\nu} T^{\mu\nu} = 0$. Furthermore, using Maxwell's equations, we obtain

$$\partial_\mu T^{\mu\nu} = \epsilon_0 j_\mu F^{\mu\nu} = -f^\nu, \tag{0.30}$$

where $f = (f^\mu) = (\boldsymbol{j} \cdot \boldsymbol{E}/c, \rho \boldsymbol{E} + \boldsymbol{j} \times \boldsymbol{B})$ is the Lorentz force density generated by the 4-current j. Without (external) sources, i.e., when $j = 0$, T is conserved, i.e., $\partial_\mu T^{\mu\nu} = 0$.

General Relativity

In a curved spacetime, the *metric* is defined as

$$ds^2 = g_{\mu\nu} dx^\mu dx^\nu, \tag{0.31}$$

where the *metric tensor* and its inverse, i.e., the *inverse metric tensor*, are given by

$$g = (g_{\mu\nu}), \quad g^{-1} = (g^{\mu\nu}), \tag{0.32}$$

respectively. In the special case that the spacetime is flat (see Special Relativity), we obtain

$$g_{\mu\nu} = \eta_{\mu\nu}, \quad g^{\mu\nu} = \eta^{\mu\nu}, \tag{0.33}$$

in Minkowski coordinates.

The *covariant derivatives* of a covariant vector A_ν and a contravariant vector A^ν are given by

$$\nabla_\mu A_\nu = A_{\nu;\mu} = \partial_\mu A_\nu - \Gamma^\lambda_{\mu\nu} A_\lambda, \quad \nabla_\mu A^\nu = A^\nu_{;\mu} = \partial_\mu A^\nu + \Gamma^\nu_{\mu\lambda} A^\lambda, \tag{0.34}$$

respectively. In particular, it holds that $\nabla_\mu \partial_\nu = \Gamma^\lambda_{\mu\nu} \partial_\lambda$, where the coefficients $\Gamma^\lambda_{\mu\nu}$ are called the *Christoffel symbols* of the second kind. Given a metric $g_{\mu\nu} = g_{\nu\mu}$, the Christoffel symbols of the Levi-Civita connection can be directly computed from

$$\Gamma^\lambda_{\mu\nu} = \frac{1}{2} g^{\lambda\omega} \left(\partial_\mu g_{\nu\omega} + \partial_\nu g_{\mu\omega} - \partial_\omega g_{\mu\nu} \right). \tag{0.35}$$

In addition, it holds that $\Gamma^\lambda_{\mu\nu} = \Gamma^\lambda_{\nu\mu}$, i.e., the Christoffel symbols are always symmetric with respect to the two lower indices.

The *parallel transport equation* for a vector A^λ is given by

$$\dot{x}^\mu \nabla_\mu A^\lambda = \dot{A}^\lambda + \Gamma^\lambda_{\mu\nu} \dot{x}^\mu A^\nu = 0, \tag{0.36}$$

where the dot above (' ˙ ') denotes differentiation with respect to the curve parameter s. Furthermore, considering the Lagrangian density given by

$$\mathcal{L} = g_{\mu\nu} \dot{x}^\mu \dot{x}^\nu, \tag{0.37}$$

and using the Euler–Lagrange equations, i.e.,

$$\frac{\partial \mathcal{L}}{\partial x^\mu} - \frac{d}{ds} \frac{\partial \mathcal{L}}{\partial \dot{x}^\mu} = 0, \tag{0.38}$$

where $\mu = 0, 1, \ldots, n$, we obtain the *geodesic equations* as

$$\ddot{x}^\lambda + \Gamma^\lambda_{\mu\nu} \dot{x}^\mu \dot{x}^\nu = 0. \tag{0.39}$$

Given three vector fields X, Y, and Z, the *torsion* T and the *curvature* R are defined as

$$T(X,Y) = \nabla_X Y - \nabla_Y X - [X,Y], \tag{0.40}$$

$$R(X,Y)Z = [\nabla_X, \nabla_Y]\, Z - \nabla_{[X,Y]} Z. \tag{0.41}$$

Both the torsion and curvature are tensors and are therefore linear in all of the arguments X, Y, and Z, including when the arguments are multiplied by a scalar function f, e.g.,

$$T(fX,Y) = T(X,fY) = fT(X,Y), \quad T(X,Y+Z) = T(X,Y) + T(X,Z). \tag{0.42}$$

Furthermore, the torsion and curvature tensors are antisymmetric in the arguments X and Y as defined above

$$T(X,Y) = -T(Y,X) \quad \text{and} \quad R(X,Y)Z = -R(Y,X)Z. \tag{0.43}$$

In local coordinates, we have

$$T(\partial_\mu, \partial_\nu)^\lambda \partial_\lambda = T^\lambda_{\mu\nu} \partial_\lambda, \quad R(\partial_\mu, \partial_\nu)\partial_\lambda = R^\omega{}_{\lambda\mu\nu} \partial_\omega, \tag{0.44}$$

and the components of the *torsion tensor* and the *Riemann curvature tensor* may be computed as

$$T^\lambda_{\mu\nu} = \Gamma^\lambda_{\mu\nu} - \Gamma^\lambda_{\nu\mu}, \tag{0.45}$$

$$R^\omega{}_{\lambda\mu\nu} = \partial_\mu \Gamma^\omega_{\nu\lambda} - \partial_\nu \Gamma^\omega_{\mu\lambda} + \Gamma^\omega_{\mu\rho}\Gamma^\rho_{\nu\lambda} - \Gamma^\omega_{\nu\rho}\Gamma^\rho_{\mu\lambda}, \tag{0.46}$$

respectively. Note that the Levi-Civita connection is torsion free as $\Gamma^\lambda_{\mu\nu} = \Gamma^\lambda_{\nu\mu}$. For fixed μ and ν, we can write the Riemann curvature tensor in matrix form as

$$R^\bullet{}_{\bullet\mu\nu} = \partial_\mu \Gamma^\bullet_{\nu\bullet} - \partial_\nu \Gamma^\bullet_{\mu\bullet} + \left[\Gamma^\bullet_{\mu\bullet}, \Gamma^\bullet_{\nu\bullet}\right]. \tag{0.47}$$

Note that the Riemann curvature tensor is antisymmetric in μ and ν, i.e., $R^\bullet{}_{\bullet\mu\nu} = -R^\bullet{}_{\bullet\nu\mu}$, or in component form, $R^\omega{}_{\lambda\mu\nu} = -R^\omega{}_{\lambda\nu\mu}$. If the torsion vanishes, i.e., $T = 0$, then we have the *first Bianchi identity*, i.e.,

$$R(X,Y)Z + R(Y,Z)X + R(Z,X)Y = 0, \tag{0.48}$$

or in component form, we have

$$R^\omega{}_{\lambda\mu\nu} + R^\omega{}_{\mu\nu\lambda} + R^\omega{}_{\nu\lambda\mu} = 0. \tag{0.49}$$

Furthermore, we have the *second Bianchi identity* in matrix form, i.e.,

$$\partial_\mu R^\bullet{}_{\bullet\nu\lambda} + \left[\Gamma^\bullet_{\mu\bullet}, R^\bullet{}_{\bullet\nu\lambda}\right] + \partial_\nu R^\bullet{}_{\bullet\lambda\mu} + \left[\Gamma^\bullet_{\nu\bullet}, R^\bullet{}_{\bullet\lambda\mu}\right] + \partial_\lambda R^\bullet{}_{\bullet\mu\nu} + \left[\Gamma^\bullet_{\lambda\bullet}, R^\bullet{}_{\bullet\mu\nu}\right] = 0. \tag{0.50}$$

Using the Riemann curvature tensor, the *Ricci tensor* can be defined as

$$R_{\mu\nu} = R^\lambda{}_{\mu\lambda\nu}, \tag{0.51}$$

which is symmetric, i.e., $R_{\mu\nu} = R_{\nu\mu}$, and in turn, the *Ricci scalar* is defined as

$$R = g^{\mu\nu} R_{\mu\nu} = R^\mu_\mu. \tag{0.52}$$

Finally, the *Einstein tensor* is defined in terms of the Ricci tensor, the Ricci scalar, and the metric tensor as

$$G_{\mu\nu} = R_{\mu\nu} - \frac{1}{2} R g_{\mu\nu}. \tag{0.53}$$

Note that it holds that the Einstein tensor is symmetric, i.e., $G_{\mu\nu} = G_{\nu\mu}$, and conserved, i.e., its covariant divergence vanishes $\nabla_\mu G^{\mu\nu} = 0$. Under local coordinate transformations $y = y(x)$, we have

$$g'_{\mu\nu}(y) = \frac{\partial x^\alpha}{\partial y^\mu}\frac{\partial x^\beta}{\partial y^\nu} g_{\alpha\beta}(x), \tag{0.54}$$

$$\Gamma'^{\lambda}{}_{\mu\nu}(y) = \frac{\partial x^\alpha}{\partial y^\mu}\frac{\partial x^\beta}{\partial y^\nu}\frac{\partial y^\lambda}{\partial x^\gamma}\Gamma^\gamma_{\alpha\beta}(x) + \frac{\partial y^\lambda}{\partial x^\gamma}\frac{\partial^2 x^\gamma}{\partial y^\mu \partial y^\nu}, \tag{0.55}$$

$$T'^{\lambda}{}_{\mu\nu}(y) = \frac{\partial y^\lambda}{\partial x^\gamma}\frac{\partial x^\alpha}{\partial y^\mu}\frac{\partial x^\beta}{\partial y^\nu} T^\gamma_{\alpha\beta}(x), \tag{0.56}$$

$$R'^{\omega}{}_{\lambda\mu\nu}(y) = \frac{\partial y^\omega}{\partial x^\delta}\frac{\partial x^\gamma}{\partial y^\lambda}\frac{\partial x^\alpha}{\partial y^\mu}\frac{\partial x^\beta}{\partial y^\nu} R^\delta{}_{\gamma\alpha\beta}(x). \tag{0.57}$$

Symmetries of a spacetime metric are associated to so-called *Killing vector fields*. Consider a vector field X. By definition, X is a Killing vector field if

$$\nabla_\mu X_\nu + \nabla_\nu X_\mu = 0, \tag{0.58}$$

for all indices μ and ν. Given a Killing vector field X^μ and a geodesic described by coordinate functions $x^\mu(s)$, the quantity

$$Q = \dot{x}^\mu X_\mu = g_{\mu\nu}\dot{x}^\mu X^\nu, \tag{0.59}$$

is constant along the geodesic.

The dynamics of spacetime in vacuum are described in the Lagrange formalism using the *Einstein–Hilbert action*, namely

$$\mathcal{S}_{\text{EH}} = -\frac{M_{\text{Pl}}^2}{2}\int R\sqrt{|\bar{g}|}\, d^4x, \tag{0.60}$$

where $M_{\text{Pl}} \equiv c^2/\sqrt{8\pi G}$ is the Planck mass, R is the Ricci scalar, and $\bar{g} = \det(g)$ is the determinant of the metric tensor. For the case of a spacetime not in vacuum, a matter contribution to the action is necessary

$$\mathcal{S}_{\text{matter}} = \int \mathcal{L}\sqrt{|\bar{g}|}\, d^4x, \tag{0.61}$$

where $\mathcal{L}$ is the Lagrangian density of the matter contribution.

The *Einstein gravitational field equations* (or simply Einstein's equations) follow from the Euler–Lagrange equations for the action and are given by

$$G_{\mu\nu} = \frac{8\pi G}{c^4} T_{\mu\nu}, \tag{0.62}$$

where G is Newton's gravitational constant and $T^{\mu\nu}$ is the *energy–momentum tensor* (or the stress–energy tensor) that describes the distribution of energy in spacetime. The energy–momentum tensor is generally given by

$$T_{\mu\nu} = \frac{2}{\sqrt{|\bar{g}|}}\frac{\delta\mathcal{S}_{\text{matter}}}{\delta g^{\mu\nu}}. \tag{0.63}$$

For example, an external electromagnetic field gives a contribution to $T^{\mu\nu}$ such that (see Special Relativity)

$$T_{\text{EM}}^{\mu\nu} = \epsilon_0 F^{\mu}{}_{\lambda} F^{\lambda\nu} + \frac{\epsilon_0}{4} g^{\mu\nu} F_{\lambda\omega} F^{\lambda\omega}, \tag{0.64}$$

whereas a *perfect fluid* (characterized by a 4-velocity u, a scalar density ρ_0, and a scalar pressure p) gives

$$T_{\text{pf}}^{\mu\nu} = (\rho_0 + p) u^{\mu} u^{\nu} - p g^{\mu\nu}. \tag{0.65}$$

In vacuum, Einstein's equations reduce to $G_{\mu\nu} = 0$.

In the Newtonian limit and the weak field approximation, i.e., $g_{\mu\nu} \simeq \eta_{\mu\nu} + h_{\mu\nu}$, where $h_{\mu\nu}$ is a small perturbation, the solutions to Einstein's equations are given by

$$h_{00} = h_{11} = h_{22} = h_{33} = \frac{2}{c^2}\Phi, \quad h_{\mu\nu} = 0 \;\; \forall \mu \neq \nu, \tag{0.66}$$

where Φ is the gravitational potential for the matter distribution ρ and given by $\Phi = -GM/r$, that is the solution to the Newtonian equation $\nabla^2 \Phi = 4\pi G\rho$. Furthermore, the geodesic equations of motion become

$$\frac{d^2 x^i}{dt^2} = \partial^i \Phi = -\partial_i \Phi. \tag{0.67}$$

The spherically symmetric vacuum solution to Einstein's equations is the *Schwarzschild solution* for which the *Schwarzschild metric* in spherical coordinates is given by

$$ds^2 = g_{\mu\nu} dx^{\mu} dx^{\nu} = \left(1 - \frac{2GM}{c^2 r}\right) c^2 dt^2 - \left(1 - \frac{2GM}{c^2 r}\right)^{-1} dr^2 - r^2 d\Omega^2, \tag{0.68}$$

where $d\Omega^2$ describes the metric on a sphere, i.e., $d\Omega^2 = d\theta^2 + \sin^2\theta d\phi^2$. For large r, the Schwarzschild metric approaches the Minkowski metric. The particular value $r = r_* \equiv 2GM/c^2$ represents the *Schwarzschild event horizon* (or the Schwarzschild radius) and is a coordinate singularity, i.e., it can be removed by a change of coordinates. Such a coordinate change is given by *Kruskal–Szekeres coordinates* u, v, θ, and ϕ, where θ and ϕ are the ordinary spherical coordinates on a unit sphere $\mathbb{S}^2$, the *Kruskal–Szekeres metric* is given by

$$ds^2 = \frac{16\mu^2}{r} e^{(2\mu - r)/(2\mu)} du dv - r^2 d\Omega^2, \quad uv = (2\mu - r) e^{(r-2\mu)/(2\mu)} < \frac{2GM}{c^2 e}, \tag{0.69}$$

where $\mu \equiv GM/c^2$.

For a static spacetime, the metric can be written on the form

$$ds^2 = \varphi(x)^2 dt^2 - g_{ij}(x) dx^i dx^j. \tag{0.70}$$

Given two static observers A and B in this spacetime, signals sent from A to B with frequencies f_A and f_B, respectively, will be redshifted according to

$$z = \frac{f_A}{f_B} - 1 = \frac{\varphi(x_B)}{\varphi(x_A)} - 1. \tag{0.71}$$

In particular, in the Schwarzschild spacetime, a signal sent from a static observer at r to an observer at infinity will be *gravitationally redshifted* according to

$$z_\infty \equiv \frac{1}{\sqrt{1 - \frac{2GM}{c^2 r}}} - 1 \simeq \frac{GM}{c^2 r}, \tag{0.72}$$

where M is the mass of the gravitating body. More generally, the frequency f of a light signal measured by an observer will be given by

$$f = g_{\mu\nu} U^\mu N^\nu, \tag{0.73}$$

where U is the 4-velocity of the observer and N the 4-frequency of the light signal, which is parallel transported along the worldline of the light signal.

In cosmology, the cosmological principles are encoded into the *Robertson–Walker metric*, which is given by

$$ds^2 = c^2 dt^2 - a(t)^2 \left(\frac{dr^2}{1 - kr^2} + r^2 d\Omega^2 \right), \tag{0.74}$$

where $a(t)$ is some function of the universal time t and k is a constant. By a suitable coordinate transformation $r \mapsto \lambda r$, one can always choose λ such that k takes one of the three values $k = 0, \pm 1$. If $k = 0$, then the spatial part for any fixed t becomes the Euclidean space $\mathbb{R}^3$.

From Einstein's equations, the assumption of the Robertson–Walker metric, and the universe being filled by an ideal fluid, follow the two independent *Friedmann equations*, namely

$$\frac{a'(t)^2 + kc^2}{a(t)^2} = \frac{8\pi G\rho + \Lambda c^2}{3}, \tag{0.75}$$

$$\frac{a''(t)}{a(t)} = -\frac{4\pi G}{3}\left(\rho + \frac{3p}{c^2}\right) + \frac{\Lambda c^2}{3}, \tag{0.76}$$

where the first equation is derived from the 00-component of Einstein's equations and the second equation is derived from the first one together with the trace of Einstein's equations. Here Λ is the *cosmological constant*.

Conventions

In this book, the following conventions will mainly be used in the presentation of the problems and their corresponding solutions:

- **Units:** We will mostly use units in which the speed of light in vacuum c has been set to $c = 1$, these are usually known as *natural units*. In some problems, we have also set $\hbar = 1$, if relevant. Normally, we do **not** use units in which Newton's gravitational constant G has been set to $G = 1$. In some problems, it is useful to use SI units.
- **Vectors:** In a four-dimensional spacetime, we will normally denote a 4-vector A by its contravariant components as follows

$$A = (A^\mu) = (A^0, A^1, A^2, A^3), \tag{0.77}$$

where A^0 is the temporal component and A^i ($i = 1, 2, 3$) are the spatial components, which is related to the 4-vector expressed in its covariant components as follows

$$(A_\mu) = (A_0, A_1, A_2, A_3), \tag{0.78}$$

where it holds that $A_\mu = g_{\mu\nu} A^\nu$ with $(g_{\mu\nu})$ being the given metric tensor. In some textbooks, the convention that the temporal component of a 4-vector is written as the last component of the vector is used, i.e., $A = (A^1, A^2, A^3, A^4)$, whereas in other textbooks, the convention that the standard components of a 4-vector are chosen as its covariant components might be used. We will **not** use these conventions.
- **Metrics:** In four-dimensional spacetimes, we adopt the convention that the signature is $+ - --$ and, when relevant, place the temporal direction first and denote it by 0. Therefore, in standard coordinates on Minkowski space, the metric tensor is $(\eta_{\mu\nu}) = \mathrm{diag}(1, -1, -1, -1)$ (see Special Relativity) and its inverse is given by $(\eta^{\mu\nu}) = \mathrm{diag}(1, -1, -1, -1)$. Thus, we have $A_\mu = \eta_{\mu\nu} A^\nu$, where $A_0 = \eta_{00} A^0 = A^0$ and $A_i = \eta_{ii} A^i = -A^i$ (for fixed $i = 1, 2, 3$). In a general coordinates, the metric is given by $ds^2 = g_{\mu\nu} dx^\mu dx^\nu$ (see General Relativity) and the metric tensor components represented in matrix form as $g = (g_{\mu\nu})$ from which its corresponding inverse components $g^{-1} = (g^{\mu\nu})$ can be computed. It must hold that $gg^{-1} = g^{-1}g = \mathbb{1}_4$, where $\mathbb{1}_4$ is the 4×4 identity matrix.
- **Sign convention of the Levi-Civita pseudotensor:** We define $\epsilon^{0123} = +1$ (see Special Relativity), which means that with our convention for the Minkowski metric, we have $\epsilon_{0123} = -1$.
- **Partial derivatives:** We will mostly denote covariant and contravariant partial derivatives as

$$\partial_\mu \equiv \frac{\partial}{\partial x^\mu}, \quad \partial^\mu \equiv g^{\mu\nu}\partial_\nu, \tag{0.79}$$

where $(g^{\mu\nu})$ is the inverse metric tensor.

- **Covariant derivatives and Christoffel symbols:** For covariant derivatives (see General Relativity), we will mostly use the notation ∇_μ, but the notation D_μ will sometimes be used. For Christoffel symbols of the second kind, we will only use the notation $\Gamma^\lambda_{\mu\nu}$ (see General Relativity).
- **Sign convention of the Ricci tensor:** The Ricci tensor may sometimes be defined as $R_{\mu\nu} = R^\lambda{}_{\mu\nu\lambda}$, which introduces a sign difference to our definition (see General Relativity) due to the antisymmetry of the Riemann curvature tensor as

$$R^\lambda{}_{\mu\nu\lambda} = -R^\lambda{}_{\mu\lambda\nu}. \tag{0.80}$$

- **Sign convention of Einstein's equations:** There is a sign convention in the definition of Einstein's equations, i.e., $G_{\mu\nu} = \pm 8\pi G T_{\mu\nu}/c^4$, where we use the positive sign.

In general, it is important to keep in mind that different texts may use different conventions. In particular, the sign discrepancies in different expressions will often be due to differing sign conventions of the metric, the Levi-Civita pseudotensor, the Ricci tensor, and Einstein's equations.

1

Special Relativity Theory

1.1 Basics

Problem 1.1 a) In Figure 1.1, a spacetime diagram for an observer $\mathcal{O}$ with an inertial frame is shown along with another observer $\mathcal{O}'$ with another inertial frame. The 4-vectors $\vec{A}$, $\vec{B}$, $\vec{U}$, and $\vec{V}$ are drawn. Which of the following statements are true?

1. In $\mathcal{O}$'s inertial frame, the scalar product between $\vec{A}$ and $\vec{B}$ is zero.
2. In $\mathcal{O}'$'s inertial frame, the scalar product between $\vec{A}$ and $\vec{B}$ is zero.
3. The scalar product between $\vec{A}$ and $\vec{B}$ is always nonzero.
4. In $\mathcal{O}$'s inertial frame, the scalar product between $\vec{U}$ and $\vec{V}$ is zero.
5. In $\mathcal{O}'$'s inertial frame, the scalar product between $\vec{U}$ and $\vec{V}$ is zero.
6. The scalar product between $\vec{U}$ and $\vec{V}$ is always nonzero.

b) Which of the 4-vectors in a) could be proportional to a 4-velocity? Explain why.

Problem 1.2 Show that

a) every 4-vector (i.e., vector in Minkowski space) that is orthogonal to a timelike 4-vector is spacelike.

b) the sum of two future directed time-like 4-vectors is another future directed timelike 4-vector.

c) every space-like 4-vector can be written as the difference between two future-directed lightlike 4-vectors.

d) the inner product of two future-directed timelike 4-vectors is positive.

Problem 1.3 In a particular inertial frame, two observers have the 3-velocities v_1 and v_2, respectively. Find an expression for the gamma factor of observer 2 in the rest frame of observer 1 in terms of these velocities.

Problem 1.4 a) Can a rest frame be chosen for a photon? Explain why!

1. always
2. sometimes
3. never

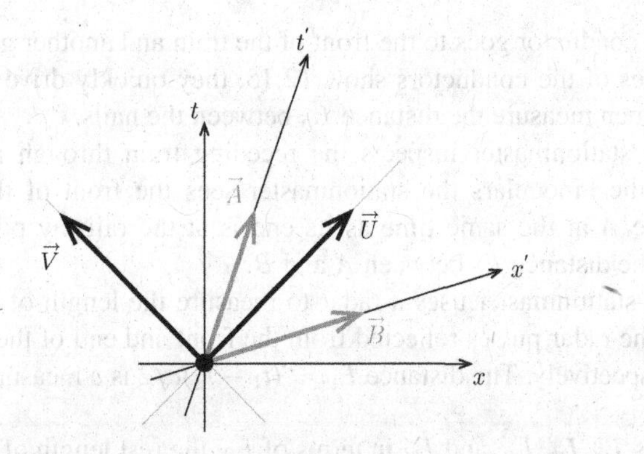

Figure 1.1 Spacetime diagram for observers $\mathcal{O}$ and $\mathcal{O}'$.

b) Can a rest frame be chosen for the center of momentum for a system of two photons? Explain why!

1. always
2. sometimes
3. never

1.2 Length Contraction, Time Dilation, and Spacetime Diagrams

Problem 1.5 a) State, explain, and derive the formula for length contraction in special relativity.

b) State, explain, and derive the formula for time dilation in special relativity.

Problem 1.6 A rod with length of 1 m is inclined 45° in the xy-plane with respect to the x-axis. An observer with the speed $\sqrt{2/3}\,c$ approaches the rod in the positive direction along the x-axis. How long does the observer measure the rod to be and at which angle does (s)he observe it to be inclined relative to its x-axis?

Problem 1.7 When the primary cosmic rays hit the atmosphere, muons are created at an altitude between 10 km and 20 km. A muon in the laboratory lives on average the time $\tau_0 = 2.2 \cdot 10^{-6}$ s before it decays into an electron (or a positron) and two neutrinos. Even though a muon can only move $\tau_0 c \approx 660$ m under the time τ_0, a large fraction of the muons will reach the surface of the Earth. How can this be explained? Make a numerical computation for a muon that moves with velocity $0.999c$.

Problem 1.8 An express train passes a station with velocity v. A measurement of the length of the train can be performed in the following different ways:

a) A "continuum" of linesmen is ordered to align along the track. The two men that see the front or the end of the train pass in front of them when their watches show 12:30 make a mark where they stand. The distance L_a between the marks is measured.

b) One conductor goes to the front of the train and another goes to the end. When the watches of the conductors show 12:15, they quickly drive a nail into the track. The linesmen measure the distance L_b between the nails.

c) The stationmaster inspects the receding train through a pair of binoculars. Through the binoculars the stationmaster sees the front of the train to be at the semaphore A at the same time as its end is at the railway point B. The linesmen measure the distance L_c between A and B.

d) The stationmaster uses a radar to measure the length of the train. The arrival times of the radar pulses reflected from the front and end of the receding train are t_1 and t_2, respectively. The distance $L_d = (t_1 - t_2)c/2$ is a measure of the length of the train.

Express L_a, L_b, L_c, and L_d in terms of L_0, the rest length of the train.

Problem 1.9 A hitchhiker in the Milky Way sits waiting on a small asteroid when a formidably long express space cruiser passes very close to the asteroid. Just as the rear end is opposite to the hitchhiker, (s)he sees lanterns in the front and in the rear end of the cruiser go on simultaneously. Actually, the rear watchman also saw them go on, but according to his hydrogen maser wristwatch he measured a small time difference of $4 \cdot 10^{-9}$ s between the lightening of the forward and rear lanterns. From the type indication on the cruiser – X2000 – our hitchhiker realized that its length was $2 \cdot 10^3$ m. Had they known what you know, they could have calculated the speed of the cruiser. What was it, according to Einstein's special theory of relativity?

Problem 1.10 Two lamps, which are separated by the distance ℓ in an inertial coordinate system K, are switched on simultaneously (in K). In another inertial coordinate system K', an observer measures the distance between the lamps to be ℓ' and observes the lamps go on with the time difference τ. Express ℓ in terms of ℓ' and τ. Assuming that the inertial coordinate system K' is moving along the axis connecting the two lamps, also find the expression for the relative velocity v between the two inertial coordinate systems.

Problem 1.11 A rod of length ℓ lies in the xz-plane of a coordinate system. If the angle between the rod and the x-axis is θ, calculate the length of the rod as seen by an observer moving with velocity v along the x-axis.

Problem 1.12 Two events A and B with coordinates x_A and x_B are simultaneous for an observer K with rest frame S. Another observer, K', moving with velocity $-u$ along the x-axis of S measures these events to not be simultaneous, but such that B is earlier than A by the amount $\Delta t'$. What is the distance L between the events A and B expressed in the frame of K if it is L' in the rest frame of K?

Problem 1.13 An observer S with rest frame K observes two events x_α and x_β. The α event takes place at the origin and the β event 2 years later at a distance of 10 light years (ly) forward along the x^1-axis. Another observer S' with rest frame K' moves with velocity v along the x^1-axis of K, passing S at the origin. The observer S' instead observes the β event 1 year later than the α event.

a) How far away does S' find the β event?

b) What is the relative velocity between S and S'?

Problem 1.14 The ratio $R(\mu/e)$ of muon neutrinos to electron neutrinos measured at ground level from the cosmic radiation is $R(\mu/e) = 2$ at low energies. These neutrinos come from the decay of pions, created by the primary cosmic radiation, which consists mostly of protons. The relevant reaction chain can be written in simplified form as follows

$$\pi \longrightarrow \mu + \nu_\mu,$$
$$\mu \longrightarrow e + \nu_\mu + \nu_e.$$

As we can see there are two muon neutrinos ν_μ produced for every electron neutrino ν_e. When the energy of the muon neutrinos, and therefore the muons, is high enough this ratio goes up, since the muons hit the Earth before they decay, and no electron neutrinos are produced. In the muon's rest frame, the muon lifetime is $\tau_0 = 2.2$ μs. The speed of light is $3 \cdot 10^8$ m/s. What is the smallest energy of the muons that hit the ground before they decay substantially if they are produced at an altitude of 10 km above ground? The rest mass of the muon is 106 MeV.

Problem 1.15 A circular accelerator has a radius of 50 m. How many turns can a muon take on average in this ring before it decays if its energy is kept constant at 1 GeV? The average lifetime of the muon in its rest frame is 2.2 μs and the muon mass is 106 MeV.

Problem 1.16 Consider a triangle at rest in the inertial system K with sides of length $a = 3\ell$, $b = 4\ell$, and $c = 5\ell$ in K.

a) Compute the lengths of the sides and the area of this triangle as measured in an inertial frame K' moving with constant velocity v parallel to the a-side of the triangle.

b) Same as in a), but now the observer K' moves parallel to the c-side of the triangle.

Problem 1.17 Consider a pole of proper length L moving along the x-axis in the negative direction with a constant velocity so that the pole is parallel to the x-axis (see Figure 1.2). At a fixed time, an observer at rest at the spatial origin sees (the optical effect is referred to) the front of the pole at an angle $\pi/3$ with the x-axis, a mark on the pole at an angle $\pi/4$, and the end of the pole at an angle $\pi/6$. What is the quotient r between the distance from the front of the pole to the mark and the full length of the pole?

Problem 1.18 Two spaceships, which are initially at rest in some common rest frame, are connected by a straight tensionless string. At time $t = 0$ in this frame, both spaceships start to accelerate in the same direction, in the direction of the string, such that their separation is constant in the initial rest frame. Both spaceships agree to stop accelerating once a predetermined time t_0 has passed in the initial rest frame.

a) Does the string break, i.e., does the distance between the two spaceships increase in the new rest frame of the spaceships?

b) If the distance between the spaceships is originally 40 km, $t_0 = 30$ s, and the spaceships have constant acceleration of 1/50 c/s in the initial rest frame, what is the

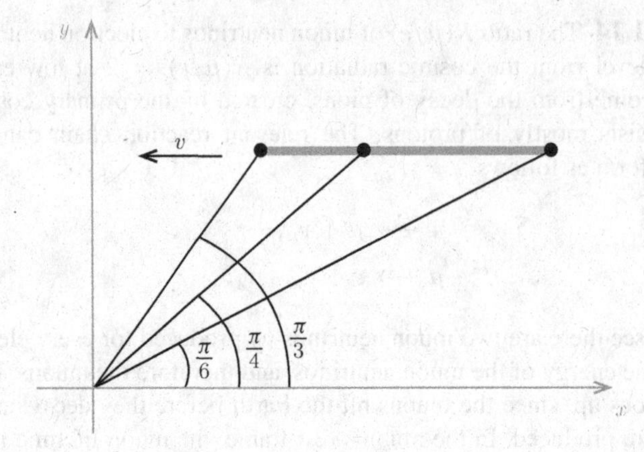

Figure 1.2 The pole is moving in the negative direction of the x-axis with a constant velocity v. The z-direction is neglected.

distance between the two spaceships in the frame of the leading spaceship after the engines are turned off?

Problem 1.19 Professor A. Einstein is traveling in a train on a rainy night. He is situated in the exact middle of the train, and suddenly lightning strikes right next to him. The train has reflectors in the rear and front, and since the reflections from rear and front reach him at the same time, he falls into slumber convinced that the reflections happened at the same time and that the speed of light is the same in both directions. What he did not see was that Professor W. Wolf was standing on the ground, also next to the lightning strike, observing the events. Draw spacetime diagrams showing how the light signals travel in each of the professors' rest frames. Use these to answer (including motivation) the following

a) Does Professor Wolf see the light reflections reaching Professor Einstein at the same time?

b) Would Professor Wolf agree with Professor Einstein that the light signals were reflected at the same time?

c) Do the reflections reach Professor Wolf at the same time?

Problem 1.20 Two rockets with rest lengths L and $2L$, respectively, move with constant velocities on an interstellar highway. Since the velocities are different, the rockets will pass each other. Call the event when the front of the faster rocket reaches the slower rocket A and the event when the end of the faster rocket reaches the front of the slower rocket B (see Figure 1.3). In each rocket there is an observer. Draw one or more spacetime diagrams describing the events, and use it/them to deduce which observer will consider time between A and B to be larger (the observer in the short rocket or the observer in the long rocket).

Problem 1.21 Muons created by cosmic rays hitting the atmosphere have a lifetime of $2.2 \cdot 10^{-6}$ s. If the muons are created at a height of 10 km, the time to reach the surface of the Earth (measured in the rest frame of the Earth) is at least $10\ \text{km}/c \simeq 3 \cdot 10^{-5}$ s, yet a large fraction of the muons can be measured at sea level.

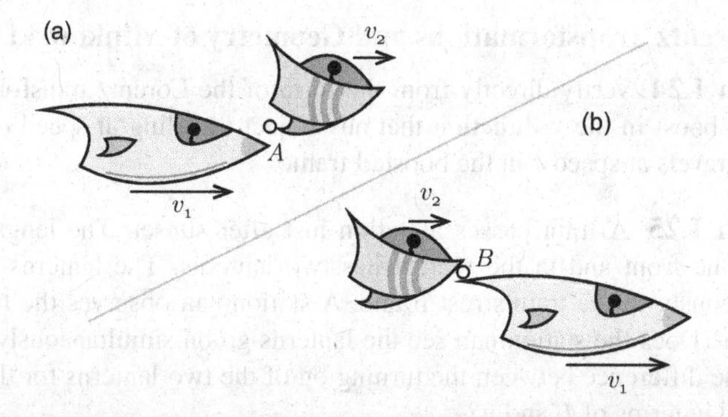

Figure 1.3 Two rockets with rest-lengths L and $2L$, respectively. Part (a) of the figure shows the event A, whereas part (b) shows the event B.

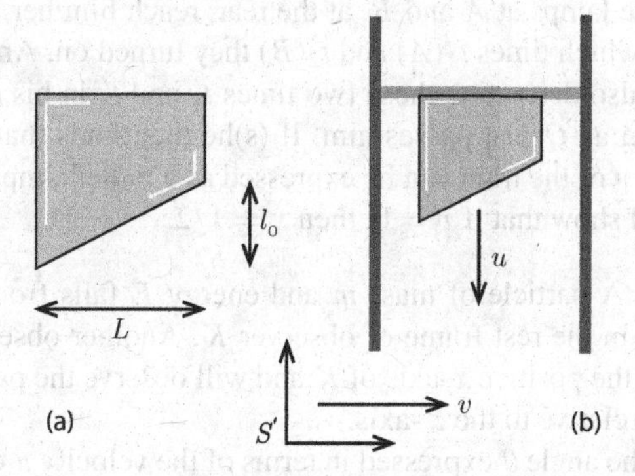

Figure 1.4 Part (a) of the figure shows the dimensions of a guillotine blade in its rest frame, whereas part (b) shows the event of the Scarlet Pimpernel riding by on his horse at velocity v (S' is the rest frame of the Scarlet Pimpernel and his horse) and the guillotine blade falls at velocity u in its own rest frame.

Explain qualitatively why this occurs by describing the situation using spacetime diagrams.

Problem 1.22 During the French Revolution, guillotines with a slanted blade were used to decapitate nobility. The guillotine blade at rest has the dimensions shown in Figure 1.4. Eager to save the nobility, the Scarlet Pimpernel rides by on his horse at velocity v. How fast does he have to ride in order for the guillotine blade to be horizontal in his rest frame S' if it falls at velocity u in the guillotine rest frame?

Problem 1.23 In an inertial frame S, two lights located on the positive x-axis are moving in the negative x-direction at speed v. An observer placed in the origin of S looks at the light signals coming from the lights. What is the distance between the *seen* positions of the lights if their separation in their common rest frame is ℓ_0? *Note:* The problem is asking for the separation *as seen* by the observer, not the actual distance between the lights at a given time.

1.3 Lorentz Transformations and Geometry of Minkowski Space

Problem 1.24 Verify directly from the form of the Lorentz transformation representing a boost in the x-direction that any object traveling at speed c in an inertial frame S travels at speed c in the boosted frame.

Problem 1.25 A train passes a station just after sunset. The length of the train is L. In the front and in the rear, it has two lanterns. The lanterns are turned on simultaneously in the train's rest frame. A stationman observes the train pass with velocity v. Does the stationman see the lanterns go on simultaneously? If not, what is the time difference between the turning on of the two lanterns for the stationman, expressed in terms of L and v?

Problem 1.26 An observer O on a train of length L and velocity v relative to the ground is standing at a distance xL $(0 \leq x \leq 1)$ from the front A of the train. When the light from the lamps at A and B, at the rear, reach him/her simultaneously, (s)he can calculate at which times $t_1(A)$ and $t_2(B)$ they turned on. Another observer O' on the ground can also determine these two times t_1' and t_2' in his rest frame, where the light reaches him as O just passes him. If (s)he then finds that $t_1' = t_2'$, it turns out that the velocity v of the train can be expressed as a rather simple function of x. Find this function and show that if $v = 0$, then $x = 1/2$.

Problem 1.27 A particle of mass m and energy E falls from zenith to the Earth along the z-axis in the rest frame of observer K. Another observer, K', moves with velocity v along the positive x-axis of K and will observe the particle to approach K' with an angle θ relative to the z'-axis.

a) Calculate the angle θ expressed in terms of the velocity u of the particle and the velocity v of K'.

b) Based on the result of a) give a description of how the starry sky would look like for a space cruiser moving with high speed in our galaxy.

Problem 1.28 Consider a particle with 4-velocity $V = \gamma(v')(c, v', 0, 0)$. By making a Lorentz transformation with velocity $-v$ along the x^1-axis, show that you can obtain the formula for relativistic addition of velocities, by expressing the velocity v'' of the particle in the new system in terms of the velocity v' in the old system and the velocity v of the motion of the observer.

Problem 1.29 Consider an equilateral triangle with sides of length ℓ, which is at rest in the inertial coordinate system K. Assume that one of the sides in the triangle is parallel to the x^1-axis of K. In an inertial coordinate system K' moving relative to K with velocity v along the positive x^1-axis of K, an observer measures the lengths of the sides and angles in the triangle. What expressions in ℓ and v for the lengths and angles does the observer find?

Problem 1.30 An observer K' is moving with constant speed v along the positive x^1-axis of an observer K. A thin rod is parallel to the x'^1-axis and moving in the direction of the positive x'^2-axis with relative velocity u. Show that according to the observer K the rod forms an angle ϕ with the x^1-axis, with

$$\tan\phi = -\frac{uv/c^2}{\sqrt{1 - v^2/c^2}}. \tag{1.1}$$

Problem 1.31 A cylinder is rotating around its axis with angular velocity ω (rad/s) in an inertial system. A straight line is drawn along the length of the cylinder. Show that the observer in an inertial system, which moves with velocity v parallel to the direction of the cylinder axis, will measure the line as twisted around the cylinder. Determine the twist angle per unit length.

Problem 1.32 A fast train (velocity v) is passing a station during the night. As the train passes the station, all compartment lights are turned on simultaneously with respect to the rest frame of the train. Relative to an observer standing at the station, the lights seem to be turned on at various times. Compute the velocity u of the line separating the illuminated and unilluminated parts of the train in the station rest frame.

Problem 1.33 A planet is moving along a circular orbit (radius R and angular velocity ω) around a star. A space ship is passing by the star, orthogonal with respect to the plane of motion of the planet, with velocity v. Compute the orbit of the planet in the rest frame coordinates of the space ship.

Problem 1.34 An observer B is moving with constant velocity v along the positive x^1-axis in the rest frame K of an observer A. An observer C is moving with constant velocity v' along the positive x'^2-axis in the rest frame K' of the observer B. Compute the absolute value of the relative velocity of C with respect to A. What is the time interval Δt between two events E_1 and E_2 that occur at the same spatial point with time difference $\Delta t''$ in the rest frame K'' of observer C.
Hint: It is sufficient to compute the time coordinate x''^0 of C as a function of the coordinates x^μ of A.

Problem 1.35 Let x be a lightlike vector in Minkowski space. Show that

$$u = N\begin{pmatrix} x^0 + x^3 \\ x^1 + ix^2 \end{pmatrix}, \tag{1.2}$$

where N is a real normalization factor, u is a spinor that satisfies $X \propto uu^*$, where X is a complex 2×2 matrix, so that $\det X = \det(uu^*) = 0$. Normalize this spinor by the requirement that $\operatorname{tr} X = 2x^0$.

A Lorentz transformation along the 3-axis is given by

$$a(v) = \begin{pmatrix} e^{-\theta/2} & 0 \\ 0 & e^{\theta/2} \end{pmatrix}, \tag{1.3}$$

where $\tanh\theta = v/c$. Show explicitly that this transformation satisfies

$$a(v)u = u(L(a(v))x), \tag{1.4}$$

where $L(a(v))x$ is the Lorentz-transformed vector and u is the normalized spinor.

Problem 1.36 Use Einstein's postulate to derive the expressions for a Lorentz boost in the x-direction.

Problem 1.37 In an inertial frame S, rockets A and B traveling with velocities v and $-v$, respectively, pass each other at time $t = 0$ at the spatial origin. A time t_0 later, light signals are sent from the origin toward each of the spaceships. Compute the time difference between the spaceships receiving the light signals in the rest frame of one of the rockets.

1.4 Relativistic Velocities and Proper Quantities

Problem 1.38 a) Explain the concept of "relativity of simultaneity." Illustrate it in a spacetime diagram.

b) The worldline of a massive particle in Minkowski space is described by the following equations in some inertial frame $(x^\mu) = (ct, x, y, z)$,

$$x(t) = \frac{3}{2}at^2, \quad y(t) = 2at^2, \quad z(t) = 0, \tag{1.5}$$

where a is constant and $0 \leq t \leq t_0$ for some value of t_0. Compute the particle's 4-velocity and 4-acceleration components. What values of t_0 are possible and why? Compute the proper time along this worldline from $t = 0$ to $t = t_0$.

Problem 1.39 A rod moves with velocity v along the positive x-axis in an inertial frame S. An observer at rest in S measures the length of the rod to be L. Another observer moves with the velocity $-v$ along the x-axis. What length, expressed as a function of L and v, will this observer measure for the rod? The measurement is done as usual with the endpoints being measured simultaneously for each observer in their respective frames.

Problem 1.40 The worldline of a particle is described by the coordinates $x^\mu(t)$ in the system S. An observer at rest in the system S', with velocity u along the positive x^2-axis relative to S, measures the velocity of the particle at time t'. Express his result as a function of the velocity of the particle in S and u.

Problem 1.41 A spaceship is moving away from Earth. The effect of the engines is regulated so that the the passengers feel the constant acceleration g. Calculate the distance between the Earth and the spaceship (measured in the rest frame of the Earth) as a function of

a) the time on Earth.

b) the time on the spaceship.

The commander of the spaceship is 40 years of age at the beginning of the voyage. How old is (s)he when the spaceship reaches the Andromeda Galaxy, which lies about 2 500 000 light years away from Earth?

Hint: 1 year $\approx \pi \cdot 10^7$ s and $g \approx 10\,\text{m/s}^2$.

Problem 1.42 A rocket (with rest mass m_0) starts from rest at the origin of a coordinate system K. Its velocity along the positive x-axis is increased by shooting matter from the rocket with constant velocity w relative to the instantaneous rest frame of the rocket in the negative x-direction. Compute the remaining mass m of the rocket as a function of its velocity v with respect to the origin of K.

Problem 1.43 You and your friend are in intergalactic space (assume the Minkowski metric). You leave simultaneously from a space station, with equal speeds v, in orthogonal directions. Neglect acceleration. After a time T has passed in your inertial frame, you want to send a message to your friend using a light signal. In which direction (in your rest frame) should you send it?

Problem 1.44 a) In an inertial frame S, an object travels with 3-velocity u. A different inertial frame S' is moving in the negative x-direction relative to S with relative speed v'. Write down the 4-velocities of the object and S' in S.

b) Show that the gamma factor of an object in any inertial frame is given by the inner product of the 4-velocity of the object and the 4-velocity of an object at rest in the frame.

c) Express the gamma factor of the object in a) in the frame S' using the result from b).

Problem 1.45 Show that the 4-velocity $V^\mu = dx^\mu/d\tau$ and 4-acceleration $A^\mu = dV^\mu/d\tau$ of an object are always perpendicular, where τ is the proper time of the object such that $V^2 = 1$.

Problem 1.46 You are traveling in your spaceship in flat intergalactic space such that special relativity can be used. You are on your way to a space station when you suddenly discover an enemy spaceship on your radar. You immediately send out a light signal for help to the space station. When you send out your signal, the distance in your coordinate system to the space station is 1 light day. You have a relative speed of $c/4$ toward the space station. When the space station receives your signal, they send out a rescue spaceship with a speed $3c/4$ relative the space station. How long does it take before you get help (as measured by your own clock)? First, how long does it take your light signal to reach the space station, and second, how long does it take for the rescue spaceship to reach you?

Problem 1.47 On an interstellar highway there is a speed limit of u relative to a reference frame S. A member of the intergalactic police force is at rest in this system when a spaceship passes at constant velocity $v > u$ (see Figure 1.5). Eager to do the job properly, the police officer starts the pursuit, accelerating with constant proper acceleration a. The pursuit ends when the police officer catches up with the criminal.

a) How long does the pursuit take according to the criminal?

b) How long does the pursuit take according to the police officer?

c) What is the relative velocity between the police officer and the criminal at the end of the pursuit?

Problem 1.48 An astronaut on an accelerated spaceship uses a coordinate system (T, X, Y, Z) related to an inertial system (t, x, y, z) as follows (we set $c = 1$)

$$t = X \sinh(aT), \quad x = X \cosh(aT), \quad y = Y, \quad z = Z. \tag{1.6}$$

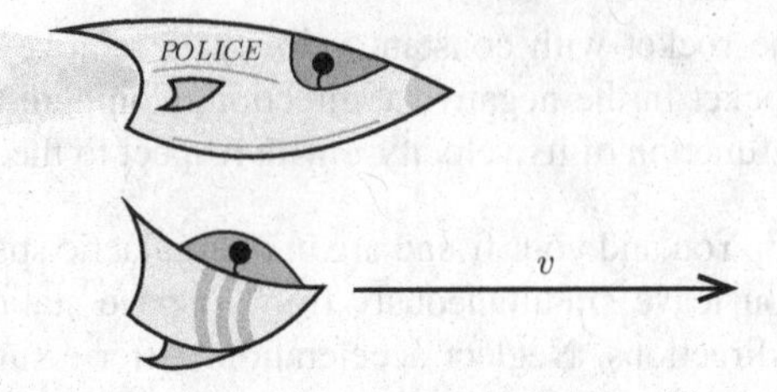

Figure 1.5 A space ship passing the intergalactic police at a relative constant velocity v.

a) Compute the metric tensor in the astronaut's coordinate system. (The metric in the inertial system is, of course, $ds^2 = dt^2 - dx^2 - dy^2 - dz^2$.)

b) Let $(k^\mu) = (\omega, \omega\cos(\theta), 0, \omega\sin(\theta))$ be the 4-wavevector of a photon emitted at time $t = t_0$ at the position $x = x_0$, $y = z = 0$ in the abovementioned inertial system. Compute the components of this 4-wavevector in the astronaut's coordinate system.

c) Compute the duration of a trip of the spaceship on the astronaut's watch (i.e., the proper time) if the trajectory of his spaceship on this trip is, in the astronaut's coordinate system, $X(T) = X_0 =$ constant, $Y(T) = vT$ for some constant $v > 0$, $Z(T) = 0$, and $0 \leq T \leq T_0$.

Problem 1.49 A rocket A is accelerating with constant proper acceleration α such that its worldline is given by $t^2 - x^2 = -1/\alpha^2$ in the reference frame S. Assume that S' is a different reference frame related to S by a Lorentz transformation in standard configuration with velocity v. What is the coordinate acceleration a' in the system S' at time $t' = 0$?

Problem 1.50 Two observers A and B are initially colocated at rest in the inertial system S. At a given time in S, observer B starts accelerating with a proper acceleration α. A time t_0 later (as measured by A), a light signal is sent from A toward B. Find an expression for the proper timed elapsed for observer B when B receives the signal. Discuss the limiting cases.

Problem 1.51 Particles in a circular accelerator are accelerated by an electromagnetic field in such a way that they are kept in a circular orbit with constant velocity. What are the corresponding 4-acceleration and proper acceleration of the particle and what is the eigentime required for the particles to complete one orbit? Introduce any quantities required to solve the problem.

Problem 1.52 An object of internal energy M moving with 4-velocity V is being acted upon by a force $F = fU$, where the known 4-vector U fulfills $U^2 = 1$ and f is a scalar. How fast is the internal energy of the object increasing (with respect to its proper time) and what is the proper acceleration of the object?

Problem 1.53 Consider a 4-force $F^\mu = (0, \boldsymbol{f})^\mu$ acting on an object of rest energy m with 3-velocity $\boldsymbol{v}$. Compute the rate of change in the rest energy $dm/d\tau$ and the

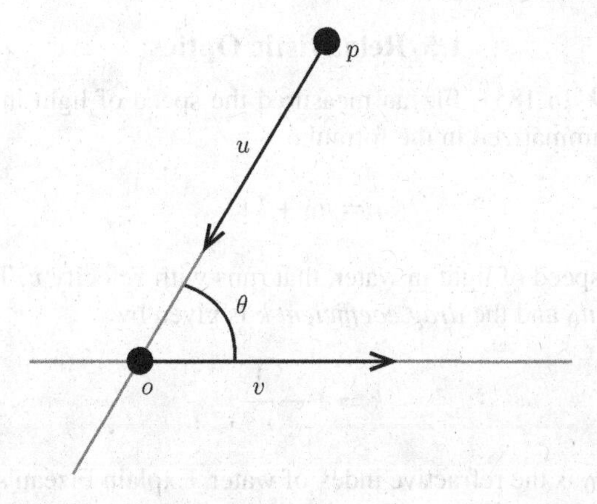

Figure 1.6 An observer o moving at velocity v and a particle p moving toward the observer o at velocity u in a direction that makes an angle θ with the direction of the velocity v.

proper acceleration α, where τ is the proper time of the object worldline. Discuss the requirements on f for the 4-force to be a pure force.

Problem 1.54 Given an object with acceleration $\boldsymbol{a}_0$ in its instantaneous rest frame S, find an expression for the acceleration in the inertial frame S', which is moving in the x-direction with velocity v relative to S. What is the maximal and minimal acceleration in S' depending on the direction of the acceleration based on your expression?

Problem 1.55 A particle has 4-momentum $P = (E, \boldsymbol{p})$ in an inertial frame S. An observer is moving by with velocity v in the x-direction. Compute the total energy this observer will measure for the particle and the velocity of the particle in the x'-direction of the observer's rest frame.

Problem 1.56 An object originally at rest with mass $m(0) = m_0$ is affected by a constant 4-force $F^\mu = f(1, \mathbf{1})^\mu$. Find the object's mass and the time elapsed in the initial rest frame as a function of the object's proper time τ.

Problem 1.57 An observer o is moving at velocity v in the x-direction in an inertial frame S. In the same inertial frame, a particle p hits the observer while traveling at a speed u in a direction that makes an angle θ with the negative x-direction, see Figure 1.6. What is the speed and angle that the observer will measure for the particle? Verify that your result is consistent with both the ultrarelativistic and nonrelativistic limits, i.e., $u \to 1$ and $u, v \ll 1$, respectively.

Problem 1.58 In an inertial frame S an object starting at rest at $t = 0$ is moving with constant *coordinate acceleration* a, i.e., $x = at^2/2$. Determine the proper time for the object to reach the speed v_0 in S and the proper acceleration of the object as a function of the time t in S.

1.5 Relativistic Optics

Problem 1.59 In 1851, Fizeau measured the speed of light in running water. His result can be summarized in the formula

$$u = u_0 + kv, \tag{1.7}$$

where u is the speed of light in water, that runs with velocity v. The speed of light in water at rest is u_0 and the *drag coefficient* k is given by

$$k = 1 - \frac{1}{n^2}, \tag{1.8}$$

where $n = c/u_0$ is the refractive index of water. Explain Fizeau's result!

Problem 1.60 See Problem 1.59. Is Fizeau's result still valid if the water runs perpendicular to the motion of light? If not, what is the correction?

Problem 1.61 In 1965, Maarten Schmidt at the Mount Palomar Observatory could identify the strongly redshift Lyman α line in the spectrum of the quasi-stellar radio source 3C 9. Normally, this line has the wavelength 1 215 Å. Schmidt instead found the value 3 600 Å for this line in this radio source. It is possible to explain the redshift in terms of the Doppler effect. This would imply that 3C 9 moves with an enormous speed relative to our galaxy. Determine a lower bound for the speed of 3C 9.

Problem 1.62 A plane electromagnetic wave moving along the x^1-axis has the form

$$E(x) = E_0 \sin\left[2\pi\left(\frac{x^1}{\lambda} - \nu t\right)\right]. \tag{1.9}$$

Introduce the angular frequency $\omega = 2\pi\nu$ and show that the argument of the wave can be written in the form $-x_\mu k^\mu$, where $k = \left(\frac{\omega}{c}, \frac{\omega}{c}, 0, 0\right)$ is the 4-wave vector of the light wave traveling along the positive x^1-axis. Show that this vector is lightlike and deduce the formula for the Doppler shift by calculating the change in angular frequency ω under a Lorentz transformation along the x^1-axis. What does the formula for the Doppler shift look like expressed in terms of the rapidity θ?

Problem 1.63 A *gamma ray burst* (GRB) observed in a cluster of faraway galaxies is time dilated and therefore has a total duration about twice as long as GRBs in nearby galaxies. According to the Hubble law, the recession speed is proportional to the distance to the GRB. Calculate the Doppler redshift $z = \Delta\lambda/\lambda_0$ of a typical spectral line from the distant GRB, where λ and λ_0 are the observed and emitted wavelengths, respectively.
Hint: All GRBs can be considered to have the same duration when measured in their respective rest frames.

Problem 1.64 A person watches two objects with constant velocities on a collision course, i.e., they approach each other on a straight line.

a) Assuming that both objects' velocities have the same absolute value $c/2$ in the person's frame of reference, compute the absolute value of the velocity with which a person traveling with the first object sees the other object approaching.

b) Assume that the first object sends a light pulse from a ruby laser, which produces visible light with a wavelength $\lambda_0 = 694.3$ nm, toward the second object. Compute the wavelength of this light pulse as seen by an observer on the second object.

Problem 1.65 A light source is moving at speed v and at an angle θ relative to the separation between the source and a stationary observer.

a) Consider a light pulse with frequency ω_0 in the rest frame of the source and determine the frequency ω measured by the observer.

b) Compute the angle θ for which $\omega = \omega_0$.

Problem 1.66 A large disk rotates at uniform angular speed Ω in an inertial frame S. Two observers, O_1 and O_2, ride on the disk at radial distances r_1 and r_2, respectively, from the center (not necessarily on the same radial line). They carry clocks, C_1 and C_2, which they adjust so that the clocks keep time with clocks in S, i.e., the clocks speed up their natural rates by the Lorentz factors

$$\gamma_1 = \frac{1}{\sqrt{1 - r_1^2\Omega^2/c^2}}, \quad \gamma_2 = \frac{1}{\sqrt{1 - r_2^2\Omega^2/c^2}}, \tag{1.10}$$

respectively. By the stationary nature of the situation, C_2 cannot appear to gain or lose relative to C_1. Deduce that, when O_2 sends a light signal to O_1, this signal is affected by a Doppler shift $\omega_2/\omega_1 = \gamma_2/\gamma_1$.

Note that, in particular, there is no relative Doppler shift between any two observers equidistant from the center.

Problem 1.67 A light source is moving with speed v through an optical medium with refractive index n. Derive an expression for the ratio between the frequency in the frame of the medium and the frequency in the frame of the source as a function of v, n, and the angle θ between the movement direction of the source and the propagation direction of the light (in the frame of the medium).

Problem 1.68 In an inertial frame S, a mirror is oriented perpendicular to the x-axis and moving with velocity v in the x-direction, see Figure 1.7. A light pulse with frequency ω approaches the mirror at an angle θ_{in} in S. What is the scattering angle θ_{out} and what frequency does the outgoing light have? Explain what happens when $v < -\cos\theta_{\text{in}}$?

Problem 1.69 In an inertial frame S, a light pulse is being directed at an optical medium with refractive index n which is moving with velocity v orthogonal to its surface, see Figure 1.8. In the rest frame of the medium, the light pulse is refracted according to Snell's law. An observer in S makes the interesting observation that the light pulse is still traveling in the same direction after entering the medium. Compute the index of refraction for the medium in terms of the velocity v and the angle θ' between the initial direction of the light pulse and the direction of motion for S in the rest frame of the medium.

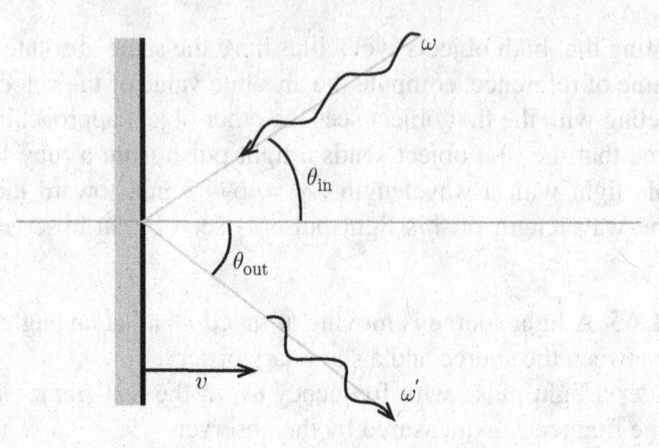

Figure 1.7 A mirror perpendicular to the x-axis moving with velocity v in the x-direction. The ingoing light has frequency ω and makes an angle θ_{in} with the x-axis, whereas the outgoing light has frequency ω' and makes an angle θ_{out} with the x-axis.

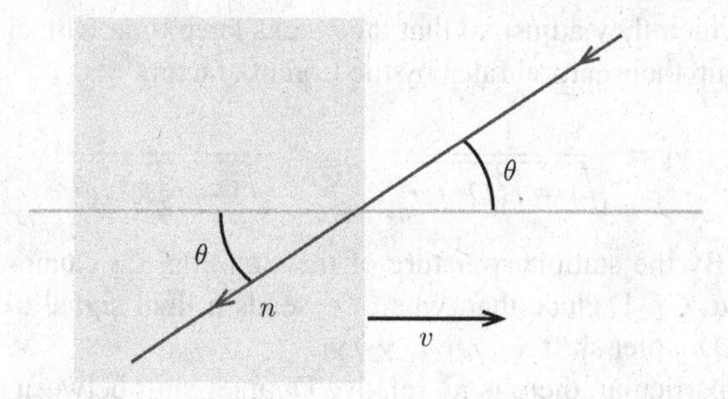

Figure 1.8 An optical medium with refractive index n moving with velocity v orthogonal to its surface. In the inertial frame S, an observer makes the observation that a light pulse is traveling in the same direction (at an angle θ relative to the velocity v) before and after entering the medium.

Problem 1.70 A sine wave propagating in a medium can be described by the function $\sin(N \cdot x)$ in the medium rest frame. Here, $(N^\mu) = (\omega, k)$, where ω is the angular frequency and k the wave number. Assuming the wave velocity in the medium is u, the relationship between k and ω is $ku = \omega$. A source with internal frequency ω_0 is moving through the medium with velocity v. Compute the Doppler shifted frequency ω in the medium rest frame when the waves are traveling in the direction of motion. Also discuss the special cases $v = u$ and $v = -u$ and make sure that your solution reduces to the classical Doppler formula when $v \ll 1$.

Problem 1.71 Using the same setup as in Problem 1.50 and assuming that A sends the light signal using a frequency ω, compute the frequency observed by B. In addition, if B carries a mirror and reflects the signal back at A, find the frequency observed by A for the reflected signal.

1.6 Relativistic Mechanics

Problem 1.72 The rest energy of an electron is about 0.51 MeV, i.e., the energy a charged particle, with charge equal to the electron charge, would receive when falling down a potential difference of 0.51 MV. Assuming that the electron is accelerated through a linear accelerator (starting from rest) with a potential difference of 10^6 V. Compute the final velocity of the electron.

Problem 1.73 a) An electron e^- (with mass m_e) collides with a positron e^+ (i.e., the antiparticle of the electron with the same mass m_e as the electron). Show that they cannot annihilate into a single photon γ (a photon has zero mass), i.e., the process $e^- + e^+ \longrightarrow \gamma$ is impossible due to conservation of energy and momentum.

b) Also show that an electron cannot spontaneously emit a photon.

c) Can an electron colliding with a positron annihilate into two photons? Justify your answer.

Problem 1.74 An elementary particle with mass M decays into two particles a and b with masses m_a and m_b, respectively. Calculate the momentum of particle a in the rest frame of particle b.

Problem 1.75 A particle A with mass m_A decays into two particles B and C with masses m_B and m_C, respectively. Assume that particle A has speed v_A before the decay and that particle B is at rest after the decay, i.e., $\mathbf{p}_B = \mathbf{0}$. Express the speed v_A in the masses m_A, m_B, and m_C.

Problem 1.76 Two particles, 1 and 2, with masses m_1 and m_2, respectively, collide and form a new particle with mass M. Calculate the mass M and the velocity v of this new particle in the rest frame of particle 2 as a function of the velocity v_1 of particle 1 in the rest frame of particle 2 and the masses m_1 and m_2.

Problem 1.77 a) Two particles with rest masses m_1 and m_2, respectively, move along the x-axis in the inertial frame of some observer at uniform velocities u_1 and u_2, respectively. They collide and form a single particle with rest mass m moving at uniform velocity u. Assuming that $c = 1$, prove that

$$m^2 = m_1^2 + m_2^2 + 2m_1 m_2 \gamma(u_1)\gamma(u_2)(1 - u_1 u_2), \tag{1.11}$$

and also find u.

b) Show that the above expression can be written as

$$m^2 = m_1^2 + m_2^2 + 2m_1 m_2 \gamma(v), \tag{1.12}$$

when v is the velocity of particle 2 as measured in the rest frame of particle 1.

c) Consider two different situations and in both of the situations the relative velocity v as defined above is the same, and thus, the rest mass m is the same in both situations, but in one $u_1 = 0$ and in the other $m_1\gamma(u_1)u_1 = -m_2\gamma(u_2)u_2$. What is the difference in total energy for the two situations in the frame of the observer?

Problem 1.78 A pion with mass m_π and energy E_π moves along the x-axis. It decays into a muon with mass m_μ and a neutrino with approximately zero mass. Calculate the energy E_μ of the muon when it moves at a right angle relative to the x-axis in terms of the velocity of the incoming pion and the masses.

Problem 1.79 A pion with mass m_π decays into an electron with mass m and an antineutrino with mass m_ν. Calculate the velocity of the antineutrino in the rest frame of the electron as a function of the masses of the particles and determine the limiting value of this velocity as the mass of the antineutrino goes to zero.

Problem 1.80 In June 1998, the Super-Kamiokande Collaboration in Japan reported that it had found evidence for massive neutrinos. Super-Kamiokande measures so-called atmospheric neutrinos, which are produced in hadronic showers resulting from collisions of cosmic rays with nuclei in the upper atmosphere. Two of the dominating processes in the production of atmospheric neutrinos are

$$\pi^+ \longrightarrow \mu^+ + \nu_\mu,$$

where π^+ is a pion, μ^+ is an antimuon, and ν_μ is a muon neutrino, followed by

$$\mu^+ \longrightarrow e^+ + \bar{\nu}_\mu + \nu_e,$$

where e^+ is a positron, $\bar{\nu}_\mu$ is an antimuon neutrino, and ν_e is an electron neutrino.

a) Calculate the kinetic energy of the antimuon, T_{μ^+}, and the absolute value of the 3-momentum of the muon neutrino, p_{ν_μ}, when the pion decays at rest according to the first decay. Despite the small mass of the muon neutrino, neglect it! The mass of the pion is m_π and the mass of the antimuon is m_μ.

b) How far will one of the antimuons, which are produced in the first decay, travel (on average) in the pion rest frame before it decays according to the second decay? The mean lifetime of an antimuon at rest is τ_μ.

Problem 1.81 The pions in the sky that are decaying into muons as in Problem 1.14 are produced in collisions between protons in the primary cosmic rays and nitrogen or oxygen in the air. When a pion with energy of 2 GeV is produced, what energy does the muon have if it continues in the same direction as the pion? The expression can be simplified due to the high energy of the pion. What is the resulting expression? What is the muon energy? The pion has a rest mass of 140 MeV, and the neutrino mass can be neglected.

Problem 1.82 A beam of protons that are accelerated to a very high energy hits a beryllium target and produces a shower of particles. Two detectors are placed in a plane behind the target symmetrically around the proton beam axis. Each detector makes an angle of 45° with this axis and detects $\mu^+\mu^-$-pairs, one type of particle in each detector. When the momentum of each muon is 2.2 GeV, one sees an enhancement in the muon rate. This is interpreted as the production of a resonance R of mass M_R that decays into the muons. What is the mass M_R of this resonance? The muon mass is 106 MeV.

Problem 1.83 A particle with mass M and 4-momentum $p = (E, \mathbf{p})$ moves toward a detector when it suddenly decays and emits a photon in the direction of motion.

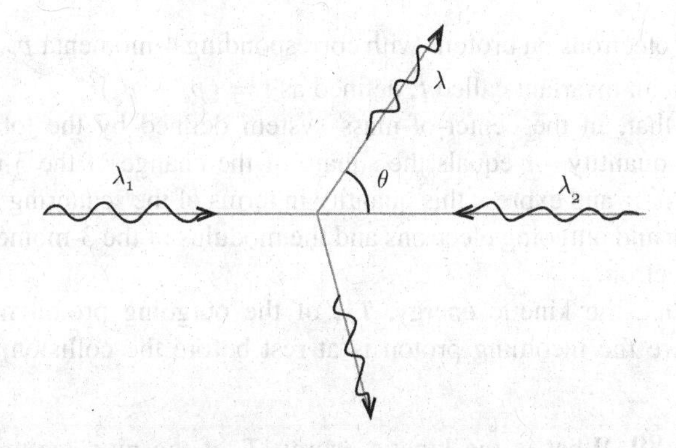

Figure 1.9 Scattering of two photons $\gamma + \gamma \longrightarrow \gamma + \gamma$.

The energy registered by the detector is ω. Determine what energy the photon had in the rest frame of the decaying particle.

Problem 1.84 An electron moves with constant velocity toward a positron at rest and they annihilate into two photons. The photons go out with angles ϕ and $-\phi$ relative to the direction of the incoming electron.

a) Calculate the angle as a function of the total energy of the electron.

b) Show that in the nonrelativistic limit the angle is given by $\cos\phi = v/(2c)$.

Problem 1.85 Two photons with wavelengths λ_1 and λ_2, respectively, are scattered against each other according to Figure 1.9. Calculate the wavelength of the photon with scattering angle θ, i.e., express λ as a function of λ_1, λ_2, and θ.

Hint: $p = \frac{h}{\lambda}$, where h is Planck's constant.

Problem 1.86 A K-meson with mass M decays at rest into two charged pions with the same mass m and a photon according to the reaction formula

$$K^0(P) \longrightarrow \pi^+(p_1) + \pi^-(p_2) + \gamma(k).$$

The momenta of the particles are given in parentheses after each particle symbol. Calculate the speed v of the pions in center-of-mass frame where (where $\mathbf{p}_1 + \mathbf{p}_2 = 0$) as a function of the masses of the particles and the photon energy $k^0 = \omega$ in the rest frame of the decaying particle.

Problem 1.87 A Σ^0 particle with speed $c/3$ in the direction toward a gamma detector suddenly decays into a Λ particle and a photon. The photon continues toward the detector.

a) What energy does the Σ^0 particle have in the system in which the detector is at rest?

b) What energy does the photon have in the rest system of the Σ^0 particle?

c) What energy will be registered in the detector?

The mass of the Λ is $m_\Lambda \approx 1\,115.7$ MeV and that of Σ^0 is $m_{\Sigma^0} \approx 1\,192.6$ MeV.

Problem 1.88 In elastic scattering of two particles onto each other, the same type of particles are present before and after the collision. Thus, in $e + p \longrightarrow e + p$ elastic scattering of electrons on protons with corresponding 4-momenta p_e, p_p, p'_e, and p'_p, one can form an invariant called t, defined as $t = (p_e - p'_e)^2$.

a) Show that, in the center-of-mass system defined by the total 3-momentum being $\mathbf{0}$, the quantity $-t$ equals the square of the change of the 3-momentum, i.e., $-t = (\mathbf{p}_e - \mathbf{p}'_e)^2$ and express this quantity in terms of the scattering angle θ between the incoming and outgoing electrons and the modulus of the 3-momentum $|\mathbf{p}_e|$ of the incoming electron.

b) Calculate the kinetic energy, T'_p, of the outgoing proton in the laboratory system, where the incoming proton is at rest before the collision, in terms of the variable t.

Problem 1.89 What is the kinetic energy T of the pion required to create the resonance $\Delta(1232)$ in the reaction

$$\pi + p \longrightarrow \pi + \Delta,$$

where π is a pion and p is a proton? The proton is at rest before the collision. The result should be expressed in terms of the masses of the particles involved.

Problem 1.90 The scattering probabilities for the reactions $\pi + d \longrightarrow p + p$ and for the reversed reaction $p + p \longrightarrow \pi + d$ are related due to so-called time reversal invariance. However, they must be compared at the same center-of-mass energy. Calculate the relation between the kinetic energy T_π of the pion (π), in the frame where the deuteron (d) is at rest before the collision in the first reaction, and the kinetic energy T_p of one of the protons (p) in reversed reaction, when the other proton is at rest, respecting the above condition on the center-of-mass energy.

Problem 1.91 Consider the reaction $\pi^+ + n \longrightarrow K^+ + \Lambda$ in the rest frame of n. The masses of the particles are m_{π^+}, m_n, m_{K^+}, and m_Λ, respectively. What is the kinetic energy T of the π^+ when the K^+ has total energy E and moves off at an angle of 90° to the direction of the incident π^+? (T should be expressed in m_{π^+}, m_n, m_{K^+}, m_Λ, and E.)

Problem 1.92 The mass of the meson π^0 can be measured by the reaction

$$p + \pi^- \longrightarrow \pi^0 + n,$$

where p is a proton, π^- is a negative pion, and n is a neutron. The uncharged π^0 meson decays very quickly into two photons and cannot be easily measured. However, the velocity of the final neutron can be measured and is found to be $v_n = (0.89418 \pm 0.00017)$ cm/ns. Derive the formula that expresses the mass of the π^0 meson as a function of the masses of the proton, the π^-, the neutron, and the velocity v_n, assuming that the reaction takes place at rest for the incoming particles. Simplify the result by showing that the velocity is small, so that we need to retain only lowest nontrivial order in v_n/c.

Problem 1.93 A thermal neutron is absorbed by a proton at rest and a deuteron is formed together with a photon. This exothermic reaction is formally

$$p + n \longrightarrow d + \gamma.$$

The binding energy B of the deuteron is about 2.23 MeV. Calculate, relativistically, the energy of the emitted photon as a function of the masses of the particles and the binding energy B.

Problem 1.94 A hydrogen atom H, consisting of an electron and a proton with binding energy $B = 13.6$ eV, can disintegrate into its two constituent particles by being hit by a photon. The reaction is

$$\gamma + \mathrm{H} \longrightarrow p + e.$$

Calculate, relativistically, the smallest photon energy in the rest frame of H required for this process to occur expressed in terms of B and the hydrogen mass m_{H}.

Problem 1.95 Similarly to the cosmic microwave background (CMB) of photons with a temperature of $T_{\mathrm{CMB}} \sim 2.7$ K, there should be a cosmic neutrino background (CNB) with a temperature of $T_{\mathrm{CNB}} \sim 1.9$ K. At these temperatures, their kinetic energy is very tiny. Suppose a very high-energy antineutrino would hit such a neutrino and annihilate it. A result of this collision could be the production of a Z^0 boson which decays hadronically. The reaction is formally

$$\bar{\nu} + \nu \longrightarrow Z^0.$$

What is the threshold energy for the antineutrino for this to occur? In particular, consider the two limits

a) The CNB neutrinos have a mass of $m_\nu = 0.15$ eV.

b) The CNB neutrinos have very small masses $(m_\nu/(k_B T) \to 0)$.

Hint: In a gas of particles at temperature T, the mean kinetic energy of the particles is given by $E_k = 3k_B T/2$, where $k_B \simeq 8.6 \cdot 10^{-5}$ eV/K is Boltzmann's constant. The Z^0 mass is $m_{Z^0} \simeq 91$ GeV.

Problem 1.96 Consider elastic scattering of photons on electrons

$$\gamma(k) + e^-(p) \longrightarrow \gamma(k') + e^-(p'),$$

where k and p are the incoming photon and electron 4-momenta and k' and p' the corresponding outgoing 4-momenta.

a) In the laboratory system, the incoming electron is at rest and the outgoing photon is scattered at an angle θ with respect to the direction of the incoming photon. Use invariants to derive the so-called "Compton formula," i.e., the difference between the outgoing and incoming photon wavelengths, as a function of θ, in units $c = 1$ and $\hbar = 1$.

b) Derive the angular frequency (energy) of the outgoing photon in the center-of-mass system in terms of the incoming photon angular frequency (energy) in the laboratory system.

Problem 1.97 In Compton scattering $\gamma + e \longrightarrow e + \gamma$, photons of a fixed energy ω are scattered against electrons, which can be considered at rest in the laboratory frame. Compute the kinetic energy of the outgoing *electron* as a function of the scattering angle θ of the outgoing *photon*.

Problem 1.98 Inverse Compton scattering occurs when low-energy photons collide with high-energy electrons. Assuming that the photon and electron are originally moving in the same direction, find an expression for the photon energy after the collision as a function of the initial photon energy, the velocity and mass of the electron, and the scattering angle θ of the photon.

Problem 1.99 An antimuon μ^+ decays into a positron e^+ and two neutrinos ν_e and $\bar{\nu}_\mu$. The reaction is

$$\mu^+ \longrightarrow e^+ + \nu_e + \bar{\nu}_\mu.$$

Give an expression for the largest possible total energy of the electron neutrino ν_e in the rest frame of the antimuon. You may assume that the neutrino masses are negligible compared to lepton masses.

Problem 1.100 A ρ-meson with mass $m_\rho \simeq 770\,\mathrm{MeV}/c^2$ sometimes decays into a pair of muons ($\mu^-$and μ^+) with mass $m_{\mu^-} = m_{\mu^+} \simeq 106\,\mathrm{MeV}/c^2$ and a photon, γ. What is the maximal *kinetic* energy that the μ^+ can have in this decay in the rest frame of the ρ-meson?

Problem 1.101 There is a possibility that neutrinos are their own antiparticles. If this is true, then the so-called neutrinoless double beta decay

$$^{76}\mathrm{Ge} \longrightarrow {}^{76}\mathrm{Se} + e^- + e^-,$$

is allowed. Derive expressions for the maximal and minimal possible values of the sum of the kinetic energy of the electrons in the rest frame of ^{76}Ge. Express your answer in terms of the particle masses.

Problem 1.102 At the LHC (Large Hadron Collider), two photons are measured with 4-momenta

$$p_1 = \omega_1(1,1,0,0) \quad \text{and} \quad p_2 = \omega_2(1, \cos\theta, \sin\theta, 0), \tag{1.13}$$

respectively. Assuming that the photon pair results from the decay of a new particle ϕ such that $\phi \longrightarrow \gamma\gamma$, what is the mass of the new particle?

Problem 1.103 In an accelerator, protons are accelerated until they reach a kinetic energy of 8 000 MeV and are then made to collide with protons at rest. If the sum of the kinetic energies of two colliding protons (measured in the center-of-mass system) is larger than the rest energy of a proton-antiproton pair, then such a pair can be formed according to the reaction formula

$$p + p \longrightarrow p + p + p + \bar{p},$$

where p is a proton and $\bar{p}$ is an antiproton.

Is the energy 8 000 MeV sufficient for the reaction to go? The rest mass of the proton is 938 MeV.

Problem 1.104 Protons at rest are bombarded with π-mesons. How large kinetic energy do the mesons need to have for the reaction

$$\pi^- + p \longrightarrow \pi^+ + \pi^- + n,$$

to take place? The rest mass of the particles are $m_{\pi^-} = m_{\pi^+} \approx 140$ MeV, $m_p \approx 938$ MeV, and $m_n \approx 940$ MeV.

Problem 1.105 In the CELSIUS ring at the The Svedberg Laboratory in Uppsala, Sweden, one would like to study the reaction

$$p + d \longrightarrow p + p + n + \eta.$$

The available kinetic energy of the protons is $T_p = 700$ MeV and the deuterons (d) can be considered to be at rest. The rest masses of the particles are $m_p \approx m_n$, $m_d \approx m_p + m_n$, $m_n = 940$ MeV, and $m_\eta = 550$ MeV.

a) Is the reaction possible?

b) If the kinetic energy of the protons in the beam is increased to $T_p = 1\,350$ MeV, what is the maximum kinetic energy that the η can get in the system in which the nucleons are at rest after the reaction, expressed in terms of the rest masses and the kinetic energies?

Problem 1.106 In neutrino detection, the quasi-elastic ($\nu_\mu + X \longrightarrow \mu + Y$, where X and Y are different nuclei) and 1π ($\nu_\mu + X \longrightarrow \mu + Y + \pi^0$) processes are relevant at relatively low energies. Compute the ratio between the neutrino threshold energies for these processes in the rest frame of the nucleus X. Express your answer in terms of the different particle masses (the neutrino may be considered massless for the purposes of this problem).

Problem 1.107 Consider the particle collision $e^- + e^- \longrightarrow e^- + e^- + e^- + e^+$. Compute the necessary total energy of one of the initial electrons in the rest frame of the other for this process to occur. Also, compute the ratio between this energy and the total required energy in the center-of-momentum frame.

Problem 1.108 We can produce neutral kaons in a proton collision through the reaction $p + p \longrightarrow p + p + K^0$. Find an expression for the threshold kinetic energy of the protons of this reaction when

a) One proton is stationary in the lab frame (find the threshold kinetic energy of the *other* proton).

b) Both protons have the same kinetic energy (quote the *total* kinetic energy of both protons).

Problem 1.109 A particle χ hits a stationary proton p and undergoes inelastic scattering to a new state χ^* while keeping the proton intact. Determine the threshold kinetic energy of χ for this scattering to occur if $m_{\chi^*} = m_\chi + \delta > m_\chi$. Discuss your result in the limit when $\delta \ll m_\chi$.

Problem 1.110 Neutrinos are emitted from core collapse supernovae. If a core collapse supernova occurs at a distance L from Earth and each neutrino has a total energy E, how much more time would pass (in the rest frame of the Earth) until the neutrinos reach us if they have a small mass $m > 0$ compared to if they were massless ($m = 0$)? Give an exact answer as well as a reasonable approximation for when $m \ll E$.

Problem 1.111 An elementary particle of charge e (e is the elementary charge) is accelerated from rest in a 100 m long straight insulated vacuum cylinder (a linear accelerator) with a constant electric field of 10^4 V/m across the endpoints.

a) What kinetic energy will the particle obtain after the acceleration?

b) How long time does it take for particle to pass through the tube if it starts from rest? *Hint:* Use the energy as an integration variable.

1.7 Electromagnetism

Problem 1.112 Show by explicit calculation, using the chain rule for derivation and the properties of the Lorentz transformations, that

$$\Box A^{\mu}(x) = 0, \tag{1.14}$$

is invariant under Lorentz transformations, i.e., if $A^{\mu}(x)$ is a solution to Eq. (1.14), then $A'^{\mu}(x')$ is a solution to the same equation in the primed variables $x' = \Lambda x$, where Λ is a Lorentz transformation.

Problem 1.113 Show that the gauge transformation $A_{\mu} \mapsto A'_{\mu} = A_{\mu} + \partial_{\mu}\psi$, where ψ is an arbitrary scalar field, does not affect the field tensor $F_{\mu\nu} = \partial_{\mu}A_{\nu} - \partial_{\nu}A_{\mu}$.

Problem 1.114 An inertial coordinate system K' is moving relative to another inertial coordinate system K with constant velocity v along the positive x^1-axis of K.

a) Assume that a stick of length ℓ is at rest in K such that $\Delta\mathbf{x} = (\ell, 0, 0)$. Calculate $\Delta\mathbf{x}'$ in K'.

b) Assume that there is a constant electric field $\mathbf{E} = (0, 0, E)$ in K (no magnetic field, i.e., $\mathbf{B} = \mathbf{0}$ in K). Calculate $\mathbf{E}'$ and $\mathbf{B}'$ in K'.

Problem 1.115 An observer at rest in a frame K experiences only an electric field $\mathbf{E}$. Another observer in another frame K', moving with velocity v along the positive x-axis, will observe a magnetic field $\mathbf{B}'$. Calculate this magnetic field for small velocities (linear terms in v) and show that this field is perpendicular to both the electric field $\mathbf{E}'$ and the velocity of K relative to K'.

Problem 1.116 Let K, K', and K'' be as in Problem 1.34. Assume that there is a constant electric field $\mathbf{E} = (0, 1, 0)$ (in some given physical units) in the coordinate system K. We assume that the magnetic field $\mathbf{B}$ vanishes in K. Compute the components of both the electric and magnetic fields in the coordinate systems K' and K''.

Problem 1.117 Compute the electric and magnetic field components due to a point charge q moving with velocity v along the positive x-axis.

Problem 1.118 A particle of mass m and electric charge q is moving in a constant electric field $\boldsymbol{E}$. Use the Lorentz force law to calculate the velocity of the particle as a function of the displacement r from the origin along the direction of motion. The particle starts off at rest.

Problem 1.119 A current I is flowing through a straight uncharged conductor. Determine the electromagnetic field in an inertial system K' that moves parallel to the conductor with velocity v

a) by transforming the electromagnetic field tensor from the rest frame K of the conductor to K',

b) by transforming the current-density 4-vector from K to K', and then, knowing the charge of the conductor and its current relative to K determine the field in K'.

Problem 1.120 Maxwell's equations can be expressed by means of the electromagnetic 4-vector potential A. When $\partial_\mu A^\mu = 0$ (i.e., the Lorenz gauge), they take on a simple form. What is this form? Assuming that Maxwell's equations are in this simple form, and furthermore, $J = 0$ (i.e., current free), show for a plane wave, $A^\mu = \varepsilon^\mu e^{ik\cdot x}$, where ε is the polarization vector, that

$$\boldsymbol{E} \cdot \boldsymbol{k} = \boldsymbol{B} \cdot \boldsymbol{k} = 0, \tag{1.15}$$

i.e., the electric and magnetic fields are perpendicular to the direction of motion.

Problem 1.121 Calculate the Lorentz invariants $F_{\mu\nu}F^{\mu\nu}$ and $\epsilon_{\mu\nu\omega\lambda}F^{\mu\nu}F^{\omega\lambda}$ for a free electromagnetic plane wave $A^\mu(x) = \epsilon^\mu e^{ik\cdot x}$, where ϵ is the polarization vector. Give a physical interpretation of your result.

Problem 1.122 a) Prove that the scalar product $\boldsymbol{E} \cdot \boldsymbol{B}$ between the electric and magnetic field vectors is invariant under Lorentz transformations.

b) Show that if the electric and magnetic fields $\boldsymbol{E}$ and $\boldsymbol{B}$ are orthogonal for one observer, they are orthogonal for any observer.

c) Show that $\boldsymbol{E}$ and $\boldsymbol{B}$ are orthogonal for free plane waves with $A^\mu(x) = \varepsilon^\mu e^{ik\cdot x}$, where ε is the polarization vector.

d) Show for the plane waves that $\boldsymbol{E} \times \boldsymbol{B} = A\boldsymbol{k}$, where $\boldsymbol{k}$ is the wave vector and A is a nonvanishing expression.

Problem 1.123 An electron with mass m_0 is moving in a homogeneous magnetic field $\boldsymbol{B} = (0,0,B)$ and no electric field. Calculate its trajectory if it has velocity $\boldsymbol{u} = (u,0,0)$ at time $t = 0$.

Problem 1.124 In an inertial coordinate system K, there is a constant electric field $\mathbf{E} = (cB,0,0)$ and a constant magnetic field $\mathbf{B} = (0,B,0)$. In another inertial system K', the same fields are measured to be $\mathbf{E}' = (0,2cB,cB)$ and the x-component $B'_x = 0$. Compute B'_y and B'_z.

Problem 1.125 Observer A measures the electric and magnetic field strengths to be $\mathbf{E} = (\alpha,-\alpha,0)$ and $\mathbf{B} = (0,0,2\alpha/c)$, respectively, where $\alpha \neq 0$. Another observer, observer B, makes the same measurements and finds $\mathbf{E}' = (0,0,2\alpha)$ and $\mathbf{B}' = (B'_x,\alpha/c,B'_z)$. Determine B'_x and B'_z.

Problem 1.126 Observer A measures the electric and magnetic field strengths to be $\mathbf{E} = (0, \beta, -\beta)$ and $\mathbf{B} = (2\beta/c, 0, 0)$, respectively, where $\beta \neq 0$. Another observer, observer B, makes the same measurements and finds $\mathbf{E}' = (2\beta, 0, 0)$ and $\mathbf{B}' = (B'_x, B'_y, \beta/c)$. Determine B'_x and B'_y.

Problem 1.127 Observer A measures the electric and magnetic field strengths to be $\mathbf{E} = (\alpha, 0, 0)$ and $\mathbf{B} = (\alpha/c, 0, 2\alpha/c)$, respectively, where $\alpha \neq 0$. Another observer, observer B, makes the same measurements and finds $\mathbf{E}' = (E'_x, \alpha, 0)$ and $\mathbf{B}' = (\alpha/c, B'_y, \alpha/c)$. Express E'_x and B'_y in terms of α and c. Finally, a third observer, observer C, is moving relative to observer B with constant velocity v along the positive x-axis of observer B. Find the electric and magnetic field strengths, $\mathbf{E}''$ and $\mathbf{B}''$, as observer C measures them.

Problem 1.128 Assume that a muon originally travels vertically down toward the ground from an altitude of 10 km. There is a magnetic field coming from the Earth of $B = 50\ \mu$T affecting the motion of the muon. To make a simple model we take the magnetic field to be constant all the way from 10 km altitude to ground level. Suppose the field lines go from south to north and we are in Japan on the northern hemisphere. How far in length and in which direction is the deviation from the point where the muon would hit the ground without magnetic field, compared to where it hits the ground due to the deviation induced by the magnetic field of the Earth, if it has the energy of 2 GeV and is negatively charged?
Hint: The combination cB, where c is the speed of light, has the value $cB = 300$ V/m, for $B = 1\ \mu$T. The trajectory of a charged particle in a homogeneous magnetic field is a circle, it is sufficient to compute the radius of the circle and then use geometric arguments.

Problem 1.129 An observer in the system S has observed an electromagnetic field tensor $F^{\mu\nu}$ with nonvanishing $\boldsymbol{E}$- and $\boldsymbol{B}$-fields. Performing a Lorentz transformation with velocity u along the positive x_1-axis to another system S' he finds that the $\boldsymbol{B}$-field is absent, i.e., all its components are equal to 0. What is the electric field in this system expressed in u and the components of the electric field in S?

Problem 1.130 In an inertial frame S there is a constant time-independent magnetic field $\boldsymbol{B}$ and no electric field ($\boldsymbol{E} = \mathbf{0}$). Consider another inertial frame S', which moves with velocity v along the positive x^1-axis of S.

a) What are the $\boldsymbol{E}'$ and $\boldsymbol{B}'$ fields in the system S' expressed in the original $\boldsymbol{B}$-field and the velocity v?

b) Verify that the Lorentz invariants are indeed invariant under this transformation.

Problem 1.131 An electron in a linear particle accelerator of length $L = 3$ km (e.g., SLAC in California, USA) is accelerated through an electric potential U.

a) Compute the trajectory $x(t)$ of this electron for $0 < |x(t)| < L$ if its motion starts at time $t = 0$ at rest at one end of the accelerator.

b) Compute the time it takes for this electron to pass through the whole accelerator.

c) Compute the time dependence of the energy of this electron in the accelerator.

Problem 1.132 a) Find the electric and magnetic fields $\boldsymbol{E}$ and $\boldsymbol{B}$ generated by a particle with charge q moving with constant velocity v parallel with the x-axis in an inertial system S, using that the electric and magnetic potentials in the particle's rest frame are

$$\phi(t',\boldsymbol{x}') = \frac{q}{4\pi\,|\boldsymbol{x}'|}, \quad \boldsymbol{A}(t',\boldsymbol{x}') = \boldsymbol{0}, \tag{1.16}$$

we use the notation $\boldsymbol{A} = (A^1, A^2, A^3)$, and similarly for $\boldsymbol{E}$, $\boldsymbol{B}$, and $\boldsymbol{x}$.

b) Explain why it is possible to check your result in a) by computing $\boldsymbol{E}\cdot\boldsymbol{B}$ and $\boldsymbol{E}^2 - \boldsymbol{B}^2$ in both inertial systems. Perform these checks!

Problem 1.133 Bubble chambers were frequently used in the 1960s in particle collision experiments. In a bubble chamber, there is a strong constant magnetic field, which bends the motion of charged particles. The charged particles give rise to bubbles, which make the trajectories of the charged particles visible.

a) In the lab frame of the bubble chamber, there is a strong magnetic field in the z-direction and no electric field. Use the Lorentz force law to show that the trajectory of a charge particle can be parametrized in the lab frame as

$$x = R\cos\omega\tau, \quad y = -R\sin\omega\tau, \tag{1.17}$$

and determine ω. Show that for a charged particle, you can obtain the 3-momentum from knowing the radius of the trajectory and the strength of the magnetic field (any energy losses can be neglect)

$$p_i p^i = q^2 R^2 B_i B^i. \tag{1.18}$$

b) In a bubble chamber, one can only see the traces of charged particles in terms of bubbles. Consider the following process

$$\Sigma^- \longrightarrow \pi^- + X^0,$$

where Σ^- and π^- are known charged particles. Here X^0 is an unknown uncharged particle, which we cannot see, since it does not give rise to bubbles. For the other two particles, we know their rest masses and their trajectory radii R_Σ and R_π (therefore, we also know their 3-momenta). From this information, derive an expression for the rest mass of the unknown particle expressed in terms of the rest masses M_Σ of Σ^- and M_π of π^-, and their respective 3-momenta, as well as the angle θ between the recorded trajectories of the charged particles close to the collision.

Problem 1.134 Starting from the plane wave solution to Maxwell's equations

$$A^\mu = \varepsilon^\mu \sin(k\cdot x), \tag{1.19}$$

show that the electric and magnetic fields are orthogonal and have the same magnitude without referring to a particular gauge condition.

Problem 1.135 Assume that the electromagnetic field in an inertial frame S satisfies $|\boldsymbol{E}| = |\boldsymbol{B}|$ and that the angle between the electric and magnetic field is α. In another inertial frame, the fields are $\boldsymbol{E}'$ and $\boldsymbol{B}'$ with a corresponding angle α'. Show that

$$\cos\alpha' = \frac{\boldsymbol{E}^2}{\boldsymbol{E}'^2}\cos\alpha. \tag{1.20}$$

Problem 1.136 The 4-potential of a stationary point charge Q in its rest frame is given by

$$A^{\mu} = \frac{Q}{4\pi r}(1,\mathbf{0})^{\mu}, \tag{1.21}$$

where $r = \sqrt{x^2+y^2+z^2}$ is the distance to the particle. Compute the electromagnetic stress–energy tensor $T^{\mu\nu}$ in $(x,y,z) = (1,0,0)$ and the corresponding trace T^{μ}_{μ}.

Problem 1.137 Starting from Maxwell's equations and without assuming a particular gauge condition, show that the components of the electromagnetic field tensor $F^{\mu\nu}$ satisfy the sourced wave equation

$$\Box F^{\mu\nu} \equiv \partial_{\sigma}\partial^{\sigma} F^{\mu\nu} = S^{\mu\nu}, \tag{1.22}$$

and express the source tensor $S^{\mu\nu}$ in terms of the 4-current density J^{μ}.

Problem 1.138 The electromagnetic stress–energy tensor is given by

$$T^{\nu}_{\mu} = -\varepsilon_0 \left[F_{\mu\sigma}F^{\nu\sigma} - \frac{1}{4}\delta^{\nu}_{\mu}(F_{\rho\sigma}F^{\rho\sigma}) \right]. \tag{1.23}$$

Given the electromagnetic plane wave solution for the 4-potential

$$A^{\mu} = \varepsilon^{\mu}\sin(k\cdot x), \tag{1.24}$$

express T^{ν}_{μ} in terms of the 4-vector k. You also need to assure that the wave actually fulfills Maxwell's equations in the absence of a source term $\partial_{\mu}F^{\mu\nu} = 0$.
Hint: You may assume the Lorenz gauge condition $\partial_{\mu}A^{\mu} = 0$.

Problem 1.139 The 4-potential A_{μ} is not physical, but may be transformed according to $A_{\mu} \mapsto A_{\mu} + \partial_{\mu}\varphi$, where φ is a scalar field, without changing the physical observables. Show that the physical electromagnetic field tensor $F^{\mu\nu}$ is invariant under this transformation.

Problem 1.140 The electric field of an electric dipole with dipole moment $\boldsymbol{d} = d\boldsymbol{e}_z$ is given by

$$\boldsymbol{E} = \frac{d}{4\pi\varepsilon_0}\left(\frac{3xz}{r^5}\boldsymbol{e}_x + \frac{3yz}{r^5}\boldsymbol{e}_y + \frac{3z^2-r^2}{r^5}\boldsymbol{e}_z\right), \tag{1.25}$$

in its rest frame S. Compute the value of the quantity $F_{\mu\nu}\tilde{F}^{\mu\nu} = \varepsilon^{\mu\nu\sigma\rho}F_{\mu\nu}F_{\sigma\rho}$ as a function of time and position in the frame S', which is moving in the positive x-direction with velocity v relative to S'.

Problem 1.141 A particle at rest acting as an electric monopole and a magnetic dipole has the electromagnetic fields

$$\boldsymbol{E} = \frac{q}{4\pi r^2}\boldsymbol{e}_r \quad \text{and} \quad \boldsymbol{B} = \frac{1}{4\pi r^3}[3(\boldsymbol{m}\cdot\boldsymbol{e}_r)\boldsymbol{e}_r - \boldsymbol{m}], \tag{1.26}$$

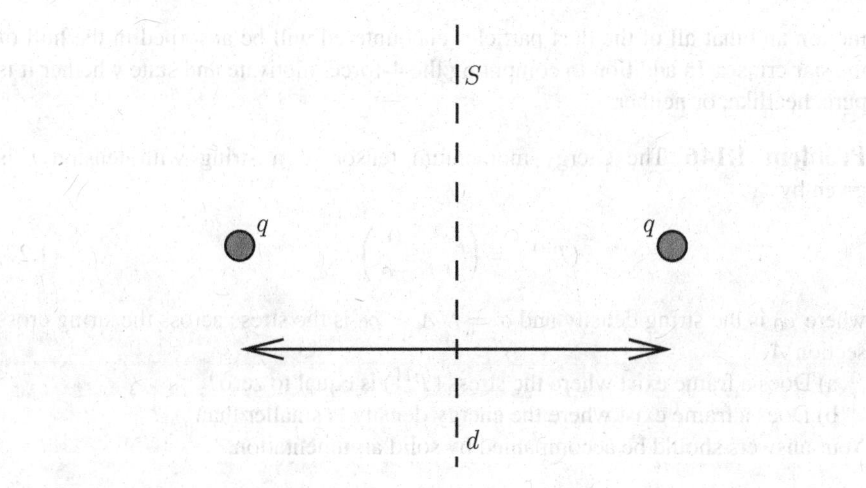

Figure 1.10 Two particles (each with charge q) and the plane S equidistant from both particles.

where $\boldsymbol{r}$ is the position vector relative to the particle, r its magnitude, q the charge of the particle, and $\boldsymbol{m}$ its magnetic dipole moment. Determine whether or not there exists a region of spacetime where the electric field is equal to zero in some inertial frame (although that frame may generally be different for different points in the region). If such a region exists, determine the shape of the region in the particle's rest frame.

Problem 1.142 Two particles with the same charge q are held fixed with a separation distance d (see Figure 1.10). Compute the stress–energy tensor of the static electric field between the charges and use your result to find the total 4-force between the electromagnetic fields on either side of the plane S that is equidistant from both charges.

Problem 1.143 Compute the Lorentz 4-force between two electrons moving in parallel with constant velocity $\boldsymbol{v}$ and a separation d orthogonal to the direction of motion.

1.8 Energy–Momentum Tensor

Problem 1.144 Determine the momentum density of a gas consisting of massless noninteracting particles in a frame which is moving with velocity $\boldsymbol{v}$ relative to the gas rest frame. Express your result in terms of $\boldsymbol{v}$ and the energy density of the gas *in the frame where the gas is moving*.

Problem 1.145 A star cruiser is moving through space with velocity v relative to the galaxy. Suddenly it encounters a gas cloud of dust particles. What is the 4-force from the dust cloud on the star cruiser at the moment it enters the cloud? You may assume that the star cruiser has a cross sectional area A relative to the direction of motion and that all of the dust particles encountered will be absorbed in the hull of the star cruiser. In addition to computing the 4-force, motivate and state whether it is pure, heatlike, or neither.

Problem 1.146 The energy–momentum tensor of a string with tension t is given by

$$(T^{\mu\nu}) = \begin{pmatrix} \rho_0 & 0 \\ 0 & -\sigma \end{pmatrix}, \tag{1.27}$$

where ρ_0 is the string density and $\sigma = t/A < \rho_0$ is the stress across the string cross section A.

a) Does a frame exist where the stress (T^{11}) is equal to zero?

b) Does a frame exist where the energy density is smaller than ρ_0?

Your answers should be accompanied by solid argumentation.

Problem 1.147 A pure photon gas such as the cosmic microwave background (CMB) can be described as a perfect fluid with pressure $p = \rho_0/3$ in its rest frame. In a frame moving with velocity $\boldsymbol{v}$ in relation to the rest frame of the CMB, compute the energy density, momentum density, and stress tensors. In addition, comment on whether the shear stress (off-diagonal elements of the stress tensor) in an arbitrary frame is zero or not.

Problem 1.148 In a perfect fluid with proper density ρ_0 and positive proper pressure p, find an expression for the energy density ρ in an arbitrary inertial frame S' and derive an upper bound on ρ/γ^2, where γ is the gamma factor between the fluid's rest frame and S'.

Problem 1.149 The energy density in the frame of an observer with 4-velocity V is given by $\rho = T_{\mu\nu}V^\mu V^\nu$. The *weak energy condition* is a condition requiring the energy density to be nonnegative for all observers, i.e., $\rho \geq 0$. For a perfect fluid, determine the condition on the equation of state parameter w in the relation $p = w\rho_0$ that the weak energy condition implies.

1.9 Lagrange's Formalism

Problem 1.150 The 4-momentum of a free particle of mass m is $p^\mu = mc\dot{x}^\mu$.

a) Show that the momentum is conserved (i.e., independent of time) by deriving the Euler–Lagrange variational equations for the Lagrangian $\mathscr{L} = p^2/(2m)$ in Minkowski space.

b) When the particle moves in an electromagnetic field, one can obtain the relevant equations of motion by using the substitution $p \mapsto p + qA/c$, where $A = A(x)$ is the electromagnetic potential and q is the charge of the particle. Show that, to lowest nontrivial order in q, the equations of motion for the particle give the equations of the Lorentz force.

2

General Relativity Theory

2.1 Some Differential Geometry

Problem 2.1 Show that the function $f(x,y) = x^2 + y$ is a smooth function on the unit sphere $\mathbb{S}^2 \subset \mathbb{R}^3$.

Problem 2.2 On the unit sphere $M = \mathbb{S}^2$, we use the spherical coordinates θ and ϕ, except at the poles $\theta = 0, \pi$. A curve can then be parametrized as $(\theta(s), \phi(s))$. A tangent vector $v \in T_p\mathbb{S}^2$ is given by its components $v = (v_\theta, v_\phi)$ with $v_\theta = \dot{\theta}(s_0)$, $v_\phi = \dot{\phi}(s_0)$, and $p = (\theta(s_0), \phi(s_0))$. How would you describe a tangent vector at the poles $\theta = 0, \pi$?

Problem 2.3 Find the metric tensor and the Christoffel symbols in the two-dimensional Euclidean plane in the following coordinates

a) s and t defined by $x = se^t$ and $y = se^{-t}$.

b) u and v defined by $x = u$ and $y = v^2$.

In both cases, discuss where in the Euclidean plane the new coordinates provide a well-defined coordinate system.

Problem 2.4 Let $\alpha(t)$ and $\beta(t)$ be a pair of smooth curves on a manifold M such that $\alpha(t_0) = \beta(t_0)$. Show that the condition

$$\left.\frac{d}{dt}x^i(\alpha(t))\right|_{t=t_0} = \left.\frac{d}{dt}x^i(\beta(t))\right|_{t=t_0} \qquad \text{for } i = 1,2,\ldots,n, \tag{2.1}$$

is independent of the choice of local coordinates x^i, i.e., if the curves are tangential in one coordinate system, then they are tangential in any other coordinate system.

Problem 2.5 Show that the system of first-order ordinary differential equations

$$\dot{x}^k \nabla_k Y^i = \dot{Y}^i(s) + \Gamma^i_{kj}(x(s))\dot{x}^k(s)Y^j(s) = 0, \quad i = 1,2,\ldots,n, \tag{2.2}$$

defining the parallel transport along a curve on a manifold M is coordinate independent in the sense that if the system is valid in one coordinate system, then it is also valid in any other coordinate system.

Problem 2.6 The distance between two points a and b on the unit sphere $\mathbb{S}^2$ along a curve $\gamma(s) = (\theta(s), \phi(s))$ is given by

$$\ell[\gamma] \equiv \int_a^b \sqrt{g_{\gamma(s)}(\dot{\gamma}(s), \dot{\gamma}(s))}\, ds = \int_a^b \sqrt{g_{\theta\theta}\dot{\theta}(s)^2 + g_{\phi\phi}\dot{\phi}(s)^2}\, ds$$
$$= \int_a^b \sqrt{\dot{\theta}(s)^2 + \sin^2\theta(s)\,\dot{\phi}(s)^2}\, ds. \tag{2.3}$$

Use Euler–Lagrange equations to derive the geodesic equations on $\mathbb{S}^2$.

Problem 2.7 Compute the Christoffel symbols on the unit sphere $\mathbb{S}^2$ with metric given by $ds^2 = d\theta^2 + \sin^2\theta\, d\phi^2$

a) directly from the metric.

b) using the general formula for the geodesic equations.

Problem 2.8 We define the Christoffel symbols on the unit sphere $\mathbb{S}^2$, using spherical coordinates (θ, ϕ). When $\theta \neq 0, \pi$, we find (see Problem 2.7)

$$\Gamma^{\theta}_{\phi\phi} = -\frac{1}{2}\sin 2\theta, \quad \Gamma^{\phi}_{\theta\phi} = \Gamma^{\phi}_{\phi\theta} = \cot\theta, \tag{2.4}$$

and all other Γ are equal to zero. Show that the apparent singularity at $\theta = 0, \pi$ can be removed by a better choice of coordinates at the poles of the sphere. Thus, the above affine connection extends to the whole $\mathbb{S}^2$.

Problem 2.9 a) Let $M = \mathbb{S}^2$ and Γ be the affine connection in Problem 2.8. The coordinates $\theta(s)$ and $\phi(s)$ of a geodesic then satisfy the geodesic equations, i.e.,

$$\ddot{\theta}(s) - \frac{1}{2}\sin 2\theta(s)\,\dot{\phi}(s)\dot{\phi}(s) = 0, \tag{2.5}$$
$$\ddot{\phi}(s) + 2\cot\theta(s)\,\dot{\theta}(s)\dot{\phi}(s) = 0. \tag{2.6}$$

Find the general solution to the geodesic equations.

b) Let M and Γ be as in a). Furthermore, let $(\theta, \phi) = (\alpha s + \beta, \phi_0)$, where s is the curve parameter and α, β, and ϕ_0 are constants. Determine the parallel transport equations for a vector field $X = (X^\theta, X^\phi)$ and solve this set of equations. In addition, if u is the tangent vector $(1,1)$ at the point $(\theta, \phi) = \left(\frac{\pi}{4}, 0\right)$, then determine the parallel transported vector v at the point $(\theta, \phi) = \left(\frac{\pi}{2}, 0\right)$.

Problem 2.10 Determine the shortest path on the conical surface $r = -az$ which connects the points $z = -h$, $\varphi = 0$ and $z = -h$, $\varphi = \pi/2$, where (r, φ, z) are cylindrical coordinates and $a > 0$ and $h > 0$ constants.

Problem 2.11 A ship starts from a position in the Atlantic Ocean with coordinates 10° N 30° W (Cape Verde). It sails directly to the north to the 45° northern latitude (Azores, Portugal) and then it turns abruptly to the west and sails until it hits the 60° western longitude (Nova Scotia, Canada). Suppose a vector is parallel transported along the route of the ship (with the help of a gyroscope). Its initial direction is 45° (north-east). What is its final direction?

Problem 2.12 A vector is first parallel transported along a great circle on a sphere from a point A on the equator to the North Pole N, then again along a great circle from N to another point B on the equator, and finally, along the equator back to the point A. Use the standard Riemannian metric on the sphere and prove that the vector is rotated in the above process by an angle θ, which is directly proportional to the area of the geodesic triangle ANB.

Problem 2.13 Let $M = \mathbb{S}^2 \subset \mathbb{R}^3$. Determine the metric g on M in terms of the spherical coordinates θ and ϕ. In particular, compute the inner product of the vectors $(1,2)$ and $(2,-1)$ at the point (θ,ϕ).

Problem 2.14 Compute the Riemann curvature tensor R of the unit sphere $\mathbb{S}^2$.

Problem 2.15 Consider the vector fields

$$X = x\frac{\partial}{\partial y} - y\frac{\partial}{\partial x} \quad \text{and} \quad Y = x\frac{\partial}{\partial x} + y\frac{\partial}{\partial y}, \tag{2.7}$$

in the xy-plane.

a) Determine the commutator $[X,Y]$.

b) Assume that an affine connection in the plane satisfies $\nabla_X X = -Y$, $\nabla_Y Y = Y$, $\nabla_Y X = X$, and that the torsion tensor T vanishes. Compute the Riemann curvature tensor R.

Problem 2.16 Let x^1 and x^2 be a pair of local coordinates and

$$X = x^2\frac{\partial}{\partial x^1} - x^1\frac{\partial}{\partial x^2}, \quad Y = x^1\frac{\partial}{\partial x^1} + x^2\frac{\partial}{\partial x^2}, \tag{2.8}$$

be a pair of vector fields in $\mathbb{R}^2\backslash\{0\}$. Assume that

$$\begin{aligned} \nabla_X X = 0, \quad \nabla_X Y = X + Y, \\ \nabla_Y X = X - Y, \quad \nabla_Y Y = 0. \end{aligned} \tag{2.9}$$

Compute the components $R^1{}_{1ij}$ in the local coordinate basis, where $i,j = 1,2$, of the Riemann curvature tensor.

Problem 2.17 A manifold $\mathcal{M}$ of dimension 3 has a basis of orthonormal vector fields $\{L_1, L_2, L_3\}$ with commutation relations

$$\left[L_i, L_j\right] = \epsilon_{ijk} L_k, \quad \text{where } i,j,k = 1,2,3. \tag{2.10}$$

Determine the Levi-Civita connection $\nabla_i = \nabla_{L_i}$ $(1 \le i \le 3)$ and its Riemann curvature tensor R.

Hint: The Levi-Civita connection is the unique metric-compatible torsion-free connection. Use the symmetry properties of the Christoffel symbols coming from this, several times, to evaluate them.

Problem 2.18 Let x and y be local coordinates on a surface S with $x + y \neq 0$. Define a metric tensor g by $g_{xx} = 1$, $g_{xy} = g_{yx} = x + y$, and $g_{yy} = 1 + (x+y)^2$. Let ∇ be an affine connection defined by

$$\nabla_x \partial_x = (x+y)\partial_x - \partial_y, \tag{2.11}$$
$$\nabla_x \partial_y = [2+(x+y)^2]\partial_x - (x+y)\partial_y, \tag{2.12}$$
$$\nabla_y \partial_x = (x+y)(x+y+1)\partial_x - (x+y+1)\partial_y, \tag{2.13}$$
$$\nabla_y \partial_y = \{(x+y+1)[1+(x+y)^2]+1\}\partial_x - (x+y)(x+y+1)\partial_y. \tag{2.14}$$

a) Compute the Christoffel symbols in the orthonormal basis

$$e_1 = \partial_x, \quad e_2 = -(x+y)\partial_x + \partial_y. \tag{2.15}$$

b) Consider the parallel transport of a pair of vectors starting from the point $(x, y) = (1, 1)$, counterclockwise along the full circle with center at $(x, y) = (2, 2)$ and radius $r = \sqrt{2}$. Assume that the initial angle between the vectors is $\pi/3$. What is the angle after the parallel transport around the loop?

Problem 2.19 Three ants are walking on a two-dimensional surface embedded in a flat three-dimensional Euclidean space as

$$x = r\cos\phi, \quad y = r\sin\phi, \quad z = \frac{2}{3}r^{3/2}. \tag{2.16}$$

Assume that the ants are walking on the surface along curves parametrized by λ such that

- Ant #1: $r = \lambda$, $\phi = 0$;
- Ant #2: $r = \lambda^{2/3} - 1$, $\phi = \pi/2$, $\quad \lambda > 1$;
- Ant #3: $r = \lambda^{1/2}$, $\phi = \ln\lambda$, $\quad \lambda > 0$.

a) Compute the induced metric on the two-dimensional surface.
b) Investigate if the ants are walking along geodesics or not.

Problem 2.20 Derive the explicit form of the geodesic equation on the hyperboloid $x^2 + y^2 - z^2 = a^2$ with x, y, and z being Cartesian coordinates on the flat Euclidean three-dimensional space (i.e., the metric is $ds^2 = dx^2 + dy^2 + dz^2$) and $a > 0$ a constant. Using the coordinates r and φ such that $x = r\cos(\varphi)$ and $y = r\sin(\varphi)$, compute also all Christoffel symbols for this hyperboloid.

Problem 2.21 Consider the surface $(ct)^2 - x^2 - y^2 = -K^2$ in the three-dimensional Minkowski space $\mathbb{R}^3$ with metric signature $+--$.

a) Compute the metric tensor on the surface in a suitable coordinate system. Is it positive definite? If not, of what type?

b) Find the geodesic equations. In particular, find a geodesic starting from the point $(0, 0, K)$ and going in the direction of the tangent vector $(c, 0, 0)$.

Problem 2.22 Consider the pseudo-Riemannian metric

$$ds^2 = (dx^1)^2 + (dx^2)^2 - (dx^3)^2 - (dx^4)^2, \tag{2.17}$$

in $\mathbb{R}^4$. This induces a pseudo-Riemannian metric g on the surface

$$S: \quad (x^1)^2 + (x^2)^2 - (x^3)^2 - (x^4)^2 = 1. \tag{2.18}$$

a) Show that the metric g on S is Lorentzian, i.e., it has one timelike and two spacelike directions at each point.

b) Construct a pair of constants of motions for freely falling bodies by integrating the geodesic equations on S once.

Problem 2.23 The flow lines generated by a vector field X are smooth curves $\gamma(t)$ such that

$$\dot{\gamma}(t) = X(\gamma(t)), \tag{2.19}$$

along the curve. Assume that all flow lines for a vector field X are geodesics with respect to a connection determined by the Christoffel symbols Γ^k_{ij}. Derive a set of partial differential equations for the components of X giving a necessary and sufficient condition for the above property of X.

Problem 2.24 Parametrize the points on the spin group SU(2) as

$$g(\boldsymbol{x}) = e^{i(x^1\sigma_1 + x^2\sigma_2 + x^3\sigma_3)} = \mathbb{1}\cos(r) + i\frac{\sin(r)}{r}\boldsymbol{x}\cdot\boldsymbol{\sigma}, \tag{2.20}$$

where $\boldsymbol{x} = (x^1, x^2, x^3) \in \mathbb{R}^3$, $r = |\boldsymbol{x}|$ and the σ_k's are the Hermitian 2×2 Pauli matrices

$$\sigma_1 = \begin{pmatrix} 0 & 1 \\ 1 & 0 \end{pmatrix}, \quad \sigma_2 = \begin{pmatrix} 0 & -i \\ i & 0 \end{pmatrix}, \quad \text{and} \quad \sigma_3 = \begin{pmatrix} 1 & 0 \\ 0 & -1 \end{pmatrix}. \tag{2.21}$$

We can identify the point $g(\boldsymbol{x})$ as a point on the 3-dimensional unit sphere $\mathbb{S}^3 \subset \mathbb{R}^4$, the first coordinate is $\cos(r)$ and the remaining three coordinates are $\sin(r)\boldsymbol{x}/r$. Show that the 1-parameter subgroups $t \mapsto e^{it\boldsymbol{a}\cdot\boldsymbol{\sigma}}$, where $\boldsymbol{a} \in \mathbb{R}^3$, are geodesics with respect to the standard metric on $\mathbb{S}^3$ coming from the Euclidean metric in $\mathbb{R}^4$.

Hint: It is more convenient to use the Euler–Lagrange equations coming from the metric element (derive the formula!) in terms of the angular coordinates θ, ϕ of the vector $\boldsymbol{x} \in \mathbb{R}^3$ and the radial coordinate r.

Problem 2.25 a) Derive the relation between the Christoffel symbols $\Gamma^\lambda_{\mu\nu}$ and the metric tensor $g_{\mu\nu}$ from the following conditions: (i) $D_\lambda g_{\mu\nu} = 0$, where D_λ is the covariant derivative, and (ii) $\Gamma^\lambda_{\mu\nu} = \Gamma^\lambda_{\nu\mu}$. (The result is the so-called "fundamental theorem" in Riemannian geometry.)

b) Consider the vector field $(V^\mu) = (x, -t)$, i.e., $V^0 = x$ and $V^1 = -t$, in two-dimensional Minkowski spacetime with coordinates $(x^\mu) = (t, x)$ and metric $ds^2 = dt^2 - dx^2$. Compute all components of the tensor $T_\mu{}^\nu = D_\mu V^\nu$ in this coordinate system. Compute also the component $T'_0{}^1$ of this tensor in Rindler coordinates $(x'^\mu) = (\lambda, a)$ defined as

$$t = a\sinh(\lambda), \quad x = a\cosh(\lambda). \tag{2.22}$$

Problem 2.26 a) Write the transformation law for a tensor with components $S^{\mu\nu}$ under a general coordinate transformation $x^\mu \mapsto x'^\mu$, i.e., give a general formula for $S'^{\mu\nu}$.

b) Write $D_\mu S^{\mu\nu}$ in terms of partial derivatives and Christoffel symbols.

c) Consider the tensor with components $S^{12} = -S^{21} = 2xy$ and $S^{11} = S^{22} = 0$ on the two-dimensional plane with coordinates $(x^\mu) = (x, y)$ and metric $ds^2 = dx^2 + dy^2$. (i) Compute the components $S'^{\mu\nu}$ of this tensor in polar coordinates $(x'^\mu) = (r, \varphi)$. (ii) Compute $D_\mu S^{\mu\nu}$.

Problem 2.27 Consider AdS_2 which is a two-dimensional curved spacetime with coordinates $(x^\mu) = (x^0, x^1) = (t, r)$ and the metric given by

$$ds^2 = v\left[(r^2 - 1)dt^2 - \frac{1}{r^2 - 1}dr^2\right], \tag{2.23}$$

where $v > 0$ is some constant. Consider also on AdS_2 and in the coordinates x^μ, a tensor field $S_{\mu\nu}$ with the following components $S_{00} = a(r^2 - 1)$, $S_{11} = -a/(r^2 - 1)$, and $S_{01} = S_{10} = 0$ for some constant $a > 0$.

a) Compute all Christoffel symbols for AdS_2 in the coordinates x^μ.

b) Compute $S^{\mu\nu}$ and $D_\mu S^{\mu\nu}$.

c) Consider another coordinate system $(x'^\mu) = (x'^0, x'^1) = (\theta, \eta)$ on AdS_2 defined by $\theta = at$ and $r = \cosh(\eta)$ with a being the same constant as above. Transform the tensor $S_{\mu\nu}$ to this new coordinate system, i.e., compute $S'_{\mu\nu}$.

Problem 2.28 Consider the so-called Rindler coordinate system $(x'^\mu) = (x'^0, x'^1) = (\lambda, a)$ in two-dimensional Minkowski space defined by

$$t = a\sinh(\lambda), \quad x = a\cosh(\lambda), \quad a > 0, \lambda \in \mathbb{R}, \tag{2.24}$$

where $(x^\mu) = (t, x)$ are the usual coordinates, i.e., $ds^2 = dt^2 - dx^2$.

a) Find the metric tensor and the Christoffel symbols in this coordinate system.

b) Determine expressions for the divergence of a vector field and the Laplacian of a scalar field in two-dimensional Minkowski space in Rindler coordinates.

c) A tensor of rank two on two-dimensional Minkowski space has the following components in the coordinates $(x^\mu) = (t, x)$: $T^0{}_0 = -T^1{}_1 = x^2 - t^2$ and $T^1{}_0 = T^0{}_1 = 0$. Compute the component $T'^0{}_0$, i.e., $T'^\mu{}_\nu$ for $\mu = \nu = 0$, of this tensor in Rindler coordinates.

Problem 2.29 Consider the metric

$$ds^2 = dt^2 - dr^2 - r^2 d\phi^2. \tag{2.25}$$

Find expressions for the covariant derivative $\nabla_\mu V_\nu$ and the divergence $\nabla_\mu V^\mu$.

Problem 2.30 The parallel transport of a vector V^μ along a curve parametrized by λ is given by the condition

$$\frac{dx^\mu}{d\lambda}\nabla_\mu V^\nu = 0. \tag{2.26}$$

a) Obtain the geodesic equation

$$\ddot{x}^\mu + \Gamma^\mu_{\nu\lambda}\dot{x}^\nu\dot{x}^\lambda = 0, \tag{2.27}$$

from parallel transporting a tangent vector.

b) If one allows the tangent vector T^μ to change in size, one can generalize the condition $\frac{dx^\mu}{d\lambda}\nabla_\mu V^\nu = 0$ to

$$T^\mu \nabla_\mu T^\nu = \alpha T^\nu, \quad \text{where } T^\mu = \frac{dx^\mu}{d\lambda}. \tag{2.28}$$

Show that one can get back the original condition, where one has zero on the right-hand side, by making a reparametrization $\lambda \to \tau(\lambda)$. Show also how the geodesic equation is modified by the extra term on the right-hand side.

Problem 2.31 A conformal transformation of the metric tensor is defined as

$$g_{\mu\nu} \longrightarrow f(x) g_{\mu\nu}, \tag{2.29}$$

for an arbitrary positive function f.

a) Show that a conformal transformation preserves the angle between any two vectors.

b) A null curve is a curve for which all tangent vectors are null vectors. Show that all null curves remain null curves after performing a conformal transformation.

Problem 2.32 A sphere is described locally by the two coordinates θ and φ is embedded in $\mathbb{R}^3$ according to

$$x = R\cos(\varphi)\sin(\theta), \quad y = R\sin(\varphi)\sin(\theta), \quad z = R\alpha\cos(\theta), \tag{2.30}$$

where R and $0 < \alpha < 1$ are constants (note that while the topology of this manifold is a sphere, this is not the standard embedding of the sphere in Euclidean space). Compute the induced metric tensor and the Christoffel symbols.

2.2 Christoffel Symbols, Riemann and Ricci Tensors, and Einstein's Equations

Problem 2.33 a) Let $\Gamma^\lambda_{\mu\nu}$ be the Levi-Civita connection associated to a metric tensor $g_{\mu\nu}$. Show that $\Gamma^\mu_{\mu\nu} = \frac{1}{2}\bar{g}^{-1}\partial_\nu \bar{g}$, where $\bar{g} = \det(g_{\mu\nu})$.

b) Derive from the definition of covariant differentiation the transformation rule for the Christoffel symbols with respect to general coordinate transformations.

c) Show directly from the definition of parallel transport that in a parallel transport defined by the Levi-Civita connection, $\Gamma^\lambda_{\mu\nu} = \frac{1}{2} g^{\lambda\omega}(\partial_\mu g_{\nu\omega} + \partial_\nu g_{\mu\omega} - \partial_\omega g_{\mu\nu})$, the length of a parallel transported vector is constant.

Problem 2.34 For any two vector fields U^μ and V^μ on a manifold with metric $g_{\mu\nu}$ equipped with the Levi-Civita connection, show that if

$$W^\nu = U^\mu \nabla_\mu V^\nu, \tag{2.31}$$

then

$$W_\nu \equiv g_{\nu\sigma} W^\sigma = U^\mu \nabla_\mu V_\nu. \tag{2.32}$$

Problem 2.35 The nonzero Christoffel symbols on a unit sphere $\mathbb{S}^2$ in spherical coordinates are

$$\Gamma^{\theta}_{\phi\phi} = -\frac{1}{2}\sin 2\theta, \quad \Gamma^{\phi}_{\theta\phi} = \Gamma^{\phi}_{\phi\theta} = \cot\theta. \tag{2.33}$$

a) Compute the Christoffel symbols $\Gamma^{j}_{\theta i}$ and $\Gamma^{j}_{\phi i}$ in the orthonormal basis $e_1 = \partial_\theta$, $e_2 = \frac{1}{\sin\theta}\partial_\phi$, $\nabla_\theta e_i = \Gamma^{j}_{\theta i} e_j$, and $\nabla_\phi e_i = \Gamma^{j}_{\phi i} e_j$.

b) Prove that the parallel transport of a vector $u = u_1 e_1 + u_2 e_2 = \begin{pmatrix} u_1 \\ u_2 \end{pmatrix}$ around a closed loop $\gamma(t)$ on $\mathbb{S}^2$ is given by the operation $u' = Ru$, where R is a rotation by an angle Ω equal to the area of the region bounded by the loop γ.
Hint: First, write the solution as a line integral of the Christoffel symbols around the loop, and then, apply Green's formula in the plane.

Problem 2.36 The Christoffel symbols for the flat Euclidean metric in $\mathbb{R}^3$ vanish. Compute the Christoffel symbols in the spherical coordinates (r,θ,φ).

Problem 2.37 A sphere can be projected onto a plane using stereographic projection. We use the metric

$$ds^2 = R^2(d\theta^2 + \sin^2\theta d\phi^2), \tag{2.34}$$

for the sphere. The x and y coordinates of the plane can be expressed in terms of the spherical coordinates as

$$x = 2R\tan(\theta/2)\cos\phi, \quad y = 2R\tan(\theta/2)\sin\phi. \tag{2.35}$$

a) Express the metric of the sphere in terms of the x and y coordinates.
b) Compute the Christoffel symbols using the metric obtained in a).
c) Draw a picture illustrating the stereographic projection.

Problem 2.38 A two-dimensional hyperbolic subspace $x^2 + y^2 - t^2 = 1$, $z = 0$ is embedded into the four-dimensional Minkowski space.
a) Parametrize the surface using only two parameters.
b) Compute the induced metric on the subspace.
c) Compute the Christoffel symbols in the subspace.

Problem 2.39 a) Starting from the definition of the curvature tensor

$$R(X,Y)Z = [\nabla_X, \nabla_Y]Z - \nabla_{[X,Y]}Z, \tag{2.36}$$

derive the formula for the components $R^{\omega}{}_{\mu\nu\lambda}$ in terms of the Christoffel symbols. Prove the first Bianchi identity

$$R^{\omega}{}_{\mu\nu\lambda} + R^{\omega}{}_{\nu\lambda\mu} + R^{\omega}{}_{\lambda\mu\nu} = 0, \tag{2.37}$$

in the case when the torsion $T = 0$.
b) Prove the second Bianchi identity

$$R_{\alpha\beta\mu\nu;\lambda} + R_{\alpha\beta\nu\lambda;\mu} + R_{\alpha\beta\lambda\mu;\nu} = 0, \tag{2.38}$$

and use this to show that the covariant derivative of the energy–momentum tensor $T^{\mu\nu}$ in Einstein's equations

$$G^{\mu\nu} = \frac{8\pi G}{c^4} T^{\mu\nu}, \tag{2.39}$$

vanishes.

Motivate that the vanishing of the covariant derivative of $T^{\mu\nu}$ coincides with local energy–momentum conservation for flat spacetime.

Problem 2.40 Derive the formula relating the Riemann curvature tensor to the parallel transport around an infinitesimal parallelogram.

Problem 2.41 Fix a metric on the paraboloid $z = x^2 + y^2$ induced by the standard Euclidean metric in $\mathbb{R}^3$. Compute the components of the Riemann curvature tensor on the paraboloid.
Hint: Use polar coordinates in the xy-plane.

Problem 2.42 Consider the two-dimensional metric

$$ds^2 = r^2(dr^2 + r^2 d\phi^2). \tag{2.40}$$

a) Calculate the component $R^r{}_{\phi r \phi}$ of the Riemann tensor.

b) In flat Euclidean space, the relation between area and circumference of a circle is $C^2 = 4\pi A$. What is the relation for a circle around the origin for the above metric? The area is given by the integral $\int \sqrt{\det(g)}\, dr d\phi$.

Problem 2.43 Let M be a Lorentzian manifold of dimension $n = 3$. Assume that there is an orthogonal basis of vector fields X, Y, Z such that

1. $g(X,X) = g(Y,Y) = -g(Z,Z) = -1$,
2. $[X,Y] = -Z$, $[Y,Z] = X$, $[Z,X] = Y$,

where g is the metric tensor. Compute the Christoffel symbols of the Levi-Civita connection and the Riemann curvature tensor in this basis.
Hint: Use the symmetry properties of the Christoffel symbols coming from the torsion-free property of the connection together with $\nabla_X g = \nabla_Y g = \nabla_Z g = 0$.

Problem 2.44 a) Show that the Ricci tensor $R_{\mu\nu} = R^\lambda{}_{\mu\lambda\nu}$ (and thus also the Einstein tensor) is symmetric when the Riemann curvature tensor $R^\alpha{}_{\mu\beta\nu}$ has been constructed from a metric.

b) Show that in two spacetime dimensions the tensor $R_{\mu\nu} - k g_{\mu\nu} R$ vanishes for some number k. Determine k.

c) Show that any metric in a 1+1-dimensional spacetime satisfies Einstein's equations in vacuum ($T_{\mu\nu} = 0$), i.e, $G_{\mu\nu} = 0$.
Hint: Use the (anti)symmetries

$$R_{\alpha\beta\mu\nu} = -R_{\beta\alpha\mu\nu} = -R_{\alpha\beta\nu\mu} = R_{\mu\nu\alpha\beta}, \tag{2.41}$$

of the Riemann curvature tensor.

Problem 2.45 Consider the two-dimensional curved spacetime with the metric given by

$$ds^2 = \frac{1}{y^2}(dt^2 - dy^2), \tag{2.42}$$

in coordinates $(x^\mu) = (x^0, x^1) = (t, y)$ with $t \in \mathbb{R}$ and $y \geq 0$.

a) Find the geodesic equations and the Christoffel symbols.

b) Compute the Riemann curvature tensor and the Ricci scalar.

Problem 2.46 Calculate the Christoffel symbols, the Riemann curvature tensor, the Ricci tensor, and the Ricci scalar for the metric

$$ds^2 = d\rho^2 + (a^2 + \rho^2)d\phi^2, \tag{2.43}$$

where $a > 0$ is a constant and the coordinates ρ and ϕ vary in the intervals $-\infty < \rho < \infty$ and $0 \leq \phi < 2\pi$, respectively.

Problem 2.47 Show by direct computation of the Riemann curvature tensor that the curvature of the Rindler space with coordinates λ and a and line element

$$ds^2 = a^2 d\lambda^2 - da^2, \tag{2.44}$$

is zero.

Problem 2.48 Consider the curved two-dimensional spacetime $t^2 - x^2 - y^2 = -1$ embedded in three-dimensional Minkowski spacetime with coordinates $(x^\mu) = (t, x, y)$ and the metric $ds^2 = dt^2 - dx^2 - dy^2$. Compute the metric tensor $g_{\mu\nu}$ and the Ricci tensor $R_{\mu\nu}$ for this two-dimensional spacetime and thus prove that

$$R_{\mu\nu} = -\Lambda g_{\mu\nu}, \tag{2.45}$$

for some constant Λ to be determined. Perform this computation in the coordinate system $(x^\mu) = (\lambda, \varphi)$, where $t = \sinh(\lambda)$, $x = \cosh(\lambda)\cos(\varphi)$, and $y = \cosh(\lambda)\sin(\varphi)$.

Problem 2.49 Compute the Ricci tensor for the two-dimensional spacetime AdS_2 and in the coordinates x^μ as defined in Problem 2.27.

Problem 2.50 Consider the 2-dimensional manifold M defined by being the surface $t^2 + u^2 - x^2 = \alpha^2$ (with $\alpha > 0$ being a constant) embedded in a 3-dimensional flat manifold with coordinates t, u, and x, and line element $ds^2 = dt^2 + du^2 - dx^2$.

a) Introduce suitable coordinates on M and compute the line element for M in terms of those coordinates.

b) Compute the Christoffel symbols in M in the coordinates introduced in a).

c) Compute the Ricci scalar in M.

Problem 2.51 a) Derive the geodesic equations and determine the metric tensor, the Christoffel symbols, the Riemann curvature tensor, the Ricci tensor, and the Ricci scalar for the spherically symmetric metric

$$ds^2 = g_{tt}(t,r)c^2dt^2 + g_{rr}(t,r)dr^2 - r^2\left(d\theta^2 + \sin^2\theta d\phi^2\right), \tag{2.46}$$

using the following parametrization

$$g_{tt} = e^{\nu}, \quad g_{rr} = -e^{\rho}, \tag{2.47}$$

with the arbitrary functions $\nu = \nu(t,r)$ and $\rho = \rho(t,r)$.

b) Derive the so-called Schwarzschild solution to Einstein's equations in empty space, i.e., solve $G_{\alpha\beta} = 0$ with the spherically symmetric metric given in a). *Hint:* Assume that Birkhoff's theorem holds, which states that any spherically symmetric solution to $G_{\alpha\beta} = 0$ must be static (and asymptotically flat), i.e., $\mathring{\nu} = 0$ and $\mathring{\rho} = 0$, where a circle denotes partial differentiation with respect to time t.

Problem 2.52 Prove Birkhoff's theorem, i.e., prove that any spherically symmetric solution to Einstein's equations in empty space must be static.

2.3 Maxwell's Equations and Energy–Momentum Tensor

Problem 2.53 The energy–momentum tensor associated with the electromagnetic field strength tensor $F^{\mu\nu}$ is

$$T^{\mu\nu} = \epsilon_0 F^{\mu}{}_{\lambda} F^{\lambda\nu} + \frac{\epsilon_0}{4} g^{\mu\nu} F_{\lambda\omega} F^{\lambda\omega}, \tag{2.48}$$

where $g^{\mu\nu}$ is the inverse of the metric tensor $g_{\mu\nu}$. Maxwell's equations in general relativity are written as in Minkowski space, except that partial derivatives are replaced by covariant derivatives, i.e., $\nabla_\mu F^{\mu\nu} = J^\nu$. Show that

$$\nabla_\mu T^{\mu\nu} = \epsilon_0 J_\mu F^{\mu\nu}. \tag{2.49}$$

Note that this does not violate the relation $\nabla_\mu T^{\mu\nu}_{\text{tot}} = 0$, since the $T^{\mu\nu}$ considered in this problem is just the electromagnetic part of the total energy–momentum tensor.

Problem 2.54 Show that half of Maxwell's equations, i.e.,

$$\partial_\alpha F_{\beta\gamma} + \partial_\beta F_{\gamma\alpha} + \partial_\gamma F_{\alpha\beta} = 0, \tag{2.50}$$

can be written precisely in the same form in general relativity; the equations transform covariantly in general coordinate transformations. Why is it unnecessary to write

$$\nabla_\alpha F_{\beta\gamma} + \nabla_\beta F_{\gamma\alpha} + \nabla_\gamma F_{\alpha\beta} = 0? \tag{2.51}$$

Problem 2.55 Show that the covariant form $\nabla_\mu j^\mu = 0$ of the current conservation law can be written as $\bar{g}^{-\frac{1}{2}} \partial_\mu \left(\bar{g}^{\frac{1}{2}} j^\mu\right) = 0$, where $\bar{g} = -\det(g_{\mu\nu})$; $g_{\mu\nu}$ is a Lorentzian metric. Show that this is compatible with the generally covariant form $\nabla_\mu F^{\mu\nu} = j^\nu$ of Maxwell's equations.

Problem 2.56 Assume that in a three-dimensional spacetime, there is a basis of vector fields $\{X_0, X_1, X_2\}$ with orthogonality relations $g(X_\mu, X_\nu) = 0$ for $\mu \neq \nu$ and $g(X_0, X_0) = -g(X_1, X_1) = -g(X_2, X_2) = 1$. In this basis (which is not a coordinate basis!), we define an affine connection by

$$\nabla_{X_0} X_1 = -\nabla_{X_1} X_0 = \frac{1}{2} X_2, \tag{2.52}$$

$$\nabla_{X_1} X_2 = -\nabla_{X_2} X_1 = -\frac{1}{2} X_0, \tag{2.53}$$

$$\nabla_{X_2} X_0 = -\nabla_{X_0} X_2 = \frac{1}{2} X_1. \tag{2.54}$$

We also assume that $[X_0, X_1] = X_2$, $[X_1, X_2] = -X_0$, and $[X_2, X_0] = X_1$.

a) Show that ∇ is the Levi-Civita connection associated to the metric g.

b) Compute the energy–momentum tensor $T_{\mu\nu}$ corresponding to the metric g from Einstein's equations in the above basis.

Problem 2.57 The action for a point particle of mass M in a curved spacetime is given by

$$\mathscr{S}_M = M \int \sqrt{g(\dot{\gamma}, \dot{\gamma})}\, d\tau, \tag{2.55}$$

where τ is the proper time and $\dot{\gamma}$ the 4-velocity of the particle. What is the energy–momentum tensor corresponding to such a point particle? Check that your expression takes the expected form in the case of standard coordinates in Minkowski space.

Problem 2.58 The Lagrangian density for an electromagnetic field A_μ is given by

$$\mathscr{L} = -\frac{1}{4} F_{\mu\nu} F^{\mu\nu} + J^\mu A_\mu, \tag{2.56}$$

where $F_{\mu\nu} = \partial_\mu A_\nu - \partial_\nu A_\mu$ and J^μ is an external 4-current density that does not depend on A_μ. Use this to derive the equations of motion for an electromagnetic field in a general spacetime.

Problem 2.59 The Lagrangian for a free massive scalar field ϕ in a general spacetime is given by

$$\mathcal{L}_\phi = \frac{1}{2}\left[g^{\mu\nu}(\partial_\mu \phi)(\partial_\nu \phi) - m^2 \phi^2\right]. \tag{2.57}$$

Find the equation of motion for ϕ in the curved spacetime based on the principle of stationary action.

Problem 2.60 A massless scalar field ϕ can be described by the Lagrangian density

$$\mathcal{L} = \frac{1}{2} g^{\mu\nu}(\partial_\mu \phi)(\partial_\nu \phi) - V(\phi), \tag{2.58}$$

where $V(\phi)$ is the potential density, which is a function of ϕ only (i.e., it does not depend on the metric). Compute the components of the stress–energy tensor $T_{\mu\nu}$

and then simplify your expression in the case where $g^{\mu\nu}(\partial_\mu \phi)(\partial_\nu \phi)$ is negligible compared to $V(\phi)$.

Problem 2.61 Consider the Robertson–Walker spacetime with metric

$$ds^2 = dt^2 - \cosh^2(Ht)d\boldsymbol{x}^2, \tag{2.59}$$

where $d\boldsymbol{x}^2$ is the standard Euclidean line element in three dimensions. Show that this spacetime is *not* a vacuum solution to Einstein's field equations.

Problem 2.62 The Lagrangian density of a free electromagnetic field A_μ is given by

$$\mathscr{L} = -\frac{1}{4} F_{\mu\nu} F^{\mu\nu}. \tag{2.60}$$

Starting from this Lagrangian density, determine the stress–energy tensor related to the electromagnetic field.

2.4 Killing Vector Fields

Problem 2.63 A Killing vector field X by definition satisfies the differential equation

$$\nabla_i X_j + \nabla_j X_i = 0. \tag{2.61}$$

For a geodesic $x(s)$, show that there is a conserved quantity

$$W(s) = X_\mu(x(s)) \frac{dx^\mu}{ds}, \tag{2.62}$$

i.e., the derivative of W with respect to the curve parameter s vanishes.
Hint: Use the symmetry of the Christoffel symbols from the Levi-Civita connection.

Problem 2.64 The symmetries of a spacetime metric are associated to so-called Killing vector fields. A vector field X is a Killing vector field if $\mathcal{L}_X g = 0$; this means that

$$X^\lambda \partial_\lambda g_{\mu\nu} = -g_{\lambda\nu} \partial_\mu X^\lambda - g_{\mu\lambda} \partial_\nu X^\lambda, \tag{2.63}$$

for all indices μ, ν.

a) Show that this condition can be written without reference to any specific choice of local coordinates as

$$X \cdot g(A, B) = g([X, A], B) + g(A, [X, B]), \tag{2.64}$$

for all vector fields A, B.

b) Show that the coordinate vector field $X = \partial_\lambda$ is a Killing vector field if and only if

$$\partial_\lambda g_{\mu\nu} = 0, \quad \forall \mu, \nu. \tag{2.65}$$

c) Show that the vector fields X_μ in Problem 2.56 are all Killing vector fields.

Problem 2.65 Find the flows of the following vector fields and determine if they are Killing vector fields or not.

a) The field $K = y\partial_x - x\partial_y$ in the Euclidean plane with Cartesian coordinates x and y.

b) The field $K = x\partial_t - t\partial_x$ in two-dimensional Minkowski space with standard coordinates t and x.

Problem 2.66 Consider the paraboloid $z = \alpha(x^2 + y^2)$ as a submanifold embedded in Euclidean three-dimensional space ($\mathbb{R}^3$).

a) Using coordinates r and φ such that $x = r\cos(\varphi)$ and $y = r\sin(\varphi)$, compute the induced metric tensor on the paraboloid.

b) Verify that the vector field $K = \partial_\varphi$ is a Killing vector field and find an expression for the corresponding conserved quantity for a geodesic in terms of the coordinates and their derivatives along the geodesic.

Problem 2.67 A torus can be parametrized using two angles θ and φ. The metric induced by a typical embedding in $\mathbb{R}^3$ corresponds to the line element

$$ds^2 = [R + \rho\sin(\varphi)]^2 d\theta^2 + \rho^2 d\varphi^2. \tag{2.66}$$

a) Find the Christoffel symbols corresponding to the Levi-Civita connection of this metric.

b) Find a Killing vector field for the torus and the corresponding conserved quantity along geodesics of the Levi-Civita connection.

c) It is possible to introduce a flat connection $\tilde{\nabla}$ with all connection coefficients $\tilde{\Gamma}^c_{ab} = 0$ in the θ-φ coordinate system. This connection is not metric compatible. Compute the components of the derivative $\tilde{\nabla}_a g$ for this connection.

Problem 2.68 A wavy two-dimensional surface locally described by the coordinates ρ and φ is embedded in $\mathbb{R}^3$ according to

$$x = \rho\cos(\varphi), \quad y = \rho\sin(\varphi), \quad z = R_0\cos(\rho/R_0), \tag{2.67}$$

where $R_0 > 0$ is a constant.

a) Compute the induced metric tensor and the Christoffel symbols.

b) Find the flow for each of the following vector fields and determine whether or not they are Killing vector fields:

$$K = \partial_\rho, \quad Q = \partial_\varphi. \tag{2.68}$$

Problem 2.69 The 2-dimensional de Sitter space dS_2 may be defined as the surface $t^2 - x^2 - y^2 = -r_0^2$ in 1+2-dimensional Minkowski space, where $r_0 > 0$ is a constant.

a) Introduce suitable coordinates on dS_2.

b) Compute the components of the metric tensor induced by the embedding in Minkowski space in your selected coordinates.

c) Find (at least) two Killing vector fields on dS_2.

2.5 Schwarzschild Metric

Problem 2.70 The Schwarzschild metric, when restricted to the plane $\theta = \frac{\pi}{2}$, is given by

$$ds^2 = \left(1 - \frac{\alpha}{r}\right)(dx^0)^2 - \left(1 - \frac{\alpha}{r}\right)^{-1} dr^2 - r^2 d\phi^2. \tag{2.69}$$

Derive the geodesic equations of motion for a test particle in this metric.

Problem 2.71 The Schwarzschild metric is normally written in terms of time and spherical coordinates. Transform this metric to coordinates $(x^1, x^2, x^3) = r(\sin\theta\cos\phi,\ \sin\theta\sin\phi,\ \cos\theta)$.

Problem 2.72 Consider the Schwarzschild metric

$$ds^2 = c^2 d\tau^2 = \left(1 - \frac{2GM}{c^2 r}\right)\left(dx^0\right)^2 - \left(1 - \frac{2GM}{c^2 r}\right)^{-1} dr^2 - r^2 d\Omega^2, \tag{2.70}$$

where τ is the proper time, $x^0 = ct$, and $d\Omega^2 = d\theta^2 + \sin^2\theta d\phi^2$.

a) Assuming circular motion in the equatorial plane, i.e., $r = r_0$, where r_0 is a constant, derive Kepler's third law

$$\Delta t = 2\pi\sqrt{\frac{r_0^3}{GM}}, \tag{2.71}$$

where Δt is the period. Compare with the classical result.

b) Compute the proper time $\Delta\tau$ for one period of circular motion.

Problem 2.73 The Schwarzschild metric is given by

$$ds^2 = \left(1 - \frac{r_*}{r}\right)(dx^0)^2 - \left(1 - \frac{r_*}{r}\right)^{-1} dr^2 - r^2 d\Omega^2, \tag{2.72}$$

where $d\Omega^2 = d\theta^2 + \sin^2\theta\, d\phi^2$ and $r_* \equiv 2GM/c^2$ is the Schwarzschild radius. Find the worldlines for bodies, outside of the Schwarzschild horizon, radially freely falling toward the black hole.

Problem 2.74 Show that there are no circular free fall orbits inside of the radius $r = 3r_*/2$ in the Scwharzschild spacetime.

Problem 2.75 For the Schwarzschild solution in the limit $r \gg r_*$ and approximately circular orbits such that $r = r_0 + \rho$, where $\rho \ll r_0$, determine the ratio between the period of oscillations in ρ to the orbital period.

Problem 2.76 The optical size of a black hole is given by $4\pi b^2$, where b is the minimal impact parameter such that the past-null geodesics originate at $r \to \infty$.

Find the optical size of the Schwarzschild black hole for which the line element is given by

$$ds^2 = \left(1 - \frac{R}{r}\right) dt^2 - \left(1 - \frac{R}{r}\right)^{-1} dr^2 - r^2 d\Omega^2. \tag{2.73}$$

Hint: For null geodesics in the Schwarzschild spacetime, the angular momentum is equal to the impact parameter (i.e., $b = L$) for $\dot{r} = 1$ at $r \to \infty$.

Problem 2.77 Show that the Schwarzschild solution of Einstein's equation

$$ds^2 = \left(1 - \frac{r_*}{r}\right) dt^2 - \left(1 - \frac{r_*}{r}\right)^{-1} dr^2 - r^2 \left[d\theta^2 + \sin(\theta)^2 d\varphi^2\right], \tag{2.74}$$

where $r_* = 2GM$ and $c = 1$, can be embedded in 5+1-dimensional Minkowski spacetime with coordinates $(Z_1, Z_2, \ldots, Z_6)$ and metric

$$ds^2 = dZ_1^2 - dZ_2^2 - dZ_3^2 - dZ_4^2 - dZ_5^2 - dZ_6^2. \tag{2.75}$$

Hint: Make the ansatz $Z_1 = 2r_* \sinh(t/(2r_*)) f(r)$, $Z_2 = 2r_* \cosh(t/(2r_*)) f(r)$, $Z_3 = g(r)$, $Z_4 = r\sin(\theta)\cos(\varphi)$, $Z_5 = r\sin(\theta)\sin(\varphi)$, and $Z_6 = r\cos(\theta)$ and determine the functions $f(r)$ and $g(r)$ [see C. Fronsdal, Phys. Rev. **116**, 778 (1959)]. The answer may contain an integral.

2.6 Metrics, Geodesic Equations, and Proper Quantities

Problem 2.78 The restriction of the Minkowski metric $(\eta_{\mu\nu})$ to the three-dimensional hyperboloid M_3, i.e.,

$$(x^0)^2 - (x^1)^2 - (x^2)^2 - (x^3)^2 = -a^2, \tag{2.76}$$

defines a curved metric on M_3. Determine the lightlike geodesics with constant spherical angle ϕ (where $x^1 = r\sin\theta\cos\phi$, $x^2 = r\sin\theta\sin\phi$, and $x^3 = r\cos\theta$ as usual).

Problem 2.79 The Minkowski metric $ds^2 = (dx^0)^2 - (dx^1)^2 - (dx^2)^2$ in $\mathbb{R}^3$ induces a nonflat Lorentzian metric on the surface $S = \left\{(x^0, x^1, x^2) : (x^0)^2 - (x^1)^2 - (x^2)^2 = -1\right\}$. Let ϕ be the polar angle in the (x^1, x^2)-plane. Compute the global time difference Δx^0 needed for a light signal to travel from a point $\phi_0 = 0$ to a point $\phi = \pi/2$ on S.

Problem 2.80 Let $(x^0(s), r(s), \theta(s), \phi(s))$ be a lightlike geodesic for the Schwarzschild metric, expressed in the spherical coordinates (r, θ, ϕ). Derive a differential equation for $r(s)$ in the form

$$\frac{dr}{ds} = f(r), \tag{2.77}$$

when restricted to the plane $\theta = \frac{\pi}{2}$.

Hint: The following nonzero Christoffel symbols for the Schwarzschild metric when $\theta = \frac{\pi}{2}$ might be useful: $\Gamma^0_{0r} = -\Gamma^r_{rr} = \frac{1}{2\alpha}\frac{d\alpha}{dr}$, $\Gamma^r_{00} = \frac{\alpha}{2}\frac{d\alpha}{dr}$, $\Gamma^r_{\theta\theta} = \Gamma^r_{\phi\phi} = -r\alpha$, and $\Gamma^\theta_{r\theta} = \Gamma^\phi_{r\phi} = \frac{1}{r}$, where $\alpha = \alpha(r) = 1 - \frac{2GM}{c^2 r}$.

Problem 2.81 Consider the metric $ds^2 = c^2dt^2 - S(t)^2(dx^2 + dy^2 + dz^2)$, where $S(t)$ is an increasing function of time t with $S(0) = 0$. Find the geodesic equations of motion. In particular, construct explicitly the lightlike geodesic when $S(t) = t/t_0$ for some constant $t_0 > 0$. What are the points $(ct, x, y, z) \in \mathbb{R}^4$ for a fixed $t > t_0$, which are causally related to the event $p = (ct_0, ct_0, 0, 0)$, i.e., the points which are connected to p by a future-directed timelike (or lightlike) curve?

Problem 2.82 The line element on the unit sphere $\mathbb{S}^2$ is given by

$$ds^2 = d\theta^2 + \sin^2(\theta)d\varphi^2, \tag{2.78}$$

and the nonzero Christoffel symbols are

$$\Gamma^\theta_{\varphi\varphi} = -\sin(\theta)\cos(\theta), \qquad \Gamma^\varphi_{\theta\varphi} = \Gamma^\varphi_{\varphi\theta} = \cot(\theta). \tag{2.79}$$

Consider two geodesics separated by a small distance δ and both orthogonal to the equator $\theta = \pi/2$, compute the rate of acceleration of the geodesics toward each other through the geodesic deviation equation

$$A^a = R^a{}_{bcd}\dot{\chi}^b\dot{\chi}^c X^d, \tag{2.80}$$

where X^d is the infinitesimal separation of the geodesics at the equator and χ^a are the coordinates.

Problem 2.83 The metric for the de Sitter universe can be expressed in the form

$$ds^2 = dt^2 - e^{2t/R}(dx^2 + dy^2 + dz^2), \tag{2.81}$$

where $R > 0$ is a constant and x, y, and z can be treated as rectangular coordinates and t as time.

a) Show that the trajectories of freely falling particles and photons are straight lines.

b) A body at a point $x = X > 0$ on the x-axis emits a photon toward the origin at time $t = 0$. Show that, if $X < R$, the photon arrives at the origin at $t = -R\log(1 - X/R)$.

Problem 2.84 The de Sitter universe dS_4 is defined as the hyperboloid

$$t^2 - (x^1)^2 - (x^2)^2 - (x^3)^2 - (x^4)^2 = -T_0^2, \tag{2.82}$$

in units with $c = 1$ and where $T_0 > 0$ is a constant. The metric on dS_4 is defined by restricting the five-dimensional Minkowski metric to the hyperboloid.

a) Find an explicit expression for the four-dimensional metric in a suitable coordinate system on dS_4.

b) Derive the geodesic equations in dS_4.

c) Compare the metric tensor with the Robertson–Walker metric

$$ds^2 = dt^2 - S(t)^2 d\Omega^2, \tag{2.83}$$

by writing down your metric in a coordinate system where the coefficient in front of the timelike coordinate is identically equal to one.

Problem 2.85 Consider AdS_2 and the coordinates x^μ defined in Problem 2.27.

a) Find the trajectory of a light ray in this spacetime.

b) A particle at rest in $r = r_0 > 2$ starts to fall freely at $t = 0$. What is the proper time it takes for the particle to freely fall and reach $r = r_1$, where $1 < r_1 < r_0$? Also compute the coordinate time for this fall, i.e., the time t at which the particle reaches r_1. Discuss your result in the limit $r_1 \to 1$, and in particular, a possible physical interpretation of what you find. (Your answers may contain integrals and functions defined by implicit equations.)

Problem 2.86 Four-dimensional anti-de Sitter space, which is also called AdS_4, is a four-dimensional curved space that can be defined as follows: Consider the five-dimensional space with coordinates $(X^a) = (U, V, X, Y, Z)$, where $a = 1, 2, 3, 4, 5$, and metric

$$ds^2 = dU^2 + dV^2 - dX^2 - dY^2 - dZ^2 \equiv dX^a dX_a. \tag{2.84}$$

Then, AdS_4 is the subspace of this space defined by the relation

$$X^a X_a = U^2 + V^2 - X^2 - Y^2 - Z^2 = 1. \tag{2.85}$$

Hint: Note that $(X_a) = (U, V, -X, -Y, -Z)$.

a) Compute the metric of AdS_4 in the coordinate system $(x^\mu) = (\alpha, \lambda, \theta, \varphi)$, where $\mu = 0, 1, 2, 3$, defined as follows: Let (r, θ, φ) be spherical coordinates for (X, Y, Z), i.e., $X = r\sin(\theta)\cos(\varphi)$, $Y = r\sin(\theta)\sin(\varphi)$, and $Z = r\cos(\theta)$, and (t, α) polar coordinates for (U, V), i.e., $U = t\cos(\alpha)$ and $V = t\sin(\alpha)$. Then, $t = \cosh(\lambda)$ and $r = \sinh(\lambda)$.

b) Compute all possible trajectories $\lambda(\alpha)$ of light rays on AdS_4 moving along the subspace $Y = Z = 0$.

c) Prove that all lightlike geodesics on AdS_4 are straight lines in the embedding space, i.e., they obey the equations

$$\ddot{X}^a = 0, \tag{2.86}$$

with the dot indicating differentiation with respect to proper time τ.

Hint: You can find these trajectories by extremizing the functional $\int \mathcal{L}\, d\tau$ with

$$\mathcal{L} = \frac{1}{2}\dot{X}^a \dot{X}_a + \lambda(X^a X_a - 1), \tag{2.87}$$

and λ a Lagrange multiplier (explain why this is so!). One key step in the proof is to show that $\dot{X}^a X_a$ is conserved and you can assume that it is identically equal to zero for a lightlike trajectory.

Problem 2.87 Consider the $1+1$-dimensional Robertson–Walker spacetime described by the metric

$$ds^2 = dt^2 - a(t)^2 dx^2, \tag{2.88}$$

for some function $a(t)$.

a) Compute the Ricci tensor $R_{\mu\nu}$ for this spacetime.

b) Derive the trajectory $x(t)$ of a light ray on this spacetime for $a(t) = 1/(A^2 + B^2t^2)$ and some constants A and B. Assume that the light ray is emitted at $t = 0$ from $x = x_0$ with $dx/dt > 0$.

Problem 2.88 Consider the $1+2$-dimensional Robertson–Walker spacetime described by the metric

$$ds^2 = dt^2 - a(t)^2\left(\frac{dr^2}{1-kr^2} + r^2 d\phi^2\right), \tag{2.89}$$

where $a(t)$ is some function and k is a constant. Derive the geodesic equations and determine the Christoffel symbols.

Problem 2.89 The Robertson–Walker metric describing a particular closed universe is given by

$$ds^2 = dt^2 - e^{-2t/a}\left[\frac{1}{1+(r/a)^2}dr^2 + r^2(\sin^2\theta d\varphi^2 + d\theta^2)\right], \tag{2.90}$$

for some parameter $a > 0$.

a) Determine all nonzero Christoffel symbols $\Gamma^r_{\mu\nu}$, $\mu, \nu = t, r, \theta, \varphi$ for this metric. Moreover, for the vector field with the components $A^t = t/a$, $A^r = r/a$, $A^\theta = A^\varphi = 0$ in the coordinate vector basis, compute the components $\nabla_\mu A^r$, $\mu = t, r$, of the covariant derivative.

b) Find the trajectory $r(t)$ of a light pulse emitted at time $t = 0$ at $r = a$ and moving radially outward such that $\theta = \pi/2$ and $\varphi = 0$ at all times.

Problem 2.90 The Robertson–Walker metric is defined by

$$ds^2 = c^2dt^2 - S(t)^2\left(\frac{dr^2}{1-kr^2} + r^2 d\Omega^2\right), \tag{2.91}$$

for some smooth function $S(t)$ and $d\Omega^2 = d\theta^2 + \sin^2\theta d\phi^2$. We consider the case $k = 1$. After a coordinate transformation $\chi = \arcsin r$ (with $0 \le \chi \le \pi/2$), this can be written as

$$ds^2 = c^2dt^2 - S(t)^2(d\chi^2 + \sin^2\chi d\Omega^2). \tag{2.92}$$

a) Derive first integrals for the geodesic equations when $d\Omega = 0$.

b) Derive a formula expressing the distance a light ray emitted at $r = 0$ at universal time t_0 travels (in the r coordinate) in the time interval $[t_0, t_0 + T]$.

Problem 2.91 A spaceship is freely falling (along a geodesic) toward the true singularity at $r = 0$ in a Schwarzschild black hole. The initial velocity is $\dot\theta = \dot\phi = 0$, $\dot r = \alpha$, and $\dot t = \beta$, where the dot means differentiation with respect to the path parameter (which can be taken to be the proper time) and the standard Schwarzschild metric is used, i.e.,

$$ds^2 = \left(1 - \frac{2GM}{c^2 r}\right) c^2 dt^2 - \left(1 - \frac{2GM}{c^2 r}\right)^{-1} dr^2 - r^2 d\Omega^2. \tag{2.93}$$

The proper time τ needed to reach the singularity $r = 0$, when starting from $r = r_0 < 2GM/c^2$, can be written as

$$\tau = \int_0^{r_0} f(r)\, dr. \tag{2.94}$$

What is the function $f(r)$?

Problem 2.92 A Schwarzschild black hole has a mass $M = 13.5 \cdot 10^{30}$ kg (about seven times the solar mass). An observer is freely falling (along a geodesic) radially toward the black hole. The initial radial coordinate is $r = r_0 = 10^{10}$ km and the initial coordinate velocity is $c(dr/dx^0) = -v_0 = -10$ km/s. Derive the formula for the proper time needed to reach the Schwarzschild horizon and give the order of magnitude of this time. Newton's gravitational constant is $G \approx 6.67 \cdot 10^{-11}\ \mathrm{m}^3/(kg \cdot s^2)$ and the speed of light is $c \approx 3 \cdot 10^8$ m/s.
Hint: The following integral can be useful

$$\int \frac{dx}{\sqrt{a + \frac{b}{x}}} = \frac{1}{\sqrt{a}} \left[\sqrt{x^2 + \frac{bx}{a}} - \frac{b}{2a} \ln\left(x + \frac{b}{2a} + \sqrt{x^2 + \frac{bx}{a}} \right) \right] + C, \tag{2.95}$$

where a, b, and C are constants.

Problem 2.93 A particle of mass $m > 0$ is freely falling radially toward the horizon of a Schwarzschild black hole of mass M. Show that $p_0 = mcg_{00}\dot{x}^0$ is a constant of motion. Find the proper time Δs (as a function of $p_0 = E/c$) needed for the particle to reach $r = 2GM/c^2$ from $r = 3GM/c^2$. Show that the result can be written as

$$\Delta s = \int_{r_*}^{3r_*/2} \frac{dr}{\sqrt{\left(\frac{E}{mc^2}\right)^2 - \left(1 - \frac{r_*}{r}\right)}}, \tag{2.96}$$

where $r_* \equiv 2GM/c^2$.

Problem 2.94 An observer in the Schwarzschild spacetime moves with fixed radial coordinate $r = r_0$ and fixed angular velocity $\dot{\varphi} = \omega$ in the plane $\theta = \pi/2$. Compute the 4-acceleration A and the proper acceleration α of the observer as a function of the proper period ω.

Problem 2.95 A particle of mass $m > 0$ is freely falling radially toward the event horizon $r = 2GM$ of a Schwarzschild black hole of mass M (we set $c = 1$), i.e., θ and φ are constant in the standard coordinates where the metric is

$$ds^2 = \left(1 - \frac{2GM}{r}\right) dt^2 - \left(1 - \frac{2GM}{r}\right)^{-1} dr^2 - r^2 \left[d\theta^2 + \sin(\theta)^2 d\varphi^2\right]. \tag{2.97}$$

a) Compute the trajectory of this particle.

b) Find also formulas for the coordinate time Δt and proper time $\Delta\tau$ it takes for the particle to reach the event horizon from some position $r = r_0$. Determine for each of these times if it is finite or infinite, and discuss the physical significance of your results.

Integration constants can be fixed so that the results have a simple form, but if so a physical interpretation of what the choice amounts to should be given.

Problem 2.96 Consider an observer in the Schwarzschild spacetime with line element

$$ds^2 = \left(1 - \frac{r_*}{r}\right) dt^2 - \left(1 - \frac{r_*}{r}\right)^{-1} dr^2 - r^2 d\Omega^2. \tag{2.98}$$

The observer starts out at $r = r_0$ and initially moves tangentially with a local velocity v_0 relative to the stationary frame.

a) Determine the minimal value of v_0 such that the observer does not fall into the black hole region of the solution.

b) Assuming that $v_0 = 0$, compute the proper time it takes for the observer to reach the singularity.

Hint: The local velocity v of an observer relative to the stationary frame has a γ factor of $\gamma = V \cdot U$, where V is the 4-velocity of the observer, $U = \alpha\partial_t$, and $U^2 = 1$.

Problem 2.97 You are sending up a satellite around the Earth. You want it to be directed such that when you turn off its engines, it will follow a geodesic around the Earth with fixed radius. The metric around the Earth is

$$ds^2 = \left(1 - \frac{r_*}{r}\right) dt^2 - \left(1 - \frac{r_*}{r}\right)^{-1} dr^2 - r^2 d\theta^2 - r^2 \sin^2\theta d\phi^2, \quad R_0 > r_*, \tag{2.99}$$

where R_0 is the radius of the Earth. Your initial data when you turn off the engines at $\tau = 0$ are the following

$$\left.\frac{dr}{d\tau}\right|_{\tau=0} = 0, \quad \left.\frac{d\theta}{d\tau}\right|_{\tau=0} = 0, \quad \left.\frac{d\phi}{d\tau}\right|_{\tau=0} = B, \tag{2.100}$$
$$r|_{\tau=0} = R, \quad \theta|_{\tau=0} = \pi/2, \quad \phi|_{\tau=0} = 0.$$

Is this possible? If so, determine how your initial condition B, which you have to choose, depends on R and r_*.

Problem 2.98 A satellite moves at a constant radial distance from the Earth with a constant orbital coordinate speed $v = r d\phi/dt$. Assume that the metric is the Schwarzschild metric and let the orbit be in the plane with angle $\theta = \pi/2$ such that

$$ds^2 = \left(1 - \frac{r_*}{r}\right) dt^2 - \left(1 - \frac{r_*}{r}\right)^{-1} dr^2 - r^2 d\phi^2, \tag{2.101}$$

where $r_* \equiv 2GM$ is the Schwarzschild radius.

a) Calculate the proper time τ for the satellite to complete one orbit around the Earth. Express the answer in terms of the coordinate speed v and the radius r.

b) Use the result in a) to calculate t/τ and show that if this is series expanded to first order in v and the gravitational potential, it holds that

$$\frac{t}{\tau} - 1 \simeq \frac{v^2}{2} - \Phi_s, \tag{2.102}$$

where Φ_s is the gravitational potential at the satellite.

Problem 2.99 The spacetime outside of the Earth may be approximately described by the Schwarzschild line element

$$ds^2 = \left(1 - \frac{r_*}{r}\right) dt^2 - \left(1 - \frac{r_*}{r}\right)^{-1} dr^2 - r^2 d\Omega^2, \tag{2.103}$$

where r_* is the Schwarzschild radius of the Earth (approximately 9 mm). A GPS satellite is orbiting the Earth in free fall at a stationary radius $r = r_0$. The motion is assumed to occur in the plane $\theta = \pi/2$.

a) Since r is constant, the motion will have a 4-velocity $U = \alpha\partial_t + \beta\partial_\varphi$. Find the values of the constants α and β.

b) Find an expression for the proper time it takes for the satellite to complete a full orbit around the Earth.

c) An observer is stationary at $r = r_0$ (note that this requires proper acceleration of this observer). At what speed will the satellite pass by the observer?
Hint: The relative speed v between two objects with 4-velocities U and V, respectively, has a γ factor of $\gamma = V \cdot U$.

Problem 2.100 Consider two observers in the exterior Schwarzschild spacetime with line element

$$ds^2 = \left(1 - \frac{r_*}{r}\right) dt^2 - \left(1 - \frac{r_*}{r}\right)^{-1} dr^2 - r^2 d\Omega^2. \tag{2.104}$$

Both observers can be assumed to move in the plane $\theta = \pi/2$. The first observer is a stationary observer with fixed spatial coordinates $r = r_0 > 3r_*$ and $\varphi = \varphi_0$, whereas the second observer is moving on a circular geodesic with radius $r = r_0$.

a) Give a parametrization of the worldline for each observer and use it to find the proper acceleration of the observers.

b) The observers meet and synchronize their clocks when they pass each other. Find the ratio between the times shown by the respective clocks when they pass each other the next time.

c) Find the relative velocity of the observers as they pass each other.

Problem 2.101 You have reached a fast rotating neutron star with your spaceship. You decide that you want to go around the neutron star once. Let your orbit be at constant radial distance and your coordinate speed $v = r d\phi/dt$.

a) Calculate the proper time τ it takes you to go around. Express your answer in terms of the radius R and the speed v (set $c = 1$). The metric describing this neutron star is given by the Kerr metric, i.e.,

$$ds^2 = \left(1 - \frac{rr_*}{\rho^2}\right)dt^2 + \frac{2arr_*\sin^2\theta}{\rho^2}dtd\phi - \frac{\rho^2}{\Delta}dr^2 - \rho^2 d\theta^2$$

$$- \left(r^2 + a^2 + \frac{a^2 rr_*\sin^2\theta}{\rho^2}\right)\sin^2\theta d\phi^2, \tag{2.105}$$

where $\rho^2 \equiv r^2 + a^2\cos^2\theta$, $\Delta \equiv r^2 - rr_* + a^2$, $a \equiv J/M$, and M and J are the mass and the angular momentum of the neutron star, respectively. You choose to make the orbit at a fixed angle of $\theta = \pi/2$.

b) Now, use your result in a) to calculate T/τ, where T is the coordinate time of your orbit, and show if you expand to first approximation in v and the gravitational potential, you obtain

$$\frac{T}{\tau} \simeq 1 + \frac{v^2}{2}\left[1 + \left(\frac{a}{R}\right)^2\right] + \frac{r_*}{2R}, \quad \text{if } \tfrac{r_*}{R} \sim v^2. \tag{2.106}$$

Problem 2.102 At the time of inflation, consider a massive free-falling particle. Assume that the metric of spacetime is

$$ds^2 = dt^2 - a(t)^2\left[d\rho^2 + \rho^2(d\theta^2 + \sin^2\theta d\phi^2)\right], \quad a(t) = a_0 e^{Ct}, \tag{2.107}$$

where a_0 and C are constants. The initial values for the free-falling particle are given by

$$\left.\frac{d\rho}{d\tau}\right|_{\tau=0} = \frac{1}{4}, \quad \left.\frac{d\theta}{d\tau}\right|_{\tau=0} = \frac{1}{4}, \quad \left.\frac{d\phi}{d\tau}\right|_{\tau=0} = \frac{1}{4}, \tag{2.108}$$

$$t|_{\tau=0} = 0, \quad \rho|_{\tau=0} = 1, \quad \theta|_{\tau=0} = 0, \quad \phi|_{\tau=0} = 0.$$

a) Is the spatial geometry curved or not (at each fixed value of time)? Justify your answer.

b) Calculate the proper time for the free-falling particle between the coordinate times $t = 0$ and $t = t_1$.

Hint: Calculations might become simpler if another coordinate system is used! It is also okay to give the answer expressed as an integral.

Problem 2.103 Assume a three-dimensional version of the Robertson–Walker metric (with $k = 0$):

$$ds^2 = dt^2 - a(t)^2(dr^2 + r^2 d\phi^2). \tag{2.109}$$

You are traveling in your spaceship in this universe. You decide to travel in a circle around $r = 0$ on a fixed radius $r = R_0$.

a) Calculate the proper time for the spaceship in this geometry going around one time (let ϕ go from zero to 2π) if you have the following constant velocity $v = a(t)R_0\frac{d\phi}{dt}$ and $a(t) = \exp(t)$. Let the initial time be $t = 0$.

b) Is it always possible to get around in a finite time?

Problem 2.104 You are traveling in your spaceship in outer intergalactic space. The metric can be assumed to be the Robertson–Walker metric with zero curvature:

$$ds^2 = dt^2 - a(t)^2(dx^2 + dy^2 + dz^2). \tag{2.110}$$

To save fuel, you do not use the spaceship's engines. You are moving according to the following initial conditions

$$\left.\frac{dx}{d\tau}\right|_{\tau=0} = A, \quad \left.\frac{dy}{d\tau}\right|_{\tau=0} = 0, \quad \left.\frac{dz}{d\tau}\right|_{\tau=0} = 0,$$
$$x|_{\tau=0} = X_0, \quad y|_{\tau=0} = 0, \quad z|_{\tau=0} = 0. \tag{2.111}$$

Calculate the proper time it takes you to reach $x = X_D$. It is sufficient to give your answer in terms of an integral, which depends on the initial value A and the scale factor $a(t)$.

Problem 2.105 An observer (A) is stationary in Schwarzschild spacetime at a radius r_0. A second observer (B) is initially stationary at $r = r_1$, but at some event (which can be assigned $t = \tau = 0$) suffers from a rocket failure and becomes freely falling. On the way into the black hole, B passes right by A.

a) What is the relative velocity of A and B as they pass by each other?

b) What proper time has passed for B since the rocket failure when they pass by each other?

Problem 2.106 For the two-dimensional spacetime with coordinates t and x and line element $ds^2 = x^2dt^2 - dx^2$:

a) Compute the proper acceleration of an observer with worldline $x = x_0$, where x_0 is a constant.

b) Compute the proper time for a free-falling observer starting at $x = x_0$ with $dx/dt = 0$ to reach the coordinate singularity at $x = 0$.

2.7 Kruskal–Szekeres Coordinates

Problem 2.107 Show that for $r > 2\mu$ the Kruskal–Szekeres metric

$$ds^2 = \frac{16\mu^2}{r}e^{(2\mu-r)/2\mu}dudv - r^2(d\theta^2 + \sin^2\theta d\phi^2), \tag{2.112}$$

is equivalent to the standard Schwarzschild metric through the relations

$$uv = (2\mu - r)e^{(r-2\mu)/2\mu}, \quad t = \frac{x^0}{c} = 2\mu \ln\left(-\frac{v}{u}\right). \tag{2.113}$$

Here $u < 0$ and $v > 0$, and we use units with $c = 1$.

Problem 2.108 Show that a spaceship cannot get out from the black hole region $u > 0$ and $v > 0$ in Kruskal–Szekeres coordinates.

Problem 2.109 An observer is freely falling to the true singularity $r = 0$ of a Schwarzschild black hole. We assume that the fall follows a radial ray $d\Omega = 0$. Since the standard (spherical) coordinates become singular at the Schwarzschild event horizon $r = 2GM/c^2 \equiv 2\mu$, we express the initial condition of the observer in terms of the Kruskal–Szekeres coordinates

$$t = \frac{x^0}{c} = 2\mu \ln\left(\frac{v}{u}\right), \quad uv = (2\mu - r)e^{\frac{r-2\mu}{2\mu}}, \tag{2.114}$$

for $u, v > 0$. The initial conditions are

$$u(0) = 0, \quad v(0) = v_0, \quad \dot{u}(0) = \frac{E}{cv_0}, \tag{2.115}$$

and $\dot{v}(0)$ is determined by the requirement of the 4-velocity being future-directed with modulus one. Compute the proper time Δs for the fall, expressed as an integral

$$\int_0^{2\mu} f(r)\, dr$$

for a certain function $f(r)$ of the radius r expressed in terms of E and v_0.
Hint: From the equations of motion, it can be shown that the quantity

$$\frac{\dot{u}v - \dot{v}u}{r} \exp\left(\frac{2\mu - r}{2\mu}\right),$$

is a constant of motion.

Problem 2.110 Consider a Schwarzschild black hole with metric

$$ds^2 = \left(1 - \frac{r_*}{r}\right) dt^2 - \left(1 - \frac{r_*}{r}\right)^{-1} dr^2 - r^2 d\Omega^2. \tag{2.116}$$

a) What conclusion can you draw about the singularity $r = 0$ of this metric from the fact that

$$R_{\mu\nu\alpha\beta} R^{\mu\nu\alpha\beta} = \frac{3r_*^2}{r^6}? \tag{2.117}$$

Do not forget to motivate your answer.

b) Determine what radial light cones look like using the Schwarzschild metric, i.e., how t depends on r.

c) Instead of using the Schwarzschild coordinates to describe black holes, it can be useful to use Kruskal–Szekeres coordinates. The metric then takes the form

$$ds^2 = \frac{4r_*^3}{r} e^{-r/r_*}(dV^2 - dU^2) - r^2 d\Omega^2. \tag{2.118}$$

Determine what radial light cones look like in these coordinates.

2.8 Weak Field Approximation and Newtonian Limit

Problem 2.111 a) What are the equations of motion for a massive particle in a gravitational potential according to Newton's mechanics and general relativity, respectively? Derive the former from the latter in the Newtonian limit.

b) What are tidal forces in Newton's theory of gravity? How are they related to the gravitational potential? Why are solar tidal forces slightly weaker than lunar tidal forces? In general relativity, explain how the tidal forces are identified with the curvature of spacetime.

Problem 2.112 a) Compute the Ricci tensor $R_{\mu\nu}$ in the linear approximation for a metric $g_{\mu\nu} = \eta_{\mu\nu} + h_{\mu\nu}$, i.e., you can ignore all but the first-order terms in $h_{\mu\nu}$.

b) Show that a coordinate transformation

$$x^{\mu} \mapsto x^{\mu} + \chi^{\mu}(x), \quad |\partial_{\nu}\chi^{\mu}| \ll 1, \tag{2.119}$$

in the linear approximation described in a) corresponds to a gauge transformation of $h_{\mu\nu}$ given by

$$h_{\mu\nu} \mapsto h_{\mu\nu} - \partial_{\mu}\chi_{\nu} - \partial_{\nu}\chi_{\mu}. \tag{2.120}$$

c) Show that, by imposing the gauge condition

$$\partial^{\mu}\bar{h}_{\mu\nu} = 0, \quad \text{where } \bar{h}_{\mu\nu} = h_{\mu\nu} - \frac{h}{2}\eta_{\mu\nu}, \; h = \eta^{\mu\nu}h_{\mu\nu}, \tag{2.121}$$

the linearized Einstein equations $R_{\mu\nu} = 0$ reduce to the wave equation for $\bar{h}_{\mu\nu}$.

Problem 2.113 The spacetime metric corresponding to a weak gravitational potential $|\Phi(\mathbf{x})| \ll c^2$ is

$$ds^2 = \left[c^2 - 2\Phi(\mathbf{x})\right]dt^2 - \left[1 + \frac{2\Phi(\mathbf{x})}{c^2}\right](dx^2 + dy^2 + dz^2). \tag{2.122}$$

a) Find the geodesic equation for this metric in the nonrelativistic and weak field (where you only keep the lowest-order terms in Φ) limits. Discuss your result.

b) Compute the redshift of a photon with angular frequency ω moving in the Earth's gravitational field $\Phi(\mathbf{x}) = -gz$ (independent of x and y) from $z = 0$ to $z = h > 0$ as observed by a stationary observer. You should give a detailed derivation of your result. Discuss also how one can understand the result using the equivalence principle.

Problem 2.114 Consider two massive particles moving freely on two close paths on a curved spacetime with metric $ds^2 = g_{\mu\nu}dx^{\mu}dx^{\nu}$ and assume that the positions of these two particles at proper time τ are $x^{\mu}(\tau)$ and $x^{\mu}(\tau) + s^{\mu}(\tau)$, respectively, with s^{μ} small (i.e., only terms linear in s^{μ} need to be taken into account and higher-order terms can be ignored).

a) Derive the geodesic deviation equation

$$\frac{D^2 s^{\mu}}{D\tau^2} = -R^{\mu}{}_{\alpha\nu\beta}s^{\nu}\frac{dx^{\alpha}}{d\tau}\frac{dx^{\beta}}{d\tau}. \tag{2.123}$$

b) Show that in the Newtonian limit the geodesic deviation equation reduces to the equation for tidal acceleration in Newton's theory of gravitation, i.e.,

$$\frac{d^2 s^i}{dt^2} = -\frac{\partial^2 \Phi}{\partial x^i \partial x^j} s^j. \tag{2.124}$$

Hint: Recall that

$$\frac{DV^\mu}{D\tau} = \frac{dV^\mu}{d\tau} + \Gamma^\mu_{\alpha\beta} \frac{dx^\alpha}{d\tau} V^\beta. \tag{2.125}$$

To derive the equation in a) it is convenient, but not necessary, to work in a local inertial frame.

Problem 2.115 Derive a relativistic generalization of the centrifugal force as follows: Consider the motion of a free particle in Minkowski space in a coordinate system rotating with constant angular velocity ω around the z-coordinate axis.

a) Compute the Christoffel symbols and the geodesic equations in this coordinate system.

Hint: It is convenient to use cylindrical coordinates (r, φ, z).

b) From your result in a), derive the equations of motion in this rotating frame in the nonrelativistic limit (for $\omega r \ll 1$).

Problem 2.116 a) Find the trajectory of a planet with mass m moving on a circle in the gravitational potential $V(\mathbf{r}) = -GMm/|\mathbf{r}|$, according to Newton's mechanics.

b) There is a natural generalization of the trajectory in a) to general relativity. Explain what this generalization is. Find this generalized trajectory.

Hint: The trajectory can be computed from Hamilton's principle

$$\delta \int \left(\frac{1}{2} m \dot{\mathbf{r}}^2 + \frac{GMm}{r} \right) dt = 0, \tag{2.126}$$

using spherical coordinates (r, θ, φ) and assuming $\theta(t) = \pi/2$ and $r(t) = r_0 =$ constant. Recall that $dx^2 + dy^2 + dz^2 = dr^2 + r^2(d\theta^2 + \sin^2\theta d\varphi^2)$. Find the relation between r and the angular momentum $L = mr^2\dot{\varphi}$.

Problem 2.117 Consider the Einstein field equations.

a) What three approximations should be made to obtain the Newtonian limit?

b) Show that, in the Newtonian limit, the Einstein field equations reduce to

$$\nabla^2 \Phi \propto \rho, \tag{2.127}$$

where Φ is the gravitational potential and ρ the mass density.

Problem 2.118 Consider a satellite in circular orbit around the Earth at a distance R_1 from the surface. The metric outside of the Earth can be considered to be

$$ds^2 = (1 + 2\Phi)dt^2 - (1 + 2\Phi)^{-1} dr^2 - r^2 d\Omega^2, \tag{2.128}$$

where $\Phi = -GM/r$ is the classical gravitational potential and $d\Omega^2 = d\theta^2 + \sin^2\theta d\phi^2$. What is the eigentime required for the satellite to complete a full orbit around the Earth? How does this compare with the global time t required for the same orbit?

2.9 Gravitational Lensing

Problem 2.119 Consider a spherical body with radius R_0, constant density, and total mass M_0. Neutrinos traveling through this body have such small masses that their worldlines can be roughly approximated as null geodesics. Find an expression for the angular deflection of a neutrino with an impact parameter (smallest distance to the body's center) $b < R_0$. Verify that your expression has the expected limit when $b \to R_0$. You may work in the low-velocity and weak field limits for computing the metric.

Problem 2.120 For large distances from the center of the halo, the Navarro–Frenk–White (NFW) dark matter halo profile assumes that the matter density varies as $\rho(r) = k/r^3$. Find the deflection angle α due to gravitational lensing of light that passes such a halo at a minimum distance r_0. You may assume that the NFW density profile is valid from some radius $r = r_s < r_0$ and that the mass inside this radius is given by M_0.

2.10 Frequency Shifts

Problem 2.121 a) Derive the formula for the gravitational redshift between stationary observers in a static spacetime with line element such that the metric components $g_{\mu\nu}$ are independent of a global time coordinate t.

b) Explain the origin of the gravitational redshift in the case of the Schwarzschild metric and derive the approximative formula

$$z \equiv \frac{\lambda_B - \lambda_A}{\lambda_A} \simeq \frac{GM}{c^2 r_A}, \tag{2.129}$$

for the redshift observed far away at B, from a source at a radial coordinate $r = r_A$.

Problem 2.122 A 1+1-dimensional universe is defined as the surface

$$(ct)^2 - x^2 - y^2 = -K^2 \quad \text{(where } K > 0\text{)}, \tag{2.130}$$

in $\mathbb{R}^3$. The metric on the surface is induced by the Minkowski metric $ds^2 = c^2dt^2 - dx^2 - dy^2$ in $\mathbb{R}^3$. Analyze the frequency shift in this mini-universe between comoving observers.

Problem 2.123 A spaceship is moving radially toward a center of mass M with a coordinate velocity $dr/dt = -0.1c$, where t is the Schwarzschild universal time and $c \simeq 3 \cdot 10^8$ m/s^2. An observer in the spaceship is measuring the wavelength of a light signal from a distant star at rest. The light signal travels along the same radius as the observer. The wavelength at $r \to \infty$ is assumed to be 4 000 Å. What is the observed wavelength when $GM = 10^{20}$ m$^3/s^2$ and $r = 10^6$ m?

Problem 2.124 Compute the redshift of starlight emitted from the surface of a star with $r_{\text{star}} = 7 \cdot 10^8$ m and mass $M = 2 \cdot 10^{30}$ kg. Use the approximate values $G \approx 6.67 \cdot 10^{-11}$ m$^3/(kg \cdot s^2)$ and $c \approx 3 \cdot 10^8$ m/s.

Problem 2.125 Elements in the chromosphere of the Sun emit sharp spectral lines. A student in relativity theory observes one such known spectral line in a spectrometer on Earth. According to general relativity, the emitted light is affected by the mass of the Sun. Calculate, using the general theory of relativity and to lowest order in the gravitational constant, the magnitude and sign of the relative frequency shift $\Delta\nu/\nu$ of this spectral line. The solar mass is about $2.0 \cdot 10^{30}$ kg, Newton's gravitational constant is $G \approx 6.7 \cdot 10^{-11}\ \mathrm{m}^3/(\mathrm{kg}\cdot\mathrm{s}^2)$, the speed of light is $c \approx 3.0 \cdot 10^8$ m/s, the solar radius is about $7.8 \cdot 10^8$ m, and the average distance Sun-Earth is about $1.5 \cdot 10^{11}$ m.

Problem 2.126 A spaceship is launched from the ground station on Earth and is moving radially upward. When it is at an altitude of 1 000 km, its velocity is only about 0.1 km/s. At that moment, a light signal is sent from the spaceship and is observed at the ground station. Compute the red/blue shifts of the signal from the two most important physical effects. Newton's gravitational constant is $G \approx 6.67 \cdot 10^{-11}\ \mathrm{m}^3/(\mathrm{kg}\cdot\mathrm{s}^2)$ and the radius and the mass of the Earth are $R \approx 6.3 \cdot 10^3$ km and $M \approx 5.98 \cdot 10^{24}$ kg, respectively.

Problem 2.127 Compute the blueshift of a light signal sent from a very distant spaceship and observed at the Earth. Assume that the spaceship is stationary in an approximately static spacetime. Useful information: The distance between the Sun and the Earth is approximately $1.5 \cdot 10^{11}$ m, the speed of light is $c \simeq 3 \cdot 10^8$ m/s, $GM_\odot \simeq 1.3 \cdot 10^{20}\ \mathrm{m}^3/\mathrm{s}^2$ and the gravitational potential of the Earth on its surface (normalized to zero at infinity) is $-6.24 \cdot 10^7\ \mathrm{m}^2/\mathrm{s}^2$.

Problem 2.128 A free-falling observer is moving radially away from a black hole with a local velocity that is just large enough to escape the gravitational pull. The free-falling observer is emitting light signals radially toward an observer stationary at infinity. Compute the frequency of the light signal received by the second observer if it was emitted with frequency f_0 at radius r by the first observer.

Problem 2.129 In the two-dimensional spacetime introduced in Problem 2.106, a series of light signals is emitted from a free-falling observer starting at $x = x_0$ and received by an observer with a worldline $x = x_1$, where x_1 is a constant. Find the redshift z of the light signals as a function of the position $x = x_e$ where the signal was emitted by the free-falling observer.

Problem 2.130 Consider the Robertson–Walker metric written as

$$ds^2 = c^2 dt^2 - S(t)^2 \left[\frac{dr^2}{1 - kr^2} + r^2 d\Omega^2 \right], \tag{2.131}$$

for some fixed parameter $k > 0$. We project the metric to two dimensions by setting $d\Omega = 0$. An observer A, located at (t_0, r_0) and at rest with respect to the coordinate r, sends a light signal. Another observer, located at (t_1, r_1) and also at rest with respect to r, receives the light signal. After a short time ϵ, A sends another light signal, which is received by B at the time $t_1 + \epsilon'$. Compute the ratio ϵ'/ϵ in terms of the unknown function $S(t)$ and deduce from this the cosmological redshift.

Problem 2.131 The Robertson–Walker metric (for $k = 1$) can be written as

$$ds^2 = c^2dt^2 - S(t)^2\left[d\chi^2 + \sin^2\chi\,(d\theta^2 + \sin^2\theta\,d\phi^2)\right], \tag{2.132}$$

where $0 \le \chi \le \pi/2$, $0 \le \theta \le \pi$, and $0 \le \phi < 2\pi$. Derive the differential equations for the geodesics.

2.11 Gravitational Waves

Problem 2.132 Consider the weak field limit in the harmonic gauge, where $g_{ab} = \eta_{ab} + \varepsilon h_{ab}$ and $g^{ab}\nabla_a\nabla_b\chi^c = 0$ for the coordinates χ^a.

a) Show that the harmonic gauge leaves a residual gauge freedom $\chi^a \mapsto \chi'^a = \chi^a + \varepsilon\xi^a$, which also preserves the weak field approximation, as long as $g^{ab}\nabla_a\nabla_b\xi^c = 0$ and find an expression for h'_{ab} in the coordinates χ'^a.

b) Defining $\bar{h}_{ab} = h_{ab} - g_{ab}h/2$, where $h = h^a_a$, show that

$$\partial^a\bar{h}_{ab} = 0, \tag{2.133}$$

in the harmonic gauge to leading order in ε.

c) A gravitational plane wave satisfies $\bar{h}_{ab} = A_{ab}\exp(ik_c\chi^c)$, where the wave equation can be used to conclude that $\eta_{ab}k^ak^b = 0$. Using the residual gauge transformation $\xi^a = iC^a\exp(ik_c\chi^c)$, find an expression for the amplitude A'_{ab} of the gravitational wave in the χ'^a coordinates.

d) Choosing $k_0 = k_3 = 1$, compute the values of the constants C^a for which $A'_{0a} = A'_{3a} = 0$ and $A'^a_a = 0$.

Hint: Note that the harmonic gauge condition in itself puts some constraints on the form of A_{ab}.

Problem 2.133 Consider gravitational waves described by perturbations $h_{\mu\nu}$ of the metric $g_{\mu\nu}$ such that $g_{\mu\nu} = \eta_{\mu\nu} + h_{\mu\nu}$, where $\eta_{\mu\nu}$ is the Minkowski metric and $|h_{\mu\nu}| \ll 1$.

a) In the transverse traceless gauge (TT gauge), the conditions $h_{0i} = 0$ and $\eta^{\mu\nu}h_{\mu\nu} = 0$ hold, which means that the Lorenz gauge condition reduces to $\partial_\nu h^\nu{}_\mu = 0$. Using the TT gauge conditions, show that the number of independent components of $h_{\mu\nu}$ is two.

b) Consider two particles at rest at $(x, y, z) = (0, 0, 0)$ and $(x, y, z) = (L, 0, 0)$, respectively. A plus-polarized gravitational wave h_+ of frequency f and amplitude $h_0 \ll 1$ passes by, propagating in the z-direction, such that

$$(h_{\mu\nu}(t, x, y, z)) = \begin{pmatrix} 0 & 0 & 0 & 0 \\ 0 & h_+ & h_\times & 0 \\ 0 & h_\times & -h_+ & 0 \\ 0 & 0 & 0 & 0 \end{pmatrix}$$

$$= h_0 \sin\left[2\pi f\left(t - \frac{z}{c}\right)\right]\begin{pmatrix} 0 & 0 & 0 & 0 \\ 0 & 1 & 0 & 0 \\ 0 & 0 & -1 & 0 \\ 0 & 0 & 0 & 0 \end{pmatrix}. \tag{2.134}$$

Show that the distance d measured along the x-axis between the two particles (i.e., the spatial separation along the equal t hypersurface), as the wave passes, is given by

$$d = \left[1 - \frac{1}{2}h_0 \sin(2\pi f t)\right] L. \tag{2.135}$$

Problem 2.134 Assuming a source described by the energy–momentum tensor $T^{\mu\nu}$, the linearized Einstein equations are given by

$$\Box \bar{h}_{\mu\nu} = 16\pi T_{\mu\nu}, \tag{2.136}$$

and the solutions in terms of Green's functions can be written as

$$\bar{h}_{\mu\nu}(t,\mathbf{x}) = 4 \int \frac{T_{\mu\nu}(t - |\mathbf{x} - \mathbf{x}'|, \mathbf{x}')}{|\mathbf{x} - \mathbf{x}'|} d^3x' \sim \frac{4}{r} \int T_{\mu\nu}(t - r, \mathbf{x}')\, d^3x', \tag{2.137}$$

where $r \equiv |\mathbf{x}|$ is far away from the source. Using the conservation law for the energy–momentum tensor, i.e., $\nabla_\nu T^{\mu\nu} = 0$, show that the spatial components are

$$\bar{h}_{ij}(t,\mathbf{x}) \sim \frac{2}{r}\frac{d^2}{dt^2} \int \rho(t - r, \mathbf{x}') x'_i x'_j \, d^3x', \tag{2.138}$$

where $\rho = T^{00}$ is the mass–energy density of the source.

Problem 2.135 a) For a neutron-star binary (with total mass $M \simeq 2.8 M_\odot$) at a distance of 5 kpc with orbital period $P = 1$ h, estimate the amplitude h of the gravitational waves.

b) Again, for the same system, but now with $P = 0.02$ s (giving $f_{\rm GW} = 2 f_{\rm orbit} = 100$ Hz, which lies in the sensitive band of the gravitational wave observatory LIGO), estimate h at a distance of 15 Mpc, which is approximately the distance to the Virgo cluster of galaxies.

c) Estimate the orbital separation R when $P = 0.02$ s.

d) For a neutron star at 1 kpc with a nonspherical deformation of mass $\delta M = 10^{-6} M_\odot$, a spin frequency of 50 Hz, and a stellar radius of 10 km, determine the gravitational wave amplitude h at Earth.

Problem 2.136 The first direct observation of gravitational waves was performed on September 14, 2015. This was presented by the LIGO and Virgo Collaborations on February 11, 2016, and was awarded the Nobel Prize in Physics in 2017. The original signal was named GW150914 and it was also the first observation of a binary black hole merger.

a) GW150914 had a maximal amplitude of $h \simeq 10^{-21}$ at a frequency of $f \simeq 200$ Hz. Compute the corresponding energy flux at the Earth. The binary source of GW150914 is situated at an estimated distance of about 400 Mpc.

b) Estimate the energy flux in electromagnetic waves that is received at Earth from a full moon. Compare this energy flux to the gravitational wave energy flux of GW150914.

Note:

$$\frac{c^5}{G} \simeq 3.63 \cdot 10^{52}\ \mathrm{W}, \tag{2.139}$$

is equal to 1 in geometric units, i.e., $c = 1$ and $G = 1$.

2.12 Cosmology and Friedmann–Lemaître–Robertson–Walker Metric

Problem 2.137 Consider the linearly expanding spacetime with metric $ds^2 = dt^2 - H^2t^2dx^2$. You start at $t = t_0$, $x = 0$ and want to arrive at $x = L$ at $t = t_1$ without accelerating at any time. Find an expression for your position $x(t)$ as a function of the global time t. What is the largest L you can arrive at in finite global time?

Problem 2.138 In a 2-dimensional mini-universe the metric element is given by

$$ds^2 = c^2dt^2 - S(t)^2d\chi^2, \tag{2.140}$$

where $S(t)$ is some positive function of the time t, constant in the "space" variable χ. Explain the cosmological redshift and derive an expression for it by studying the emission and detection of light signals by comoving observers at two different locations χ_0 and χ_1.

Problem 2.139 Consider the two-dimensional de Sitter universe with the metric $(c = 1)$

$$ds^2 = dt^2 - e^{2t/R}dx^2, \tag{2.141}$$

where $R > 0$ is a constant. Find an expression for the cosmological redshift between comoving observers at $x_0 > 0$ and $x_1 > x_0$.

Problem 2.140 Consider the two-dimensional de Sitter universe as defined in Problem 2.139.

a) Compute all nonzero Christoffel symbols for this metric.

b) Find the explicit form of the wave equation $g_{\mu\nu}\nabla^\mu\nabla^\nu\Phi = 0$, where Φ is a scalar field, in this universe.

Problem 2.141 Consider the Robertson–Walker spacetime described by the metric

$$ds^2 = dt^2 - e^{2t/t_H}\left[dr^2 + r^2(d\theta^2 + \sin^2\theta d\varphi^2)\right], \tag{2.142}$$

with the coordinates $(x^\mu) = (t, r, \theta, \varphi)$, where t is the universal time and $t_H \approx 14$ Gyr is the Hubble time.

a) Compute the path $r(t)$ of a light pulse emitted at time $t = t_0$ at the origin $r = 0$. You may assume that the light ray moves along the line $\theta = \pi/2$ and $\varphi = 0$ so that $dr/dt > 0$.

b) Compute the proper distance between the origin $r = 0$ and a point $r > 0$ on the line $\theta = \pi/2$ and $\varphi = 0$ at fixed universal time t.

c) Compute the spectral shift of the light pulse in a) during the time between emission at the origin and arrival at a point $r > 0$.

Hint: The spectral shift is defined as $z = \lambda_{\text{rec}}/\lambda_{\text{em}} - 1$, where λ_{em} is the wavelength of the light at emission and λ_{rec} its wavelength when it arrives.

Problem 2.142 a) Derive Hubble's law

$$v_p = \frac{\dot{a}(t)}{a(t)} d_p, \tag{2.143}$$

where $\dot{a}(t) = da/dt$ and v_p and d_p are the proper velocity and the proper distance, respectively, from the Robertson–Walker metric

$$ds^2 = dt^2 - a(t)^2 G_{ij} dx^i dx^j. \tag{2.144}$$

b) What are the physical consequences of Hubble's law?

Problem 2.143 a) The first Friedmann equation can be written as

$$\frac{\dot{a}^2}{a^2} + \frac{k}{a^2} = \frac{8\pi G}{3}\rho, \tag{2.145}$$

which can also be written as $1 - \Omega = -k/\dot{a}^2$, where $\Omega = \rho/\rho_c$. Assume now that k is small compared to the energy density ρ, which mainly consists of the cosmological constant ρ_Λ, thus leading to an inflationary universe. Show that the longer this scenario is assumed to last, the closer Ω gets to one.

b) Describe the flatness problem in words and how it is solved by inflation.

Problem 2.144 Our present universe can roughly be described by a spatially flat Friedmann–Lemaître–Robertson–Walker spacetime with $\Omega_\Lambda = 0.7$, $\Omega_m = 0.3$, $\Omega_r \simeq 10^{-4}$, being the density parameters of the cosmological constant (dark energy), matter, and radiation, respectively.

a) How much smaller was the scale factor when the energy density of the dark energy was equal to the energy density of matter?

b) At what redshift z did the matter–radiation equality (equal amounts of radiation and matter energy density) occur?

Problem 2.145 Our current universe is roughly described by a dark energy component $\Omega_\Lambda = 0.7$ and a matter component $\Omega_m = 0.3$. Determine an integral expression for the future behavior of the scale factor $a(t)$. You may assume that the scale factor today is $a_0 = 1$ and that the current Hubble parameter is H_0. Plot the result of your integral for $0 \le a(t) \le 100$. Compare your result to the analytic result $a(t) = \exp(H_0(t - t_0))$ for $\Omega_\Lambda = 1$ and $\Omega_m = 0$ and determine the age of the universe t_0 if $a(0) = 0$.

Hint: You may use numerical integration.

Problem 2.146 A toy model 1+1-dimensional circular Robertson–Walker universe has the line element

$$ds^2 = dt^2 - a(t)^2 d\varphi^2, \tag{2.146}$$

where φ and $\varphi + 2n\pi$ ($n \in \mathbb{N}$) correspond to the same spatial point. An object is thrown from $\varphi = \varphi_0$ at time t_0 with a velocity v relative to a comoving observer.

a) Find a condition that must be satisfied in order for the object to complete a full lap around the universe to reach the comoving observer again from the other direction.

b) What is the relative velocity between the object and a comoving observer at an arbitrary time t?
You may assume that the scale factor $a(t)$ has a known dependence on t.

Problem 2.147 Consider a flat universe containing only matter and radiation components such that the radiation density today corresponds to $\Omega_{\rm rad} = x \ll 1$. Find an expression for the time that has passed since matter and radiation had equal energy densities.

Problem 2.148 Consider a Friedmann–Lemaître–Robertson–Walker universe with curvature parameter $\kappa \neq 0$. Determine the condition on the equation-of-state parameter $w = p/\rho$ such that the curvature parameter $|\Omega_K| = \left|-1/(Ha)^2\right|$ decreases with cosmological time t for an expanding universe ($\dot{a} > 0$).

Problem 2.149 For a flat single-component Friedmann–Lemaître–Robertson–Walker universe with an arbitrary, but fixed, equation-of-state parameter $w = p/\rho$:

a) Compute the scale factor $a(t)$ as a function of cosmological time t.

b) Compute the Hubble parameter $H(t)$ as a function of t.

Hint: You may assume that, for some cosmological time $t = t_0$, we normalize our parameters such that $a_0 = a(t_0) = 1$ and $H_0 = H(t_0)$ are known. Your answers should be given in terms of ω, t_0, and H_0.

Problem 2.150 A scalar field ϕ with potential energy density $V(\phi)$ has a Lagrangian density given by

$$\mathscr{L} = \frac{1}{2} g^{\mu\nu}(\partial_\mu \phi)(\partial_\nu \phi) - V(\phi). \tag{2.147}$$

a) Derive the equation of motion for the scalar field ϕ.

b) Assuming that the N+1-dimensional spacetime has a metric given by

$$ds^2 = g_{\mu\nu} dx^\mu dx^\nu = dt^2 - a(t)^2 G_{ij} dx^i dx^j, \tag{2.148}$$

where G_{ij} are the metric components on an N-dimensional Riemannian manifold, and that the scalar field ϕ only depends on the time coordinate t, show that the scalar field is an ideal fluid and find the (time-dependent) equation-of-state parameter $w = p/\rho_0$.

3
Solutions to Problems

3.1 Solutions to Problems in Special Relativity Theory

1.1

a) A scalar is always the same independent of the inertial frame. The scalar product between the two 4-vectors $\vec{A}$ and $\vec{B}$ is zero in the inertial frame of the observer $\mathcal{O}'$ since they point in different coordinate directions. Thus, it is always zero. The two 4-vectors $\vec{U}$ and $\vec{V}$ are lightlike in different directions. The scalar product between two lightlike 4-vectors pointing in different directions is always nonzero. Assume that $U^0 = |\alpha|$, $U^1 = -|\alpha|$ and $V^0 = |\beta|$, $V^1 = |\beta|$, where $\alpha, \beta \neq 0$, then we have

$$\vec{U} \cdot \vec{V} = \eta_{\mu\nu} U^\mu V^\nu = \eta_{00} U^0 V^0 + \eta_{11} U^1 V^1 = |\alpha||\beta| - (-|\alpha|)|\beta| = 2|\alpha\beta| > 0. \tag{3.1}$$

In conclusion, the statements 1, 2, and 6 are true.

b) The 4-vector $\vec{A}$ can be proportional to a 4-velocity vector for a massive particle, since $\vec{A}$ is equal to $(A^0, 0, 0, 0)$ in S' a 4-velocity vector of a massive particle is given by $(c, 0, 0, 0)$, where c is the speed of light. The two 4-vectors $\vec{U}$ and $\vec{V}$ are proportional to the worldline tangent vector of massless particles, since they are null vectors. Vector $\vec{B}$ is spacelike and cannot be the tangent of a worldline.

1.2

a) Let A^μ be the timelike 4-vector. There exists an inertial coordinate system K' such that $A'^\mu = (A'^0, \mathbf{0})$, where $A'^0 \neq 0$. If the 4-vector B^μ is orthogonal to A^μ, then $A'^\mu B'_\mu = A'^0 B'_0 = 0$, which means that $B'_0 = B'^0 = 0$. Hence, it holds that

$$B^2 = B'^2 = B'^\mu B'_\mu = -\mathbf{B}'^2 < 0, \tag{3.2}$$

which means that B^μ is spacelike. (We assume that $B^\mu \not\equiv 0$.)

b) Let A^μ and B^μ be the 4-vectors. There exists an inertial coordinate system K' such that $A'^\mu = \left(A'^0, 0\right)$ and $B'^\mu = \left(B'^0, \boldsymbol{B}'\right)$, where $A'^0 > 0$ and $B'^0 > 0$. Therefore, it holds that

$$\left(A'^\mu + B'^\mu\right)\left(A'_\mu + B'_\mu\right) = \left(A'^0 + B'^0\right)^2 - \boldsymbol{B}'^2$$
$$= \left(A'^0\right)^2 + 2A'^0 B'^0 + \left(B'^0\right)^2 - \boldsymbol{B}'^2 > 0, \tag{3.3}$$

because $\left(B'^0\right)^2 - \boldsymbol{B}'^2 > 0$, since B^μ is timelike.

c) There exists an inertial coordinate system K' such that the spacelike 4-vector has the components $A'^\mu = \left(0, A'^1, 0, 0\right)$ relative to K'. Thus, we can set

$$A'^\mu = \frac{1}{2}\left(A'^1, A'^1, 0, 0\right) - \frac{1}{2}\left(A'^1, -A'^1, 0, 0\right). \tag{3.4}$$

d) Introduce K' in the same way as in b). Then, it holds that

$$A^\mu B_\mu = A'^\mu B'_\mu = A'^0 B'_0 > 0. \tag{3.5}$$

1.3

The relative gamma factor between two observers is given by

$$\gamma_{12} = V_1 \cdot V_2, \tag{3.6}$$

where V_1 and V_2 are the 4-velocities of the observers. (This may be seen by going to the rest frame of one of the observers, where the 4-velocity of two other is given by $V = \gamma_{12}(1, \boldsymbol{v}_{\text{rel}})$.) We find that

$$V_1 = \gamma_1(1, \boldsymbol{v}_1), \quad V_2 = \gamma_2(1, \boldsymbol{v}_2), \tag{3.7}$$

where

$$\gamma_i = \frac{1}{\sqrt{1 - v_i^2}}. \tag{3.8}$$

Consequently, we obtain

$$\gamma_{12} = V_1 \cdot V_2 = \gamma_1\gamma_2(1 - \boldsymbol{v}_1 \cdot \boldsymbol{v}_2) = \frac{1}{\sqrt{1 - v_1^2}\sqrt{1 - v_2^2}}(1 - \boldsymbol{v}_1 \cdot \boldsymbol{v}_2). \tag{3.9}$$

1.4

a) The correct answer is: Statement 3 (never). This is due to the fact that a photon is traveling at the speed of light relative to all inertial frames. Hence, it is impossible to perform a Lorentz transformation to a frame where it is does not move.

b) The correct answer is: Statement 2 (sometimes). In order to have a rest frame for the center of momentum for a system of two photons, it cannot be moving with the speed of light. This happens when the two photons are moving in directions parallel to each other. However, in all other configurations, the center of momentum will move at a speed slower than c, and then, we can find a rest frame for it.

1.5

a) The formula for *length contraction* (or *Lorentz contraction*) in special relativity is given by

$$\ell' = \frac{\ell}{\gamma(v)}, \quad \gamma(v) \equiv \frac{1}{\sqrt{1 - \frac{v^2}{c^2}}}, \tag{3.10}$$

where ℓ is the length of an object in its rest frame $\mathcal{O}$ and ℓ' is the corresponding length of the object in a frame $\mathcal{O}'$ moving with constant velocity v ($0 \leq v < c$) relative to the rest frame $\mathcal{O}$ of the object. Note that $0 < \ell' \leq \ell$, since $1 \leq \gamma(v) < \infty$, so the object appears to be shortened in the frame $\mathcal{O}'$ moving relative to the rest frame $\mathcal{O}$.

The length contraction formula can be derived using the fact that the length of the object in a given frame is the spatial distance between the ends of the object measured *at the same time* in that frame. Consider the time t' in S' to be the time at which we measure the position of the ends of the object, which are then separated by ℓ'. Calling the events at the ends of the object A and B, respectively, the spatial distance between them in S' is therefore ℓ' and the temporal separation is zero. At the same time, the spatial and temporal separations between the events in S are ℓ and Δt, respectively, as the object is at rest in S. For the events to be simultaneous in S', the Lorentz transformation requires

$$c\Delta t' = \gamma\left(c\Delta t - \frac{v\ell}{c}\right) = 0, \tag{3.11}$$

and therefore,

$$\Delta t = \frac{v\ell}{c^2}. \tag{3.12}$$

As the spacetime interval between the events is invariant, we now find that

$$\ell'^2 - c^2\Delta t'^2 = \ell'^2 = \ell^2 - c^2\Delta t^2 = \ell^2 - \frac{v^2}{c^2}\ell^2, \tag{3.13}$$

or, once simplified,

$$\ell' = \ell\sqrt{1 - \frac{v^2}{c^2}} = \frac{\ell}{\gamma}, \tag{3.14}$$

which is the length contraction formula.

b) The formula for *time dilation* in special relativity is given by

$$t = \gamma(v)t', \quad \gamma(v) \equiv \frac{1}{\sqrt{1 - \frac{v^2}{c^2}}}, \tag{3.15}$$

where t is the time in the frame $\mathcal{O}$ and t' is the corresponding time in the frame $\mathcal{O}'$ moving with constant velocity v $(0 \leq v < c)$ relative to the frame $\mathcal{O}$. Note that $t > t'$, since $1 \leq \gamma(v) < \infty$, so an observer in the frame $\mathcal{O}$ measures a longer time than an observer in $\mathcal{O}'$ measures between two events which occur at the same spatial point in the frame $\mathcal{O}'$.

In order to derive the time dilation formula, we consider the worldline of an observer at rest in S'. In S', the spatial and temporal differences between two events on the worldline are given by zero and t', respectively. In the frame S, the observer is instead moving at speed v and the spatial and temporal separations are therefore t and vt, respectively. As the spacetime interval between the events is invariant, we find that

$$c^2t'^2 - 0^2 = c^2t^2 - v^2t^2 \implies t' = t\sqrt{1 - \frac{v^2}{c^2}}, \tag{3.16}$$

or, equivalently,

$$t = \gamma(v)t', \tag{3.17}$$

which is the time dilation formula.

1.6

Length contraction in the x-direction yields (see Figure 3.1)

$$a' = a\sqrt{1 - \frac{v^2}{c^2}} = \{v = \sqrt{2/3}c\} = a\sqrt{1 - \frac{2}{3}} = \frac{a}{\sqrt{3}}, \tag{3.18}$$

where $a = 1/\sqrt{2}$ m. Note that there is no length contraction in the y-direction.

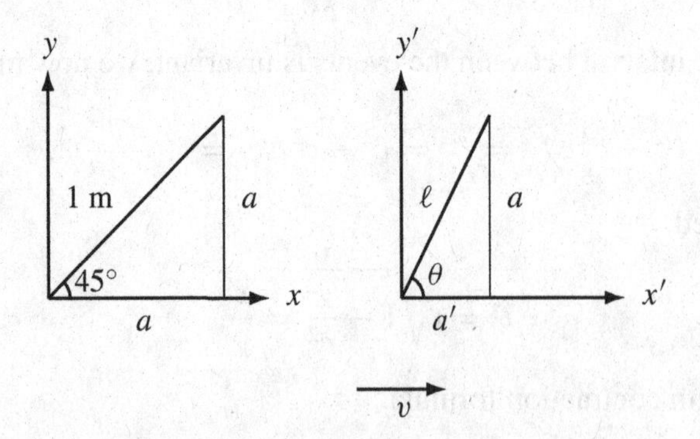

Figure 3.1 A rod inclined 45° in the xy-plane with respect to the x-axis and the angle θ in the $x'y'$-plane with respect to the x'-axis. The relative speed between the two planes is v.

Now, Pythagoras' theorem gives

$$\ell = \sqrt{a'^2 + a^2} = \{a' = a/\sqrt{3}\} = \sqrt{\frac{a^2}{3} + a^2} = \frac{2a}{\sqrt{3}} = \{a = 1/\sqrt{2}\,\text{m}\} = \sqrt{\frac{2}{3}}\,\text{m}. \tag{3.19}$$

Thus, the angle θ is given by

$$\theta = \arctan\frac{a}{a'} = \{a' = a/\sqrt{3}\} = \arctan\sqrt{3} = \frac{\pi}{3} = 60^\circ. \tag{3.20}$$

In conclusion, the answer is $\ell = \sqrt{\frac{2}{3}}$ m and $\theta = 60^\circ$.

1.7

The lifetime τ_0 is measured in the rest frame of the muon. Due to time dilation, the lifetime in the rest frame of the Earth will be

$$\tau = \tau_0 \gamma(v), \tag{3.21}$$

where v is the velocity of the muon relative to Earth. With $v = 0.999c$, one obtains

$$\tau = \frac{\tau_0}{\sqrt{1 - \frac{v^2}{c^2}}} \approx 22\tau_0. \tag{3.22}$$

During this time, the muon will move the distance

$$\tau v \approx 22 \cdot 660 \cdot 0.999\ \text{m} \approx 15\ \text{km}, \tag{3.23}$$

in the rest frame of the Earth, which explains that many muons arrive at the surface.

This can also be derived in the muons' rest frame: Due to length contraction, the thickness of the atmosphere (10 km) in the rest frame of the Earth will correspond to $10^4/22$ m $\approx$ 450 m in the rest frame of the muon. During the lifetime of the muon, the Earth will move the distance $\tau_0 v = 660 \cdot 0.999$ m ≈ 660 m in the rest frame of the muon, which is larger than 450 m.

1.8

Introduce the rest frames of the station and the train as K and K', respectively.

a) For the two events that involve the markings on the track, we have

$$\Delta t = 0, \quad \Delta x = L_a, \quad \text{and} \quad \Delta x' = L_0. \tag{3.24}$$

When this is inserted into

$$\Delta x' = \gamma(v)(\Delta x - v\Delta t), \tag{3.25}$$

we obtain

$$L_a = L_0/\gamma(v). \tag{3.26}$$

b) For the nailings, we have

$$\Delta t' = 0, \quad \Delta x = L_b, \quad \text{and} \quad \Delta x' = L_0. \tag{3.27}$$

Inserting into

$$\Delta x = \gamma(v)(\Delta x' + v\Delta t'), \tag{3.28}$$

gives

$$L_b = \gamma(v)L_0. \tag{3.29}$$

c) We introduce the events E_1: the end passes B, and E_2: the front passes A. For these events, we have

$$\Delta t = -L_c/c, \quad \Delta x = L_c, \quad \text{and} \quad \Delta x' = L_0. \tag{3.30}$$

The relation $\Delta x' = \gamma(v)(\Delta x - v\Delta t)$ gives

$$L_c = \sqrt{\frac{1 - v/c}{1 + v/c}} L_0. \tag{3.31}$$

d) We let the events E_1 and E_2 be the radar pulses reflection against the front and end, respectively. For these events, we have

$$\Delta t = (t_2 - t_1)/2, \quad \Delta t' = -L_0/c, \quad \text{and} \quad \Delta x' = -L_0. \tag{3.32}$$

Inserting into $\Delta t = \gamma(v)(\Delta t' + v\Delta x'/c^2)$ gives

$$L_d = (t_1 - t_2)\frac{c}{2} = \sqrt{\frac{1 + v/c}{1 - v/c}} L_0. \tag{3.33}$$

1.9

Let the coordinates of the front and rear end of the express space cruiser be $x_F = (-\tau c, L, 0, 0)$ and $x_R = (0, 0, 0, 0)$, respectively, in the rest frame of the cruiser at the time when the rear watchman sees the rear light go on. The difference between these are $\Delta x = (-\tau c, L, 0, 0)$. In the rest frame of the hitchhiker, this difference is given by an inverse Lorentz transformation to the rest frame of the asteroid with velocity $-v$, v being the velocity of the cruiser in the rest frame of the asteroid. His/her time difference for the lightening of the lanterns is then $\Delta x'^0/c = -\tau \cosh\theta + (L/c)\sinh\theta$. However, this time difference is equal to zero. Thus, we have

$$\tau = \frac{L}{c}\tanh\theta = \frac{vL}{c^2}, \tag{3.34}$$

from which we obtain

$$v = \frac{\tau c^2}{L} = 1.8 \cdot 10^5 \text{ m/s} \approx 0.0006c. \tag{3.35}$$

1.10

Using the fact that the interval between two points in spacetime is invariant, i.e., $\Delta s^2 = \Delta s'^2$ or $c^2\Delta t^2 - \Delta x^2 = c^2\Delta t'^2 - \Delta x'^2$, gives together with the information in the problem text

$$c^2 \cdot 0^2 - \ell^2 = c^2\tau^2 - \ell'^2. \tag{3.36}$$

Thus, we find

$$\ell = \sqrt{\ell'^2 - c^2\tau^2}. \tag{3.37}$$

Now, using the length contraction formula $\ell' = \ell\gamma(v)$, where $\gamma(v) = \frac{1}{\sqrt{1-v^2/c^2}}$, one obtains

$$\ell' = \sqrt{\ell'^2 - c^2\tau^2}\frac{1}{\sqrt{1 - v^2/c^2}}, \tag{3.38}$$

from which it follows that

$$v = \pm\frac{c^2\tau}{\ell'}. \tag{3.39}$$

1.11

Lorentz contraction only takes place for the projection of the rod that lies along the x-axis. This projection is $\ell\cos\theta$. The orthogonal component is $\ell\sin\theta$. Lorentz contraction of the x-component is then $\ell\cos\theta\sqrt{1 - v^2/c^2}$. Therefore, to the moving observer, the length of the rod is

$$\ell' = \sqrt{\left(\ell\cos\theta\sqrt{1 - \frac{v^2}{c^2}}\right)^2 + (\ell\sin\theta)^2} = \ell\sqrt{1 - \left(\frac{v}{c}\right)^2\cos^2\theta}. \tag{3.40}$$

Thus, the answer is $\ell' = \ell\sqrt{1 - \left(\frac{v}{c}\right)^2\cos^2\theta}$.

1.12

The spacetime interval of the two events A and B with coordinates x_A and x_B is $s = (x_A - x_B)^2$. For observer K in S, this is $s = -L^2$. In the rest frame of K' we have $s = c^2\Delta t'^2 - L'^2$. Therefore, we obtain $L = \sqrt{L'^2 - c^2\Delta t'^2}$. Thus, the answer is $L = \sqrt{L'^2 - c^2\Delta t'^2}$.

1.13

The invariant interval is $(x_\alpha - x_\beta)^2 = (x'_\alpha - x'_\beta)^2$.

a) Denote the distance to the β event for S' by L'. We set $c = 1$ and calculate the length in ly. Then, we have

$$2^2\ \mathrm{ly}^2 - 10^2\ \mathrm{ly}^2 = 1^2\ \mathrm{ly}^2 - L'^2. \tag{3.41}$$

Solving this equation gives

$$L' = \sqrt{100 + 1 - 4}\ \text{ly} = \sqrt{97}\ \text{ly} \approx 9.85\ \text{ly}. \tag{3.42}$$

b) The Lorentz transformation from K to K' gives

$$1 = 2\gamma - 10v\gamma, \tag{3.43}$$
$$L'/\text{ly} = -2v\gamma + 10\gamma, \tag{3.44}$$

where $\gamma = 1/\sqrt{1 - v^2}$. Solving this equation yields $v \approx 0.1$ as the permissible root.

1.14

The smallest energy required for a muon to hit the ground within its mean life in the rest frame of the Earth is given when the direction of the muon is vertical. In this case, the muon must travel the distance $\ell = 10$ km in the time τ, which is the time dilated mean life τ_0 of the muon in its rest frame. It follows that

$$\ell \leq v\tau = v\gamma(v)\tau_0. \tag{3.45}$$

The velocity of the muon is given by $v = pc^2/E$, and thus, we have

$$\gamma(v) = \frac{1}{\sqrt{1 - (v/c)^2}} = \frac{E}{\sqrt{E^2 - p^2c^2}} = \frac{E}{mc^2}. \tag{3.46}$$

Inserting this into the above inequality, we obtain

$$\ell \leq \frac{pc}{mc^2} c\tau_0. \tag{3.47}$$

Inserting the numerical values, we find that $pc \gtrsim 15mc^2 \simeq 1.6$ GeV, and thus, it holds that $E \simeq pc \gtrsim 1.6$ GeV.

1.15

A time interval for a muon and a time interval for an observer in the lab frame are related through the time dilation formula

$$\gamma(v)d\tau = dt, \tag{3.48}$$

where $d\tau$ is the time interval for the muon, dt is the time interval for the observer in the lab frame, and v is the muon velocity. The muon velocity is constant (since the total energy is constant) and given by

$$v = \frac{p}{E} = \frac{\sqrt{E^2 - m^2}}{E} = \sqrt{1 - \frac{m^2}{E^2}}, \tag{3.49}$$

where p is the muon momentum, E is the muon energy, and m is the muon mass. It follows that

$$\gamma(v) = \frac{E}{m}, \tag{3.50}$$

and thus, we obtain

$$t = \frac{E}{m}\tau = \frac{1\text{ GeV}}{106\text{ MeV}}\tau \simeq 10\,\tau. \tag{3.51}$$

The average lifetime of the muon in the lab frame is therefore 22 μs. Since the muon is highly relativistic ($E \gg m$), the length traveled by the muon in the lab frame in the average lifetime is given by

$$\ell = vt \simeq ct \simeq 3 \cdot 10^8 \text{ m/s} \cdot 22 \cdot 10^{-6} \text{ s} = 6\,600 \text{ m}. \tag{3.52}$$

The circumference of the circular accelerator is given by

$$L = 2\pi r \simeq 300 \text{ m}. \tag{3.53}$$

Thus, the average number of turns taken by a muon is given by

$$N = \frac{\ell}{L} \simeq 22. \tag{3.54}$$

1.16

We let $\ell = 1$ to simplify the notation.

a) We use the standard formulas for Lorentz contractions, and thus, we have

$$a' = a/\gamma = 3/\gamma, \quad b' = b = 4, \quad c' = \sqrt{(a')^2 + (b')^2} = \sqrt{9/\gamma^2 + 16}, \tag{3.55}$$

where $\gamma = 1/\sqrt{1 - (v/c)^2}$, and thus, the area becomes

$$A' = a'b'/2 = 6/\gamma = A/\gamma, \tag{3.56}$$

where $A = 6$ is the area in K.

b) Let h be the height of the triangle in K'. Since $A = 6 = ch/2$, we get $h = 12/5$. We have $c' = c/\gamma$, $h' = h$, and thus, the area becomes $A' = c'h'/2 = A/\gamma$ as in a). Let c_1 be the distance in K from the corner, where the a- and c-sides of the triangle meet in the point, where the height intersects the c-side in a right angle. Obviously, $c_1 = \sqrt{a^2 - h^2}$. In K', the same distance is $c_1' = c_1/\gamma$, and thus, we obtain

$$a' = \sqrt{\left(c_1'\right)^2 + (h')^2} = \sqrt{(a^2 - h^2)/\gamma^2 + h^2} = \sqrt{9/\gamma^2 + (144/25)(1 - 1/\gamma^2)}, \tag{3.57}$$

and similarly, we find that

$$b' = \sqrt{16/\gamma^2 + (144/25)(1 - 1/\gamma^2)}. \tag{3.58}$$

1.17

The rest frame of the pole is denoted S with coordinates $X = (t, x, y, z)$, where the pole has length L. The lab frame is denoted S' with coordinates $X' = (t', x', y', z')$, in which the pole length is $L' = L/\gamma$, where $\gamma = 1/\sqrt{1 - v^2/c^2}$. In the lab frame, the pole is moving in the negative x-direction with speed v, parallel to the x-axis at some distance y. Everything occurs in the spatial plane $z = 0$, so we can ignore the z-direction. Denote the events at which light is emitted from front of the pole by A, from the mark on the pole by B, and from the end of the pole by C. The three rays of light all reach the origin event in S', i.e., the spatial origin at $t' = 0$. Since the three light rays are emitted from different positions, but all reach the origin at the same time, the light rays were emitted at different times t'. We now consider each of the three light rays separately.

Event A: Let $X'_{\mathrm{A}} = (t', x', y')$ denote the coordinates in S' of the event A. Since the line element along the 4-path of a light signal vanishes, we have

$$t'^2 = x'^2 + y'^2. \tag{3.59}$$

The light ray defines an angle $\pi/3$ with the x-axis, and therefore, we find

$$\frac{y'}{x'} = \tan\frac{\pi}{3} = \sqrt{3}. \tag{3.60}$$

Combining the two equations, we get

$$t'^2 = 4x'^2, \quad y' = \sqrt{3}x'. \tag{3.61}$$

Defining the time at which the event A occurs to be $t' = -\delta_{\mathrm{A}}$, we obtain the coordinates of A in S' as

$$X'_{\mathrm{A}} = \left(-1, \frac{1}{2}, \frac{\sqrt{3}}{2}\right)\delta_{\mathrm{A}}. \tag{3.62}$$

Event B: Let $X'_{\mathrm{B}} = (t', x', y')$ denote the coordinates in S' of the event B. In this case, the angle is $\pi/4$, so we have

$$t'^2 = x'^2 + y'^2, \quad \frac{y'}{x'} = \tan\frac{\pi}{4} = 1, \tag{3.63}$$

which means that

$$t'^2 = 2x'^2, \quad y' = x'. \tag{3.64}$$

Thus, defining the time of the event B as $t' = -\delta_{\mathrm{B}}$, we obtain

$$X'_{\mathrm{B}} = \left(-1, \frac{1}{\sqrt{2}}, \frac{1}{\sqrt{2}}\right)\delta_{\mathrm{B}}. \tag{3.65}$$

Event C: Let $X'_{\mathrm{C}} = (t', x', y')$ denote the coordinates in S' of the event C. Similar to events A and B, defining the time of the event C as $t' = -\delta_{\mathrm{C}}$, we obtain

$$X'_{\mathrm{C}} = \left(-1, \frac{\sqrt{3}}{2}, \frac{1}{2}\right) \delta_{\mathrm{C}}. \tag{3.66}$$

Since the values of y' are all equal for the three events A, B, and C, we find that

$$\frac{\sqrt{3}}{2}\delta_{\mathrm{A}} = \frac{1}{\sqrt{2}}\delta_{\mathrm{B}} = \frac{1}{2}\delta_{\mathrm{C}}, \tag{3.67}$$

which means that the coordinates of all three events can be expressed in the time δ_{A} only, i.e.,

$$X'_{\mathrm{A}} = \left(-1, \frac{1}{2}, \frac{\sqrt{3}}{2}\right) \delta_{\mathrm{A}}, \quad X'_{\mathrm{B}}\left(-\sqrt{\frac{3}{2}}, \frac{\sqrt{3}}{2}, \frac{\sqrt{3}}{2}\right) \delta_{\mathrm{A}}, \quad X'_{\mathrm{C}} = \left(-\sqrt{3}, \frac{3}{2}, \frac{\sqrt{3}}{2}\right) \delta_{\mathrm{A}}. \tag{3.68}$$

The event A occurs at the front of the pole at $x' = \delta_{\mathrm{A}}/2$ at time $-\delta_{\mathrm{A}}$, whereas the event C occurs at the end of the pole at $x' = 3\delta_{\mathrm{A}}/2$ at time $-\sqrt{3}\delta_{\mathrm{A}}$. The pole moves at speed v in the negative x-direction. Therefore, at time $-\delta_{\mathrm{A}}$, the end of the pole is located at

$$x'_{\mathrm{C}} - v(t'_{\mathrm{A}} - t'_{\mathrm{C}}) = \left[\frac{3}{2} - v(\sqrt{3} - 1)\right] \delta_{\mathrm{A}}. \tag{3.69}$$

In S', we find that the length of the pole is the difference between this and $x'_{\mathrm{A}} = \delta_{\mathrm{A}}/2$, which occurs simultaneously at the front, so

$$L' = \left[1 - v(\sqrt{3} - 1)\right] \delta_{\mathrm{A}}. \tag{3.70}$$

However, in S', we know that the length of the pole is $L' = L/\gamma$, which means that we have determined δ_{A} to be

$$\delta_{\mathrm{A}} = \frac{L'}{1 - v(\sqrt{3} - 1)} = \frac{L}{\gamma\left[1 - v(\sqrt{3} - 1)\right]}, \tag{3.71}$$

and therefore, we have determined all coordinates of the events A, B, and C in S'.

Now, we determine the distance between the front of the pole and the mark on the pole. At time $-\delta_{\mathrm{A}}$, the x'-coordinate of the mark is given by

$$x'_{\mathrm{B}} - v(t'_{\mathrm{A}} - t'_{\mathrm{B}}) = \left[\frac{\sqrt{3}}{2} - v\left(\sqrt{\frac{3}{2}} - 1\right)\right] \delta_{\mathrm{A}}. \tag{3.72}$$

Thus, the distance between the front and the mark is

$$\tilde{L}' = \left[\frac{\sqrt{3}}{2} - \frac{1}{2} - v\left(\sqrt{\frac{3}{2}} - 1\right)\right]\delta_{\mathrm{A}}. \tag{3.73}$$

Therefore, in S', the ratio r between the distance from the front of the pole to the mark on the pole and the length of the whole pole is

$$r \equiv \frac{\tilde{L}'}{L'} = \frac{\frac{\sqrt{3}}{2} - \frac{1}{2} - v\left(\sqrt{\frac{3}{2}} - 1\right)}{1 - v(\sqrt{3} - 1)} \approx 0.37 - \frac{0.22v}{1 - 0.73v}. \tag{3.74}$$

Note that r is the same in all inertial frames due to the linearity of Lorentz transformations. For example, in the rest frame of a pole with length L and the front at the origin, the mark would be located at the x-coordinate rL.

1.18

a) In the initial rest frame of the spaceships, they are at a constant distance L from each other during the entire accelerating phase. However, this length is the length contraction of the string in the new rest frame and the proper length will then be $L' = \gamma L$. Trying to force the string to be of length L in the initial rest frame will therefore break it.

b) The initial distance between the two spaceships is 40 km. The spaceships accelerate with a constant acceleration $a = 1/50\ c/\mathrm{s}$. When the spaceships stop after 30 s in the initial rest frame, the speed of the spaceships will then be $v = at = 3c/5$. Hence, the gamma factor can be computed to be

$$\gamma = \frac{1}{\sqrt{1 - \frac{v^2}{c^2}}} = \frac{1}{\sqrt{1 - \left(\frac{3}{5}\right)^2}} = \frac{5}{4}. \tag{3.75}$$

Thus, the distance between the two spaceships is $L' = \gamma L = 5/4 \cdot 40\ \mathrm{km} = 50\ \mathrm{km}$.

1.19

Spacetime diagrams showing the events described in the problem are shown in Figure 3.2. Part (a) of the diagram depicts the events using the time and x axes of the rest frame of Professor Einstein as orthogonal, while part (b) is the same spacetime diagram with the axes of Professor Wolf orthogonal. The coordinate axes of Professor Einstein are marked by t_{E} and x_{E}, respectively, while those of Professor Wolf are marked by t_{W} and x_{W}, respectively. The light gray shaded region is the train, the thick black lines and the thick dark gray lines represent the worldlines of Professor Einstein and Professor Wolf, respectively, while the thick white lines represent the light signals.

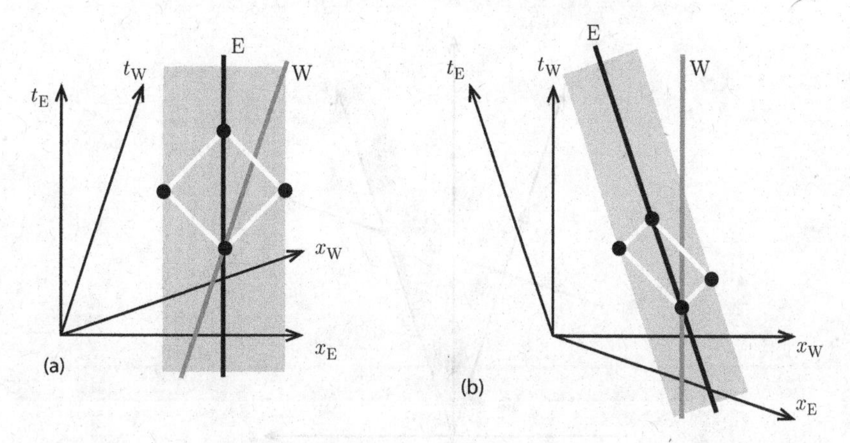

Figure 3.2 Spacetime diagrams describing the events of Problem 1.19.

From the diagrams it is clear that

a) The light signals will always reach Professor Einstein at the same time. This is something physically observable and cannot depend on the set of coordinates used to describe the events. (In fact, a light signal reaching Professor Einstein is an event and both light signals reach him at the same event).
b) Professor Wolf sees the light signals being reflected at different times. This can be seen as the line between the reflection points is not parallel to the spatial axis in Professor Wolf's rest frame, i.e., it is not a simultaneity in that frame.
c) The signals will reach Professor Wolf at different times. The signal from the back of the train has already reached Professor Wolf before it reaches Professor Einstein, while the signal from the front reflection will reach him after it reaches Professor Einstein.

1.20

Let us consider the situation seen in the frame S where the two spaceships have equal but opposite velocities. The situation can then be described through the spacetime diagram in Figure 3.3. Since the velocities of both spaceships relative to this frame are the same, they are equally Lorentz contracted and the ship with rest length $2L$ is twice as long (i.e., light gray shaded region) as the ship with rest length L (i.e., dark gray shaded region). The coordinates have been chosen such that A occurs in the spacetime origin and the event B has been marked in the diagram. The time axes of both the frame S', in which the shorter ship is at rest, and S'', in which the longer is at rest, are also shown. The dark (light) gray line represents the surface simultaneous with B in S' (S'') and the intersection with the t' (t'')-axis has been marked. Since the t' and t'' axes are tilted with the same angle relative to the t-axis,

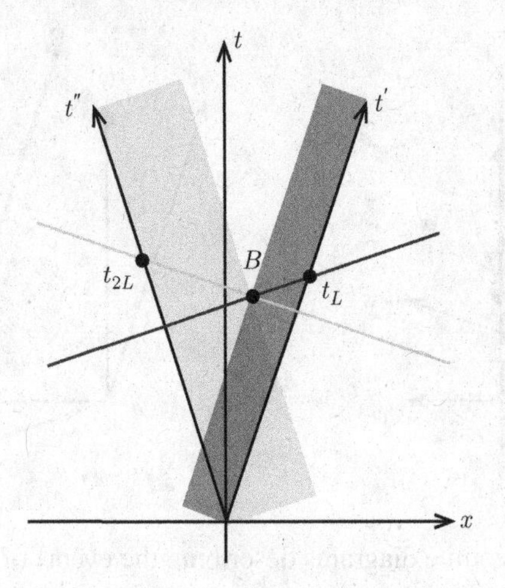

Figure 3.3 The spacetime diagram describing the situation in Problem 1.20.

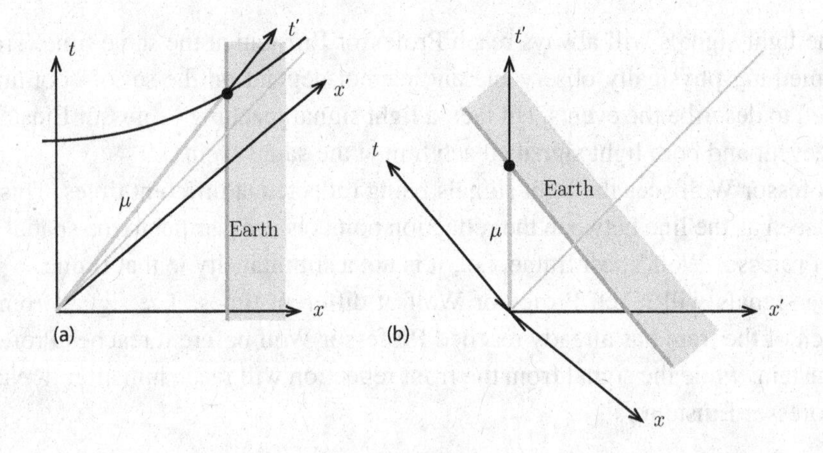

Figure 3.4 Spacetime diagrams describing the situation in Problem 1.21.

they both have the same normalization. Thus, $t_{2L} > t_L$, i.e., the time measured in the rest frame of the shorter rocket is shorter.

1.21

The situation is described by the spacetime diagrams in Figure 3.4. In part (a) of the diagram, we see the situation described in S, the rest frame of the Earth. The light gray thick line (marked by μ) corresponds to the muon worldline, the length of which is the muon eigentime to reach the Earth's surface. The black solid curve is a hyperbolic curve, which means that straight worldlines from the origin to that curve will have the same length, i.e., eigentime. From this we draw the conclusion that the eigentime experienced by the muons will be significantly smaller than the time taken to perform the travel from the top of the atmosphere to the surface and the muons thus will not have sufficient time to decay. More quantitative statements cannot be made without further information. The t'- and x'-axes of S', the muon rest frame, are also shown along with the light gray thin line, which shows the light cone. For completeness, we also show how the situation looks in S' in part (b) of the figure.

1.22

The problem is most easily solved by considering the two events when the ends of the blade reaches the $y' = y = 0$ plane. Letting the tip of the guillotine cutting the plane be the origin $x = x' = t = t' = 0$, the other end of the guillotine cutting the plane will have the coordinates

$$t = \frac{\ell}{u} = \frac{\ell_0}{u\gamma_u}, \quad x = L, \tag{3.76}$$

where we have taken into account that the blade is length contracted in the y-direction in S with a factor $\gamma_u = 1/\sqrt{1-u^2}$. Transforming this to the S' frame, we find that

$$t' = \gamma_v(t - vx) = \gamma_v\left(\frac{\ell_0}{u\gamma_u} - vL\right). \tag{3.77}$$

If the blade edge is horizontal in S', then all points on the edge will cut the $y' = 0$ plane at the same time and since $t' = 0$ for the point cutting it, we find that

$$\frac{\ell_0}{u\gamma_u} - vL = 0 \quad \Longrightarrow \quad v = \frac{\ell_0}{uL\gamma_u}. \tag{3.78}$$

1.23

The spacetime diagram in Figure 3.5 shows the worldlines of the observer, the two lights, and a light signal from the lights that arrives to the observer at the same time.

The positions where the lights were when this signal was emitted will be the positions *seen* by the observer at $t = t^*$ and the *seen* separation L will be given by the difference of these positions.

The worldlines of the lights are given by

$$x_1(t) = x_{1,0} - vt, \quad x_2(t) = x_{2,0} - vt. \tag{3.79}$$

If the time coordinate is chosen such that the signal is emitted from $x_2(0)$ at $t = 0$, the signal worldline is given by

$$x_s(t) = x_{2,0} - t, \tag{3.80}$$

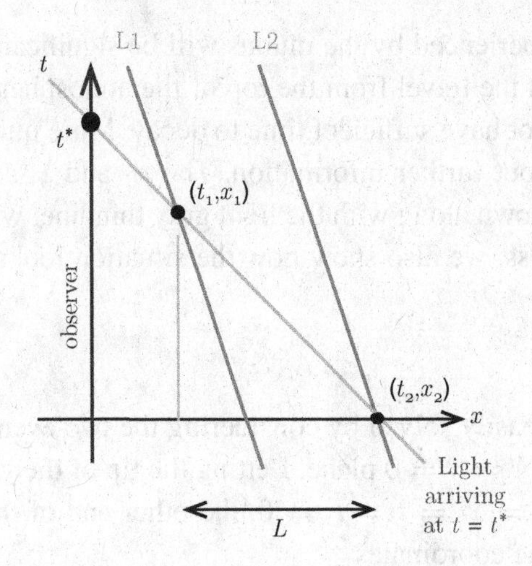

Figure 3.5 Spacetime diagram showing the worldlines of an observer, two lights (L1 and L2), and a light signal arriving to the observer at $t = t^*$.

and intersects the worldlines of light 2 at $x_{2,0}$. The intersection of the worldline of light 1 is given by

$$x_1(t_1) = x_s(t_1) \quad \Rightarrow \quad x_{1,0} - vt_1 = x_{2,0} - t_1 \quad \Rightarrow \quad x_{2,0} - x_{1,0} = (1-v)t_1$$

$$\Rightarrow \quad t_1 = \frac{x_{2,0} - x_{1,0}}{1-v} = \frac{\ell_0}{\gamma(1-v)} = \ell_0\sqrt{\frac{1+v}{1-v}}, \tag{3.81}$$

where we have used that $x_{2,0} - x_{1,0} = \frac{\ell_0}{\gamma}$ by Lorentz contraction. The *seen* distance between the lights is therefore

$$L = x_s(0) - x_s(t_1) = x_{2,0} - x_{2,0} + t_1 = \ell_0\sqrt{\frac{1+v}{1-v}}. \tag{3.82}$$

1.24

The Lorentz transformation representing a boost in the x-direction is given by

$$ct' = \gamma\left(ct - \frac{v}{c}x\right), \quad x' = \gamma(x - vt), \quad y' = y, \quad z' = z. \tag{3.83}$$

An object traveling at speed c in the frame S has a position given by $\boldsymbol{x} = c\boldsymbol{n}t + \boldsymbol{x}_0$, where $\boldsymbol{n}^2 = 1$. The Lorentz transform then becomes

$$ct' = \gamma(c - vn_1)t + ct_0', \quad x' = \gamma(cn_1 - v)t + x_0',$$
$$y' = cn_2 t + y_0', \quad z' = cn_3 t + z_0'. \tag{3.84}$$

The velocity in the new inertial frame S' is given by

$$\boldsymbol{v}' = \frac{d\boldsymbol{x}'}{dt'} = \frac{d\boldsymbol{x}'/dt}{dt'/dt} = \frac{(\gamma(cn_1 - v), cn_2, cn_3)}{\gamma(1 - vn_1/c)}. \tag{3.85}$$

Squaring this relation leads to

$$v'^2 = \frac{\gamma^2(c^2n_1^2 - 2cn_1v + v^2) + c^2(n_2^2 + n_3^2)}{\gamma^2(1 - 2vn_1/c + v^2n_1^2/c^2)}$$
$$= \frac{\gamma^2[c^2n_1^2 - 2cn_1v + v^2 + (c^2 - v^2)(n_2^2 + n_3^2)]}{\gamma^2(1 - 2vn_1/c + v^2n_1^2/c^2)}$$
$$= \frac{c^2 - 2cn_1v + v^2n_1^2}{1 - 2vn_1/c + v^2n_1^2/c^2} = c^2, \tag{3.86}$$

where we have used that $n_1^2 + n_2^2 + n_3^2 = 1$. Thus, the object travels at speed c also in S'.

1.25

Let F and R denoted the front and the rear of the train, respectively. In the rest frame of the train at the time of passing of the rear in front of the station man, we take the coordinates to be $x'_F = (0, L)$ and $x'_R = (0, 0)$. For the stationman, these coordinates are instead $x_F = (x_F^0, x_F^1)$ and $x_R = (x_R^0, x_R^1)$, which are obtained from the first ones by means of an inverse Lorentz transformation along the x-axis, the direction of motion of the train, given by

$$\begin{cases} x_i^0 = x_i'^0 \cosh\theta + x_i'^1 \sinh\theta \\ x_i^1 = x_i'^0 \sinh\theta + x_i'^1 \cosh\theta \end{cases}, \quad \text{where } i = F, R \text{ and } \tanh\theta = \tfrac{v}{c}. \tag{3.87}$$

It then holds that

$$\Delta x^0 = x_F^0 - x_R^0 = L\sinh\theta = L\beta\gamma = L\frac{v}{c}\frac{1}{\sqrt{1 - v^2/c^2}}. \tag{3.88}$$

The time difference, in the rest frame of the stationman, is therefore

$$\Delta t = \frac{\Delta x^0}{c} = \frac{vL}{c^2\sqrt{1 - v^2/c^2}}. \tag{3.89}$$

1.26

Let A and B be the front and rear end of the train, respectively. Since the velocity of light is c for all observers, the times are given by $t_1 = -Lx/c$ and $t_2 = -L(1 - x)/c$. These times are related to the times t'_1 and t'_2 determined by the observer on the ground by a Lorentz transformation

$$ct'_2 = ct_2\cosh\theta + [-L(1 - x)]\sinh\theta, \tag{3.90}$$
$$ct'_1 = ct_1\cosh\theta + xL\sinh\theta, \tag{3.91}$$

where we have put the origin at $O = O'$. This gives $c(t'_2 - t'_1) = c(t_2 - t_1)\cosh\theta - L\sinh\theta$. If $t'_2 - t'_1 = 0$, we obtain $v = c\tanh\theta = (2x - 1)c$, and $v = 0$ therefore implies $x = 1/2$. This means that if $x < 1/2$ the train has to move in opposite direction to when $x > 1/2$, for the situation to occur, i.e., A and B change roles of being rear and front, respectively.

One can also calculate the invariant interval $s^2 = c^2(t_2 - t_1)^2 - L^2 = c^2(t_2' - t_1')^2 - L'^2$ in the two frames and use the length contraction formula to obtain the velocity. In this treatment, the sign of the velocity must be discussed separately.

1.27

a) Suppose the particle moves through the origin of K. Then, the event $A = (ct, 0, 0, -ut)$ belongs to the worldline of the particle. Transforming A to K' using the standard Lorentz transformation, we find that $A' = (ct\gamma(v), -vt\gamma(v), 0, -ut)$. The angle is therefore given by

$$\tan\theta = \gamma(v)\frac{v}{u} \quad \Rightarrow \quad \theta = \arctan\left(\gamma(v)\frac{v}{u}\right). \tag{3.92}$$

b) The stars will all seem to gather in front of the spaceship.

1.28

The Lorentz transformation of the 4-velocity to the new system gives for the non-trivial components

$$V'^0 = c\gamma(v'') = c\gamma(v')\cosh\theta + \gamma(v')v'\sinh\theta, \tag{3.93}$$

$$V'^1 = v''\gamma(v'') = c\gamma(v')\sinh\theta + \gamma(v')v'\cosh\theta, \tag{3.94}$$

where $v = c\tanh\theta$. Calculating $v'' = V'^1/v'^0$ yields

$$v'' = c\frac{\sinh\theta + \frac{v'}{c}\cosh\theta}{\cosh\theta + \frac{v'}{c}\sinh\theta} = \frac{v + v'}{1 + \frac{vv'}{c^2}}, \tag{3.95}$$

which is the desired formula for relativistic addition of velocities.

1.29

The standard configuration Lorentz transformation (in one temporal and two spatial dimensions) is given by $x' = \Lambda x$, where

$$x = \begin{pmatrix} x^0 \\ x^1 \\ x^2 \end{pmatrix} \quad \text{and} \quad \Lambda = \begin{pmatrix} \gamma(v) & -\frac{v}{c}\gamma(v) & 0 \\ -\frac{v}{c}\gamma(v) & \gamma(v) & 0 \\ 0 & 0 & 1 \end{pmatrix}. \tag{3.96}$$

Here $\gamma(v) = \frac{1}{\sqrt{1-(\frac{v}{c})^2}}$. This means that

$$x'^0 = \gamma(v)\left(x^0 - \frac{v}{c}x^1\right), \tag{3.97}$$

$$x'^1 = \gamma(v)\left(x^1 - \frac{v}{c}x^0\right), \tag{3.98}$$

$$x'^2 = x^2. \tag{3.99}$$

The observer in K' measures the triangle at time $x'^0 = 0$. Using Eq. (3.97) together with $x'^0 = 0$, implies that $x^0 = \frac{v}{c}x^1$. Inserting $x^0 = \frac{v}{c}x^1$ into Eq. (3.98), yields

$$x'^1 = \gamma(v)\left(x^1 - \frac{v}{c}\cdot\frac{v}{c}x^1\right) = \gamma(v)\left(1 - \frac{v^2}{c^2}\right)x^1 = \gamma(v)\frac{1}{\gamma(v)^2}x^1 = \frac{1}{\gamma(v)}x^1, \tag{3.100}$$

i.e., $x'^1 = \frac{1}{\gamma(v)}x^1$, which is the Lorentz length contraction formula.

In K:

All three sides of the triangle have length ℓ and all three angles in the triangle are 60° (i.e., $\frac{\pi}{3}$). One of the sides in the triangle (the base $b = \ell$) is parallel to the x^1-axis. Using Pythagoras' theorem, $\ell^2 = \left(\frac{b}{2}\right)^2 + h^2$, one finds the length of the altitude (the height) of the triangle to be $h = \frac{\sqrt{3}}{2}\ell$.

In K':

The length of base of the triangle is: $b' = \frac{1}{\gamma(v)}b = \frac{1}{\gamma(v)}\ell$. The length of the altitude (the height) of the triangle is: $h' = h = \frac{\sqrt{3}}{2}\ell$ [using Eq. (3.99)]. Assume that the length of the two other sides of the triangle is ℓ'. Again, using Pythagoras' theorem, $\ell'^2 = \left(\frac{b'}{2}\right)^2 + h'^2$, one finds the length of the two other sides as

$$\ell' = \sqrt{\left(\frac{b'}{2}\right)^2 + h'^2} = \sqrt{\left[\frac{\ell}{2\gamma(v)}\right]^2 + \left(\frac{\sqrt{3}}{2}\ell\right)^2} = \frac{\ell}{2}\sqrt{3 + \frac{1}{\gamma(v)^2}}. \tag{3.101}$$

The base angle α can be obtained from the relation $\ell' \cdot \cos\alpha = \frac{b'}{2}$ and the apex angle β from the relation $\ell' \cdot \sin\frac{\beta}{2} = \frac{b'}{2}$. The results are: $\alpha = \arccos\frac{1}{\sqrt{1+3\gamma(v)^2}}$ and $\beta = 2\arcsin\frac{1}{\sqrt{1+3\gamma(v)^2}}$.

1.30

For two events, E_1 and E_2 that occur at the left and right endpoints of the rod, respectively, we have

$$\Delta x'^2 = u\Delta t'. \tag{3.102}$$

By using the standard configuration Lorentz transformation, this can reformulated as

$$\Delta x^2 = u\gamma(v)\left(\Delta t - \frac{v}{c^2}\Delta x^1\right). \tag{3.103}$$

When $\Delta t = 0$, we obtain

$$\tan\phi = \frac{\Delta x^2}{\Delta x^1} = -\frac{uv}{c^2}\gamma(v) = -\frac{uv/c^2}{\sqrt{1 - v^2/c^2}}. \tag{3.104}$$

1.31

Let the axis of the cylinder coincide with the x-axis in K. The straight line on the cylinder surface is described by the equation

$$\varphi = \omega t, \tag{3.105}$$

where φ is the angle of rotation around the x-axis. We now transform the equation $\varphi = \omega t$ to K', that moves with velocity v in the x-direction. We find that

$$\varphi = \varphi' \quad \text{and} \quad t = \gamma(v)\left(t' + \frac{vx'}{c^2}\right), \tag{3.106}$$

which means that the equation of motion of the straight line relative to K' is described by the equation

$$\varphi' = \omega\gamma(v)\left(t' + \frac{vx'}{c^2}\right). \tag{3.107}$$

The twist per unit length for a fixed t' is therefore given by

$$\frac{\partial\varphi'}{\partial x'} = \omega\gamma(v)\frac{v}{c^2}, \tag{3.108}$$

so that to the observer in K', the straight line appears as a twisted line around the cylinder.

1.32

Let E_1 and E_2 be the events of turning on two compartment lights. If K and K' are the rest frames of the station and the train, respectively, then

$$\Delta t' = \gamma(v)\left(\Delta t - \frac{v}{c^2}\Delta x\right) = 0. \tag{3.109}$$

However, it holds that $\Delta x = u\Delta t$, which gives $u = c^2/v$.

1.33

Let K and K' be the rest frames of the star and spaceship, respectively. Furthermore, let the planet have its orbit in the xy-plane in the coordinate system K, i.e., $z = 0$ for the planet. The spacetime trajectory of the planet in K is then $x = (x^0, \boldsymbol{x}) = (ct, \boldsymbol{x})$, where $\boldsymbol{x} = (R\cos\omega t, R\sin\omega t, 0)$. The orbit of the planet in K' is now given by the Lorentz transformation:

$$\begin{cases} t' = \gamma(v)\left(t - \frac{v}{c^2}z\right) \\ x' = x \\ y' = y \\ z' = \gamma(v)(z - vt) \end{cases}, \tag{3.110}$$

where $\gamma(v) = \frac{1}{\sqrt{1-v^2/c^2}}$. The spaceship is moving along the positive z-axis with velocity v. Therefore, we have

$$t' = \gamma(v)t, \tag{3.111}$$

$$x' = R\cos\omega t = R\cos\omega\frac{t'}{\gamma(v)} = R\cos\omega' t', \tag{3.112}$$

$$y' = R\sin\omega t = R\sin\omega\frac{t'}{\gamma(v)} = R\sin\omega' t', \tag{3.113}$$

$$z' = -\gamma(v)vt = -vt', \tag{3.114}$$

where $\omega' = \omega/\gamma(v)$. Thus, the spacetime trajectory of the planet in K' is given by $x' = \left(x'^0, \boldsymbol{x}'\right) = (ct', \boldsymbol{x}')$, where

$$\boldsymbol{x}'(t') = \left(R\cos\frac{\omega t'}{\gamma(v)}, R\sin\frac{\omega t'}{\gamma(v)}, -vt'\right). \tag{3.115}$$

1.34

Let x^μ, x'^μ, and x''^μ be the rest coordinates of the observers A, B, and C, respectively.

The Lorentz transformation between A and B (B is moving with velocity v along the positive x^1-axis in K) is given by

$$\begin{cases} x'^0 = x^0\cosh\theta - x^1\sinh\theta \\ x'^1 = -x^0\sinh\theta + x^1\cosh\theta \\ x'^2 = x^2 \\ x'^3 = x^3 \end{cases}, \tag{3.116}$$

where $\tanh\theta = \frac{v}{c}$, and the Lorentz transformation between B and C (C is moving with velocity v' along the positive x'^2-axis in K') is given by

$$\begin{cases} x''^0 = x'^0 \cosh\theta' - x'^2 \sinh\theta' \\ x''^1 = x'^1 \\ x''^2 = -x'^0 \sinh\theta' + x'^2 \cosh\theta' \\ x''^3 = x'^3 \end{cases}, \tag{3.117}$$

where $\tanh\theta' = \frac{v'}{c}$.

Inserting the equations for x'^0 and x'^2 into the equation for x''^0, we obtain

$$x''^0 = x^0 \cosh\theta \cosh\theta' - x^1 \sinh\theta \cosh\theta' - x^2 \sinh\theta' \equiv x^0 \cosh\theta'' - \cdots,$$

where $\tanh\theta'' \equiv \frac{v''}{c}$. The velocity v'' is the (magnitude of) the relative velocity between A and C. Using the hint, this means that

$$\cosh\theta'' = \cosh\theta \cosh\theta', \tag{3.118}$$

i.e.,

$$\theta'' = \operatorname{arcosh}(\cosh\theta \cosh\theta'). \tag{3.119}$$

Thus, we have

$$v'' = c \tanh\theta'' = c \tanh \operatorname{arcosh}(\cosh\theta \cosh\theta'), \tag{3.120}$$

or with the rapidities inserted

$$v'' = c \tanh \operatorname{arcosh}\left(\cosh \operatorname{artanh}\frac{v}{c} \cosh \operatorname{artanh}\frac{v'}{c}\right). \tag{3.121}$$

We know that $\gamma(v'') = \cosh\theta''$. It follows that $\gamma(v'') = \cosh\theta \cosh\theta' = \gamma(v)\gamma(v')$, and thus, the time dilation formula between the time intervals $\Delta t \equiv t_{E_2} - t_{E_1}$ and $\Delta t'' \equiv t''_{E_2} - t''_{E_1}$ is given by

$$\Delta t'' = \frac{\Delta t}{\gamma(v)\gamma(v')} = \sqrt{1 - \frac{v^2}{c^2}}\sqrt{1 - \frac{v'^2}{c^2}}\Delta t. \tag{3.122}$$

1.35

Using

$$u = N\begin{pmatrix} x^0 + x^3 \\ x^1 + ix^2 \end{pmatrix} \quad \Leftrightarrow \quad u^* = N\begin{pmatrix} x^0 + x^3 & x^1 - ix^2 \end{pmatrix}, \tag{3.123}$$

so we have

$$\begin{aligned} uu^* &= N^2\begin{pmatrix} (x^0 + x^3)^2 & (x^0 + x^3)(x^1 - ix^2) \\ (x^1 + ix^2)(x^0 + x^3) & (x^1)^2 + (x^2)^2 \end{pmatrix} \\ &= N^2(x^0 + x^3)\begin{pmatrix} x^0 + x^3 & x^1 - ix^2 \\ x^1 + ix^2 & \frac{(x^1)^2 + (x^2)^2}{x^0 + x^3} \end{pmatrix}. \end{aligned} \tag{3.124}$$

However, $x^2 = 0$ implies that $(x^0)^2 = (x^1)^2 + (x^2)^2 + (x^3)^2$, so we can write $(x^1)^2 + (x^2)^2 = (x^0)^2 - (x^3)^2 = (x^0 - x^3)(x^0 + x^3)$. Therefore, we obtain

$$uu^* = N^2(x^0 + x^3)\begin{pmatrix} x^0 + x^3 & x^1 - ix^2 \\ x^1 + ix^2 & x^0 - x^3 \end{pmatrix}, \tag{3.125}$$

and we find that

$$\mathrm{tr}\,(uu^*) = N^2(x^0 + x^3) \cdot 2x^0 \equiv 2x^0 \quad \Leftrightarrow \quad N = \pm\frac{1}{\sqrt{x^0 + x^3}}. \tag{3.126}$$

Thus, it holds that

$$u = \begin{pmatrix} \frac{x^0+x^3}{\sqrt{x^0+x^3}} \\ \frac{x^1+ix^2}{\sqrt{x^0+x^3}} \end{pmatrix} \quad \text{is normalized,}$$

$$uu^* = \begin{pmatrix} x^0 + x^3 & x^1 - ix^2 \\ x^1 + ix^2 & x^0 - x^3 \end{pmatrix}, \tag{3.127}$$

and

$$\det(uu^*) = (x^0)^2 - (x^1)^2 - (x^2)^2 - (x^3)^2 = 0. \tag{3.128}$$

Consider the Lorentz transformation of u, i.e.,

$$a(v)u = \begin{pmatrix} e^{-\theta/2} & 0 \\ 0 & e^{\theta/2} \end{pmatrix}\begin{pmatrix} \frac{x^0+x^3}{\sqrt{x^0+x^3}} \\ \frac{x^1+ix^2}{\sqrt{x^0+x^3}} \end{pmatrix} = \begin{pmatrix} e^{-\theta/2}\frac{x^0+x^3}{\sqrt{x^0+x^3}} \\ e^{\theta/2}\frac{x^1+ix^2}{\sqrt{x^0+x^3}} \end{pmatrix} = \begin{pmatrix} \frac{e^{-\theta}(x^0+x^3)}{\sqrt{e^{-\theta}(x^0+x^3)}} \\ \frac{x^1+ix^2}{\sqrt{e^{-\theta}(x^0+x^3)}} \end{pmatrix}. \tag{3.129}$$

Now, the Lorentz transformation of x is given by

$$\begin{cases} x'^0 = x^0\cosh\theta - x^3\sinh\theta \\ x'^1 = x^1 \\ x'^2 = x^2 \\ x'^3 = -x^0\sinh\theta + x^3\cosh\theta \end{cases}. \tag{3.130}$$

Therefore, we find that

$$\begin{aligned} x'^0 + x'^3 &= x^0(\cosh\theta - \sinh\theta) + x^3(-\sinh\theta + \cosh\theta) \\ &= (\cosh\theta - \sinh\theta)(x^0 + x^3) \end{aligned} \tag{3.131}$$

and so we have

$$\cosh\theta - \sinh\theta = \frac{e^\theta + e^{-\theta}}{2} - \frac{e^\theta - e^{-\theta}}{2} = e^{-\theta}. \tag{3.132}$$

Thus, we obtain

$$a(v)u = \begin{pmatrix} \frac{x'^0 + x'^3}{\sqrt{x'^0 + x'^3}} \\ \frac{x'^1 + ix'^2}{\sqrt{x'^0 + x'^3}} \end{pmatrix} = u(x') = u(L(a(v))x). \tag{3.133}$$

1.36

A Lorentz transformation is linear and we will derive the expressions for a boost in the x-direction. The general linear transformation is given by

$$x' = \gamma x + bt, \tag{3.134}$$

$$t' = Ax + Bt, \tag{3.135}$$

for a transformation from an inertial frame S to an inertial frame S' moving with a speed v relative to S. In the case that $x = vt$, i.e., moving along with S', we must have $x' = 0$. Hence, we have

$$x' = \gamma vt + bt = (\gamma v + b)t = 0 \quad \Rightarrow \quad b = -\gamma v, \tag{3.136}$$

and thus, we obtain

$$x' = \gamma(x - vt), \tag{3.137}$$

$$x = \gamma(x' + vt'), \tag{3.138}$$

where for the second equation we have assumed that the situation must be symmetric to the first one. Furthermore, we must have $t = x/c$ and $t' = x'/c$ for a light ray. Therefore, we find that

$$x' = \gamma\left(1 - \frac{v}{c}\right)x, \tag{3.139}$$

$$x = \gamma\left(1 + \frac{v}{c}\right)x'. \tag{3.140}$$

Thus, inserting the first equation into the second one, we obtain

$$x = \gamma^2\left(1 - \frac{v^2}{c^2}\right)x \quad \Rightarrow \quad \gamma = \frac{1}{\sqrt{1 - \frac{v^2}{c^2}}}. \tag{3.141}$$

In addition, we can determine the parameters A and B to be $A = -\gamma\frac{v}{c^2}$ and $B = \gamma$ by requiring the inverse transformation to take a similar form.

1.37

Calling the events where the light signals reach the rockets 1 and 2, respectively (with 2 being the event for the rocket traveling with velocity v), event 2 is given by the intersection of the worldlines

$$x = vt \quad \text{and} \quad x = t - t_0. \tag{3.142}$$

This leads to

$$t_2 = \frac{t_0}{1-v} \quad \text{and} \quad x_2 = vt_2 = \frac{vt_0}{1-v}. \tag{3.143}$$

By symmetry, $t_1 = t_2$ and $x_1 = -x_2$. It follows that the spacetime separation between the events is given by

$$\Delta x = x_2 - x_1 = \frac{2vt_0}{1-v} \quad \text{and} \quad \Delta t = t_2 - t_1 = 0. \tag{3.144}$$

Lorentz transforming this to the rest frame S' of rocket 2, we find that

$$\Delta t' = t_2' - t_1' = \gamma(\Delta t - v\Delta x) = -\frac{2v^2\gamma t_0}{1-v}. \tag{3.145}$$

Thus, in the rest frame of one of the rockets, the other rocket receives the signal a time

$$-\Delta t' = \frac{2v^2\gamma t_0}{1-v}, \tag{3.146}$$

later.

1.38

a) Relativity of simultaneity means that it is possible for two events that are simultaneous in one inertial frame to not be simultaneous in a different frame. This occurs if there is a spatial separation between the events in the direction of the relative velocity between the frames.

The concept of relativity of simultaneity can be illustrated by the spacetime diagram shown in Figure 3.6. In this spacetime diagram, there are two events A and B, where A occurs at the origin of both frames $\mathcal{O}$ and $\mathcal{O}'$, i.e., $x_A = x_A' = 0$, at equal times $t_A = t_A' = 0$, whereas B occurs at x_B in $\mathcal{O}$ at time $t_B = t_A = 0$, but at x_B'

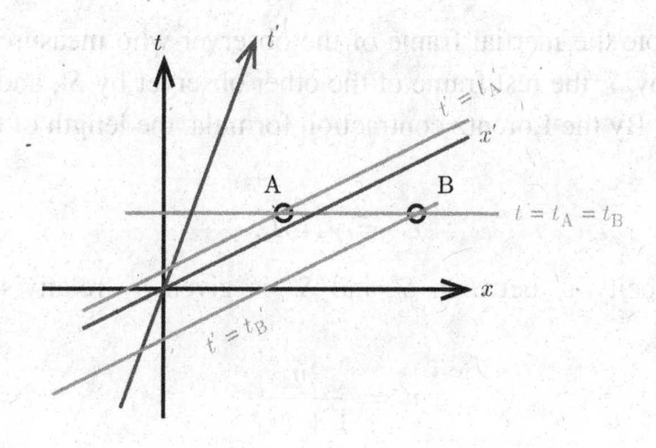

Figure 3.6 Relativity of simultaneity: $t_A = t_B$, but $t_A' > t_B'$.

in $\mathcal{O}'$ at time $t'_B \neq t'_A$. In fact, the two events are simultaneous in $\mathcal{O}$, since $t_A = t_B$, but they are **not** simultaneous in $\mathcal{O}'$, since $t'_A > t'_B$. In mathematical language, using the Lorentz transformation between $\mathcal{O}$ and $\mathcal{O}'$ (assuming that the frames are moving with constant speed v relative to each other), we obtain the relations for B as $x'_B = \gamma(v)x_B$ and $t'_B = -\gamma(v)vx_B/c^2$, where $\gamma(v) \equiv 1/\sqrt{1-v^2/c^2}$. Note that $x'_B \geq x_B$, since $\gamma(v) \geq 1$, and $t'_B \leq 0$ and directly proportional to the spatial coordinate x_B.

b) We compute

$$\boldsymbol{v} = \left(\frac{dx}{dt}, \frac{dy}{dt}, \frac{dz}{dt}\right) = at(3,4,0), \tag{3.147}$$

and thus, we find that $\boldsymbol{v}^2 = (5at)^2$. The condition $|\boldsymbol{v}| < c$ implies $t_0 < c/(5a)$. We also find

$$\gamma = \frac{1}{\sqrt{1-(v/c)^2}} = \frac{1}{\sqrt{1-(5at/c)^2}}. \tag{3.148}$$

Thus, we can compute the 4-velocity and the 4-acceleration to be

$$(V^\mu) = \left(c\gamma, \gamma\frac{dx}{dt}, \gamma\frac{dy}{dt}, \gamma\frac{dz}{dt}\right) = \gamma(c, 3at, 4at, 0), \tag{3.149}$$

$$(A^\mu) = \gamma\frac{d}{dt}(V^\mu) = \gamma\left(c\frac{d\gamma}{dt}, 3a\frac{d(\gamma t)}{dt}, 4a\frac{d(\gamma t)}{dt}, 0\right) = a\gamma^4(25at/c, 3, 4, 0), \tag{3.150}$$

respectively. The proper time along the trajectory is

$$\begin{aligned}\tau &= \int_0^{t_0} \sqrt{1-(v(t)/c)^2}\,dt = \int_0^{t_0} \sqrt{1-(5at/c)^2}\,dt \\ &= \frac{t_0}{2}\sqrt{1-(5at_0/c)^2} + \frac{c}{10a}\arcsin(5at_0/c).\end{aligned} \tag{3.151}$$

1.39

Solution 1: Denote the inertial frame of the observer who measures the length of the rod to be L by S, the rest frame of the other observer by S', and the rest frame of the rod by S''. By the Lorentz contraction formula, the length of the rod in S'' is given by

$$L'' = \gamma(v)L. \tag{3.152}$$

The relative velocity v' between S' and S'' is given by relativistic addition of velocities

$$v' = \frac{2v}{1+v^2}. \tag{3.153}$$

By the Lorentz contraction formula, the length of the rod in S' is given by

$$L' = \frac{L''}{\gamma(v')} = \frac{\gamma(v)L}{\gamma(v')} = \frac{\sqrt{1-v^2}}{1+v^2}L. \tag{3.154}$$

Solution 2: Let $x_A(t_A) = (t_A, vt_A)$ and $x_B(t_B) = (t_B, vt_B + L)$ be the worldlines of the ends of the rod. Also, let x'_A and x'_B be events on those worldlines which are simultaneous in S'. Without loss of generality, we can choose $x'_A = (0,0)$. By Lorentz transformation, we find that the worldline x_B is given by

$$x'_B(t_B) = \gamma(v)(t_B(1+v^2) + Lv, 2vt_B + L), \tag{3.155}$$

in the S' frame. Using that x'_A and x'_B are simultaneous in S', we obtain that $x'_B = x'_B(\tau)$ where $\tau = -Lv/(1+v^2)$. Thus, after simplification, the expression for the length of the rod for an observer at rest in S' is given by

$$L' = x'^{\,1}_B - x'^{\,1}_A = \frac{\sqrt{1-v^2}}{1+v^2}L. \tag{3.156}$$

1.40

The 4-velocity of the particle in S can be written as

$$V^\mu = \frac{dx^\mu}{ds} = \gamma(v)(1, \boldsymbol{v}) = \gamma(v)(1, v^1, v^2, v^3). \tag{3.157}$$

This is related to the 4-velocity in S' by the Lorentz transformation

$$V'^\mu = \Lambda^\mu{}_\nu V^\nu, \tag{3.158}$$

where

$$(\Lambda^\mu{}_\nu) = \begin{pmatrix} \gamma(u) & 0 & -u\gamma(u) & 0 \\ 0 & 1 & 0 & 0 \\ -u\gamma(u) & 0 & \gamma(u) & 0 \\ 0 & 0 & 0 & 1 \end{pmatrix}. \tag{3.159}$$

Thus, we find that

$$V'^\mu = \gamma(v)(\gamma(u)(1-uv^2), v^1, \gamma(u)(v^2-u), v^3). \tag{3.160}$$

By using the relation $\boldsymbol{v}' = \boldsymbol{V}'/V'^0$, we obtain the velocity of the particle in S' as

$$\boldsymbol{v}' = \frac{1}{1-uv^2}\left(v^1\sqrt{1-u^2}, v^2 - u, v^3\sqrt{1-u^2}\right). \tag{3.161}$$

1.41

Let K and be the rest frame of the Earth and let K' be the momentary rest frame of the spaceship. Relative to K', the space ship has the velocity $u' = 0$ at the considered time, which, according to transformations of velocities and accelerations, i.e.,

$$u' = \frac{dx'}{dt'} = \frac{u - v}{1 - \frac{uv}{c^2}}, \quad a' = \frac{du'}{dt'} = \frac{d^2x'}{dt'^2} = \frac{a}{\gamma(v)^3 \left(1 - \frac{uv}{c^2}\right)^3}, \quad \gamma(v) \equiv \frac{1}{\sqrt{1 - \frac{v^2}{c^2}}}, \tag{3.162}$$

where $u = \frac{dx}{dt}$ and $a = \frac{du}{dt} = \frac{d^2x}{dt^2}$, means that

$$u = v \quad \text{and} \quad a = a'\left(1 - \frac{v^2}{c^2}\right)^{3/2}. \tag{3.163}$$

Since $a = \frac{du}{dt}$ and $a' = g$, it follows that

$$\frac{du}{dt} = g\left(1 - \frac{u^2}{c^2}\right)^{3/2}, \tag{3.164}$$

which with the initial condition $u(0) = 0$ has the solution

$$u = \frac{dx}{dt} = \frac{gt}{\sqrt{1 + \left(\frac{gt}{c}\right)^2}}. \tag{3.165}$$

a) Integration of Eq. (3.165) with the initial condition $x(0) = 0$ gives

$$x = \frac{c^2}{g}\left[\sqrt{1 + \left(\frac{gt}{c}\right)^2} - 1\right], \tag{3.166}$$

which is a hyperbola in the Minkowski diagram; the motion is said to be *hyperbolic*.

b) According to *the clock hypothesis*, it holds that $d\tau = dt'$, where τ is the proper time of the spaceship and t' is the time relative to K'. This means that

$$\frac{d\tau}{dt} = \frac{dt'}{dt} = \frac{1}{\gamma(u)} = \{\text{Eq. } (3.165)\} = \frac{1}{\sqrt{1 + \left(\frac{gt}{c}\right)^2}}. \tag{3.167}$$

After the substitution $\frac{gt}{c} = \sinh\phi$, Eq. (3.167) is easily integrated. If $\tau(0) = 0$, then the solution is

$$t = \frac{c}{g}\sinh\frac{g\tau}{c}. \tag{3.168}$$

Inserting Eq. (3.168) into Eq. (3.166) gives

$$x = \frac{c^2}{g}\left(\cosh\frac{g\tau}{c} - 1\right). \tag{3.169}$$

If we measure distances in light years and times in years, then $c = 1$ light year/year and $g \simeq 1.05$ light years/(year)^2. For $x_A = 2\,500\,000$ light years, Eq. (3.169) gives the proper time

$$\tau_A = \frac{c}{g}\,\text{arcosh}\left(1 + \frac{g x_A}{c^2}\right) \simeq \frac{c}{g}\ln\frac{2 g x_A}{c^2}$$
$$\approx \frac{1}{1.05}\ln(2 \cdot 1.05 \cdot 2.5 \cdot 10^6)\text{ years} \approx 14.7\text{ years.} \tag{3.170}$$

Thus, the commander of the spaceship will be almost 55 years old when the spaceship reaches the Andromeda Galaxy.

1.42

With $dm < 0$ being the decrease in mass of the rocket and dm' the mass of the ejecta, the conservation of energy and momentum in the instantaneous rest frame of the rocket takes the form

$$mc^2 = (m + dm)\gamma(du)c^2 + dm'\gamma(-w)c^2$$
$$\simeq (m + dm)c^2 + dm'c^2\gamma(w), \tag{3.171}$$
$$0 = (m + dm)\gamma(du)du - w\,dm'\gamma(-w)$$
$$\simeq m\,du - w\,dm'\gamma(w), \tag{3.172}$$

where we have kept only terms to linear order in small quantities and du is the change in velocity of the rocket in the instantaneous rest frame. Solving for du in terms of dm, we find that

$$du = -\frac{w}{m}dm. \tag{3.173}$$

With the relative velocity of the instantaneous rest frame and K being v, we find that the change in velocity is given by relativistic addition of velocity

$$v + dv = \frac{v + du}{1 + \frac{v\,du}{c^2}} \simeq (v + du)\left(1 - \frac{v\,du}{c^2}\right) \simeq v + du\left(1 - \frac{v^2}{c^2}\right)$$
$$= v - w\left(1 - \frac{v^2}{c^2}\right)\frac{dm}{m}, \tag{3.174}$$

where we again keep only linear terms in the small quantities. It follows that

$$\frac{dm}{m} = -\frac{dv}{w(1 - v^2/c^2)}. \tag{3.175}$$

Integrating this equation with the condition $m = m_0$ when $v = 0$ leads to

$$m(v) = m_0\left(\frac{c - v}{c + v}\right)^{\frac{c}{2w}}. \tag{3.176}$$

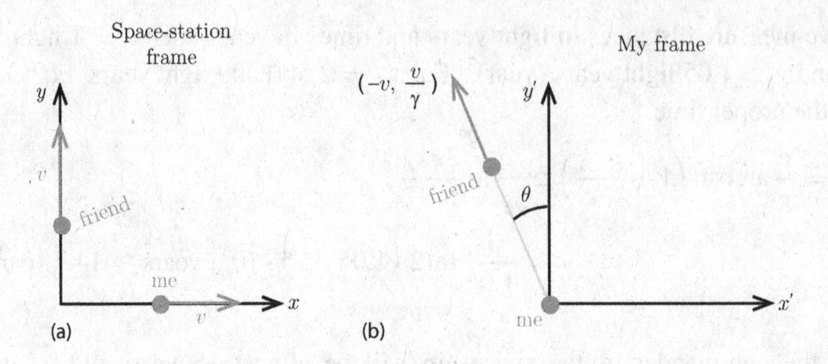

Figure 3.7 Space-station frame (a) and my frame (b).

1.43

In order to obtain the velocity that my friend has in my frame, which will give me the direction, use the formula for relativistic addition of velocities with $u_x = 0$ and $u_y = v$ (which are measured in the space-station frame), i.e.,

$$u'_x = \frac{u_x + (-v)}{1 + u_x(-v)/c^2} = -v, \quad u'_y = \frac{u_y}{\gamma(1 + u_x(-v)/c^2)} = \frac{v}{\gamma}, \tag{3.177}$$

where $-v$ is also the velocity (along the x-axis) of the space station in my frame and $\gamma = (1 - v^2/c^2)^{-1/2}$. The angle θ is given by (see Figure 3.7)

$$\tan\theta = \frac{u'_x}{u'_y} = \frac{-v}{v/\gamma} = -\gamma. \tag{3.178}$$

Thus, I should send the message to my friend in the direction

$$\theta = -\arctan\gamma = -\text{arccot}\sqrt{1 - v^2/c^2}, \tag{3.179}$$

defined in my frame. Note that in the nonrelativistic limit (i.e., $v \ll c$), the direction is $\theta_{\text{nr}} = -\pi/4 = -45°$.

1.44

a) The 4-velocity of the object is given by $U = \gamma(u)(c, \mathbf{u})$ while the 4-velocity of the frame S' is given by $V = \gamma(v)(c, -v, 0, 0)$.

b) In any inertial frame, the 4-velocity of an object traveling at velocity $\mathbf{u}$ is given by $U = \gamma(u)(c, \mathbf{u})$ as given in a). An object at rest in the inertial frame therefore has 4-velocity $V = (c, \mathbf{0})$ and as a result

$$U \cdot V = c^2\gamma(u). \tag{3.180}$$

As this inner product is a Lorentz invariant, it may be computed in any frame.

c) Taking the inner product of U and V as defined in a) we obtain

$$U \cdot V = c^2\gamma(u') = \gamma(u)\gamma(v)(c^2 - vu_1) \quad \Longrightarrow \quad \gamma(u') = \gamma(u)\gamma(v)\left(1 - \frac{vu_1}{c^2}\right). \tag{3.181}$$

1.45

From the definition of the 4-velocity, we know that $V^2 = 1$. Differentiating this relation with respect to the proper time τ leads to

$$0 = \frac{d1}{d\tau} = \frac{dV^2}{d\tau} = V \cdot \frac{dV}{d\tau} = V \cdot A. \tag{3.182}$$

It follows that the 4-velocity V and the 4-acceleration A are always perpendicular as $V \cdot A = 0$.

1.46

a) With the distance to the space station being 1 light day in my rest frame and the station moving toward me with speed $c/4$ in the same frame. The distance between the signal and the station will decrease at $5c/4$. Hence, it will take

$$\frac{1 \text{ light day}}{5c/4} = 0.8 \text{ days}, \tag{3.183}$$

for the signal to reach the station in my rest frame. The station will then be at a distance of 0.8 light days away from me as this is the distance the signal traveled. The rescue ship then travels toward me with speed $3c/4$ in the station's rest frame. Using relativistic addition of velocities, the rescue ship's speed relative to me is

$$v' = \frac{3c/4 + c/4}{1 + 3/16} = \frac{16c}{19}. \tag{3.184}$$

The time taken for the rescue ship to reach me is therefore

$$\frac{0.8 \text{ light days}}{16c/19} = 0.95 \text{ days}. \tag{3.185}$$

In total, it therefore takes the rescue ship $0.8 + 0.95 = 1.75$ days to reach me according to my clock.

1.47

Let us assume that the criminal passes the police officer at $t = 0$ and describe the events from the inertial frame S. We use a coordinate system such that the worldline of the police ship is given by

$$x_p^2 - t^2 = \frac{1}{a^2}, \tag{3.186}$$

as this corresponds to a worldline with proper acceleration of a (this may be shown by studying the dependence of x_p on t for small t). As this worldline has $x_p(t = 0) = \frac{1}{a}$, the worldline of the criminal must also fulfill $x_c(t = 0) = 0$ and correspond to the worldline of an object with constant velocity v. Thus, the criminal's worldline is

$$x_c = vt + \frac{1}{a}. \tag{3.187}$$

In order to deduce where the two worldlines intersect, we set $x_c = x_p$ and solve for the time t at which this happens, we obtain

$$t\left[2\frac{v}{a} - (1 - v^2)t\right] = 0 \quad \Rightarrow \quad t_1 = 0 \quad \text{or} \quad t_2 = 2\frac{v\gamma^2}{a}. \tag{3.188}$$

The solution t_1 corresponds to the pass where the pursuit starts and t_2 to that where it ends.

a) The criminal is moving with velocity v relative to S and is therefore simply time dilated by a factor $1/\gamma$. Thus, the pursuit takes

$$\Delta t_c = \frac{t_2 - t_1}{\gamma} = 2\frac{v\gamma}{a}, \tag{3.189}$$

according to the criminal.

b) We can parametrize the police officer's worldline using hyperbolic functions as

$$x_p = \frac{1}{a}\cosh(a\tau), \quad t = \frac{1}{a}\sinh(a\tau), \tag{3.190}$$

which fulfills $x_p^2 - t^2 = 1/a^2$. The line element ds is then given by

$$ds^2 = dt^2 - dx^2 = d\tau^2[\cosh^2(a\tau) - \sinh^2(a\tau)] = d\tau^2. \tag{3.191}$$

It follows that the parameter τ can be taken as the proper time of the police officer. We obtain

$$\Delta t_p = \tau_2 - \tau_1 = \frac{1}{a}\operatorname{arsinh}(at_2) - \frac{1}{a}\operatorname{arsinh}(at_1) = \frac{1}{a}\operatorname{arsinh}\left(2v\gamma^2\right). \tag{3.192}$$

c) By time reversal symmetry in the criminal's rest frame, the relative velocity between the two at the end of the pursuit must be v (in the opposite direction to when the pursuit started).

For peace of mind, we note that for $v \to 0$, we find that

$$\Delta t_c \simeq 2\frac{v}{a} \simeq \frac{1}{a}\operatorname{arsinh}\left(2v\gamma^2\right), \tag{3.193}$$

as expected in the nonrelativistic limit.

1.48

a) The relations $dt = dX\sinh(aT) + Xa\cosh(aT)dT$, $dx = dX\cosh(aT) + Xa\sinh(aT)dT$, $dy = dY$, $dz = dZ$ give

$$ds^2 = dt^2 - dx^2 - dy^2 - dz^2 = \cdots = (Xa)^2 dT^2 - dX^2 - dY^2 - dZ^2, \tag{3.194}$$

i.e., the nonzero components of the metric tensor are $g_{TT} = (Xa)^2$ and $g_{XX} = g_{YY} = g_{ZZ} = -1$.

b) The components of this vector in the astronauts coordinate system are

$$(k')^{\mu'} = \frac{\partial(x')^{\mu'}}{\partial x^\nu} k^\nu, \tag{3.195}$$

with $(x')^{\mu'} = (T, X, Y, Z)$ and $(x^\nu) = (t, x, y, z)$. We compute the matrix

$$\Lambda = \left(\frac{\partial(x')^{\mu'}}{\partial x^\nu}\right) = \begin{pmatrix} \frac{\partial T}{\partial t} & \cdots & \frac{\partial T}{\partial z} \\ \vdots & \ddots & \vdots \\ \frac{\partial Z}{\partial t} & \cdots & \frac{\partial Z}{\partial z} \end{pmatrix}. \tag{3.196}$$

To find this we note that $x^2 - t^2 = X^2$ and $t/x = \tanh(aT)$, i.e.,

$$X = \sqrt{x^2 - t^2}, \quad T = \frac{1}{a}\operatorname{artanh}(t/x), \quad Y = y, \quad Z = z. \tag{3.197}$$

Using this we find that

$$\Lambda = \begin{pmatrix} \frac{x}{a(x^2-t^2)} & -\frac{t}{a(x^2-t^2)} & 0 & 0 \\ -\frac{t}{\sqrt{x^2-t^2}} & \frac{x}{\sqrt{x^2-t^2}} & 0 & 0 \\ 0 & 0 & 1 & 0 \\ 0 & 0 & 0 & 1 \end{pmatrix} = \begin{pmatrix} \frac{\cosh(aT)}{aX} & -\frac{\sinh(aT)}{aX} & 0 & 0 \\ -\sinh(aT) & \cosh(aT) & 0 & 0 \\ 0 & 0 & 1 & 0 \\ 0 & 0 & 0 & 1 \end{pmatrix}, \tag{3.198}$$

and thus, we obtain

$$\begin{pmatrix} (k')^T \\ (k')^X \\ (k')^Y \\ (k')^Z \end{pmatrix} = \Lambda \begin{pmatrix} \omega \\ \omega\cos(\theta) \\ 0 \\ \omega\sin(\theta) \end{pmatrix} = \omega \begin{pmatrix} (1/aX)[\cosh(aT) - \sinh(aT)\cos(\theta)] \\ -\sinh(aT) + \cosh(aT)\cos(\theta) \\ 0 \\ \sin(\theta) \end{pmatrix} \tag{3.199}$$

c) Inserting $t(T) = X_0\sinh(aT)$, $x(T) = X_0\cosh(aT)$, $y(T) = vT$, and $z(T) = 0$, we obtain

$$\tau = \int_0^{T_0} \sqrt{t'(T)^2 - x'(T)^2 - y'(T)^2}\, dT = T_0\sqrt{(X_0 a)^2 - v^2}. \tag{3.200}$$

1.49

By Lorentz invariance of the quantity $x^2 - t^2 = 1/\alpha^2$, it follows that $x'^2 - t'^2 = 1/\alpha^2$. Differentiating this once with respect to t' gives

$$2x'\frac{dx'}{dt'} - 2t' = 0 \quad \Longrightarrow \quad v' = \frac{t'}{x'}. \tag{3.201}$$

A second differentiation with respect to t' results in

$$v'^2 + x'a' - 1 = 0 \quad \Longrightarrow \quad a' = \frac{1 - v'^2}{x'}. \tag{3.202}$$

For $t' = 0$, we find that $x' = 1/\alpha$ and $v' = 0$. It follows that

$$a' = \alpha, \tag{3.203}$$

for $t' = 0$. Thus, the acceleration in S' at time $t' = 0$ is the proper acceleration regardless of the relative velocity between the frames S and S'.

1.50

The worldline of observer B may be described as

$$x^2 - t^2 = \frac{1}{\alpha^2} \quad \Longrightarrow \quad x(\tau) = \frac{1}{\alpha}\cosh(\alpha\tau), \quad t(\tau) = \frac{1}{\alpha}\sinh(\alpha\tau), \tag{3.204}$$

where τ is the proper time since the start of acceleration and the acceleration starts at time $t = 0$. We may of course add an arbitrary constant to the x-coordinate, but this is irrelevant for our purposes as we can simply put observer A at $x = 1/\alpha$. With the light signal being sent from A at $t = t_0$, the worldline of the signal is given by

$$x = \frac{1}{\alpha} + t - t_0. \tag{3.205}$$

The signal is received by B at the event where these worldlines cross and so we find that

$$\left(\frac{1}{\alpha} + t - t_0\right)^2 - t^2 = \frac{1}{\alpha^2} \quad \Longrightarrow \quad t = \frac{t_0}{2}\frac{2 - \alpha t_0}{1 - \alpha t_0}. \tag{3.206}$$

Using the expression for t in terms of the proper time, this leads to

$$\tau = \frac{1}{\alpha}\operatorname{arsinh}\left(\frac{\alpha t_0}{2}\frac{2 - \alpha t_0}{1 - \alpha t_0}\right). \tag{3.207}$$

For the case where $\alpha t_0 \ll 1$, we can find a good approximation for the proper time τ by expanding in this parameter with the result

$$\tau \simeq t_0\left(1 + \frac{\alpha t_0}{2}\right). \tag{3.208}$$

This is no surprise to us as the physical interpretation of the condition is that observer B has not had enough time to accelerate to a significant velocity, thus the main contribution to the proper time is given by the time t_0.

In the case $\alpha t_0 \to 1$, the time t as well as the proper time τ for the signal arriving at B diverge. This is also relatively straightforward to understand as the asymptote of $x^2 - t^2 = 1/\alpha^2$ is the line $x = t$. Therefore, if $\alpha t_0 \geq 1$, then the signal will never reach B.

Note that the solutions obtained for $\alpha t_0 > 1$ are unphysical. They correspond to the intersections of the straight line with the *other* branch of $x^2 - t^2 = 1/\alpha^2$.

1.51

The particles are moving in a circle with constant angular velocity ω. Thus, we can write down an expression for the worldline (in the lab frame) of the particles by using the time t in the lab frame as the parameter

$$x^\mu = (t, R\cos(\omega t), R\sin(\omega t))^\mu, \tag{3.209}$$

where we have assumed the accelerator to have radius R and suppressed the z-coordinate which is constant. The velocity of the particles in the lab frame is

$$\boldsymbol{v} = \frac{d\boldsymbol{x}}{dt} = R\omega(-\sin(\omega t), \cos(\omega t)) \implies v = |\boldsymbol{v}| = R\omega = \text{const.} \tag{3.210}$$

We now use the differential relation between the time in the lab frame and the eigentime of the particles

$$\frac{dt}{d\tau} = \gamma(v) = \frac{1}{\sqrt{1-v^2}} = \frac{1}{\sqrt{1-R^2\omega^2}}, \tag{3.211}$$

to obtain an expression for the 4-velocity

$$V^\mu = \frac{dx^\mu}{d\tau} = \frac{dt}{d\tau}\frac{dx^\mu}{dt} = \gamma(1, -R\omega\sin(\omega t), R\omega\cos(\omega t))^\mu. \tag{3.212}$$

Repeating the procedure to obtain the 4-acceleration, we get

$$A^\mu = \frac{dV^\mu}{d\tau} = \frac{dt}{d\tau}\frac{dV^\mu}{dt} = -\gamma^2 R\omega^2(0, \cos(\omega t), \sin(\omega t))^\mu, \tag{3.213}$$

since γ is a constant. Note that this may also be written in terms of the orbital velocity v as

$$A^\mu = -\gamma^2\frac{v^2}{R}(0, \cos(vt/R), \sin(vt/R))^\mu, \tag{3.214}$$

if using v as a parameter instead of ω. The proper acceleration a is given by $-a^2 = A^2$ and thus

$$a = \sqrt{-A^2} = \gamma^2 R\omega^2. \tag{3.215}$$

The eigentime required for the particles to complete one orbit is simply given by the relation between the eigentime and the lab time

$$\tau_0 = \frac{T}{\gamma} = \frac{2\pi}{\gamma\omega}, \tag{3.216}$$

where $T = 2\pi/\omega$ is the time to complete the orbit as measured in the lab, by integrating the differential relation taking into account the fact that γ is constant.

1.52

The definition of the 4-force is

$$F = \frac{dP}{d\tau}, \tag{3.217}$$

where $P = MV$ is the 4-momentum and τ the proper time of the object. The internal energy can be found by squaring the 4-momentum $P^2 = M^2V^2 = M^2$, since the square of the 4-velocity is equal to one. It follows that

$$\frac{dP^2}{d\tau} = 2P \cdot \frac{dP}{d\tau} = 2M\frac{dM}{d\tau}, \tag{3.218}$$

where $dP/d\tau = F$ is the 4-force and $dM/d\tau$ is the sought derivative of the internal energy with respect to the proper time. Using $P = MV$, we can solve for this quantity as

$$\frac{dM}{d\tau} = V \cdot F = fV \cdot U = f\gamma, \tag{3.219}$$

where $\gamma = V \cdot U$ is the gamma factor for the relative velocity v between objects moving with the 4-velocities U and V. (Alternatively, we note that there is a frame where $U = (1, \mathbf{0})$. In this frame, $V = \gamma(1, \boldsymbol{v})$, where $\boldsymbol{v}$ is the velocity of the object in this frame.)

By using the product rule on the derivative $F = dP/d\tau$, we find that

$$F = \frac{dM}{d\tau}V + M\frac{dV}{d\tau} = f\gamma V + MA \quad \Longrightarrow \quad A = \frac{f}{M}(U - \gamma V), \tag{3.220}$$

where $A = dV/d\tau$ is the 4-acceleration of the object. The relation between the 4-acceleration A and the proper acceleration α is

$$\begin{aligned} \alpha^2 = -A^2 &= -\frac{f^2}{M^2}(U - \gamma V)^2 = -\frac{f^2}{M^2}(U^2 - 2\gamma V \cdot U + \gamma^2 V^2) \\ &= -\frac{f^2}{M^2}(1 - 2\gamma^2 + \gamma^2) = \frac{f^2}{M^2}(\gamma^2 - 1) = \frac{f^2}{M^2}\gamma^2 v^2. \end{aligned} \tag{3.221}$$

Thus, the proper acceleration is

$$\alpha = \frac{f}{M}\gamma v. \tag{3.222}$$

In particular, we note that $\alpha = 0$ if $v = 0$, which is stating that the object is not accelerating if the 4-force is parallel to the 4-velocity, as expected.

1.53

With the 4-force being the derivative of the 4-momentum with respect to the proper time, we find that

$$F^\mu = \frac{dP^\mu}{d\tau} = \frac{d(mV^\mu)}{d\tau} = \frac{dm}{d\tau}V^\mu + mA^\mu, \tag{3.223}$$

where V^μ is the 4-velocity and A^μ the 4-acceleration. We can find the rate of change in the rest energy by taking the inner product with the 4-velocity, resulting in

$$F \cdot V = \frac{dm}{d\tau}V^2 + mA \cdot V = \frac{dm}{d\tau}, \tag{3.224}$$

since $V^2 = 1$ and $A \cdot V = 0$. Using the given expression for F^μ and that $V^\mu = \gamma(1, \boldsymbol{v})^\mu$, we now obtain

$$\frac{dm}{d\tau} = (0, \boldsymbol{f}) \cdot \gamma(1, \boldsymbol{v}) = -\gamma \boldsymbol{f} \cdot \boldsymbol{v}. \tag{3.225}$$

Squaring the 4-force, we find the relation

$$F^2 = \left(\frac{dm}{d\tau}V + mA\right)^2 = \left(\frac{dm}{d\tau}\right)^2 + m^2A^2 = \gamma^2(\boldsymbol{f} \cdot \boldsymbol{v})^2 - m^2\alpha^2. \tag{3.226}$$

Obviously, we also have $F^2 = -\boldsymbol{f}^2$ and so we can solve for the proper acceleration α as

$$\alpha^2 = \frac{1}{m^2}\left[\boldsymbol{f}^2 + \gamma^2(\boldsymbol{f} \cdot \boldsymbol{v})^2\right]. \tag{3.227}$$

The force is a pure force only if the rate $dm/d\tau = 0$, corresponding to $\boldsymbol{f} \cdot \boldsymbol{v} = 0$. Therefore, the force is a pure force if $\boldsymbol{f}$ and $\boldsymbol{v}$ are orthogonal.

1.54

The 4-acceleration is generally given by

$$A = \frac{dV}{d\tau} = \gamma\left[\frac{d\gamma}{dt}(1, \boldsymbol{v}) + \gamma(0, \boldsymbol{a})\right] = \gamma^4 \boldsymbol{v} \cdot \boldsymbol{a}(1, \boldsymbol{v}) + \gamma^2(0, \boldsymbol{a}). \tag{3.228}$$

In the instantaneous rest frame, the 4-acceleration is therefore given by

$$A = (0, \boldsymbol{a}_0) = (0, a_{0x}, a_{0y}), \tag{3.229}$$

where we have suppressed the z-direction, which will behave just as the y-direction. By Lorentz transforming this to S', we find that

$$A' = (-v\gamma a_{0x}, \gamma a_{0x}, a_{0y}). \tag{3.230}$$

We now use that the 4-velocity in S' is given by $V = \gamma(1, -v, 0)$ to identify this with the general expression for the 4-acceleration

$$A' = -v\gamma^4 a'_x(1, -v, 0) + \gamma^2(0, a'_x, a'_y), \tag{3.231}$$

which gives us

$$a'_x = \gamma^{-3} a_{0x}, \quad a'_y = \gamma^{-2} a_{0y}. \tag{3.232}$$

For a fixed proper acceleration α, the acceleration in the instantaneous rest frame is given by $a_{0x} = \alpha \cos(\theta)$ and $a_{0y} = \alpha \sin(\theta)$, where θ is the angle between the acceleration and the x-direction. This results in an acceleration

$$a'^2 = \boldsymbol{a}'^2 = a'^2_x + a'^2_y = \gamma^{-4}\alpha^2[\gamma^{-2}\cos^2(\theta) + \sin^2(\theta)]. \tag{3.233}$$

Since $\gamma \geq 1$, the expression in the parenthesis varies between one [when $\sin^2(\theta) = 1$] and γ^{-2} [when $\cos^2(\theta) = 1$]. Consequently, we find that

$$a'_{\text{max}} = \alpha\gamma^{-2}, \quad a'_{\text{min}} = \alpha\gamma^{-3}. \tag{3.234}$$

1.55

In the S' frame, we can extract the energy of the particle by taking the inner product of the 4-momentum with the 4-vector

$$V' = (1, 0, 0, 0), \tag{3.235}$$

which is the 4-velocity of an object at rest in S'. Similarly, the x'-component of the 3-momentum can be obtained by taking the inner product with the 4-vector

$$T' = (0, -1, 0, 0). \tag{3.236}$$

To see this, we note that the 4-momentum in S' is given by $P' = (E', \boldsymbol{p}')$, resulting in

$$V' \cdot P' = (1, 0, 0, 0) \cdot (E', \boldsymbol{p}') = E', \tag{3.237}$$

$$T' \cdot P' = (0, -1, 0, 0) \cdot (E', \boldsymbol{p}') = -(-1)p'_x = p'_x. \tag{3.238}$$

Lorentz transforming V' and T' to the S frame, we find that

$$V = \gamma(1, v, 0, 0), \quad T = -\gamma(v, 1, 0, 0). \tag{3.239}$$

Since the inner products are Lorentz invariant, we can directly compute that

$$E' = V \cdot P = \gamma(1, v, 0, 0) \cdot (E, \boldsymbol{p}) = \gamma(E - vp_x). \tag{3.240}$$

In the same fashion, we also find that

$$p'_x = T \cdot P = -\gamma(v, 1, 0, 0) \cdot (E, \boldsymbol{p}) = \gamma(p_x - vE). \tag{3.241}$$

Furthermore, the velocity of a particle in an arbitrary frame is given by $\boldsymbol{v} = \boldsymbol{p}/E$, where $\boldsymbol{p}$ is the 3-momentum and E the energy. In particular, in S', this results in the velocity component in the x'-direction being

$$v'_x = \frac{p'_x}{E'} = \frac{p_x - vE}{E - vp_x}. \tag{3.242}$$

1.56

By definition of the 4-force, we know that

$$F^\mu = \frac{dP^\mu}{d\tau} = f(1, \mathbf{1})^\mu. \tag{3.243}$$

Integrating this relation with the initial condition $P^\mu(0) = (m_0, \mathbf{0})^\mu$ leads to

$$P^\mu(\tau) = (f\tau + m_0, f\tau)^\mu. \tag{3.244}$$

The mass of the object at proper time τ is therefore given by

$$m(\tau) = \sqrt{P(\tau)^2} = \sqrt{(f\tau + m_0)^2 - f^2\tau^2} = \sqrt{m_0(2f\tau + m_0)}. \tag{3.245}$$

The relation between the coordinate time t and the proper time τ is given by

$$\frac{dt}{d\tau} = \gamma = \frac{E}{m} = \frac{f\tau + m_0}{\sqrt{m_0(2f\tau + m_0)}}. \tag{3.246}$$

Integrating this differential equation with the initial condition $t(0) = 0$ leads to

$$t(\tau) = \left(\frac{\tau}{3m_0} + \frac{2}{3f}\right)\sqrt{m_0(2f\tau + m_0)} - \frac{2m_0}{3f}. \tag{3.247}$$

Note that the integral can be performed by making the ansatz that a primitive function of the integrand is given by

$$g(\tau) = (A\tau + B)\sqrt{m_0(2f\tau + m_0)}, \tag{3.248}$$

and fixing A and B by comparing $g'(\tau)$ with the integrand.

1.57

The 4-velocity of p is given by

$$U^\mu = \gamma(u)(1, -u\cos(\theta), -u\sin(\theta))^\mu, \tag{3.249}$$

in S. Lorentz transforming this to the rest frame S' of the observer o, we find that

$$U^{\mu'} = \gamma(v)\gamma(u)(1 + vu\cos(\theta), -u\cos(\theta) - v, -u\sin(\theta)\gamma(v)^{-1})^{\mu'}. \tag{3.250}$$

The speed u' of p in S' can be found by considering the time component of the 4-velocity U in S' as

$$\gamma(u') = \frac{1}{\sqrt{1-u'^2}} \implies u' = \sqrt{1 - \frac{1}{\gamma(u')^2}}. \tag{3.251}$$

From the expression for U in S', we therefore have

$$\gamma(u')^2 = \frac{[1+vu\cos(\theta)]^2}{(1-v^2)(1-u^2)} \implies u' = \frac{\sqrt{[1+vu\cos(\theta)]^2 - (1-v^2)(1-u^2)}}{1+vu\cos(\theta)}. \tag{3.252}$$

The tangent of the angle θ', can be expressed as the ratio between $U^{2'}$ and $U^{1'}$, i.e.,

$$\tan(\theta') = \frac{U^{2'}}{U^{1'}} = \frac{u\sin(\theta)}{\gamma(v)[u\cos(\theta)+v]}. \tag{3.253}$$

In the nonrelativistic limit $u, v \ll 1$, the above expressions become

$$u' = \sqrt{v^2+u^2+2vu\cos(\theta)} \quad \text{and} \quad \tan(\theta') = \frac{u\sin(\theta)}{u\cos(\theta)+v}, \tag{3.254}$$

to leading order. The relation for the speed u' is just the cosine theorem for classical addition of velocities and the expression for $\tan(\theta')$ is the classical aberration formula.

In the limit $u \to 1$, we find that

$$u' \to 1 \quad \text{and} \quad \tan(\theta') \to \frac{\sin(\theta)}{\cos(\theta)+v}. \tag{3.255}$$

The first relation represents the invariance of the speed of light and the second is the relativistic aberration formula for light.

1.58

In S, the object's worldline is given by

$$x(t) = \frac{at^2}{2}, \tag{3.256}$$

and its velocity by $v(t) = \frac{dx}{dt} = at$. The speed v_0 is obtained at time $t_0 = \frac{v_0}{a}$. The proper time to reach this speed is therefore

$$\tau(v_0) = \int_0^{t_0} \sqrt{1-v(t)^2}\, dt = \int_0^{t_0} \sqrt{1-a^2t^2}\, dt. \tag{3.257}$$

Substituting $at = \sin\theta$, we find that $dt = \frac{1}{a}\cos\theta\, d\theta$, and therefore,

$$\begin{aligned}\tau(v_0) &= \frac{1}{a}\int_0^{\theta_0} \cos^2\theta\, d\theta = \frac{1}{2a}\int_0^{\theta_0} (1+\cos 2\theta)\, d\theta = \frac{\theta_0}{2a} + \frac{1}{4a}\sin 2\theta_0 \\ &= \frac{1}{2a}(\theta_0 + \sin\theta_0\cos\theta_0),\end{aligned} \tag{3.258}$$

where $\theta_0 = \arcsin(at_0) = \arcsin v_0$. Inserting the expression for θ_0 leads to

$$\tau(v_0) = \frac{1}{2a}\left(\arcsin v_0 + v_0\sqrt{1 - v_0^2}\right). \tag{3.259}$$

The proper acceleration of the object is given by the square of its 4-acceleration. We find that

$$\begin{aligned} A = \frac{dV}{d\tau} = \frac{d^2X}{d\tau^2} &= \frac{dt}{d\tau}\frac{d}{dt}\left(\frac{dt}{d\tau}\frac{dX}{dt}\right) \\ &= \left(\frac{dt}{d\tau}\right)^2\frac{d^2X}{dt^2} + \frac{dX}{dt}\frac{dt}{d\tau}\frac{d}{dt}\left(\frac{dt}{d\tau}\right) = \gamma^2(0,a) + V\frac{d\gamma}{dt}. \end{aligned} \tag{3.260}$$

This leads to $A - V\frac{d\gamma}{dt} = \gamma^2(0,a)$ and squaring this expression, we find

$$A^2 + V^2\left(\frac{d\gamma}{dt}\right)^2 = -\alpha^2 + \left(\frac{d\gamma}{dt}\right)^2 = -\gamma^4 a^2, \tag{3.261}$$

where α is the proper acceleration and we have used that $A \cdot V = 0$. It follows that

$$\alpha^2 = \gamma^4 a^2 + \left(\frac{d\gamma}{dt}\right)^2. \tag{3.262}$$

Inserting the expression for γ into the derivative leads to

$$\frac{d\gamma}{dt} = \frac{d}{dt}\frac{1}{\sqrt{1 - a^2t^2}} = -\frac{1}{\sqrt{1 - a^2t^2}^3}a^2 t = -\gamma^3 va. \tag{3.263}$$

We therefore find

$$\begin{aligned} \alpha^2 &= \gamma^4 a^2 + \gamma^6 v^2 a^2 = \gamma^4 a^2(1 + \gamma^2 v^2) \\ &= \gamma^4 a^2 \frac{1 - v^2 + v^2}{1 - v^2} = \gamma^4 a^2 \frac{1}{1 - v^2} = \gamma^6 a^2. \end{aligned} \tag{3.264}$$

Taking the square root of this, we obtain

$$\alpha = \gamma^3 a = \frac{a}{\sqrt{1 - a^2t^2}^3}. \tag{3.265}$$

1.59

The speed of a light signal relative to the water is given by $u_0 = c/n$, where n is the refractive index of water. Meanwhile, the speed of the water relative to the lab frame is v and application of the formula for relativistic addition of velocities then results in the speed of the light signal relative to the lab frame

$$u = \frac{u_0 + v}{1 + \frac{u_0 v}{c^2}} \simeq u_0 + kv, \tag{3.266}$$

where $k = 1 - \frac{u_0^2}{c^2} = 1 - \frac{1}{n^2}$ and only linear terms in $v \ll c$ have been kept, which is Fizeau's result.

1.60

See the solution to Problem 1.59, i.e., Fizeau's result. In the case of the water moving perpendicular to the light, we assume that the water is moving in the x-direction with speed v and the light is moving in the y-direction with speed c/n (when the water is standing still). In this case, the formula for addition of velocities reads

$$u_2 = \frac{u'}{\gamma(1 + u_1 v/c^2)} = \{u_1 = 0\} = \frac{u'}{\gamma} \simeq u'\left(1 + \frac{v^2}{2c^2}\right) \approx u' = \frac{c}{n}. \tag{3.267}$$

In other words, the correction enters only at second order in the water speed.

1.61

The redshift is maximal for when the source moves directly away from the observer. A lower bound for the speed of 3C 9 is therefore determined as

$$u_{\min} = \frac{x^2 - 1}{x^2 + 1} c \simeq 0.80c, \quad \text{where } x = \frac{3\,600}{1\,215}, \tag{3.268}$$

is the ratio of the observed and emitted wavelengths, by solving for x from the relativistic Doppler shift formula.

1.62

The electromagnetic wave is $E(x) = E_0 \sin 2\pi \left(\frac{x^1}{\lambda} - \nu t\right) = \sin\left(\frac{2\pi x^1}{\lambda} - 2\pi \nu t\right)$. The argument can be rewritten as follows

$$\begin{aligned}
\frac{2\pi x^1}{\lambda} - 2\pi\nu t &= -2\pi\nu t + \frac{2\pi x^1}{\lambda} = -\left(2\pi\nu t - \frac{2\pi x^1}{\lambda}\right) = -\left(2\pi\nu t - \frac{2\pi\nu}{\lambda\nu} x^1\right) \\
&= \{\omega = 2\pi\nu \text{ and } c = \lambda\nu\} = -\left(\omega t - \frac{\omega}{c} x^1\right) = -\left(\frac{\omega}{c} ct - \frac{\omega}{c} x^1\right) \\
&= \{x^0 = ct\} = -\left(\frac{\omega}{c} x^0 - \frac{\omega}{c} x^1\right) \\
&= -\left(\frac{\omega}{c}, \frac{\omega}{c}, 0, 0\right) \cdot (x^0, x^1, x^2, x^3) = -k_\mu x^\mu,
\end{aligned} \tag{3.269}$$

where $k = \left(\frac{\omega}{c}, \frac{\omega}{c}, 0, 0\right)$ and $x = (x^0, x^1, x^2, x^3)$. Thus, we can write $E(x) = \sin(-k_\mu x^\mu) = -\sin k_\mu x^\mu$.

The wave vector k is lightlike, since $k^2 = k_\mu k^\mu = \left(\frac{\omega}{c}, \frac{\omega}{c}, 0, 0\right) \cdot \left(\frac{\omega}{c}, \frac{\omega}{c}, 0, 0\right) = \left(\frac{\omega}{c}\right)^2 - \left(\frac{\omega}{c}\right)^2 - 0^2 - 0^2 = 0$.

In K', we have $E'(x') = -E_0 \sin k'_\mu x'^\mu = -E_0 \sin k_\mu x^\mu = E(x)$. Since this is Lorentz invariant, we find that $k' = \Lambda k$, where Λ is a Lorentz transformation, and

$$\frac{\omega'}{c} = k'^0 = k^0 \cosh\theta - k^1 \sinh\theta = \frac{\omega}{c}\cosh\theta - \frac{\omega}{c}\sinh\theta = \frac{\omega}{c}(\cosh\theta - \sinh\theta). \tag{3.270}$$

Using the definitions of the hyperbolic functions and the fact that $\omega = 2\pi\nu$ and $\omega' = 2\pi\nu'$, we obtain the answer

$$\nu' = \nu e^{-\theta}, \tag{3.271}$$

which is the formula for the Doppler shift. If we instead use the relations $\cosh\theta = \gamma(v)$ and $\sinh\theta = \frac{v}{c}\gamma(v)$, where $\gamma(v) \equiv \frac{1}{\sqrt{1-(\frac{v}{c})^2}}$, we instead obtain

$$\nu' = \nu\left[\gamma(v) - \frac{v}{c}\gamma(v)\right] = \nu\gamma(v)\left(1 - \frac{v}{c}\right) = \nu\sqrt{\frac{c-v}{c+v}}, \tag{3.272}$$

which is the usual formula for the Doppler shift.

1.63

From the statement that the GRB has twice the duration as a nearby GRB, we conclude that the GRB is time dilated due to its motion and that $\gamma = 2$. From this relation, we can solve for the velocity of the GRB and obtain

$$v = \frac{\sqrt{3}}{2}. \tag{3.273}$$

The quotient between the observed and emitted wavelengths is then given by the formula for the Doppler shift, i.e.,

$$\frac{\lambda}{\lambda_0} = \sqrt{\frac{1+v}{1-v}} = 2 + \sqrt{3}. \tag{3.274}$$

Computing the redshift z, we obtain

$$z \equiv \frac{\lambda - \lambda_0}{\lambda_0} = 1 + \sqrt{3}. \tag{3.275}$$

1.64

The trajectories of objects 1 and 2 in the person's frame of reference $K(t,x,y,z)$ are

$$x_1(t) = \frac{ct}{2}, \quad x_2(t) = L - \frac{ct}{2}, \tag{3.276}$$

where we assume, without loss of generality, that the objects move parallel to the x-axis and that the first object is at the origin at time $t = 0$; $y_j(t) = z_j(t) = 0$, for $j = 1, 2$, and $L > 0$ is the distance between the objects at time $t = 0$, of course. The frame of reference K' of the observer on the first object is related to K by a Lorentz transformation

$$t' = \gamma(t - vx/c^2), \quad x' = \gamma(x - vt), \quad \gamma = 1\sqrt{1 - v^2/c^2}, \tag{3.277}$$

and $y' = y$, $z' = z$ with $v = c/2$ so that $x_1'(t) = 0$.

a) We parametrize the trajectory of the second object in K as

$$t(s) = s, \quad x(s) = L - \frac{cs}{2}, \tag{3.278}$$

and Lorentz transform this to K' such that

$$t'(s) = \gamma\left[t(s) - (c/2)x(s)/c^2\right], \quad x'(s) = \gamma\left[x(s) - (c/2)t(s)\right],$$
$$\gamma = \frac{1}{\sqrt{1-(1/2)^2}} = \frac{2}{\sqrt{3}}, \tag{3.279}$$

i.e.,

$$t'(s) = \frac{2}{\sqrt{3}}\left[\frac{5s}{4} - \frac{L}{2c}\right], \quad x'(s) = \frac{2}{\sqrt{3}}(L - cs). \tag{3.280}$$

We now can compute the velocity of the second object in K' such that

$$\frac{dx'}{dt'} = \frac{dx'/ds}{dt'/ds} = \frac{-c}{5/4} = -\frac{4c}{5}. \tag{3.281}$$

Note that a faster way to obtain this answer is to use relativistic addition of velocities

$$v = \frac{v_1 + v_2}{1 + v_1 v_2/c^2}, \tag{3.282}$$

with $v_1 = v_2 = -c/2$.

b) The 4-wavevector of the photon in the first object's rest frame is

$$(k'^{\mu}) = \left(k'^0, k'^1, 0, 0\right), \tag{3.283}$$

where $k'^0 = k'^1 = 2\pi/\lambda_0$. The corresponding 4-wavevector in the second object's rest frame is

$$(k''^{\mu}) = \left(k''^0, k''^1, 0, 0\right), \tag{3.284}$$

where

$$k''^0 = \tilde{\gamma}\left(k'^0 + \tilde{v}k'^1/c\right), \quad k''^1 = \tilde{\gamma}\left(k'^1 + \tilde{v}k'^0/c\right), \quad \tilde{\gamma} = 1/\sqrt{1-(\tilde{v}/c)^2}, \tag{3.285}$$

with $\tilde{v} = 4c/5$, i.e.,

$$(k''^{\mu}) = (k, k, 0, 0), \quad k = \frac{2\pi}{\lambda_0}\tilde{\gamma}(1 + \tilde{v}/c) = \frac{2\pi}{\lambda_0}\sqrt{\frac{1+\tilde{v}/c}{1-\tilde{v}/c}} = \frac{2\pi}{\lambda_0}\cdot 3 = \frac{2\pi}{\lambda}, \tag{3.286}$$

Thus, the observer on the second object assigns the wavelength

$$\lambda = \lambda_0 \sqrt{\frac{1 - \tilde{v}/c}{1 + \tilde{v}/c}} = \frac{\lambda_0}{3} \simeq 231.4 \text{ nm}, \tag{3.287}$$

to the photon. The wavelength is shorter: it has been shifted from red to UV, i.e., the frequency of the photon is higher.

1.65

a) In the rest frame of the observer, we can choose coordinates such that the light source is on the positive x-axis. The 4-velocity of the light source can then be written as $V = \gamma(v)(1, v\cos(\theta), v\sin(\theta), 0)$ and the 4-frequency of the light pulse is given by $N = \omega(1, -1, 0, 0)$. The frequency in the rest frame of the source is then obtained as

$$\omega_0 = V \cdot N = \gamma(v)\omega[1 + v\cos(\theta)]. \tag{3.288}$$

Consequently, we find that

$$\omega = \omega_0 \frac{\sqrt{1 - v^2}}{1 + v\cos(\theta)}. \tag{3.289}$$

In particular, when the light source is moving directly away from the observer, we have $\cos(\theta) = 1$ and therefore recover the relativistic Doppler formula

$$\omega = \omega_0 \frac{\sqrt{(1 - v)(1 + v)}}{1 + v} = \omega_0 \sqrt{\frac{1 - v}{1 + v}}. \tag{3.290}$$

b) Requiring that $\omega = \omega_0$ results in

$$1 = \frac{\sqrt{1 - v^2}}{1 + v\cos(\theta)}. \tag{3.291}$$

Solving for $\cos(\theta)$ yields the result

$$\cos(\theta) = \frac{\sqrt{1 - v^2} - 1}{v}. \tag{3.292}$$

Note that for $v \ll 1$, this may be approximated by

$$\cos(\theta) \simeq \frac{1 - v^2/2 - 1}{v} = -\frac{v}{2}, \tag{3.293}$$

while for $v = 1 - \varepsilon$, where $\varepsilon \ll 1$, leads to

$$\cos(\theta) \simeq \frac{\sqrt{1 - (1 - 2\varepsilon)} - 1}{1 - \varepsilon} \simeq -1 + \sqrt{2\varepsilon}. \tag{3.294}$$

Thus, for small velocities, the angle θ is close to 90°, but the source must move slightly toward the observer, whereas for velocities close to the speed of light, the source must move *almost* straight toward the observer.

1.66

We have a rotating disk with two observers O_1 and O_2 at different radii r_1 and r_2, respectively. Setting $c = 1$ and omitting the trivial z-direction, introducing polar coordinates on the inertial frame leads to

$$ds^2 = dt^2 - dr^2 - r^2 d\phi^2. \tag{3.295}$$

We now introduce a rotating frame with $\phi_r = \phi + \Omega t$, which leads to $\phi = \phi_r - \Omega t$ and find that

$$\begin{aligned} ds^2 &= dt^2 - dr^2 - r^2 d\phi^2 = dt^2 - dr^2 - r^2(d\phi_r - \Omega dt)^2 \\ &= (1 - r^2\Omega^2)dt^2 + r^2\Omega dt d\phi_r - dr^2 - r^2 d\phi_r^2. \end{aligned} \tag{3.296}$$

Thus, we observe that we have a nontrivial time–time component of the metric, i.e.,

$$g_{00} = 1 - r^2\Omega^2. \tag{3.297}$$

For a light wave sent from O_2 to O_1, the observed proper times between consecutive wave fronts are therefore given by

$$\Delta\tau_1 = \sqrt{g_{00}(r_1)}\Delta t, \quad \Delta\tau_2 = \sqrt{g_{00}(r_2)}\Delta t, \tag{3.298}$$

where Δt is the time difference between the emission of the wave fronts in the inertial frame. Therefore, using that the observed angular frequency ω is inversely proportional to the time period, we obtain

$$\frac{\omega_2}{\omega_1} = \frac{\Delta\tau_1}{\Delta\tau_2} = \frac{\sqrt{g_{00}(r_1)}}{\sqrt{g_{00}(r_2)}} = \frac{\sqrt{1 - r_1^2\Omega^2}}{\sqrt{1 - r_2^2\Omega^2}} = \frac{\gamma_2}{\gamma_1}. \tag{3.299}$$

1.67

In the rest frame of the medium S, the waves in direction θ have a 4-frequency given by

$$(N^\mu) = \nu(1, \cos(\theta)/n, \sin(\theta)/n), \tag{3.300}$$

where the coordinate system has been arranged such that the third spatial component is zero (which is why it has been omitted). Here, ν is the frequency of the wave in the rest frame of the medium. In the rest frame of the source S', the frequency is given by the zeroth component of the 4-frequency in that frame

$$\nu_0 = (1,0,0)_\mu N'^\mu = V'_\mu N'^\mu, \tag{3.301}$$

where V is the 4-velocity of the source itself. This expression is a Lorentz scalar and may be computed in any frame. In particular, in S, the 4-velocity of the source is

$$(V^{\mu}) = \gamma(1, v, 0), \quad \nu_0 = V \cdot N = \gamma\nu\left(1 - v\frac{\cos\theta}{n}\right). \tag{3.302}$$

It follows that

$$\frac{\nu}{\nu_0} = \sqrt{1 - v^2}\frac{n}{n - v\cos\theta}. \tag{3.303}$$

In particular, when $n \to 1$, we recover

$$\frac{\nu}{\nu_0} = \frac{\sqrt{1 - v^2}}{1 - v\cos\theta} \underset{\theta \to 0}{\longrightarrow} \sqrt{\frac{1 + v}{1 - v}}, \tag{3.304}$$

which is the usual Doppler formula in vacuum.

1.68

In the frame S, where the mirror is moving, the 4-frequency of the incoming light p and that of the outgoing light k are given by

$$p = \omega(1, -c_i, -s_i), \quad k = \omega'(1, c_o, -s_o), \tag{3.305}$$

where we have introduced $c_i = \cos\theta_{\text{in}}$, $s_i = \sin\theta_{\text{in}}$, $c_o = \cos\theta_{\text{out}}$, $s_o = \sin\theta_{\text{out}}$, and used a coordinate system such that the mirror is moving in the x-direction and the light is not propagating in the z-direction, which we therefore have omitted. In the rest frame of the mirror S', the incident angle is equal to the reflected angle. Furthermore, the frequencies of the incoming and reflected light are the same, which means that

$$p' = \omega''(1, -c', -s'), \quad k' = \omega''(1, c', -s'), \tag{3.306}$$

where $c' = \cos\theta'$, $s' = \sin\theta'$. Since p and k are related to p' and k' by Lorentz transformation, we also have

$$p' = \omega\gamma(1 + vc_i, -c_i - v, s_i/\gamma), \quad k = \omega''\gamma(1 + vc', c' + v, s'/\gamma), \tag{3.307}$$

where $\gamma = 1/\sqrt{1 - v^2}$. By identification, it follows (using p') that

$$\omega'' = \omega\gamma(1 + vc_i), \quad c' = \frac{c_i + v}{1 + vc_i}, \tag{3.308}$$

and (using k) that

$$\omega' = \omega''\gamma(1 + vc') = \omega\frac{1 + 2vc_i + v^2}{1 - v^2}, \tag{3.309}$$

$$c_o = \frac{\omega''\gamma}{\omega'}(c' + v) = \frac{c' + v}{1 + vc'} = \frac{c_i + 2v + v^2 c_i}{1 + 2vc_i + v^2}. \tag{3.310}$$

For $v = -c_i = -\cos\theta_{\text{in}}$, the mirror is moving away from the light at the same speed that the light is approaching the mirror. As a result, the light never reaches the mirror. In S', the light is moving parallel to the mirror. The reflection angle in S approaches $c_o = -c_i$, which simply means that the light continues in a straight line. For values $v \le c_i$, the mirror outruns the light and there is no reflection.

1.69

In the rest frame S' of the medium, the 4-frequencies of the light waves are given by

$$N^{\mu'} = f_0(1, -c', -s')^{\mu'}, \quad \mathcal{N}^{\mu'} = f_0\left(1, -\frac{c''}{n}, -\frac{s''}{n}\right)^{\mu'}, \tag{3.311}$$

where $c' = \cos\theta'$, $s' = \sin\theta'$, $s'' = \sin\theta''$, $c'' = \cos\theta''$ and the relation between θ' and θ'' is given by Snell's law $\sin\theta' = n\sin\theta''$. Here N is the 4-frequency of the wave before entering the medium and $\mathcal{N}$ is the 4-frequency after entering the medium. Expressing θ'' in terms of θ', we obtain

$$\mathcal{N}^{\mu'} = f_0\left(1, -\frac{1}{n}\sqrt{1 - \frac{s'^2}{n^2}}, -\frac{s'}{n^2}\right)^{\mu'}. \tag{3.312}$$

Lorentz transforming this to the frame S, we obtain

$$N^{\mu} = f_0\gamma\left(1 - vc', -c' + v, -\frac{s'}{\gamma}\right)^{\mu}, \tag{3.313}$$

$$\mathcal{N}^{\mu} = f_0\gamma\left(1 - \frac{v}{n}\sqrt{1 - \frac{s'^2}{n^2}}, +v - \frac{1}{n}\sqrt{1 - \frac{s'^2}{n^2}}, -\frac{s'}{n^2\gamma}\right)^{\mu}. \tag{3.314}$$

In order for the observer in S to observe the light traveling in the same direction after entering the medium, the relation

$$\frac{\mathcal{N}^1}{N^1} = \frac{\mathcal{N}^2}{N^2}, \tag{3.315}$$

must be fulfilled (the spatial part of the 4-frequencies must be in the same direction). We obtain

$$\frac{\mathcal{N}^1}{N^1} = \frac{\frac{1}{n}\sqrt{1 - \frac{s'^2}{n^2}} - v}{c' - v}, \tag{3.316}$$

$$\frac{\mathcal{N}^2}{N^2} = \frac{\frac{s'}{n^2\gamma}}{\frac{s'}{\gamma}} = \frac{1}{n^2}. \tag{3.317}$$

This leads to the relation

$$c' - v = \sqrt{n^2 - s'^2} - n^2 v \quad \Rightarrow \quad c' + v(n^2 - 1) = \sqrt{n^2 - 1 + c'^2}. \tag{3.318}$$

Squaring this, we obtain

$$(n^2 - 1)[2c'v - (1 - v^2)(n^2 - 1)] = 0, \tag{3.319}$$

which has the solutions $n^2 = 1$ and $n^2 = 1 + 2c'v\gamma^2$. Since $n > 1$ was given, the latter of these is the sought solution and we obtain

$$n = \sqrt{1 + 2c'v\gamma^2}. \tag{3.320}$$

Note that the solution $n = 1$ corresponds to the scenario where the medium has the same refractive index as vacuum, i.e., is vacuum. In this case, it does not matter how fast the medium moves and the light wave will continue undisturbed.

1.70

We can find the angular frequency ω_0 of the wave in the source rest frame by multiplying the 4-frequency N by the 4-velocity of the source. This results in

$$\omega_0 = N \cdot V = \gamma(\omega, k) \cdot (1, v) = \gamma\omega\left(1 - \frac{v}{u}\right). \tag{3.321}$$

Solving for the Doppler shifted frequency ω results in

$$\omega = \frac{\omega_0 u}{(u - v)\gamma}. \tag{3.322}$$

When $v \to u$, the source is moving essentially at the same velocity as the wave speed in the medium and in the same direction as the wave. As a result, the wave train is compressed and the frequency approaches infinity. This is the same behavior as obtained for the classical Doppler shift. When $v \to -u$, the source moves in the opposite direction to the wave, resulting in a frequency $\omega \to \omega_0/(2\gamma)$. The γ factor appears due to the time dilation of the source, while the factor of two results from the wavelength being doubled due to the motion of the source. The only difference here to the classical Doppler shift is the appearance of the factor γ, describing the time dilation due to the motion of the source.

1.71

From the form of the worldline for observer B, we find that

$$x\dot{x} - t\dot{t} = 0, \tag{3.323}$$

by differentiating with respect to the proper time. It directly follows that $v = \dot{x}/\dot{t} = t/x$. Solving for x in terms of t and inserting it into this expression now gives

$$v^2 = \frac{\alpha^2 t^2}{1 + \alpha^2 t^2}. \tag{3.324}$$

The 4-frequency of the signal is given by $\Omega = \omega(1,1)$ in S and the frequency observed by B can be found by taking the inner product of this 4-frequency with the 4-velocity of B. We find that

$$\omega_B = \Omega \cdot V_B = \omega\gamma(1-v). \tag{3.325}$$

Inserting the expression we have found for v into this results in

$$\omega_B = \omega\left(\sqrt{1+\alpha^2 t^2} - \alpha t\right). \tag{3.326}$$

In general, this expression can also be found as

$$\omega_B = \omega\sqrt{\frac{1-v}{1+v}}, \tag{3.327}$$

while the frequency ω_A of the reflected signal is given by

$$\omega_A = \omega\frac{1-v}{1+v}. \tag{3.328}$$

Direct comparison therefore results in

$$\omega_A = \omega\left(\sqrt{1+\alpha^2 t^2} - \alpha t\right)^2. \tag{3.329}$$

1.72

The rest energy of an electron is $E_0 = m_0 c^2 \simeq 0.51$ MeV, where m_0 is the rest mass of the electron. Thus, the total energy after acceleration is therefore given by

$$E = \frac{m_0 c^2}{\sqrt{1-\frac{v^2}{c^2}}} = \frac{E_0}{\sqrt{1-\frac{v^2}{c^2}}} = E_0 + 1\,\text{MeV} \simeq 1.51\,\text{MeV}. \tag{3.330}$$

Solving for v, we obtain

$$v = c\sqrt{1-\left(\frac{E_0}{E}\right)^2} = c\sqrt{1-\left(\frac{0.51}{1.51}\right)^2} \simeq 0.94\,c, \tag{3.331}$$

i.e., the final velocity of the electron is $v \simeq 0.94\,c$.

1.73

a) If the equation $p_{e^-} + p_{e^+} = k_\gamma$ i.e., conservation of 4-momentum (where p_{e^-} is the 4-momentum of the electron, p_{e^+} is the 4-momentum of the positron, and k_γ is the 4-momentum of the photon), is squared and the left-hand side is calculated in the rest frame of the electron, then the relation

$$2m_e^2 + 2m_e E_{e^+} = 0, \tag{3.332}$$

is obtained, where m_e is the rest mass of the electron (or positron) and E_{e^+} is the total energy of the positron relative to the rest frame of the electron. This proves that this process is incompatible with conservation of 4-momentum (i.e., conservation of energy and momentum) as all of the quantities on the left-hand side are strictly positive.

Alternatively, this can also be seen in an inertial frame where the spatial parts of the total momenta (i.e., the 3-momenta) of the electron and the positron are zero before the collision. In this frame, the 4-momenta of the electron and the positron before the collision are

$$p_{e^-} = \left(\sqrt{m_e^2 + \boldsymbol{p}^2}, \boldsymbol{p}\right) \quad \text{and} \quad p_{e^+} = \left(\sqrt{m_e^2 + \boldsymbol{p}^2}, -\boldsymbol{p}\right), \tag{3.333}$$

where $\boldsymbol{p} = (p_1, p_2, p_3)$. Let $k_\gamma = (|\boldsymbol{k}|, \boldsymbol{k})$ be the lightlike 4-momentum of the photon. Then, conservation of 4-momentum $p_{e^-} + p_{e^+} = k_\gamma$ implies that

$$2\sqrt{m_e^2 + \boldsymbol{p}^2} = |\boldsymbol{k}| \quad \text{and} \quad \boldsymbol{0} = \boldsymbol{k}. \tag{3.334}$$

These two conditions clearly contradict each other.

b) Now, let p_{in} and p_{out} be the 4-momentum of an electron before and after emitting a photon, respectively. In addition, let the photon have 4-momentum k. For this process to conserve the total 4-momentum, the relation

$$p_{\text{in}} = p_{\text{out}} + k, \tag{3.335}$$

must hold. However, squaring this expression gives

$$m_e^2 = m_e^2 + 2p_{\text{out}} \cdot k = m_e^2 + 2m_e\omega, \tag{3.336}$$

where ω is the photon energy in the rest frame of the electron after emitting the photon. This cannot hold for any nonzero photon energy ω.

c) In this case, we have $p_\mu + p'_\mu = k_\mu + k'_\mu$, where $(k_\mu) = (|\boldsymbol{k}|, \boldsymbol{k})$ and $(k'_\mu) = (|\boldsymbol{k}'|, \boldsymbol{k}')$ are the 4-momenta of the two photons, respectively. This implies that

$$2\sqrt{m_e^2 + \boldsymbol{p}^2} = |\boldsymbol{k}| + |\boldsymbol{k}'| \quad \text{and} \quad \boldsymbol{0} = \boldsymbol{k} + \boldsymbol{k}', \tag{3.337}$$

which clearly has a nontrivial solution $\boldsymbol{k}' = -\boldsymbol{k}$ and $|\boldsymbol{k}| = \sqrt{m_e^2 + \boldsymbol{p}^2}$. Thus, the answer to the question is "yes."

1.74

It holds that $M^2 = P_{\text{before}}^2 = P_{\text{after}}^2 = (p_a + p_b)^2$, where $p_a = (E_a, \boldsymbol{p}_a)$ and $p_b = (m_b, \boldsymbol{0})$ in the rest frame of b. Solving this equation for $|\boldsymbol{p}_a|$, using $E_a = \sqrt{m_a^2 + \boldsymbol{p}_a^2}$, gives

$$|\boldsymbol{p}_a| = \frac{1}{2m_b}\sqrt{\left[M^2 - (m_a - m_b)^2\right]\left[M^2 - (m_a + m_b)^2\right]}. \tag{3.338}$$

1.75

Consider the reaction $A \longrightarrow B + C$. Conservation of 4-momentum yields

$$P_A = P_B + P_C. \tag{3.339}$$

The redundant information about particle C can be removed by rewriting this as

$$P_C = P_A - P_B, \tag{3.340}$$

and squaring both sides, leading to

$$P_C^2 = (P_A - P_B)^2 = P_A^2 + P_B^2 - 2P_A \cdot P_B. \tag{3.341}$$

Using the fact that $P^2 = m^2$ we find

$$m_C^2 = m_A^2 + m_B^2 - 2P_A \cdot P_B. \tag{3.342}$$

In the rest frame of particle B (i.e., $\boldsymbol{p}_B = \boldsymbol{0}$), it holds that

$$P_A = (E_A, \boldsymbol{p}_A) \quad \text{and} \quad P_B = (m_B, \boldsymbol{0}), \tag{3.343}$$

which gives

$$m_C^2 = m_A^2 + m_B^2 - 2E_A m_B. \tag{3.344}$$

Using $E_A = \sqrt{m_A^2 + \boldsymbol{p}_A^2}$ and rearranging the above equation, we find that

$$\boldsymbol{p}_A^2 = \left(\frac{m_A^2 + m_B^2 - m_C^2}{2m_B} \right)^2 - m_A^2. \tag{3.345}$$

Particle A has speed v_A before the decay (relative to the rest frame of particle B after the decay), which means that

$$\boldsymbol{p}_A = m\boldsymbol{v}_A = \frac{m_A}{\sqrt{1 - v_A^2}} \boldsymbol{v}_A, \tag{3.346}$$

i.e.,

$$\boldsymbol{p}_A^2 = \frac{m_A^2 v_A^2}{1 - v_A^2}. \tag{3.347}$$

Combining the two expressions for $\boldsymbol{p}_A^2$ yields

$$\frac{m_A^2 v_A^2}{1 - v_A^2} = \left(\frac{m_A^2 + m_B^2 - m_C^2}{2m_B} \right)^2 - m_A^2. \tag{3.348}$$

Solving for v_A^2, we obtain

$$v_A^2 = \frac{m_A^4 + m_B^4 + m_C^4 - 2m_A^2 m_B^2 - 2m_A^2 m_C^2 - 2m_B^2 m_C^2}{\left(m_A^2 + m_B^2 - m_C^2\right)^2}. \tag{3.349}$$

1.76

Conservation of 4-momentum tells us that $P = p_1 + p_2$, where P is the 4-momentum of the new particle and p_1 and p_2 that of the initial two particles. Squaring this relation we find that

$$M^2 = P^2 = (p_1 + p_2)^2 = m_1^2 + m_2^2 + 2p_1 \cdot p_2. \tag{3.350}$$

In the rest frame of particle 2, we have $p_1 = m\gamma(1, \boldsymbol{v}_1)$ and $p_2 = (m, \mathbf{0})$, leading to

$$M = \sqrt{m_1^2 + m_2^2 + \frac{2m_1 m_2}{\sqrt{1 - v_1^2}}}. \tag{3.351}$$

In addition, using that $\boldsymbol{v} = \boldsymbol{p}/E$ for any particle and that the final 3-momentum is equal to the 3-momentum of particle 1 in the rest frame of particle 2, we obtain

$$\boldsymbol{v} = \frac{\boldsymbol{p}}{E} = \frac{\boldsymbol{p}_1}{E} = \frac{E_1 \boldsymbol{v}_1}{E_1 + m_2} = \frac{\boldsymbol{v}_1}{1 + m_2/E_1} = \frac{m_1}{m_1 + m_2\sqrt{1 - v_1^2}} \boldsymbol{v}_1. \tag{3.352}$$

1.77

a) Conservation of 4-momentum yields

$$p = p_1 + p_2 \quad \Rightarrow \quad p^2 = (p_1 + p_2)^2 = p_1^2 + p_2^2 + 2p_1 \cdot p_2, \tag{3.353}$$

which implies that

$$m^2 = m_1^2 + m_2^2 + 2\left(\sqrt{m_1^2 + \mathbf{p}_1^2}\sqrt{m_2^2 + \mathbf{p}_2^2} - \mathbf{p}_1 \cdot \mathbf{p}_2\right). \tag{3.354}$$

Assuming that the two particles move along the x-axis in the frame of some observer, we can express the 3-momenta of the two particles as

$$\mathbf{p}_1 = m_1\gamma(u_1)u_1\mathbf{e}_x, \quad \mathbf{p}_2 = m_2\gamma(u_2)u_2\mathbf{e}_x, \tag{3.355}$$

where $\gamma(u_i) \equiv 1/\sqrt{1 - u_i^2}, i = 1, 2$. The energies of the two particles can then be rewritten as

$$\begin{aligned}\sqrt{m_i^2 + \mathbf{p}_i^2} &= \sqrt{m_i^2 + m_i^2\gamma(u_i)^2u_i^2} = m_i\sqrt{1 + \gamma(u_i)^2u_i^2} = m_i\sqrt{1 + \frac{u_i^2}{1 - u_i^2}} \\ &= m_i\sqrt{\frac{1 - u_i^2 + u_i^2}{1 - u_i^2}} = \frac{m_i}{\sqrt{1 - u_i^2}} = m_i\gamma(u_i).\end{aligned} \tag{3.356}$$

Inserting the 3-momenta and the energies, we obtain

$$\begin{aligned}m^2 &= m_1^2 + m_2^2 + 2\left[m_1\gamma(u_1) \cdot m_2\gamma(u_2) - m_1\gamma(u_1)u_1 \cdot m_2\gamma(u_2)u_2\right] \\ &= m_1^2 + m_2^2 + 2m_1m_2\gamma(u_1)\gamma(u_2)(1 - u_1u_2),\end{aligned} \tag{3.357}$$

which is what we wanted to prove.

Now, we want to find u. Conservation of 3-momentum yields

$$\mathbf{p} = \mathbf{p}_1 + \mathbf{p}_2. \tag{3.358}$$

Assuming that the two particles move along the x-axis in the frame of the observer and using the expressions for the 3-momenta of the two particles, we find that

$$m\gamma(u)u = m_1\gamma(u_1)u_1 + m_2\gamma(u_2)u_2$$
$$\Rightarrow \quad [m\gamma(u)u]^2 = m^2\frac{u^2}{1-u^2} = [m_1\gamma(u_1)u_1 + m_2\gamma(u_2)u_2]^2, \tag{3.359}$$

where we have used $\gamma(u) \equiv 1/\sqrt{1-u^2}$. This implies that

$$\frac{u^2}{1-u^2} = \frac{[m_1\gamma(u_1)u_1 + m_2\gamma(u_2)u_2]^2}{m^2} \equiv M^2. \tag{3.360}$$

Solving for u, we obtain

$$u^2 = \frac{M^2}{1+M^2} \quad \Rightarrow \quad u = \pm\sqrt{\frac{M^2}{1+M^2}}, \tag{3.361}$$

and reinserting M, we find the velocity

$$u = \pm\sqrt{\frac{\frac{[m_1\gamma(u_1)u_1+m_2\gamma(u_2)u_2]^2}{m^2}}{1+\frac{[m_1\gamma(u_1)u_1+m_2\gamma(u_2)u_2]^2}{m^2}}} = \pm\frac{m_1\gamma(u_1)u_1 + m_2\gamma(u_2)u_2}{\sqrt{m^2 + [m_1\gamma(u_1)u_1 + m_2\gamma(u_2)u_2]^2}}. \tag{3.362}$$

Note that only the positive root is permissible.

b) In the rest frame of particle 1, we have

$$\mathbf{u}_1 = \mathbf{0}, \quad \mathbf{u}_2 = v\mathbf{e}_x. \tag{3.363}$$

Inserting this into the expression for m from a), we find that

$$m^2 = m_1^2 + m_2^2 + 2m_1m_2\gamma(0)\gamma(v)(1-0v) = m_1^2 + m_2^2 + 2m_1m_2\gamma(v), \tag{3.364}$$

which is the required expression in terms of the relative velocity v.

c) In situation 1 ($u_1 = 0$) we have $u_2 = v$. The total energy is therefore

$$E_{\text{tot},1} = E_1 + E_2 = m_1 + \frac{m_2}{\sqrt{1-v^2}}. \tag{3.365}$$

In situation 2 ($m_1\gamma(u_1)u_1 = -m_2\gamma(u_2)u_2$), our frame is instead the rest frame of the new particle and the total energy is therefore

$$E_{\text{tot},2} = m = \sqrt{m_1^2 + m_2^2 + 2m_1m_2\gamma(v)}. \tag{3.366}$$

We find the difference between the energies to be

$$\Delta E = E_{\text{tot},1} - E_{\text{tot},2} = m_1 + \frac{m_2}{\sqrt{1-v^2}} - \sqrt{m_1^2 + m_2^2 + 2\frac{m_1 m_2}{\sqrt{1-v^2}}}. \qquad (3.367)$$

This may be rewritten as

$$\Delta E = m_1 + m_2\gamma - \sqrt{(m_1 + m_2\gamma)^2 - \frac{m_2^2 v^2}{1-v^2}} \geq 0, \qquad (3.368)$$

where $\gamma = \gamma(v)$, with equality only if $v = 0$. Therefore, more energy will be required in the rest frame of one of the particles than in the center-of-mass frame. For $v \ll 1$, keeping only terms up to second order in v, we find that

$$\Delta E \simeq (m_2 - \mu)\frac{v^2}{2} = \frac{m_2^2 v^2}{2(m_1 + m_2)}, \qquad (3.369)$$

where μ is the reduced mass $\mu = m_1 m_2/(m_1 + m_2)$ of the two-particle system.

1.78

Conservation of 4-momentum gives $p_\pi = p_\mu + p_\nu$, where $p_\pi = (E_\pi, p, 0, 0)$ and $p_\mu = (E_\mu, 0, \hat{p}, 0)$. Here $(p, 0, 0)$ and $(0, \hat{p}, 0)$ are the 3-momenta of the pion (in the x-direction) and the muon (in the y-direction), respectively, which are, however, not important for this problem. Taking the square of the 4-momentum relation after moving p_μ to the left-hand side, we find that

$$(p_\pi - p_\mu)^2 = p_\pi^2 + p_\mu^2 - 2p_\pi \cdot p_\mu = p_\nu^2 = m_\nu^2 = 0$$
$$\Rightarrow \quad m_\pi^2 + m_\mu^2 - 2E_\pi E_\mu = 0. \qquad (3.370)$$

Hence, we obtain the energy of the muon as

$$E_\mu = \frac{m_\pi^2 + m_\mu^2}{2E_\pi} = \frac{m_\pi^2 + m_\mu^2}{2m_\pi \gamma(v)}, \qquad (3.371)$$

where $\gamma(v) \equiv 1/\sqrt{1-v^2}$ and v is the velocity of the incoming pion.

1.79

Let the 4-momenta of the pion, the electron, and the antineutrino be p_π, p_e, and p_ν, respectively. We find from energy–momentum conservation the relation

$$m_\pi^2 = {p_\pi}^2 = (p_e + p_\nu)^2. \qquad (3.372)$$

In the rest frame of the electron, we have $p_e = (m, \mathbf{0})$ and $p_\nu = (E_\nu, \mathbf{p})$, where m is the mass of the electron and $E_\nu = \sqrt{m_\nu^2 + \mathbf{p}^2}$ and $\mathbf{p}$ are the total energy and the

3-momentum of the antineutrino, respectively, m_ν being the mass of the antineutrino. Thus, we obtain

$$m_\pi^2 = (p_e + p_\nu)^2 = p_e^2 + p_\nu^2 + 2p_e \cdot p_\nu = m^2 + m_\nu^2 + 2mE_\nu, \tag{3.373}$$

and hence, we have $E_\nu = \Delta/(2m)$, where $\Delta \equiv m_\pi^2 - m^2 - m_\nu^2$. Using that the absolute value of the 3-momentum of the antineutrino is

$$|\mathbf{p}| = \sqrt{E_\nu^2 - m_\nu^2} = \sqrt{\frac{\Delta^2}{4m^2} - m_\nu^2}, \tag{3.374}$$

we can calculate the velocity of the antineutrino as $v = |\mathbf{p}|/E_\nu$. The result is given by

$$v = \frac{|\mathbf{p}|}{E_\nu} = \sqrt{1 - \frac{m_\nu^2}{E_\nu^2}} = \sqrt{1 - \frac{4m^2 m_\nu^2}{\Delta^2}} = \sqrt{1 - \frac{4m^2 m_\nu^2}{(m_\pi^2 - m^2 - m_\nu^2)^2}}. \tag{3.375}$$

For the limiting value of v as the rest mass of the antineutrino goes to zero, since $\lim_{m_\nu \to 0} \Delta = m_\pi{}^2 - m^2$, we find that $v \to 1$ as $m_\nu \to 0$.

1.80

a) Consider the reaction $\pi^+ \longrightarrow \mu^+ + \nu_\mu$. Conservation of 4-momentum gives that

$$P_{\pi^+} = P_{\mu^+} + P_{\nu_\mu}. \tag{3.376}$$

Subtracting P_{μ^+} from both sides and squaring results in

$$0 = P_{\nu_\mu}^2 = (P_{\pi^+} - P_{\mu^+})^2 = m_\pi^2 + m_\mu^2 - 2P_{\pi^+} \cdot P_{\mu^+}. \tag{3.377}$$

In the rest frame of the pion, we find that $P_{\pi^+} \cdot P_{\mu^+} = m_\pi E_\mu$ and therefore

$$m_\pi^2 + m_\mu^2 - 2m_\pi E_\mu = 0. \tag{3.378}$$

Solving for the total energy E_μ of the muon in the pion rest frame now results in

$$E_\mu = \frac{m_\pi^2 + m_\mu^2}{2m_\pi}. \tag{3.379}$$

The kinetic energy of the muon is the difference between its total energy and its mass and we therefore find

$$T_{\mu^+} = E_\mu - m_\mu = \frac{m_\pi^2 + m_\mu^2}{2m_\pi} - m_\mu = \frac{(m_\pi - m_\mu)^2}{2m_\pi}. \tag{3.380}$$

Since the pion decays at rest, the absolute value of the neutrino momentum must equal that of the muon momentum. The energy–momentum relation for the pion therefore yields

$$E_\mu^2 = m_\mu^2 + p_\nu^2 \quad \Longrightarrow \quad p_\nu = \frac{m_\pi^2 - m_\mu^2}{2m_\pi}, \tag{3.381}$$

after inserting the expression for E_μ and solving for p_ν.

b) In the rest frame of the pion, the muon will travel the distance $s \equiv \gamma(v_\mu)v_\mu\tau_\mu = p_\mu\tau_\mu/m_\mu$ before it decays. We therefore find that

$$s = \frac{m_\pi^2 - m_\mu^2}{2m_\pi m_\mu}\tau_\mu, \tag{3.382}$$

since the muon and neutrino momenta are equal in magnitude.

1.81

In the rest frame of the decaying pion, the solution to Problem 1.80 resulted in

$$E_\mu = \frac{m_\pi^2 + m_\mu^2}{2m_\pi} \quad \text{and} \quad p = \frac{m_\pi^2 - m_\mu^2}{2m_\pi}, \tag{3.383}$$

for the muon energy and momentum, respectively. It follows that the 4-momentum of the muon in the rest frame of the pion is given by $p_\mu = (E_\mu, p)$. However, we want to compute the energy of the muon in the rest frame of the Earth, and thus, we must Lorentz transform p_μ to this frame. The Lorentz transformation is in the opposite direction of the motion of the muon in the rest frame of the pion with velocity $v = \sqrt{E_\pi^2 - m_\pi^2}/E_\pi$, where E_π is the energy of the pion in the rest frame of the Earth. It follows that in the rest frame of the Earth, we have

$$E'_\mu = \gamma(v)(E_\mu + vp) = \frac{E_\pi}{m_\pi}\left(\frac{m_\pi^2 + m_\mu^2}{2m_\pi} + \frac{m_\pi^2 - m_\mu^2}{2m_\pi}\sqrt{1 - \frac{m_\pi^2}{E_\pi^2}}\right). \tag{3.384}$$

If we series expand the square root in the small quantity m_π/E_π and keep the zeroth order term only, we obtain

$$E'_\mu \simeq E_\pi = 2\,\text{GeV}. \tag{3.385}$$

1.82

We study the decay $R \longrightarrow \mu^+ + \mu^-$. In this decay, the total 4-momentum must be preserved, and thus, we have

$$p_R = p_{\mu^+} + p_{\mu^-}. \tag{3.386}$$

The square of the mass M_R of the resonance is given by the square of its 4-momentum, i.e.,

$$M_R^2 = p_R^2 = p_{\mu^+}^2 + p_{\mu^-}^2 + 2p_{\mu^+} \cdot p_{\mu^-} = 2\left(m_\mu^2 + p_{\mu^+} \cdot p_{\mu^-}\right). \tag{3.387}$$

If we place our coordinate system in such a way that the μ^+ is traveling in the x-direction and the μ^- in the y-direction (this is possible, since the angle between the directions is 90°), then the 4-momentum of the muons will be

$$p_{\mu^+} = (E, p, 0, 0) \quad \text{and} \quad p_{\mu^-} = (E, 0, p, 0), \tag{3.388}$$

respectively, where $p = 2.2$ GeV and $E = \sqrt{p^2 + m_\mu^2}$. It follows that

$$p_{\mu^+} \cdot p_{\mu^-} = E^2 = p^2 + m_\mu^2, \tag{3.389}$$

and thus, we obtain

$$M_R^2 = 2(p^2 + 2m_\mu^2) \simeq 2p^2 \quad \Rightarrow \quad M_R \simeq \sqrt{2}p \simeq 3 \text{ GeV}, \tag{3.390}$$

since $p \gg m_\mu$.

1.83

There are several methods of solving this problem.

Method 1. We use the Doppler effect. In the system of the detector, the frequency of the photon is ω/h. This frequency is related to the frequency ω'/h of the photon in the system of the decaying particle, from which it is emitted, according to the formula for the Doppler shift as

$$\frac{\omega}{h} = \frac{\omega'}{h}\sqrt{\frac{1+v}{1-v}}, \tag{3.391}$$

where v is the velocity of the decaying particle before it emits the photon. The frequency must be blueshifted, since the emitting particle moves toward the detector. Now, we have $v = |\boldsymbol{p}|/E$. Inserting this into the Doppler formula above gives, after some simplifications,

$$\frac{\omega}{h} = \frac{\omega'}{h}\frac{E + |\boldsymbol{p}|}{M}, \tag{3.392}$$

since $E^2 - \boldsymbol{p}^2 = M^2$. Thus, solving for ω' gives the answer

$$\omega' = \omega\frac{M}{E + |\boldsymbol{p}|}. \tag{3.393}$$

Method 2. Conservation of energy and momentum says that if the decaying particle emits a photon with 4-momentum $k = (w, \boldsymbol{k})$ and a rest product (which can be several particles) of momentum $p' = (E', \boldsymbol{p}')$, then we have, in the rest frame of the decaying particle,

$$M = E' + \omega'. \tag{3.394}$$

In this system, conservation of momentum reads $\boldsymbol{p}' = -\boldsymbol{k}$. We then obtain

$$(M - \omega')^2 = E'^2 = p'^2 + (\boldsymbol{p}')^2 = p'^2 + \omega'^2, \tag{3.395}$$

where p'^2 is the 4-momentum squared of p'. However, conservation of 4-momentum also gives

$$p = p' + k, \tag{3.396}$$

whence $p'^2 = (p-k)^2 = M^2 - 2p \cdot k$, where we have used $k^2 = 0$. Inserting this into the first relation gives

$$(M - \omega')^2 = M^2 - 2p \cdot k + \omega'^2. \tag{3.397}$$

If we solve for ω', we obtain $\omega' = p \cdot k / M$. By inserting $p = (E, \boldsymbol{p})$ and $k = (\omega, \boldsymbol{k})$ with $\boldsymbol{p}$ parallel to $\boldsymbol{k}$, we obtain the same answer as with *Method 1*.

Method 3. The previous method suggests that one can solve the problem by studying the relativistic invariant $p \cdot k$. In the rest frame of the decaying particle, its value is $M\omega'$. In the frame of the detector, its value is

$$p \cdot k = E\omega - \boldsymbol{p} \cdot \boldsymbol{k} = \omega(E - |\boldsymbol{p}|). \tag{3.398}$$

Since the value of the invariant is independent of the frame, one finds $M\omega' = \omega(E - |\boldsymbol{p}|)$, which, after simplifications, leads to the same answer as with the previous two methods.

Method 4. Lastly, one can also simply make a Lorentz transformation of $k = (\omega', \omega', 0, 0)$, which is 4-wavevector of the photon in the rest frame of the decaying particle and where we have put the direction to the detector to coincide with the x-axis to the detector system in which the particle moves with speed $v = |\boldsymbol{p}|/E$ toward the detector. The detector then moves with speed $-v$ relative to the particle. For the 0-component, one then finds that

$$\omega = \omega'\gamma + \omega'\gamma v = \omega'\gamma(1 + v), \tag{3.399}$$

where $\gamma \equiv 1/\sqrt{1 - v^2}$. Inserting the value of v above gives, after simplifications, the same result as obtained by the other three methods.

1.84

In the rest frame of the positron, let the 4-momenta be $p_e = (E_e/c, 0, 0, p)$ and $p_p = (m_e c, 0, 0, 0)$ for the electron and the positron, respectively, where E_e is the total energy of the electron, $(0, 0, p)$ is the 3-momentum of the electron, and m_e is the mass of an electron or a positron. The 4-momenta of the photons are

$k_1 = (\omega_1, \omega_1 \sin\phi, 0, \omega_1 \cos\phi)$ and $k_2 = (\omega_2, -\omega_2 \sin\phi, 0, \omega_2 \cos\phi)$, respectively. Conservation of 4-momentum gives

$$p_e + p_p = k_1 + k_2, \tag{3.400}$$

i.e., in the rest frame of the positron, we have

$$\frac{E_e}{c} + m_e c = \omega_1 + \omega_2, \tag{3.401}$$

$$0 = \omega_1 \sin\phi - \omega_2 \sin\phi, \tag{3.402}$$

$$p = \omega_1 \cos\phi + \omega_2 \cos\phi. \tag{3.403}$$

Hence, we find that $\omega_1 = \omega_2$.

a) From the relations above, we obtain the angle ϕ as a function of the total energy of the electron E_e as

$$\cos\phi = \frac{p}{k_1 + k_2} = \frac{p}{\omega_1 + \omega_2} = \frac{p}{\frac{E_e}{c} + m_e c} = \frac{\sqrt{\frac{E_e^2}{c^2} - m_e^2 c^2}}{\frac{E_e}{c} + m_e c} = \sqrt{\frac{E_e - m_e c^2}{E_e + m_e c^2}}$$

$$\Rightarrow \quad \phi = \arccos\sqrt{\frac{E_e - m_e c^2}{E_e + m_e c^2}}. \tag{3.404}$$

b) In the nonrelativistic limit, i.e., $E_e \simeq m_e c^2 + p^2/(2m_e)$, with $p \ll m_2$, we find that

$$\cos\phi \simeq \sqrt{\frac{p^2}{4m_e^2 c^2 + p^2}} \simeq \frac{p}{2m_e c} = \{p \simeq m_e v\} = \frac{v}{2c}. \tag{3.405}$$

1.85

Conservation of 4-momentum gives

$$P_1 + P_2 = P + P', \tag{3.406}$$

which implies that $P' = P_1 + P_2 - P$. Now, photons are lightlike. Thus, we find that

$$P'^2 = 0 = (P_1 + P_2 - P)^2 = 2(P_1 \cdot P_2 - P_1 \cdot P - P_2 \cdot P), \tag{3.407}$$

since $P_1{}^2 = P_2{}^2 = P^2 = 0$. Therefore, we have

$$P_1 \cdot P_2 - P_1 \cdot P - P_2 \cdot P = 0. \tag{3.408}$$

The energy for a photon is $E = pc = \{p = \frac{h}{\lambda}\} = \frac{hc}{\lambda}$. Using $P_1 = \frac{h}{\lambda_1}(1,1,0)$, $P_2 = \frac{h}{\lambda_2}(1,-1,0)$, and $P = \frac{h}{\lambda}(1, \cos\theta, \sin\theta)$, we obtain

$$\frac{2h^2}{\lambda_1\lambda_2} = P_1 \cdot P_2 = P \cdot (P_1 + P_2) = \frac{h^2}{\lambda}\left[\frac{1-\cos\theta}{\lambda_1} + \frac{1+\cos\theta}{\lambda_2}\right]. \quad (3.409)$$

Solving for λ now results in

$$\lambda = \frac{1}{2}\left[\lambda_2(1-\cos\theta) + \lambda_1(1+\cos\theta)\right]. \quad (3.410)$$

1.86

First, define the Lorentz invariant total 4-momentum squared of the pions as $s = (p_1 + p_2)^2$. Using conservation of 4-momentum, i.e., $P = p_1 + p_2 + k$, we can then calculate s to be $s = (P - k)^2 = M(M - 2\omega)$. However, in the rest frame of the pions, we know that $\sqrt{s} = 2E$, where $E = \sqrt{m^2 + p^2}$ and $p^2 = p_1^2 = p_2^2$. Therefore, using the two expressions for s, we obtain $M^2 - 2M\omega = 4(m^2 + p^2)$, which can be solved for p to give

$$p = \frac{1}{2}\sqrt{M^2 - 2M\omega - 4m^2}. \quad (3.411)$$

Finally, the speed of the pions relative to their center-of-mass frame is $v = p/E$, so inserting the expression for p and $E = \frac{1}{2}\sqrt{s} = \frac{1}{2}\sqrt{M^2 - 2M\omega}$, we find that

$$v = \frac{p}{E} = \sqrt{\frac{M^2 - 2M\omega - 4m^2}{M^2 - 2M\omega}} = \sqrt{1 - \frac{4m^2}{M(M - 2\omega)}}. \quad (3.412)$$

1.87

Before the decay, the Σ^0 particle moves with 4-momentum $(E, \boldsymbol{p})$ toward the detector. After the decay, the Λ particle moves toward the detector with 4-momentum $(E', \boldsymbol{p}')$ and the photon with momentum $(\omega', \boldsymbol{k}')$.

a) The total energy of the Σ^0 particle is given by

$$E_\Sigma = \frac{m_\Sigma}{\sqrt{1 - v^2}} = \frac{m_\Sigma}{\sqrt{1 - 1/3^2}} = \frac{3m_\Sigma}{2\sqrt{2}}. \quad (3.413)$$

b) Conservation of 4-momentum gives the relation

$$P_\Sigma = P_\gamma + P_\Lambda. \quad (3.414)$$

Subtracting P_γ from both sides and squaring results in

$$m_\Lambda^2 = P_\Lambda^2 = m_\Sigma^2 - 2P_\Sigma \cdot P_\gamma = m_\Sigma(m_\Sigma - 2E_\gamma), \quad (3.415)$$

where E_γ is the energy of the photon in the rest frame of the Σ^0 particle. Solving for E_γ results in

$$E_\gamma = \frac{m_\Sigma^2 - m_\Lambda^2}{2m_\Sigma}. \quad (3.416)$$

c) Since the Σ^0 particle is moving straight toward the detector, the energy of the photon as registered by the detector will be given by the relativistic Doppler shift formula

$$E'_\gamma = E_\gamma \sqrt{\frac{1+v}{1-v}} = E_\gamma \sqrt{\frac{1+1/3}{1-1/3}} = E_\gamma \sqrt{2} = \frac{m_\Sigma^2 - m_\Lambda^2}{\sqrt{2} m_\Sigma}. \tag{3.417}$$

1.88

a) In the center-of-mass system, we have by definition $\boldsymbol{p}_e + \boldsymbol{p}_p = \boldsymbol{0}$ and conservation of 3-momentum then leads to $\boldsymbol{p}'_e + \boldsymbol{p}'_p = \boldsymbol{0}$. Due to conservation of energy, we have for elastic scattering that $|\boldsymbol{p}_e| = |\boldsymbol{p}_p| = |\boldsymbol{p}'_e| = |\boldsymbol{p}'_p| \equiv p$. Therefore, we have $E_e = E'_e$. Using these results, we find that $t = (p_e - p'_e)^2 = (E_e - E'_e)^2 - (\boldsymbol{p}_e - \boldsymbol{p}'_e)^2 = -(\boldsymbol{p}_e - \boldsymbol{p}'_e)^2$, and thus, we obtain $-t = (\boldsymbol{p}_e - \boldsymbol{p}'_e)^2$. Introducing the scattering angle θ by $\boldsymbol{p}_e \cdot \boldsymbol{p}'_e = p^2 \cos\theta$, we find that

$$\begin{aligned} -t &= \boldsymbol{p}_e^2 + \boldsymbol{p}_e'^2 - 2\boldsymbol{p}_e \cdot \boldsymbol{p}'_e = p^2 + p^2 - 2p^2 \cos\theta \\ &= 2p^2(1 - \cos\theta) = 4p^2 \sin^2\frac{\theta}{2}. \end{aligned} \tag{3.418}$$

Thus, the result is $-t = 4p^2 \sin^2(\theta/2)$.

b) Since conservation of 4-momentum holds, we also have $p_e - p'_e = p'_p - p_p$, which means that $t = (p'_p - p_p)^2$. In the laboratory system, we have $p_p = (m_p, \boldsymbol{0})$ and $p'_p = (m_p + T'_p, \boldsymbol{p}_p)$. Therefore, we find that

$$\begin{aligned} t &= (p'_p - p_p)^2 = p_p'^2 + p_p^2 - 2p'_p \cdot p_p = 2m_p^2 - 2p_p \cdot p'_p \\ &= 2m_p^2 - 2m_p(m_p + T'_p) = -2m_p T'_p. \end{aligned} \tag{3.419}$$

Thus, the result is $T'_p = -t/(2m_p)$.

1.89

The energy of the pion before the collision is $E = m_\pi + T$. Now, $s_0 = (m_\pi + m_\Delta)^2$ gives the minimal center-of-mass energy squared for production of the Δ. However, this is also given by $s_0 = ((E, \boldsymbol{p}) + (m_p, \boldsymbol{0}))^2 = (E + m_p)^2 - \boldsymbol{p}^2 = (T + m_\pi + m_p)^2 - \boldsymbol{p}^2$. From $E^2 = (m + T)^2 = m^2 + \boldsymbol{p}^2$ follows that $\boldsymbol{p}^2 = T^2 + 2mT$. Thus, the kinetic energy T of the pion required to create the Δ is given by

$$\begin{aligned} & (m_\pi + m_\Delta)^2 = (T + m_\pi + m_p)^2 - (T^2 + 2m_\pi T) \\ \Longrightarrow \quad & T = \frac{(m_\pi + m_\Delta)^2 - (m_\pi + m_p)^2}{2m_p}. \end{aligned} \tag{3.420}$$

1.90

The square of the center-of-mass energy is given by

$$P_{\text{tot}}^2 = (p_\pi + p_d)^2 = (p_{p_1} + p_{p_2})^2. \tag{3.421}$$

In the case when the pion hits a deuteron at rest, we have

$$p_\pi = (m_\pi + T_\pi, \mathbf{p}_\pi) \quad \text{and} \quad p_d = (m_d, \mathbf{0}). \tag{3.422}$$

This gives the square of the center-of-mass energy, i.e.,

$$P_{\text{tot}}^2 = p_\pi^2 + p_d^2 + 2p_\pi \cdot p_d = (m_\pi + m_d)^2 + 2m_d T_\pi. \tag{3.423}$$

Similarly, we have, in the case of one proton hitting another proton at rest,

$$p_{p_1} = (m_p + T_p, \mathbf{p}_{p_1}) \quad \text{and} \quad p_{p_2} = (m_p, \mathbf{0}), \tag{3.424}$$

resulting in

$$P_{\text{tot}}^2 = 4m_p^2 + 2m_p T_p. \tag{3.425}$$

Requiring the two expressions for the square of the center-of-mass energy to be the same, we obtain

$$T_p = \frac{(m_\pi + m_d)^2}{2m_p} - 2m_p + \frac{m_d}{m_p} T_\pi. \tag{3.426}$$

1.91

Consider the reaction $\pi^+ + n \longrightarrow K^+ + \Lambda$. Let $\theta = [\pi^+, K^+] = 90°$. Conservation of 4-momentum gives

$$p_{\pi^+} + p_n = p_{K^+} + p_\Lambda. \tag{3.427}$$

The unknown kinematics of the Λ particle can be removed by isolating p_Λ and squaring such that

$$p_\Lambda = p_{\pi^+} + p_n - p_{K^+}, \tag{3.428}$$

$$\begin{aligned} p_\Lambda^2 &= (p_{\pi^+} + p_n - p_{K^+})^2 \\ &= p_{\pi^+}^2 + p_n^2 + p_{K^+}^2 + 2p_{\pi^+} \cdot p_n - 2p_{\pi^+} \cdot p_{K^+} - 2p_n \cdot p_{K^+}. \end{aligned} \tag{3.429}$$

Using the fact that $p^2 = m^2$ implies that

$$m_\Lambda^2 = m_{\pi^+}^2 + m_n^2 + m_{K^+}^2 + 2p_{\pi^+} \cdot p_n - 2p_{\pi^+} \cdot p_{K^+} - 2p_n \cdot p_{K^+}. \tag{3.430}$$

In the rest frame of n, one has

$$p_{\pi^+} = (E_{\pi^+}, \mathbf{p}_{\pi^+}), \quad p_n = (m_n, \mathbf{0}), \quad \text{and} \quad p_{K^+} = (E_{K^+}, \mathbf{p}_{K^+}). \tag{3.431}$$

This leads to

$$m_\Lambda^2 = m_{\pi^+}^2 + m_n^2 + m_{K^+}^2 + 2E_{\pi^+} \cdot m_n - 2\,(E_{\pi^+}E_{K^+} - \boldsymbol{p}_{\pi^+} \cdot \boldsymbol{p}_{K^+}) - 2m_n \cdot E_{K^+}. \tag{3.432}$$

Using $\boldsymbol{p}_{\pi^+} \cdot \boldsymbol{p}_{K^+} = |\boldsymbol{p}_{\pi^+}||\boldsymbol{p}_{K^+}| \cos\theta = |\boldsymbol{p}_{\pi^+}||\boldsymbol{p}_{K^+}| \cdot 0 = 0$ yields

$$m_\Lambda^2 = m_{\pi^+}^2 + m_n^2 + m_{K^+}^2 + 2E_{\pi^+} \cdot m_n - 2E_{\pi^+}E_{K^+} - 2m_n \cdot E_{K^+}. \tag{3.433}$$

Solving the above equation for E_{π^+}, we find that

$$E_{\pi^+} = \frac{m_\Lambda^2 - m_{\pi^+}^2 - m_n^2 - m_{K^+}^2 + 2E_{K^+}m_n}{2\,(m_n - E_{K^+})}. \tag{3.434}$$

The kinetic energy of π^+ is given by $T_{\pi^+} = E_{\pi^+} - m_{\pi^+}$. Thus, one obtains

$$T_{\pi^+} = \frac{m_\Lambda^2 - m_{\pi^+}^2 - m_n^2 - m_{K^+}^2 + 2E_{K^+}m_n}{2\,(m_n - E_{K^+})} - m_{\pi^+}. \tag{3.435}$$

Using the given quantities T and E, i.e., $T_{\pi^+} = T$ and $E_{K^+} = E$, this gives the result

$$T = \frac{m_\Lambda^2 - m_{\pi^+}^2 - m_n^2 - m_{K^+}^2 + 2Em_n}{2\,(m_n - E)} - m_{\pi^+}. \tag{3.436}$$

1.92

Let the 4-momenta of the different particles be indexed by their respective symbols. Then, conservation of 4-momentum gives

$$p_p + p_{\pi^-} = p_n + p_{\pi^0}. \tag{3.437}$$

Since we have no information on the π^0 meson, we solve for its 4-momentum and square it, which leads to

$$m_{\pi^0}^2 = (p_p + p_{\pi^-} - p_n)^2. \tag{3.438}$$

Thus, in the common rest frame of the incoming particles, we obtain

$$m_{\pi^0} = \sqrt{(m_p + m_{\pi^-} - E_n)^2 - \boldsymbol{p}_n^2} = \sqrt{(m_p + m_{\pi^-})^2 + m_n^2 - 2E_n(m_p + m_{\pi^-})}. \tag{3.439}$$

Now, we have $v_n \approx 3 \cdot 10^{-2}$, which is small compared to 1. Therefore, we can approximate

$$E_n = \frac{m_n}{\sqrt{1 - v_n^2}} \simeq m_n\left(1 + \frac{1}{2}v_n^2\right), \tag{3.440}$$

which yields

$$m_{\pi^0} \simeq \sqrt{(m_p + m_{\pi^-})^2 + m_n^2 - 2(m_p + m_{\pi^-})m_n\left(1 + \frac{1}{2}v_n^2\right)}$$
$$= \sqrt{(m_p + m_{\pi^-} - m_n)^2 - m_n(m_p + m_{\pi^-})v_n^2}$$
$$= (m_p + m_{\pi^-} - m_n)^2\sqrt{1 - \frac{m_n(m_p + m_{\pi^-})}{(m_p + m_{\pi^-} - m_n)^2}v_n^2}, \quad (3.441)$$

and finally, we find that

$$m_{\pi^0} \simeq m_p + m_{\pi^-} - m_n - \frac{1}{2}\frac{m_n(m_p + m_{\pi^-})}{m_p + m_{\pi^-} - m_n}v_n^2. \quad (3.442)$$

1.93

The kinetic energy of the neutron is negligible (the kinetic energy of a thermal neutron is of the order 25 meV), so both the neutron and the proton can be considered at rest before the reaction. By conservation of 4-momentum, we have the relation

$$p_p + p_n - p_\gamma = p_d \quad \Rightarrow \quad (p_p + p_n)^2 - 2(p_p + p_n)\cdot p_\gamma = p_d^2. \quad (3.443)$$

If we let $M = m_p + m_n$, then $p_p + p_n = (M, \mathbf{0})$ and $p_\gamma = (E_\gamma, \mathbf{p}_\gamma)$ in the lab frame, while $p_d = (M - B, \mathbf{0})$ in the rest frame of the deutron. It follows from the Lorentz invariance of the Minkowski product that

$$M^2 - 2ME_\gamma = M^2 - 2MB + B^2$$
$$\Rightarrow \quad E_\gamma = B\left(1 - \frac{B}{2M}\right) = B\left(1 - \frac{B}{2(m_p + m_n)}\right). \quad (3.444)$$

The difference between the energies B and E_γ is due to the recoil energy of the deutron.

1.94

Let the 4-momenta for the particles in the reaction be k, p_{H}, p_p, and p_e. Since 4-momentum is conserved, we have $k + p_{\mathrm{H}} = p_p + p_e$. From Lorentz invariance, we also have $(k + p_{\mathrm{H}})^2 = (p_p + p_e)^2$. Now, since both sides of this relation are Lorentz invariants, we can calculate them in different inertial systems. In the system where H is at rest, we can take $k = (\omega, \omega, 0, 0)$ and $p_{\mathrm{H}} = (m_{\mathrm{H}}, 0, 0, 0)$. Thus, we obtain $(k + p_{\mathrm{H}})^2 = 2\omega m_{\mathrm{H}} + m_{\mathrm{H}}^2$. For the final particles, we choose the center-of-mass frame, where the particles are at rest at threshold for the reaction to occur.

Thus, we find that $(p_p + p_e)^2 = (m_p + m_e)^2 = (m_{\mathrm{H}} + B)^2$. Finally, combining the above expressions and solving for ω, we obtain

$$\omega = B\left(1 + \frac{B}{2m_{\mathrm{H}}}\right). \tag{3.445}$$

1.95

In order for the collision of the two neutrinos to produce a Z^0 boson, the center-of-momentum energy must be equal to the Z^0 mass, i.e.,

$$(p_1 + p_2)^2 = m_{Z^0}^2, \tag{3.446}$$

where p_1 is the 4-momentum of the CNB neutrino and p_2 is the 4-momentum of the UHE neutrino. The two 4-momenta are given by

$$p_1 = \left(m_\nu + E_k, \sqrt{E_k(E_k + 2m_\mu)}\boldsymbol{a}\right), \tag{3.447}$$

$$p_2 = (E, E\boldsymbol{b}), \tag{3.448}$$

where m_ν is the neutrino mass, E_k is the thermal energy of the CNB neutrino, E is the energy of the UHE neutrino, and $\boldsymbol{a}$ and $\boldsymbol{b}$ are 3-vectors with modulus one ($|\boldsymbol{a}| = |\boldsymbol{b}| = 1$). Note that we have neglected the neutrino mass compared to the UHE neutrino energy. In general, we now need to take care if we obtain an expression, where terms including E cancel (since the corrections of the small parameter m_μ/E would then become the leading terms). However, this will not be the case in this problem.

Summing and then squaring the 4-vectors results in

$$m_{Z^0}^2 = (p_1 + p_2)^2 = 2p_1 \cdot p_2 + \underbrace{p_1^2 + p_2^2}_{\ll p_1 \cdot p_2} \simeq 2p_1 \cdot p_2$$

$$= E\left[m_\nu + E_k - \sqrt{E_k(E_k + 2m_\nu)}\boldsymbol{a} \cdot \boldsymbol{b}\right]. \tag{3.449}$$

Furthermore, since $|\boldsymbol{a} \cdot \boldsymbol{b}| \le 1$, we obtain

$$m_{Z^0}^2 \le E\left[m_\nu + E_k - \sqrt{E_k(E_k + 2m_\nu)}\right]$$

$$\Rightarrow \quad E \ge \frac{m_{Z^0}^2}{2} \frac{1}{m_\nu + E_k + \sqrt{E_k(E_k + 2m_\nu)}}. \tag{3.450}$$

For the case when $m_\nu = 0.15$ eV $\gg E_k$, this simplifies to

$$E \ge \frac{m_{Z^0}^2}{2m_\nu} \simeq 28 \cdot 10^{21} \text{ eV} = 28 \text{ ZeV}. \tag{3.451}$$

Similarly, in the case when $E_k \simeq 3k_B T/2 \gg m_\nu$, we have

$$E \ge \frac{m_{Z^0}^2}{4E_k} \simeq 8 \cdot 10^{24} \text{ eV} = 8 \text{ YeV}. \tag{3.452}$$

1.96

a) Conservation of 4-momentum yields $k + p = k' + p'$. We can remove the redundant information about p' as follows

$$(p + k - k')^2 = p'^2 = m^2. \tag{3.453}$$

Using $k^2 = k'^2 = 0$ and simplifying gives

$$p \cdot k - p \cdot k' - k \cdot k' = 0. \tag{3.454}$$

Inserting $p = (m, \mathbf{0})$, $k = (\omega, \boldsymbol{k})$, and $k' = (\omega', \boldsymbol{k}')$, this leads to

$$\omega - \omega' = \frac{2}{m}\omega\omega' \sin^2\frac{\theta}{2}, \tag{3.455}$$

in the rest frame of the initial electron. We then use $\omega = 2\pi\nu = \{\nu = 1/\lambda\} = 2\pi/\lambda$ and similar for ω' to obtain the Compton formula, i.e.,

$$\lambda' - \lambda = \frac{4\pi}{m}\sin^2\frac{\theta}{2}. \tag{3.456}$$

b) We use the Mandelstam variable s defined as

$$s \equiv (p + k)^2 = p^2 + k^2 + 2p \cdot k = m^2 + 2m\omega. \tag{3.457}$$

On the other hand, due to conservation of 4-momentum, we also have, in the center-of-mass system (i.e., $\boldsymbol{p}' + \boldsymbol{k}' = \mathbf{0}$), the expression

$$s = (p' + k')^2 = (E' + \omega')^2 = \left(\sqrt{m^2 + \omega'^2} + \omega'\right)^2, \tag{3.458}$$

where $E' = \sqrt{m^2 + \boldsymbol{p}'^2}$ is the energy of the outgoing electron and $\boldsymbol{p}'^2 = \boldsymbol{k}'^2 = \omega'^2$. Now, since s is Lorentz invariant, equating the two expressions for s and solving for ω', we obtain

$$m^2 + 2m\omega = \left(\sqrt{m^2 + \omega'^2} + \omega'\right)^2 \quad \Leftrightarrow \quad \omega' = \frac{\omega}{\sqrt{1 + \frac{2\omega}{m}}}. \tag{3.459}$$

1.97

Denoting the incoming 4-momenta of the electron and photon p_e and p_γ, respectively, and the corresponding outgoing quantities by k_e and k_γ, we have

$$p_e = (m_e, \mathbf{0}),\ p_\gamma = \omega(1, 1, 0),\ k_e = (E_e, \boldsymbol{p}_e),\ k_\gamma = \omega'(1, \cos\theta, \sin\theta), \tag{3.460}$$

in the laboratory frame, where we have oriented our coordinate system such that there is no momentum in the z-direction either before or after the collision. Thus, we have also omitted the z-components of the 4-vectors. Conservation of 4-momentum yields

$$p_e + p_\gamma = k_e + k_\gamma \quad \Longrightarrow \quad p_e + p_\gamma - k_\gamma = p_e. \tag{3.461}$$

Squaring this relation, we obtain

$$m^2 + 2p_e \cdot p_\gamma - 2p_e \cdot k_\gamma - 2p_\gamma \cdot k_\gamma = m^2 \quad \Longrightarrow \quad m\omega = m\omega' + \omega\omega'(1 - \cos\theta). \tag{3.462}$$

Solving for ω', we have

$$\omega' = \frac{\omega}{1 + \frac{\omega}{m}(1 - \cos\theta)}. \tag{3.463}$$

Furthermore, the total energy of the outgoing electron E_e is given by $E_e = m + \omega - \omega'$ from the time component of the 4-momentum conservation (i.e., energy conservation). Since the kinetic energy is $T_e = E_e - m = \omega - \omega'$, we insert our result for ω' and simplify

$$T_e = \omega - \omega' = \omega \left[1 - \frac{1}{1 + \frac{\omega}{m}(1 - \cos\theta)} \right] = \frac{\omega^2(1 - \cos\theta)}{m + \omega(1 - \cos\theta)}. \tag{3.464}$$

1.98

The setup is shown in Figure 3.8. The 4-momenta of the particles are given by (suppressing the z-component):

$$k = E_0(1, 1, 0), \quad p = m\gamma(1, v, 0), \quad k' = E(1, \cos\theta, -\sin\theta), \quad p' = mV_e', \tag{3.465}$$

where E_0 is the initial photon energy, v is the initial speed of the electron, and E is the sought energy of the photon after scattering. By conservation of 4-momentum, we find that

$$p + k - k' = p' \quad \Rightarrow \quad (p + k - k')^2 = m^2. \tag{3.466}$$

This leads to

$$\begin{aligned} m^2 + 2p \cdot k - 2(p + k) \cdot k' = m^2 \quad &\Rightarrow \quad p \cdot k = (p + k) \cdot k' \\ \Rightarrow \quad m\gamma E_0(1 - v) &= m\gamma E(1 - v\cos\theta) + EE_0(1 - \cos\theta). \end{aligned} \tag{3.467}$$

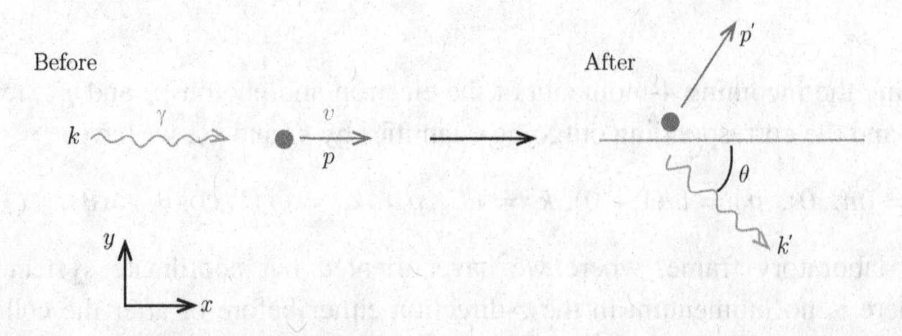

Figure 3.8 Setup of inverse Compton scattering $\gamma + e^- \longrightarrow \gamma + e^-$ before and after scattering.

Solving for E leads to

$$E = \frac{m\gamma E_0(1 - v)}{m\gamma(1 - v\cos\theta) + E_0(1 - \cos\theta)}. \tag{3.468}$$

1.99

Consider the reaction $\mu^+ \longrightarrow e^+ + \nu_e + \bar{\nu}_\mu$. Let p_μ, p_e, p_{ν_e}, and $p_{\bar{\nu}_\mu}$ be the 4-momenta of the antimuon, positron, electron neutrino, and antimuon neutrino, respectively. Conservation of 4-momentum gives $p_\mu = p_e + p_{\nu_e} + p_{\bar{\nu}_\mu}$, so that $p_\mu - p_{\nu_e} = p_e + p_{\bar{\nu}_\mu} \equiv P$. Squaring both sides yields

$$\begin{aligned} P^2 = p_\mu^2 - 2p_\mu \cdot p_{\nu_e} + p_{\nu_e}^2 &= m_\mu^2 - 2p_\mu \cdot p_{\nu_e} + m_{\nu_e}^2 \\ &= (p_e + p_{\bar{\nu}_\mu})^2 = m_e^2 + 2p_e \cdot p_{\bar{\nu}_\mu} + m_{\bar{\nu}_\mu}^2. \end{aligned} \tag{3.469}$$

Neglecting the neutrino masses compared to lepton masses yields the relation

$$m_\mu^2 - 2p_\mu \cdot p_{\nu_e} \simeq m_e^2 + 2p_e \cdot p_{\bar{\nu}_\mu}. \tag{3.470}$$

All of the terms in this relation are Lorentz invariant and may be calculated in any inertial system. Calculating $p_e \cdot p_{\bar{\nu}_\mu}$ in the rest frame of the positron, it is easy to obtain $p_e \cdot p_{\bar{\nu}_\mu} \geq m_e m_{\bar{\nu}_\mu}$, which is negligible compared to m_e^2. In the rest frame of the antimuon, we have $p_\mu = (m_\mu, \mathbf{0})$ and $p_{\nu_e} = (E, \mathbf{p})$, where $|\mathbf{p}| = \sqrt{E^2 - m_{\nu_e}^2}$ and E being the total energy of the electron neutrino. Inserting this into the equation above results in the inequality

$$m_\mu^2 - 2m_\mu E \geq m_e^2 \quad \Rightarrow \quad E \leq \frac{m_\mu^2 - m_e^2}{2m_\mu}. \tag{3.471}$$

Thus, the largest possible total energy of the electron neutrino in the rest frame of the antimuon is given by

$$E_{\max} = \frac{m_\mu^2 - m_e^2}{2m_\mu}. \tag{3.472}$$

1.100

Consider the decay $\rho \longrightarrow \mu^- + \mu^+ + \gamma$. Conservation of 4-momentum gives the relation

$$p_\rho - p_{\mu^+} = p_{\mu^-} + p_\gamma. \tag{3.473}$$

Squaring this expression yields

$$(p_\rho - p_{\mu^+})^2 = (p_{\mu^-} + p_\gamma)^2, \tag{3.474}$$

where both sides are clearly Lorentz invariant. In the rest frame of the ρ-meson, the left-hand side of this expression is given by

$$(p_\rho - p_{\mu^+})^2 = m_\rho^2 + m_\mu^2 - 2p_\rho \cdot p_{\mu^+} = m_\rho^2 + m_\mu^2 - 2m_\rho(T_\mu + m_\mu), \tag{3.475}$$

where $m_\mu \equiv m_{\mu^-} = m_{\mu^+}$ and T_μ is the kinetic energy of the μ^+. In the rest frame of the μ^-, the right-hand side becomes

$$(p_{\mu^-} + p_\gamma)^2 = m_\mu(m_\mu + 2k) \geq m_\mu^2, \tag{3.476}$$

where k is the energy of the γ. It follows that

$$m_\rho^2 + m_\mu^2 - 2m_\rho(T_\mu + m_\mu) \geq m_\mu^2 \quad \Rightarrow \quad T_\mu \leq \frac{m_\rho}{2} - m_\mu \simeq 279 \text{ MeV}. \tag{3.477}$$

Alternatively, one may realize that the maximal energy of the muons is given when the energy of the photon goes to zero. In that case, since the muons have identical masses, the *total* energy of the ρ-meson will be evenly divided to the *total* energy of the muons. The kinetic energy of one of the muons is then given by $T_\mu = E_\mu - m_\mu$, where $E_\mu = m_\rho/2$ is the total energy of one of the muons. Thus, the maximal kinetic energy that one of the muons can have in this decay in the rest frame of the ρ-meson is $T_\mu = m_\rho/2 - m_\mu$.

1.101

Giving all quantities in the rest frame of the ^{76}Ge, we have that

$$p_{\text{Ge}} = (m_{\text{Ge}}, \mathbf{0}), \tag{3.478}$$

$$p_{\text{Se}} = (E_{\text{Se}}, \mathbf{p}_{\text{Se}}), \tag{3.479}$$

$$p_{e1} = (E_1, \mathbf{p}_1), \tag{3.480}$$

$$p_{e2} = (E_2, \mathbf{p}_2). \tag{3.481}$$

The conservation of 4-momentum states that

$$p_{\text{Ge}} = p_{\text{Se}} + p_{e1} + p_{e2}. \tag{3.482}$$

In particular, the time component of this relation states that

$$m_{\text{Ge}} - E_{\text{Se}} = E_1 + E_2 = E = T + 2m_e, \tag{3.483}$$

where E is the total energy of the two electrons and T is their total kinetic energy, the quantity in which we are interested. Since the masses are invariant, we deduce that

$$T = m_{\text{Ge}} - E_{\text{Se}} - 2m_e \tag{3.484}$$

is a function of E_{Se} only. In order to maximize T, $E_{\text{Se}} = \sqrt{m_{\text{Se}}^2 + \mathbf{p}_{\text{Se}}^2}$ must take its minimum allowed value, i.e., $E_{\text{Se}} = m_{\text{Se}}$. It is necessary to check that this is kinematically allowed, which is the case since the momentum conservation is solved by $\boldsymbol{p}_1 = -\boldsymbol{p}_2$, which leaves the energy of the electrons as a free parameter, which may be adjusted to solve the energy conservation. It follows that

$$T \leq m_{\text{Ge}} - m_{\text{Se}} - 2m_e \equiv T_{\text{max}}. \tag{3.485}$$

The minimal energy of the electrons is instead obtained when the ^{76}Se obtains its maximal energy. Conservation of 4-momentum gives

$$(p_{\text{Ge}} - p_{\text{Se}})^2 = (p_{e1} + p_{e2})^2 \geq 4m_e^2. \tag{3.486}$$

The left-hand side of this equation evaluates to

$$(p_{\text{Ge}} - p_{\text{Se}})^2 = m_{\text{Ge}}^2 + m_{\text{Se}}^2 - 2m_{\text{Ge}}E_{\text{Se}}. \tag{3.487}$$

Solving for E_{Se}, we obtain

$$E_{\text{Se}} \leq \frac{m_{\text{Ge}}^2 + m_{\text{Se}}^2 - 4m_e^2}{2m_{\text{Ge}}}, \tag{3.488}$$

and thus, we have

$$T \geq \frac{(m_{\text{Ge}} - 2m_e)^2 - m_{\text{Se}}^2}{2m_{\text{Ge}}} \equiv T_{\text{min}}. \tag{3.489}$$

It is of interest to note that

$$\frac{T_{\text{max}} - T_{\text{min}}}{T_{\text{max}}} = \frac{m_{\text{Ge}} - m_{\text{Se}} + 2m_e}{2m_{\text{Ge}}}, \tag{3.490}$$

which is typically a very small number since the the difference between the masses of the nuclei is relatively small. Thus, the electron spectrum for the neutrinoless double beta decay is very peaked. This is in sharp contrast to the case of the more common double beta decay ($X \longrightarrow Y + 2e^- + 2\bar{\nu}_e$), where the electron spectrum is continuous and broad due to the possibility of the neutrinos taking some of the energy.

1.102

According to the conservation of 4-momentum, the 4-momentum of the new particle ϕ must be given by

$$p_\phi = p_1 + p_2. \tag{3.491}$$

In general, the magnitude of a particle's 4-momentum is its mass, and therefore, we find that

$$m_\phi^2 = p_\phi^2 = (p_1 + p_2)^2 = 2p_1 \cdot p_2, \tag{3.492}$$

since $p_1^2 = p_2^2 = 0$. Computing the remaining inner product in the lab frame, we find that

$$m_\phi^2 = 2\omega_1\omega_2(1 - \cos\theta). \tag{3.493}$$

1.103

The total 4-momentum is:

1. In the laboratory (lab) system: $p_{\text{lab}} = (E + m_p, p, 0, 0)$,
2. In the center-of-mass (CM) system: $p_{\text{CM}} = (E_*, 0, 0, 0)$,

where E and p are the energy and the momentum of the incoming proton, respectively, E_* is the energy in the CM system, and m_p is the rest energy (mass) for a proton (or antiproton).

The 4-momentum squared is an invariant, and thus, the same in the two systems. Therefore, we have

$$p_{\text{lab}}^2 = p_{\text{CM}}^2, \tag{3.494}$$

which gives

$$(E + m_p)^2 - p^2 = E_*{}^2. \tag{3.495}$$

Inserting $E^2 = m_p^2 + p^2$ and $T = E - m_p$, i.e., the kinetic energy of the incoming proton, we obtain

$$E_* = \sqrt{2m_p(T + 2m_p)}. \tag{3.496}$$

A necessary condition for production of a proton-antiproton pair is $E_* \geq 4m_p$, i.e.,

$$T \geq 6m_p \simeq 6 \cdot 938\ \text{MeV} \approx 5\,628\ \text{MeV}. \tag{3.497}$$

The kinetic energy 8 000 MeV is therefore sufficient.

1.104

The same method as in Problem 1.103 gives that the kinetic energy of the negative pion must satisfy the inequality

$$T_\pi > \frac{1}{2m_p}\left(3m_\pi^2 + 4m_\pi m_n + m_n^2 - m_p^2\right) - m_\pi \approx 174\ \text{MeV}. \tag{3.498}$$

1.105

a) Before the reaction, we have

$$P_{\text{in}}^2 = (p_p + p_d)^2 = (m_p + m_d)^2 + 2T_p m_d \simeq 9m^2 + 4mT_p, \tag{3.499}$$

where m is the nucleon mass. On the other hand, after the reaction, we have

$$P_{\text{out}}^2 = (p_p + p_{p'} + p_n + p_\eta)^2 \geq (m_p + m_p + m_n + m_\eta)^2 \simeq (3m + m_\eta)^2$$
$$= 9m^2 + 6mm_\eta + m_\eta^2. \tag{3.500}$$

Using $P_{\text{in}}^2 = P_{\text{out}}^2$, this gives

$$T_p \simeq \frac{m_\eta(6m + m_\eta)}{4m} \geq \frac{13}{8} m_\eta \approx 900\text{ MeV} > 700\text{ MeV}. \tag{3.501}$$

The reaction is therefore not possible at this kinetic energy.

b) Let the 4-momentum of the η be $p_\eta = (m_\eta + T, \boldsymbol{p}_\eta)$, where T is the kinetic energy and $\boldsymbol{p}_\eta$ is the 3-momentum in the rest frame of the nucleons. Squaring the relation from conservation of 4-momentum now results in

$$9m^2 + 4mT_p = [(3m, 0) + p_\eta]^2 = 9m^2 + 6m(m_\eta + T) + m_\eta^2. \tag{3.502}$$

Solving for T results in

$$T = \frac{2T_p}{3} - m_\eta - \frac{m_\eta^2}{6m} \simeq 300\text{ MeV}. \tag{3.503}$$

1.106

In general, we have the relation

$$\left(\sum_i p_i\right)^2 \geq \left(\sum_i m_i\right)^2, \tag{3.504}$$

for an arbitrary sum of particle 4-momenta p_i. For our two reactions, the conservation of 4-momentum yields

$$p_\nu + p_X = p_\mu + p_Y[+p_\pi], \tag{3.505}$$

where the term in brackets only appears in the 1π reaction. Squaring this relation gives us

$$p_\nu^2 + 2p_\nu \cdot p_X + p_X^2 = 2p_\nu \cdot p_X + m_X^2 = (p_\mu + p_Y[+p_\pi])^2 \geq (m_\mu + m_Y[+m_\pi])^2. \tag{3.506}$$

In the rest frame of X, the product $p_\nu \cdot p_X$ evaluates to

$$p_\nu \cdot p_X = E_\nu m_X, \tag{3.507}$$

and we therefore obtain

$$E_\nu \geq \frac{1}{2m_X}\left\{(m_\mu + m_Y[+m_\pi])^2 - m_X^2\right\}, \tag{3.508}$$

where the lower limit is the threshold energy for the reaction to be possible. Dividing the two threshold energies, we obtain the ratio

$$\frac{E_{\nu,\text{th},1\pi}}{E_{\nu,\text{th},\text{QE}}} = \frac{(m_\mu + m_Y + m_\pi)^2 - m_X^2}{(m_\mu + m_Y)^2 - m_X^2}. \tag{3.509}$$

1.107

We denote the incoming momenta p_1 and p_2, respectively, while we call the outgoing momenta k_i, $i = 1,2,3,4$. Since all of the particles have the same mass, we have the relation $p_i^2 = k_i^2 = m^2$. By conservation of 4-momentum, we have

$$p_1 + p_2 = k_1 + k_2 + k_3 + k_4. \tag{3.510}$$

Squaring this relation and using the inequality $A \cdot B \geq \sqrt{A^2 B^2}$ for any timelike vectors A and B, we find that

$$2m^2 + 2p_1 \cdot p_2 \geq (4m)^2, \tag{3.511}$$

where the equality holds only if all of the outgoing particles are at relative rest, i.e., at the reaction threshold. At threshold, we therefore have

$$p_1 \cdot p_2 = 7m^2. \tag{3.512}$$

In the rest frame of one of the initial electrons (say the one with 4-momentum p_1), we can write down the 4-momenta as

$$p_1 = m(1,\mathbf{0}), \quad p_2 = (E,\mathbf{p}), \tag{3.513}$$

where E is the total energy of the other electron. It follows that

$$mE = 7m^2 \quad \Rightarrow \quad E = 7m. \tag{3.514}$$

On the other hand, in the center-of-momentum frame, the total energy squared is the square of the total 4-momentum. Thus, the total center-of-mass (CM) energy is given by

$$E_{\text{CM}} = \sqrt{(4m)^2} = 4m, \tag{3.515}$$

at threshold. The ratio E/E_{CM} is therefore 7/4. Note that the total energy in the frame where one of the electrons is at rest is $E + m = 8m$.

1.108

In general, conservation of 4-momentum gives

$$p_1 + p_2 = p_1' + p_2' + p_K', \tag{3.516}$$

where p_i are the 4-momenta of the incoming protons, p_i' are the 4-momenta of the outgoing protons, and p_K' is the 4-momentum of the kaon. By squaring this relation, we obtain

$$p_1^2 + p_2^2 + 2p_1 \cdot p_2 = 2m_p^2 + 2p_1 \cdot p_2 = (p_1' + p_2' + p_K')^2 \geq (2m_p + m_K)^2, \tag{3.517}$$

where the equality holds at the threshold energy. In the first case where one of the protons is at rest, we find that

$$p_1 = (m_p, \mathbf{0}), \quad p_2 = (T + m_p, \mathbf{p}), \tag{3.518}$$

where T is the kinetic energy of the moving proton and $\mathbf{p}$ is its momentum. This results in

$$p_1 \cdot p_2 = m_p^2 + Tm_p. \tag{3.519}$$

Inserting this into the relation above, we find that

$$T_{\text{th}} = \frac{m_K^2}{2m_p} + 2m_K. \tag{3.520}$$

On the other hand, in the second case where both protons are moving with the same velocity, we have

$$p_1 = (T + m_p, \mathbf{p}), \quad p_2 = (T + m_p, -\mathbf{p}), \tag{3.521}$$

where T is the kinetic energy of *one* proton. As a result, we find

$$p_1 + p_2 = 2(T + m_p, \mathbf{0}) \implies (p_1 + p_2)^2 = (2T + 2m_p)^2 \geq (2m_p + m_K)^2. \tag{3.522}$$

We conclude that

$$2T \geq m_K, \tag{3.523}$$

and, thus, the total threshold kinetic energy is given by

$$T_{\text{th, tot}} = m_K. \tag{3.524}$$

1.109

By conservation of 4-momentum, we have the relation

$$p_\chi + p_p = k_p + k_{\chi^*}, \tag{3.525}$$

where the p_i are the incoming and the k_i the outgoing 4-momenta. By squaring this relation, we find that

$$\begin{aligned} p_\chi^2 + 2p_\chi \cdot p_p + p_p^2 &= m_\chi^2 + m_p^2 + 2(m_\chi + T)m_p \\ &= (k_p + k_{\chi^*})^2 \geq (m_p + m_\chi + \delta)^2. \end{aligned} \tag{3.526}$$

Rearranging this inequality leads to

$$T \geq \left(1 + \frac{m_\chi}{m_p}\right)\delta + \frac{\delta^2}{2m_p}. \tag{3.527}$$

In the limit $\delta \ll m_\chi$, we can neglect the δ^2 term and obtain

$$T \geq \left(1 + \frac{m_\chi}{m_p}\right)\delta. \tag{3.528}$$

As long as $\delta/T \ll m_\chi/m_p$, the threshold energy T will be in the classical limit and given by $T = m_\chi v^2/2$. This leads to

$$v \geq \sqrt{2\frac{m_p + m_\chi}{m_p m_\chi}\delta} = \sqrt{\frac{2\delta}{\mu}}, \tag{3.529}$$

where $\mu \equiv m_p m_\chi/(m_p + m_\chi)$ is the reduced mass of the proton-χ system.

1.110

Since, for a particle of mass m, $p^\mu = (E, \boldsymbol{p})^\mu = mV^\mu = m\gamma(1, \boldsymbol{v})^\mu$, it follows that

$$\boldsymbol{v} = \frac{\boldsymbol{p}}{E}, \quad v = \frac{|\boldsymbol{p}|}{E}. \tag{3.530}$$

It also holds that

$$m^2 = p^2 = E^2 - \boldsymbol{p}^2 \quad \Longrightarrow \quad p = E\sqrt{1 - \frac{m^2}{E^2}}. \tag{3.531}$$

Thus, the time $t(m)$ for neutrinos of mass m and energy E to reach the Earth is given by

$$t(m) = \frac{L}{v} = \frac{L}{\sqrt{1 - \frac{m^2}{E^2}}}, \tag{3.532}$$

and the difference compared to if they were massless by

$$\Delta t \equiv t(m) - t(0) = L\left(\frac{1}{\sqrt{1 - \frac{m^2}{E^2}}} - 1\right) \simeq \frac{m^2 L}{2E^2}, \tag{3.533}$$

where the last approximation holds when $m \ll E$.

1.111

a) From conservation of energy, it follows that the total energy of the elementary particle after acceleration is given by

$$E = E_{\text{in}} + E_{\text{acc}}, \tag{3.534}$$

where E_{in} is the initial total energy, and E_{acc} is the energy added by acceleration. Since the particle was initially at rest, we have $E_{\text{in}} = mc^2$, where m is the mass of the particle. From the relation

$$E = mc^2 + T, \tag{3.535}$$

where T is the kinetic energy, it immediately follows that $T = E_{\text{acc}}$. The energy added by the accelerator is given by

$$E_{\text{acc}} = Q\Delta U, \tag{3.536}$$

where $Q = e$ is the charge of the particle and ΔU is the difference in the electric potential at the start and end of the acceleration. Since the electric field is known, we can easily compute ΔU as

$$\Delta U = E'L = 10^4 \text{ V/m} \cdot 100 \text{ m} = 10^6 \text{ V}, \tag{3.537}$$

where E' is the electric field strength. Thus, the kinetic energy after the acceleration is

$$T = e\Delta U = 10^6 \text{ eV} = 1 \text{ MeV}. \tag{3.538}$$

b) The time for the particle to travel the distance dx in the tube is given by

$$dt = \frac{dt}{dx}dx = \frac{1}{v}dx. \tag{3.539}$$

Thus, the total time for the particle to pass through the tube is

$$t = \int_0^L \frac{1}{v}\, dx, \tag{3.540}$$

where L is the detector length. However, the length traveled in the detector is related to the total energy as

$$E = eE'x + mc^2, \tag{3.541}$$

cf., problem a). Thus, we change the variable of integration to the total energy E and obtain

$$t = \int_{mc^2}^{mc^2+E_{\text{acc}}} \frac{dE}{veE'}. \tag{3.542}$$

The velocity v can be expressed as a function of the total energy through the relation

$$v = \frac{pc^2}{E} = c\sqrt{1 - \frac{m^2c^4}{E^2}}, \tag{3.543}$$

which yields

$$t = \frac{1}{ceE'} \int_{mc^2}^{mc^2+E_{\text{acc}}} \frac{E\,dE}{\sqrt{E^2 - m^2c^4}}, \tag{3.544}$$

when inserted into the integral. A new change of variables to $\alpha = E^2$ results in the formula

$$\begin{aligned} t &= \frac{1}{ceE'} \int_{m^2c^4}^{(mc^2+E_{\text{acc}})^2} \frac{d\alpha}{2\sqrt{\alpha - m^2c^4}} = \frac{1}{ceE'} \left[\sqrt{\alpha - m^2c^4}\right]_{m^2c^4}^{(mc^2+E_{\text{acc}})^2} \\ &= \frac{1}{ceE'} \sqrt{E_{\text{acc}}(2mc^2 + E_{\text{acc}})} = \frac{1}{ceE'} \sqrt{eE'L(2mc^2 + eE'L)}. \end{aligned} \tag{3.545}$$

1.112

Let $A^\mu = A^\mu(x)$ be the solution to $\Box A^\mu(x) = 0$. Calculate $\Box' A'^\mu(x')$, where $x' = \Lambda x$ and $A'^\mu(x') = \Lambda^\mu{}_\nu A^\nu(x)$. An explicit calculation yields

$$\begin{aligned} \Box' A'^\mu(x') &= \partial'_\nu \partial'^\nu A'^\mu = \Lambda_\nu{}^\alpha \partial_\alpha \Lambda^\nu{}_\beta \partial^\beta \Lambda^\mu{}_\gamma A^\gamma(x) = \Lambda_\nu{}^\alpha \Lambda^\nu{}_\beta \Lambda^\mu{}_\gamma \partial_\alpha \partial^\beta A^\gamma \\ &= (\Lambda^T)^\alpha{}_\nu \Lambda^\nu{}_\beta \Lambda^\mu{}_\gamma \partial_\alpha \partial^\beta A^\gamma = (\Lambda^T \Lambda)^\alpha{}_\beta \Lambda^\mu{}_\gamma \partial_\alpha \partial^\beta A^\gamma = \eta^\alpha{}_\beta \Lambda^\mu{}_\gamma \partial_\alpha \partial^\beta A^\gamma \\ &= \Lambda^\mu{}_\gamma \partial_\alpha \partial^\alpha A^\gamma = \Lambda^\mu{}_\gamma \Box A^\gamma = \{\Box A^\mu = 0\} = \Lambda^\mu{}_\gamma \cdot 0 = 0, \end{aligned} \tag{3.546}$$

i.e., $\Box' A'^\mu(x') = 0$. Thus, the equation $\Box A^\mu(x) = 0$ is invariant under Lorentz transformations.

1.113

Inserting the definition $A'_\mu = A_\mu + \partial_\mu \psi$ into $F'_{\mu\nu} = \partial_\mu A'_\nu - \partial_\nu A'_\mu$, we obtain

$$F'_{\mu\nu} = \partial_\mu A_\nu + \partial_\mu \partial_\nu \psi - \partial_\nu A_\mu - \partial_\nu \partial_\mu \psi = F_{\mu\nu} + \underbrace{(\partial_\mu \partial_\nu - \partial_\nu \partial_\mu)\psi}_{=0} = F_{\mu\nu}. \tag{3.547}$$

Thus, the gauge transformation results in the same field tensor $F_{\mu\nu}$.

1.114

The standard configuration Lorentz transformation is given by

$$x'^0 = x^0 \cosh\theta - x^1 \sinh\theta, \tag{3.548}$$

$$x'^1 = -x^0 \sinh\theta + x^1 \cosh\theta, \tag{3.549}$$

$$x'^2 = x^2, \tag{3.550}$$

$$x'^3 = x^3, \tag{3.551}$$

where $\tanh\theta = v/c$. This means that the Lorentz transformation in matrix from is

$$\Lambda = \begin{pmatrix} \cosh\theta & -\sinh\theta & 0 & 0 \\ -\sinh\theta & \cosh\theta & 0 & 0 \\ 0 & 0 & 1 & 0 \\ 0 & 0 & 0 & 1 \end{pmatrix}, \tag{3.552}$$

such that $x' = \Lambda x$.

a) The observer in K' must measure the stick simultaneously in both endpoints. This means at time $x'^0 = 0$ in his/her coordinate system. Without loss of generality, we can put one of the endpoints of the stick at the origin in K'. Thus, we have $(0,0,0)$ in K' at time $x'^0 = 0$ and $(\ell,0,0)$ in K' at time $x'^0 = 0$. Therefore, it holds that $\Delta x'^0 = 0 = \Delta x^0 \cosh\theta - \ell \sinh\theta$, which leads to

$$\Delta x^0 = \frac{\ell \sinh\theta}{\cosh\theta} = \ell \tanh\theta. \tag{3.553}$$

However, $\ell' = -\Delta x^0 \sinh\theta + \ell \cosh\theta$, so we have

$$\begin{aligned} \ell' &= -\frac{\ell \sinh\theta}{\cosh\theta} \sinh\theta + \ell \cosh\theta = \ell \frac{\cosh^2\theta - \sinh^2\theta}{\cosh\theta} = \{\cosh^2\theta - \sinh^2\theta = 1\} \\ &= \frac{\ell}{\cosh\theta}. \end{aligned} \tag{3.554}$$

The relation $\cosh\theta = 1/\sqrt{1 - v^2/c^2}$ implies that $\ell' = \ell\sqrt{1 - v^2/c^2}$. Thus, the result is

$$\Delta \mathbf{x}' = (\ell', 0, 0) = \left(\ell\sqrt{1 - v^2/c^2}, 0, 0\right). \tag{3.555}$$

b) The electric and magnetic fields $\mathbf{E} = (0,0,E)$ and $\mathbf{B} = (0,0,0)$ in K implies that the electromagnetic field strength tensor in K is

$$F = \begin{pmatrix} 0 & 0 & 0 & -E \\ 0 & 0 & 0 & 0 \\ 0 & 0 & 0 & 0 \\ E & 0 & 0 & 0 \end{pmatrix}. \tag{3.556}$$

The electromagnetic field strength tensor in K' is given by $F' = \Lambda F \Lambda^T$. Thus, we find

$$F' = \Lambda F \Lambda^T = \begin{pmatrix} \cosh\theta & -\sinh\theta & 0 & 0 \\ -\sinh\theta & \cosh\theta & 0 & 0 \\ 0 & 0 & 1 & 0 \\ 0 & 0 & 0 & 1 \end{pmatrix} \begin{pmatrix} 0 & 0 & 0 & -E \\ 0 & 0 & 0 & 0 \\ 0 & 0 & 0 & 0 \\ E & 0 & 0 & 0 \end{pmatrix}$$

$$\times \begin{pmatrix} \cosh\theta & -\sinh\theta & 0 & 0 \\ -\sinh\theta & \cosh\theta & 0 & 0 \\ 0 & 0 & 1 & 0 \\ 0 & 0 & 0 & 1 \end{pmatrix}$$

$$= \begin{pmatrix} 0 & 0 & 0 & -E\cosh\theta \\ 0 & 0 & 0 & E\sinh\theta \\ 0 & 0 & 0 & 0 \\ E\cosh\theta & -E\sinh\theta & 0 & 0 \end{pmatrix}$$

$$= \begin{pmatrix} 0 & -E'^1 & -E'^2 & -E'^3 \\ E'^1 & 0 & -cB'^3 & cB'^2 \\ E'^2 & cB'^3 & 0 & -cB'^1 \\ E'^3 & -cB'^2 & cB'^1 & 0. \end{pmatrix}. \tag{3.557}$$

Therefore, $E'^1 = E'^2 = 0$, $E'^3 = E\cosh\theta$, $B'^1 = B'^3 = 0$, and $B'^2 = \frac{E}{c}\sinh\theta$. Using the relations $\cosh\theta = \gamma(v)$ and $\sinh\theta = \frac{v}{c}\gamma(v)$, where $\gamma(v) = \frac{1}{\sqrt{1-(\frac{v}{c})^2}}$, we obtain the electric and magnetic fields in K' as

$$\mathbf{E}' = \left(E'^1, E'^2, E'^3\right) = (0, 0, E\gamma(v)), \tag{3.558}$$

$$\mathbf{B}' = \left(B'^1, B'^2, B'^3\right) = \left(0, E\frac{v}{c^2}\gamma(v), 0\right). \tag{3.559}$$

1.115

For small velocities, i.e., $\beta = v/c \ll 1$, we have

$$\gamma \simeq 1 + \frac{1}{2}\frac{v^2}{c^2} \simeq 1, \quad \beta\gamma \simeq \frac{v}{c}\left(1 + \frac{1}{2}\frac{v^2}{c^2}\right) \simeq \frac{v}{c}. \tag{3.560}$$

Consider the electric and magnetic fields in K, i.e., $\mathbf{E} = (E^1, E^2, E^3)$ and $\mathbf{B} = \mathbf{0}$. Thus, the corresponding electromagnetic field strength tensor in K' is given by $F' = \Lambda F \Lambda^T$, where

$$F = (F^{\mu\nu}) = \begin{pmatrix} 0 & -E^1 & -E^2 & -E^3 \\ E^1 & 0 & 0 & 0 \\ E^2 & 0 & 0 & 0 \\ E^3 & 0 & 0 & 0 \end{pmatrix}, \quad \Lambda = (\Lambda^\mu{}_\nu) \simeq \begin{pmatrix} 1 & -\beta & 0 & 0 \\ -\beta & 1 & 0 & 0 \\ 0 & 0 & 1 & 0 \\ 0 & 0 & 0 & 1 \end{pmatrix}, \tag{3.561}$$

are the electromagnetic field strength tensor in K and the Lorentz transformation in the x^1-direction with velocity v, respectively. Therefore, we find that

$$F' \simeq \begin{pmatrix} 0 & -E^1 & -E^2 & -E^3 \\ E^1 & 0 & \beta E^2 & \beta E^3 \\ E^2 & -\beta E^2 & 0 & 0 \\ E^3 & -\beta E^3 & 0 & 0 \end{pmatrix} = \begin{pmatrix} 0 & -E'^1 & -E'^2 & -E'^3 \\ E'^1 & 0 & -cB'^3 & cB'^2 \\ E'^2 & cB'^3 & 0 & -cB'^1 \\ E'^3 & -cB'^2 & cB'^1 & 0 \end{pmatrix}, \tag{3.562}$$

to linear order in β. Thus, the magnetic field in K' is

$$\mathbf{B}' = (B'^1, B'^2, B'^3) = \frac{v}{c^2}(0, E^3, -E^2). \tag{3.563}$$

In addition, the electric field in K' is $\mathbf{E}' = (E'^1, E'^2, E'^3) = (E^1, E^2, E^3) = \mathbf{E}$. Clearly, $\mathbf{B}'$ is perpendicular to both the x^1-axis (i.e., the direction of the velocity $\mathbf{v} = -v\mathbf{e}_1$ of K in K') and $\mathbf{E}'$, since it holds that

$$\mathbf{B}' \cdot \mathbf{e}_1 = \frac{v}{c^2}(0, E^3, -E^2) \cdot (1, 0, 0) = 0, \tag{3.564}$$

$$\mathbf{B}' \cdot \mathbf{E} = \frac{v}{c^2}(0, E^3, -E^2) \cdot (E^1, E^2, E^3) = 0. \tag{3.565}$$

Thus, one finds that the magnetic field in K' for small velocities is

$$\mathbf{B}' \simeq \frac{v}{c^2}(0, E^3, -E^2) = -\frac{1}{c^2}(\mathbf{v} \times \mathbf{E}). \tag{3.566}$$

It holds that $\mathbf{B}' \cdot \mathbf{E} \propto (\mathbf{v} \times \mathbf{E}) \cdot \mathbf{E} = 0$ and $\mathbf{B}' \cdot \mathbf{v} \propto (\mathbf{v} \times \mathbf{E}) \cdot \mathbf{v} = 0$.

1.116

Consider the electric and magnetic fields in a coordinate system of K, i.e., $\mathbf{E} = (0, 1, 0)$ and $\mathbf{B} = \mathbf{0}$, which means that the electromagnetic field components are

$$E^1 = 0, \quad E^2 = 1, \quad E^3 = 0, \quad B^1 = 0, \quad B^2 = 0, \quad B^3 = 0. \tag{3.567}$$

Inserting the electromagnetic field components in the coordinate system of K into the Lorentz transformation corresponding to a velocity boost v_x in the x^1-direction, namely

$$E'^1 = E^1, \tag{3.568}$$

$$E'^2 = \gamma(v_x)\left(E^2 - v_x B^3\right), \tag{3.569}$$

$$E'^3 = \gamma(v_x)\left(E^3 + v_x B^2\right), \tag{3.570}$$

$$B'^1 = B^1, \tag{3.571}$$

$$B'^2 = \gamma(v_x)\left(B^2 + \frac{v_x}{c^2}E^3\right), \tag{3.572}$$

$$B'^3 = \gamma(v_x)\left(B^3 - \frac{v_x}{c^2}E^2\right), \tag{3.573}$$

the electromagnetic field components in the coordinate system of K' are

$$E'^1 = 0, \quad E'^2 = \gamma(v_x), \quad E'^3 = 0, \quad B'^1 = 0, \quad B'^2 = 0, \quad B'^3 = -\frac{v_x}{c^2}\gamma(v_x), \tag{3.574}$$

where $\gamma(v) \equiv \frac{1}{\sqrt{1-v^2/c^2}}$. Similarly, with a velocity boost v_y in the x'^2-direction, the electromagnetic field components in the coordinate system of K'' are given by

$$E''^1 = -\frac{v_x v_y}{c^2}\gamma(v_x)\gamma(v_y), \tag{3.575}$$

$$E''^2 = \gamma(v_x), \tag{3.576}$$

$$E''^3 = 0, \tag{3.577}$$

$$B''^1 = 0, \tag{3.578}$$

$$B''^2 = 0, \tag{3.579}$$

$$B''^3 = -\frac{v_x}{c^2}\gamma(v_x)\gamma(v_y). \tag{3.580}$$

1.117

The electric field $\boldsymbol{E}$ due to a point charge q at the origin is known to be

$$\boldsymbol{E}(x) = \frac{q\boldsymbol{x}}{4\pi\epsilon_0 r^3}. \tag{3.581}$$

After giving an observer a boost to velocity v along the positive x^1-axis, the field is transformed to

$$\boldsymbol{E}'(x') = \frac{q}{4\pi\epsilon_0 r^3}\left(x^1, x^2\cosh\theta, x^3\cosh\theta\right), \tag{3.582}$$

$$c\boldsymbol{B}'(x') = \frac{q}{4\pi\epsilon_0 r^3}\left(0, x^3\sinh\theta, -x^2\sinh\theta\right), \tag{3.583}$$

where $\cosh\theta = \gamma(v) \equiv 1/\sqrt{1-v^2/c^2}$, $\sinh\theta = v\gamma(v)/c$, and $r = \sqrt{(x^1)^2+(x^2)^2+(x^3)^2}$.

Note that in order to compute the electromagnetic field strengths at the point x', we have to write the x-coordinates on the right-hand side of the equations in terms of the new x'-coordinates. After doing so, we obtain

$$\boldsymbol{E}'(x') = \frac{q}{4\pi\epsilon_0 r^3}\left(x'^1 + vt', x'^2, x'^3\right)\cosh\theta, \tag{3.584}$$

$$c\boldsymbol{B}'(x') = \frac{q}{4\pi\epsilon_0 r^3}\left(0, x'^3, -x'^2\right)\sinh\theta, \tag{3.585}$$

where $r = \sqrt{\cosh^2\theta\left(x'^1 + vt'\right)^2 + \left(x'^2\right)^2 + \left(x'^3\right)^2}$.

For small velocities, i.e., $v \ll c$, we recover the classical formulas

$$E'^1(x') = \frac{q}{4\pi\epsilon_0}\left(x'^1 + vt'\right)\left[\left(x'^1 + vt'\right)^2 + \left(x'^2\right)^2 + \left(x'^3\right)^2\right]^{-3/2}, \tag{3.586}$$

$$E'^2(x') = \frac{q}{4\pi\epsilon_0}x'^2\left[\left(x'^1 + vt'\right)^2 + \left(x'^2\right)^2 + \left(x'^3\right)^2\right]^{-3/2}, \tag{3.587}$$

$$E'^3(x') = \frac{q}{4\pi\epsilon_0}x'^3\left[\left(x'^1 + vt'\right)^2 + \left(x'^2\right)^2 + \left(x'^3\right)^2\right]^{-3/2}, \tag{3.588}$$

$$B'^1(x') = 0, \tag{3.589}$$

$$B'^2(x') = \frac{q}{4\pi\epsilon_0}x'^3\frac{v}{c^2}\left[\left(x'^1 + vt'\right)^2 + \left(x'^2\right)^2 + \left(x'^3\right)^2\right]^{-3/2}, \tag{3.590}$$

$$B'^3(x') = -\frac{q}{4\pi\epsilon_0}x'^2\frac{v}{c^2}\left[\left(x'^1 + vt'\right)^2 + \left(x'^2\right)^2 + \left(x'^3\right)^2\right]^{-3/2} \tag{3.591}$$

for the electromagnetic field of a point charge moving along the negative x'^1-axis with velocity v, i.e., we have an electric current in the negative x'^1-direction.

1.118

Maxwell's equations describe how sources (charges and currents) give rise to electric and magnetic fields, whereas the Lorentz force law describes how the field strengths determine the trajectory of a moving test particle with rest mass m and electric charge q. Let us parametrize the trajectory of the particle as $x = x(s)$, where s is the proper time parameter. The Lorentz force law is then given by

$$mc^2\ddot{x}^\mu(s) = q\dot{x}_\nu(s)F^{\mu\nu}(x(s)), \tag{3.592}$$

which is covariant under Lorentz transformations, i.e., both sides of the equation transform as 4-vectors, and where $F^{\mu\nu}$ is the electromagnetic field strength tensor. In order to understand the physical meaning of the Lorentz force law, we will first replace proper time derivatives by ordinary time derivatives, using $x^0 = ct$. For the time derivatives of the spatial components of $x = x(s)$, we have

$$\dot{\boldsymbol{x}} = \frac{d\boldsymbol{x}}{ds} = \frac{d\boldsymbol{x}}{dt}\frac{dt}{ds} = \boldsymbol{u}\frac{1}{c}\frac{dx^0}{ds} = \frac{1}{c}\boldsymbol{u}\dot{x}^0. \tag{3.593}$$

Then, from the spatial part of the Lorentz force law and using the definitions of the electromagnetic field strengths, we obtain

$$mc^2\ddot{x}^i = mc^2\frac{d\dot{x}^i}{ds} = mc^2\frac{dt}{ds}\frac{d}{dt}\left(\frac{dx^i}{dt}\frac{dt}{ds}\right)$$
$$= m\frac{dx^0}{ds}\frac{d}{dt}\left(u^i\frac{dx^0}{ds}\right) = \frac{dx^0}{ds}\frac{d}{dt}\left(mu^i\dot{x}^0\right)$$

$$= q\dot{x}_\nu F^{i\nu} = q\left(\dot{x}_0 F^{i0} + \dot{x}_j F^{ij}\right) = q\left(\frac{dx_0}{ds}E^i + \frac{dt}{ds}\left(\boldsymbol{u}\times c\boldsymbol{B}\right)^i\right)$$

$$= q\left(\frac{dx^0}{ds}E^i + \frac{dx^0}{ds}\left(\boldsymbol{u}\times\boldsymbol{B}\right)^i\right) = \frac{dx^0}{ds}q\left(\boldsymbol{E}+\boldsymbol{u}\times\boldsymbol{B}\right)^i. \quad (3.594)$$

Thus, the factors $\frac{dx^0}{ds}$ cancel of both sides of this equation to give

$$\frac{d}{dt}\left(m\boldsymbol{u}\dot{x}^0\right) = q\left(\boldsymbol{E}+\boldsymbol{u}\times\boldsymbol{B}\right). \quad (3.595)$$

Now, what is $m\boldsymbol{u}\dot{x}^0$? The relativistic 4-momentum is defined as $p^\mu = mc\dot{x}^\mu$, so we have $\boldsymbol{p} = mc\dot{\boldsymbol{x}} = m\boldsymbol{u}\dot{x}^0$ (note that $\boldsymbol{p} \neq m\boldsymbol{u}$), using $\dot{\boldsymbol{x}} = \frac{1}{c}\boldsymbol{u}\dot{x}^0$. Thus, the Lorentz force law for the spatial components is given by

$$\frac{d\boldsymbol{p}}{dt} = q\left(\boldsymbol{E}+\boldsymbol{u}\times\boldsymbol{B}\right), \quad (3.596)$$

where $\boldsymbol{p}$ is the relativistic 3-momentum and $\boldsymbol{E}$ and $\boldsymbol{B}$ are the electric and magnetic fields, respectively. Furthermore, what is $\dot{x}^0$? The proper time parameter s is defined such that $\dot{x}^2 = \left(\dot{x}^0\right)^2 - \dot{\boldsymbol{x}}^2 = 1$. Combining this with $c\dot{\boldsymbol{x}} = \boldsymbol{u}\dot{x}^0$, we deduce

$$u = |\boldsymbol{u}| = \frac{c}{\dot{x}^0}|\dot{\boldsymbol{x}}| = \frac{c|\dot{\boldsymbol{x}}|}{\sqrt{1+|\dot{\boldsymbol{x}}|^2}} \quad \Rightarrow \quad |\dot{\boldsymbol{x}}| = \frac{u/c}{\sqrt{1-u^2/c^2}}. \quad (3.597)$$

Thus, we have

$$\dot{x}^0 = \sqrt{1+\dot{\boldsymbol{x}}^2} = \frac{1}{\sqrt{1-u^2/c^2}} \equiv \gamma(u). \quad (3.598)$$

Assuming $\boldsymbol{E}$ is a constant electric field and $\boldsymbol{B} = \boldsymbol{0}$ as well as multiplying both sides with $d\boldsymbol{r}$, we find that

$$\frac{d\boldsymbol{p}}{dt}\cdot d\boldsymbol{r} = q\boldsymbol{E}\cdot d\boldsymbol{r} = q\boldsymbol{E}\cdot\boldsymbol{u}\,dt. \quad (3.599)$$

The left-hand side is then

$$\frac{d\boldsymbol{p}}{dt}\cdot d\boldsymbol{r} = \frac{d}{dt}(m\boldsymbol{u}\dot{x}^0)\cdot\boldsymbol{u}dt = d(m\boldsymbol{u}\dot{x}^0)\cdot\boldsymbol{u} = m\boldsymbol{u}\cdot d(\boldsymbol{u}\dot{x}^0). \quad (3.600)$$

Inserting $\dot{x}^0 = 1/\sqrt{1-u^2/c^2}$ and using straightforward differential calculus, gives for the left-hand side

$$\frac{d\boldsymbol{p}}{dt}\cdot d\boldsymbol{r} = m\boldsymbol{u}\cdot d\left(\frac{\boldsymbol{u}}{\sqrt{1-u^2/c^2}}\right) = \frac{m\boldsymbol{u}\cdot d\boldsymbol{u}}{(1-u^2/c^2)^{3/2}} = d\left(\frac{mc^2}{\sqrt{1-u^2/c^2}}\right). \quad (3.601)$$

Integrating both sides from the origin, where $u = 0$, to the displacement r, where the velocity momentarily is u, we find that

$$\int_{u(0)=0}^{u(r)} \frac{d\boldsymbol{p}}{dt} \cdot d\boldsymbol{r} = \int_{r=0}^{r} q\boldsymbol{E} \cdot d\boldsymbol{r} \quad \Rightarrow \quad \frac{mc^2}{\sqrt{1-u^2/c^2}} - mc^2 = qEr$$

$$\Rightarrow \quad \frac{1}{\sqrt{1-u^2/c^2}} = 1 + \frac{qEr}{mc^2}, \tag{3.602}$$

since the electric field $\boldsymbol{E}$ is constant along the trajectory. Finally, introducing $x \equiv 1 + \frac{qEr}{mc^2}$ and solving for u, we obtain

$$u(r) = c\sqrt{1-x^{-2}} = c\sqrt{1-\left(1+\frac{qEr}{mc^2}\right)^{-2}}, \tag{3.603}$$

which is the velocity $u(r)$ of the particle as a function of the displacement r from the origin along the direction of motion.

An alternative solution is to note that the electric field in the momentary rest frame $\mathring{K}$ of the electron is, according to the transformation equations for the field tensor, equal to the electric field in the laboratory system. Using units with $c = 1$, the acceleration of the electron relative to $\mathring{K}$ is therefore $\mathring{a} = eE/m_0$, where e is the electron charge and m_0 is the electron rest mass. We obtain the answer by changing $g \to eE/m_0$ in the formula [see Problem 1.41, Eq. (3.166)]

$$x = \frac{1}{g}\left[\sqrt{1+(gt)^2} - 1\right]. \tag{3.604}$$

The answer is

$$x = \frac{m_0}{eE}\left[\sqrt{1+\left(\frac{eEt}{m_0}\right)^2} - 1\right]. \tag{3.605}$$

We now want to calculate the velocity as a function of the displacement, we multiply the Lorentz force $\boldsymbol{f} = e(\boldsymbol{E} + \boldsymbol{u} \times \boldsymbol{B})$ by $\boldsymbol{u}$ and use that the change in kinetic energy is $dT = \boldsymbol{f} \cdot d\boldsymbol{r}$. We then introduce the electrostatic potential Φ ($\boldsymbol{E} = -\nabla\Phi$):

$$\frac{dT}{dt} = -e\nabla\Phi \cdot \boldsymbol{u} = -e\nabla\Phi \cdot \frac{d\boldsymbol{r}}{dt} = -e\frac{d\Phi}{dt}. \tag{3.606}$$

Integration with respect to t gives

$$T = m_0(\gamma(u) - 1) = -e\Delta\Phi = eEr, \tag{3.607}$$

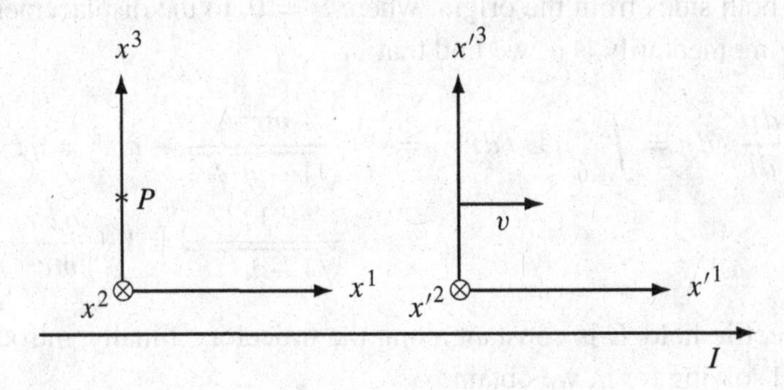

Figure 3.9 A straight uncharged conductor with current I as viewed in the inertial systems K and K'. Note that K' is moving with velocity v relative to K along the x^1-axis.

i.e.,

$$u(r) = \sqrt{1 - \left(1 + \frac{eEr}{m_0}\right)^{-2}}. \tag{3.608}$$

1.119

Introduce K and K' according to Figure 3.9.

a) The components of the electromagnetic field in K at the point $P = (0, 0, r)$ are given by

$$E^i = B^1 = B^3 = 0,$$
$$B^2 = -\frac{\mu_0 I}{2\pi r}. \tag{3.609}$$

According to the Lorentz transformation formulas for the electromagnetic field

$$E'^1 = E^1, \quad E'^2 = \gamma(v)\left(E^2 - vB^3\right), \quad E'^3 = \gamma(v)\left(E^3 + vB^2\right),$$
$$B'^1 = B^1, \quad B'^2 = \gamma(v)\left(B^2 + \frac{v}{c^2}E^3\right), \quad B'^3 = \gamma(v)\left(B^3 - \frac{v}{c^2}E^2\right), \tag{3.610}$$

the corresponding components of the electromagnetic field in K' are therefore

$$E'^1 = E'^2 = 0, \quad E'^3 = -\frac{v\gamma(v)\mu_0 I}{2\pi r},$$
$$B'^1 = B'^3 = 0, \quad B'^2 = -\frac{\gamma(v)\mu_0 I}{2\pi r}. \tag{3.611}$$

b) The current-density 4-vector in the conductor has the components $J^\mu = (0, c\mu_0 I/A, 0, 0)$ in K and $J'^\mu = (\rho'/\epsilon_0, c\mu_0 j'^1, c\mu_0 j'^2, c\mu_0 j'^3)$ in K', where $J'^\mu = \Lambda^\mu{}_\nu J^\nu$, Λ being the Lorentz transformation from K to K' and A the cross-sectional area of the conductor. We then obtain

$$\rho' = -\frac{v\gamma(v)}{c^2}\frac{I}{A} \quad \text{and} \quad j'^1 = \gamma(v)\frac{I}{A}. \tag{3.612}$$

Since the cross-sectional area A' relative to K' is A, the current relative to K' becomes

$$I' = A'j'^1 = \gamma(v)I. \tag{3.613}$$

Now, ρ' and I' generate the components of the electromagnetic field in K' as

$$E'^1 = E'^2 = 0, \quad E'^3 = \frac{\rho' A'}{2\pi r'\epsilon_0} = -\frac{v\gamma(v)I\mu_0}{2\pi r'},$$
$$B'^1 = B'^3 = 0, \quad B'^2 = -\frac{\mu_0 I'}{2\pi r'}, \tag{3.614}$$

which is the same result as we obtained in a).

1.120

Inserting the definition of the electromagnetic field strength tensor, i.e., $F^{\mu\nu} = \partial^\mu A^\nu - \partial^\nu A^\mu$, into Maxwell's equations $\partial_\mu F^{\mu\nu} = J^\nu$, one obtains

$$\begin{aligned}\partial_\mu F^{\mu\nu} &= \partial_\mu(\partial^\mu A^\nu - \partial^\nu A^\mu) = \{\partial_\mu\partial^\mu = \Box \text{ and } \partial_\mu\partial^\nu = \partial^\nu\partial_\mu\} \\ &= \Box A^\nu - \partial^\nu(\partial_\mu A^\mu) = J^\nu.\end{aligned} \tag{3.615}$$

Using the Lorenz gauge condition, i.e., $\partial_\mu A^\mu = 0$, yields

$$\Box A^\nu = J^\nu. \tag{3.616}$$

This is the simple form of Maxwell's equations and is a wave equation for A with source term J. Assuming that $J = 0$, which implies that $J^\nu = 0$, one finds $\Box A^\nu = 0$. A useful formula is given by

$$\partial^\mu A^\nu = \partial^\mu(\varepsilon^\nu e^{ik\cdot x}) = \varepsilon^\nu \partial^\mu e^{ik\cdot x} = \varepsilon^\nu e^{ik\cdot x} ik^\mu = ik^\mu \varepsilon^\nu e^{ik\cdot x} = ik^\mu A^\nu. \tag{3.617}$$

Using Eq. (3.617), one obtains the electric and magnetic field components as

$$E^i = F^{i0} = \partial^i A^0 - \partial^0 A^i = ik^i\varepsilon^0 e^{ik\cdot x} - ik^0\varepsilon^i e^{ik\cdot x} = i(k^i\varepsilon^0 - k^0\varepsilon^i)e^{ik\cdot x}, \tag{3.618}$$

$$B^i = (\nabla \times \mathbf{A})^i = \epsilon^{ijk}\partial^j A^k = i\epsilon^{ijk}k^j A^k. \tag{3.619}$$

Multiplying Eqs. (3.618) and (3.619) with k^i, one finds that

$$\begin{aligned}\boldsymbol{E}\cdot\boldsymbol{k} = E^i k^i &= i(k^i\varepsilon^0 - k^0\varepsilon^i)e^{ik\cdot x}k^i = i(k^i k^i\varepsilon^0 - k^0 k^i\varepsilon^i)e^{ik\cdot x} \\ &= i(\boldsymbol{k}^2\varepsilon^0 - k^0\boldsymbol{k}\cdot\varepsilon)e^{ik\cdot x},\end{aligned} \tag{3.620}$$

$$\begin{aligned}\boldsymbol{B}\cdot\boldsymbol{k} = B^i k^i &= \epsilon^{ijk} ik^j A^k k^i = i\epsilon^{ijk}k^i k^j A^k \\ &= \{\epsilon^{ijk} \text{ is antisymmetric and } k^i k^j \text{ is symmetric}\} = 0.\end{aligned} \tag{3.621}$$

Multiplying Eq. (3.617) with $\eta_{\mu\nu}$, one has

$$\partial_\mu A^\mu = \eta_{\mu\nu}\partial^\mu A^\nu = \eta_{\mu\nu} i k^\mu A^\nu = i k_\mu A^\mu = i k_\mu \varepsilon^\mu e^{ik\cdot x} = i(k_0\varepsilon^0 - \boldsymbol{k}\cdot\boldsymbol{\varepsilon})e^{ik\cdot x}, \tag{3.622}$$

but $\partial_\mu A^\mu = 0$ (Lorenz gauge), so $i(k_0\varepsilon^0 - \boldsymbol{k}\cdot\boldsymbol{\varepsilon})e^{ik\cdot x} = 0$, i.e.,

$$\varepsilon^0 = \frac{\boldsymbol{k}\cdot\boldsymbol{\varepsilon}}{k_0}, \tag{3.623}$$

since $e^{ik\cdot x} \neq 0$. Inserting Eq. (3.623) into Eq. (3.620) yields

$$\boldsymbol{E}\cdot\boldsymbol{k} = i\left(\boldsymbol{k}^2\frac{\boldsymbol{k}\cdot\boldsymbol{\varepsilon}}{k_0} - k^0\boldsymbol{k}\cdot\boldsymbol{\varepsilon}\right)e^{ik\cdot x} = -i\frac{\boldsymbol{k}\cdot\boldsymbol{\varepsilon}}{k_0}\left(k_0k^0 - \boldsymbol{k}^2\right)e^{ik\cdot x} = -i\frac{\boldsymbol{k}\cdot\boldsymbol{\varepsilon}}{k_0}k^2 e^{ik\cdot x}. \tag{3.624}$$

Taking the partial derivative with respect to x^μ of Eq. (3.617), one finds

$$\Box A^\nu = \partial_\mu\partial^\mu A^\nu = \partial_\mu(ik^\mu A^\nu) = ik^\mu\partial_\mu A^\nu = ik^\mu(ik_\mu A^\nu) = -k_\mu k^\mu A^\nu = -k^2 A^\nu, \tag{3.625}$$

but $\Box A^\nu = 0$, so $-k^2 A^\nu = 0$, i.e., $k^2 = 0$, since $A^\nu \neq 0$. Inserting $k^2 = 0$ into Eq. (3.624) yields

$$\boldsymbol{E}\cdot\boldsymbol{k} = -i\frac{\boldsymbol{k}\cdot\boldsymbol{\varepsilon}}{k_0}\cdot 0\cdot e^{ik\cdot x} = 0. \tag{3.626}$$

Thus, one obtains

$$\boldsymbol{E}\cdot\boldsymbol{k} = \boldsymbol{B}\cdot\boldsymbol{k} = 0. \tag{3.627}$$

1.121

Inserting the expression for the free electromagnetic plane wave $A^\mu(x) = \varepsilon^\mu e^{ik\cdot x}$ into the definition of the electromagnetic field strength tensor $F^{\mu\nu} = \partial^\mu A^\nu - \partial^\nu A^\mu$, one obtains

$$F^{\mu\nu} = ik^\mu\varepsilon^\nu e^{ik\cdot x} - ik^\nu\varepsilon^\mu e^{ik\cdot x} = i\left(k^\mu A^\nu - k^\nu A^\mu\right). \tag{3.628}$$

Thus, we find that

$$\begin{aligned} F_{\mu\nu}F^{\mu\nu} &= -\left(k_\mu A_\nu - k_\nu A_\mu\right)\left(k^\mu A^\nu - k^\nu A^\mu\right) \\ &= -k^2A^2 + (k\cdot A)^2 + (k\cdot A)^2 - k^2A^2 \\ &= 2\left[(k\cdot A)^2 - k^2A^2\right]. \end{aligned} \tag{3.629}$$

Now, Maxwell's equations $\partial_\mu F^{\mu\nu} = J^\nu$ expressed in the 4-vector potential A when the 4-current $J = 0$, i.e., the free case, are $\Box A^\nu - \partial^\nu\partial\cdot A = 0$ and in k-space

$$-k^2A^\nu + k^\nu k\cdot A = 0. \tag{3.630}$$

Therefore, multiplying the above equation with A_ν, one has

$$(k\cdot A)^2 - k^2A^2 = 0. \tag{3.631}$$

Thus, one obtains

$$F_{\mu\nu}F^{\mu\nu} = 2 \cdot 0 = 0, \tag{3.632}$$

i.e., the invariant $F_{\mu\nu}F^{\mu\nu}$ is zero. Finally, by writing out the invariant $\epsilon_{\mu\nu\omega\lambda}F^{\mu\nu}F^{\omega\lambda}$ explicitly, one finds that

$$\begin{aligned}\epsilon_{\mu\nu\omega\lambda}F^{\mu\nu}F^{\omega\lambda} &= -\epsilon_{\mu\nu\omega\lambda}(k^\mu\epsilon^\nu - k^\nu\epsilon^\mu)(k^\omega\epsilon^\lambda - k^\lambda\epsilon^\omega)e^{2ik\cdot x}\\ &= -\epsilon_{\mu\nu\omega\lambda}(k^\mu k^\omega\epsilon^\nu\epsilon^\lambda - k^\mu k^\lambda\epsilon^\nu\epsilon^\omega - k^\nu k^\omega\epsilon^\mu\epsilon^\lambda + k^\nu k^\lambda\epsilon^\mu\epsilon^\omega)e^{2ik\cdot x}\\ &= 0,\end{aligned} \tag{3.633}$$

since $\epsilon_{\mu\nu\omega\lambda}$ is totally antisymmetric and each term inside the parenthesis has two pairs of symmetric indices, i.e., the invariant $\epsilon_{\mu\nu\omega\lambda}F^{\mu\nu}F^{\omega\lambda}$ is also zero.

The result shows that for a plane wave solution of Maxwell's equations, the electric and magnetic fields oscillate in such a way that their magnitude is always equal (if multiplying the magnetic field with c^2 to get the correct units, the invariant $F_{\mu\nu}F^{\mu\nu}$ is proportional to $\boldsymbol{E}^2 - c^2\boldsymbol{B}^2$) and that they are always orthogonal ($\epsilon_{\mu\nu\omega\lambda}F^{\mu\nu}F^{\omega\lambda}$ is proportional to $\boldsymbol{E}\cdot\boldsymbol{B}$).

1.122

a) Using the Lorentz invariant $\epsilon_{\mu\nu\lambda\omega}F^{\mu\nu}F^{\lambda\omega} = -8c\boldsymbol{E}\cdot\boldsymbol{B}$, one finds

$$\boldsymbol{E}\cdot\boldsymbol{B} = -\frac{1}{8c}\epsilon_{\mu\nu\lambda\omega}F^{\mu\nu}F^{\lambda\omega}, \tag{3.634}$$

where $\epsilon^{\mu\nu\lambda\omega}$ is a totally antisymmetric fourth rank tensor with $\epsilon^{0123} = 1$. For any vector index μ, the inner product $A^\mu B_\mu$ is Lorentz invariant, and for a Lorentz transformation Λ with $\det\Lambda = 1$, it holds that $\epsilon'^{\mu\nu\lambda\omega} = \epsilon^{\mu\nu\lambda\omega}$. Thus, one has

$$\epsilon'_{\mu\nu\lambda\omega}F'^{\mu\nu}F'^{\lambda\omega} = \epsilon_{\mu\nu\lambda\omega}F'^{\mu\nu}F'^{\lambda\omega} = \epsilon_{\mu\nu\lambda\omega}F^{\mu\nu}F^{\lambda\omega}. \tag{3.635}$$

Finally, using Eq. (3.634), one obtains

$$\boldsymbol{E}'\cdot\boldsymbol{B}' = \boldsymbol{E}\cdot\boldsymbol{B}. \tag{3.636}$$

b) The expression $\epsilon_{\alpha\beta\gamma\delta}F^{\alpha\beta}F^{\gamma\delta}$ is Lorentz invariant and equal to $-8c\boldsymbol{E}\cdot\boldsymbol{B}$. Thus, if $\boldsymbol{E}\cdot\boldsymbol{B} = 0$ for one observer, then it is zero for any observer in inertial frames.

c) From Problem 1.120 we have $E^i = i(k^i\varepsilon^0 - k^0\varepsilon^i)e^{ik\cdot x}$ and $B^i = i\epsilon^{ijk}k^j\varepsilon^k e^{ik\cdot x}$. This gives immediately that $\boldsymbol{E}\cdot\boldsymbol{B} = E^iB^i = 0$, since ϵ^{ijk} is antisymmetric.

d) The components of $\boldsymbol{E}\times\boldsymbol{B}$ can be written as $(\boldsymbol{E}\times\boldsymbol{B})^i = e^{2ik\cdot x}(k^iM + \varepsilon^iN)$ with $M = k^0\varepsilon^2 - \boldsymbol{k}\cdot\boldsymbol{\varepsilon}\varepsilon^0$ and $N = \boldsymbol{k}^2\varepsilon^0 - k^0\boldsymbol{k}\cdot\boldsymbol{\varepsilon}$. Choose $k = (k^0, k^0, 0, 0)$ and $\varepsilon = (0, 0, 1, 0)$ or $\varepsilon = (0, 0, 0, 1)$. Then, it holds that $\boldsymbol{E}\times\boldsymbol{B} = k^0(k^0, 0, 0)e^{2ik\cdot x}$, showing that only the 1-component of $\boldsymbol{E}\times\boldsymbol{B}$ is nonvanishing, and thus, the product is proportional to $\boldsymbol{k}=(k^0,0,0)$.

1.123

Using the Lorentz force law, we have

$$\frac{d\boldsymbol{p}}{dt}=-e(\boldsymbol{u}\times\boldsymbol{B}), \tag{3.637}$$

where e is the electron charge and $\boldsymbol{p}=m_0c\dot{\boldsymbol{x}}=m_0\boldsymbol{u}\gamma$ with $\gamma\equiv 1/\sqrt{1-u^2/c^2}$. From the Lorentz force law, we deduce that

$$\frac{d\boldsymbol{p}}{dt}\cdot\boldsymbol{p}=0, \tag{3.638}$$

which implies that $\boldsymbol{p}^2=m_0^2u^2\gamma^2=\text{const.}$ Since $|\boldsymbol{u}|$ is constant and the magnetic field $\boldsymbol{B}=(0,0,B)$ is symmetric around the z-axis, we make the ansatz

$$\boldsymbol{u}(t)=(u\cos(\alpha t),u\sin(\alpha t),0), \tag{3.639}$$

which is automatically consistent with the initial condition $\boldsymbol{u}(t=0)=(u,0,0)$. Now, the Lorentz force law for the components is

$$\frac{dp_x}{dt}=-eBu_y,\quad \frac{dp_y}{dt}=eBu_x,\quad \frac{dp_z}{dt}=0, \tag{3.640}$$

which after inserting the ansatz yields

$$\frac{dp_x}{dt}=-eBu\sin(\alpha t),\quad \frac{dp_y}{dt}=eBu\cos(\alpha t),\quad \frac{dp_z}{dt}=0. \tag{3.641}$$

Then, integrating these components leads to

$$p_x=\frac{eBu}{\alpha}\cos(\alpha t)+c_1,\quad p_y=\frac{eBu}{\alpha}\sin(\alpha t)+c_2,\quad p_z=c_3, \tag{3.642}$$

where c_1, c_2, and c_3 are integration constants. Consistency with $\boldsymbol{p}=m_0\boldsymbol{u}\gamma$ means that $c_1=c_2=c_3=0$ and $\alpha=eB/(m_0\gamma)$. Thus, integrating the expression for $\boldsymbol{u}$, we obtain the trajectory of the electron as

$$\boldsymbol{x}(t)=\frac{um_0\gamma}{eB}\left(\sin\frac{eBt}{m_0\gamma},-\cos\frac{eBt}{m_0\gamma},0\right)+\boldsymbol{x}_0, \tag{3.643}$$

where $\boldsymbol{x}_0$ is an integration constant. This trajectory is the equation for a circle perpendicular to the z-axis with center at $\boldsymbol{x}_0$ and radius $r=|\boldsymbol{x}-\boldsymbol{x}_0|=um_0\gamma/(eB)$. The time for one revolution is $t_0=2\pi m_0\gamma/(eB)$.

1.124

The two quantities $\mathbf{E}\cdot\mathbf{B}$ and $\mathbf{E}^2-c^2\mathbf{B}^2$ are Lorentz invariants, where $\mathbf{E}\cdot\mathbf{B}=0$ and $\mathbf{E}^2-c^2\mathbf{B}^2=0$, since $\mathbf{E}=(cB,0,0)$ and $\mathbf{B}=(0,B,0)$. Therefore, one has

$\mathbf{E}' \cdot \mathbf{B}' = 0$ and $\mathbf{E}'^2 - c^2\mathbf{B}'^2 = 0$. Inserting $\mathbf{E}' = (0, 2cB, cB)$ and $\mathbf{B}' = (0, B_y', B_z')$, one obtains

$$\begin{cases} B_y'^2 + B_z'^2 = 5B^2 \\ 2B_y' + B_z' = 0 \end{cases}. \tag{3.644}$$

Solving this system of equations, one finds that $B_y' = \pm B$ and $B_z' = \mp 2B$. Thus, the answer is $\mathbf{B}' = \pm(0, B, -2B)$.

1.125

The two quantities $\mathbf{E}^2 - c^2\mathbf{B}^2$ and $\mathbf{E} \cdot \mathbf{B}$ are Lorentz invariants, where $\mathbf{E}^2 - c^2\mathbf{B}^2 = -2\alpha^2$ and $\mathbf{E} \cdot \mathbf{B} = 0$, since $\mathbf{E} = (\alpha, -\alpha, 0)$ and $\mathbf{B} = (0, 0, 2\alpha/c)$, where $\alpha \neq 0$. Therefore, one has $\mathbf{E}'^2 - c^2\mathbf{B}'^2 = -2\alpha^2$ and $\mathbf{E}' \cdot \mathbf{B}' = 0$. Inserting $\mathbf{E}' = (0, 0, 2\alpha)$ and $\mathbf{B}' = (B_x', \alpha/c, B_z')$, one obtains

$$\begin{cases} B_x'^2 + B_z'^2 = \frac{5\alpha^2}{c^2} \\ 2\alpha B_z' = 0 \end{cases}. \tag{3.645}$$

Solving this system of equations, one finds that $B_x' = \pm\frac{\alpha\sqrt{5}}{c}$ and $B_z' = 0$. Thus, it follows that $\mathbf{B}' = (\pm\alpha\sqrt{5}/c, \alpha/c, 0)$.

1.126

The two quantities $\mathbf{E}^2 - c^2\mathbf{B}^2$ and $\mathbf{E} \cdot \mathbf{B}$ are Lorentz invariants, where $\mathbf{E} \cdot \mathbf{B} = 0$ and $\mathbf{E}^2 - c^2\mathbf{B}^2 = -2\beta^2$. Therefore, one has $\mathbf{E}' - c^2\mathbf{B}' = -2\beta^2$ and $\mathbf{E}' \cdot \mathbf{B}' = 0$. Inserting $\mathbf{E}' = (2\beta, 0, 0)$ and $\mathbf{B}' = (B_x', B_y', \beta/c)$, one obtains

$$\begin{cases} c^2(B_x'^2 + B_y'^2) = -5\beta^2 \\ 2\beta B_x' = 0 \end{cases}. \tag{3.646}$$

Solving this system of equations, one finds that $B_x' = 0$ and $B_y' = \pm\sqrt{5}\,\beta/c$. Thus, it follows that $\mathbf{B}' = (0, \pm\sqrt{5}\,\beta/c, \beta/c)$.

1.127

Observer A measures the electric and magnetic fields to be $\mathbf{E} = (\alpha, 0, 0)$ and $\mathbf{B} = (\alpha/c, 0, 2\alpha/c)$, respectively, where $\alpha \neq 0$. The two quantities $\mathbf{E}^2 - c^2\mathbf{B}^2$ and $\mathbf{E} \cdot \mathbf{B}$ are Lorentz invariants, where $\mathbf{E}^2 - c^2\mathbf{B}^2 = -4\alpha^2$ and $\mathbf{E} \cdot \mathbf{B} = \alpha^2/c$. Therefore, one has $\mathbf{E}'^2 - c^2\mathbf{B}'^2 = -4\alpha^2$ and $\mathbf{E}' \cdot \mathbf{B}' = \alpha^2/c$. Inserting $\mathbf{E}' = (E_x', \alpha, 0)$ and $\mathbf{B}' = (\alpha/c, B_y', \alpha/c)$, one obtains

$$\begin{cases} E_x'^2 - c^2 B_y'^2 = -3\alpha^2 \\ E_x' + cB_y' = \alpha \end{cases}. \tag{3.647}$$

Solving this system of equations, one finds that $E'_x = -\alpha$ and $B'_y = 2\alpha/c$. Thus, it holds that $\mathbf{E}' = (-\alpha,\alpha,0)$ and $\mathbf{B}' = (\alpha/c, 2\alpha/c, \alpha/c)$, which are the electric and magnetic fields as measured by observer B.

The electric and magnetic fields $\mathbf{E}' = (-\alpha,\alpha,0)$ and $\mathbf{B}' = (\alpha/c, 2\alpha/c, \alpha/c)$ imply that the electromagnetic field strength tensor as seen by observer B is given by

$$F' = \begin{pmatrix} 0 & \alpha & -\alpha & 0 \\ -\alpha & 0 & -\alpha & 2\alpha \\ \alpha & \alpha & 0 & -\alpha \\ 0 & -2\alpha & \alpha & 0 \end{pmatrix}. \tag{3.648}$$

The electromagnetic field strength tensor as seen by observer C is then given by $F'' = \Lambda F' \Lambda^T$, where

$$\Lambda = \begin{pmatrix} \gamma & -\beta\gamma & 0 & 0 \\ -\beta\gamma & \gamma & 0 & 0 \\ 0 & 0 & 1 & 0 \\ 0 & 0 & 0 & 1 \end{pmatrix}. \tag{3.649}$$

Here $\beta = \beta(v) \equiv \frac{v}{c}$ and $\gamma = \gamma(v) \equiv \frac{1}{\sqrt{1-\beta^2}} = \frac{1}{\sqrt{1-\frac{v^2}{c^2}}}$. Thus, one obtains

$$F'' = \Lambda F' \Lambda^T = \begin{pmatrix} 0 & \alpha & -\alpha(1-\beta)\gamma & -2\alpha\beta\gamma \\ -\alpha & 0 & -\alpha(1-\beta)\gamma & 2\alpha\gamma \\ \alpha(1-\beta)\gamma & \alpha(1-\beta)\gamma & 0 & -\alpha \\ 2\alpha\beta\gamma & -2\alpha\gamma & \alpha & 0 \end{pmatrix}$$
$$= \begin{pmatrix} 0 & -E''^1 & -E''^2 & -E''^3 \\ E''^1 & 0 & -cB''^3 & cB''^2 \\ E''^2 & cB''^3 & 0 & -cB''^1 \\ E''^3 & -cB''^2 & cB''^1 & 0 \end{pmatrix}. \tag{3.650}$$

Therefore, one finds that $E''^1 = -\alpha$, $E''^2 = \alpha(1-\beta)\gamma = \alpha\sqrt{\frac{c-v}{c+v}}$, $E''^3 = 2\alpha\beta\gamma = 2\alpha\frac{v}{\sqrt{c^2-v^2}}$, $B''^1 = \alpha/c$, $B''^2 = 2\alpha\gamma/c = 2\alpha\frac{1}{\sqrt{c^2-v^2}}$, and $B''^3 = \alpha(1-\beta)\gamma/c = \frac{\alpha}{c}\sqrt{\frac{c-v}{c+v}}$. Summarizing, the electric and magnetic fields as measured by observer C are

$$\mathbf{E}'' = \left(E''^1, E''^2, E''^3\right) = \left(-\alpha, \alpha\sqrt{\frac{c-v}{c+v}}, 2\alpha\frac{v}{\sqrt{c^2-v^2}}\right), \tag{3.651}$$

$$\mathbf{B}'' = \left(B''^1, B''^2, B''^3\right) = \left(\frac{\alpha}{c}, 2\alpha\frac{1}{\sqrt{c^2-v^2}}, \frac{\alpha}{c}\sqrt{\frac{c-v}{c+v}}\right). \tag{3.652}$$

1.128

For the negatively charged muon, the Lorentz force equation is

$$\frac{d\boldsymbol{p}}{dt} = -e\boldsymbol{u} \times \boldsymbol{B}, \tag{3.653}$$

where $\boldsymbol{p} = mu\gamma(u)$ is the relativistic 3-momentum, e is the elementary charge (i.e., $-e$, since the muon is negatively charged), $\boldsymbol{u}$ is the velocity of the muon, and $\boldsymbol{B}$ is the magnetic field. By scalar multiplication of the Lorentz force equation with $\boldsymbol{p}$, we obtain $d\boldsymbol{p}^2/dt = 0$, which is equivalent to have u = constant. If we assume that there is no motion in the z-direction, then we find that

$$\boldsymbol{u} = u(\cos(\omega t), \sin(\omega t), 0) \quad \Rightarrow \quad \boldsymbol{x} = \frac{u}{\omega}(\sin(\omega t), -\cos(\omega t), 0) + \boldsymbol{x}_0, \tag{3.654}$$

i.e., the particle trajectory is a circle with radius $R = u/\omega$. By solving the Lorentz force equation, we obtain

$$\omega = \frac{eB}{m\gamma(u)} \quad \Rightarrow \quad R = \frac{mu\gamma(u)}{eB}. \tag{3.655}$$

By geometric considerations, it is easy to find that the deviation is given by

$$\Delta x = R - \sqrt{R^2 - h^2}, \tag{3.656}$$

where h is the altitude of the muon production. If $R \gg h$ (which is the case for our numerical values), then this simplifies to

$$\Delta x \simeq \frac{h^2}{2R}, \tag{3.657}$$

by a simple series expansion. By applying the correct right-hand rule, we find that the deviation is westward. Inserting the numerical values, we obtain $\Delta x \simeq 380$ m.

1.129

By Lorentz transformation of the electromagnetic field tensor, we can deduce that

$$\boldsymbol{E}' = (E^1, E^2\cosh\theta - B^3\sinh\theta, E^3\cosh\theta + B^2\sinh\theta), \tag{3.658}$$

$$\boldsymbol{B}' = (B^1, B^2\cosh\theta + E^3\sinh\theta, B^3\cosh\theta - E^2\sinh\theta). \tag{3.659}$$

Since the magnetic field in S' vanishes, we must have $\boldsymbol{B}' = \boldsymbol{0}$, which implies that

$$B^1 = 0, \quad B^2 = -E^3\tanh\theta, \quad B^3 = E^2\tanh\theta. \tag{3.660}$$

Inserting this into the expression for the $\boldsymbol{E}'$ field, we obtain

$$\boldsymbol{E}' = \left(E^1, \frac{E^2}{\cosh\theta}, \frac{E^3}{\cosh\theta}\right) = \frac{1}{\gamma}(\gamma E^1, E^2, E^3). \tag{3.661}$$

1.130

a) By applying the transformation law $F'^{\mu\nu} = \Lambda^{\mu}{}_{\omega}\Lambda^{\nu}{}_{\lambda}F^{\omega\lambda}$ for an electromagnetic field tensor $F^{\omega\lambda}$ with vanishing electric field $\boldsymbol{E}$, we obtain

$$\boldsymbol{E}' = v\gamma(0, -B^3, B^2) = \boldsymbol{v}\times\boldsymbol{B}\gamma, \tag{3.662}$$

$$\boldsymbol{B}' = (B^1, \gamma B^2, \gamma B^3), \tag{3.663}$$

where $\gamma = 1/\sqrt{1-v^2}$.

b) The original Lorentz invariants are given by

$$\boldsymbol{B}^2 - \boldsymbol{E}^2 = \boldsymbol{B}^2 \quad \text{and} \quad \boldsymbol{E}\cdot\boldsymbol{B} = 0. \tag{3.664}$$

In the new frame, we obtain

$$\boldsymbol{B}'^2 - \boldsymbol{E}'^2 = (B^1)^2 + c^2\left[(B^2)^2 + (B^3)^2\right] - s^2\left[(B^2)^2 + (B^3)^2\right] = \boldsymbol{B}^2, \tag{3.665}$$

as well as

$$\boldsymbol{E}'\cdot\boldsymbol{B}' = cs(-B^3B^2 + B^2B^3) = 0. \tag{3.666}$$

Thus, the Lorentz invariants are indeed invariant under the given boost.

1.131

a) The equation of motion is

$$\frac{d}{dt}(\gamma m_e \boldsymbol{u}) = q\boldsymbol{E}, \quad \gamma = (1 - \boldsymbol{u}^2/c^2)^{-1/2}. \tag{3.667}$$

Integrating this equation gives

$$\frac{m_e\boldsymbol{u}}{\sqrt{1-\boldsymbol{u}^2/c^2}} = q\boldsymbol{E}t, \tag{3.668}$$

where we fixed the integration constants so that $\boldsymbol{u} = \boldsymbol{0}$ at $t = 0$. We find that

$$\boldsymbol{u}^2 = \frac{(q\boldsymbol{E}ct)^2}{(m_e c)^2 + (q\boldsymbol{E}t)^2}. \tag{3.669}$$

Choosing a coordinate system with $\boldsymbol{E} = (E, 0, 0)$ with $E = U/L$, we obtain $\boldsymbol{x}(t) = (x(t), 0, 0)$ and $\boldsymbol{u} = d\boldsymbol{x}/dt = (u, 0, 0)$ with $u = dx/dt$. Thus, we have

$$\frac{dx(t)}{dt} = \frac{qEct}{\sqrt{(m_e c)^2 + (qEt)^2}} \quad \text{and} \quad x(t) = \int_0^t \frac{qEcs}{\sqrt{(m_e c)^2 + (qEs)^2}}\,ds. \tag{3.670}$$

The integral can be performed and gives

$$x(t) = \frac{c}{qE}\left[\sqrt{(m_e c)^2 + (qEt)^2} - m_e c\right], \tag{3.671}$$

where we used the initial condition $x(0) = 0$. For times so small that $qEt \ll m_e c$, this gives

$$x(t) \simeq x_{\text{nr}}(t) \equiv \frac{qEt^2}{2m_e}, \tag{3.672}$$

which is the correct nonrelativistic limit.

b) The time t^* it takes to pass through the accelerator is determined by $x(t^*) = L$. By using $q = e$ and $E = U/L$, we find

$$t^* = \frac{L}{c}\sqrt{\frac{2m_e c^2 + |eU|}{|eU|}}. \tag{3.673}$$

In the nonrelativistic limit, this gives $t^* \simeq L\sqrt{2m_e/|eU|}$, identical with the solution of $x_{\text{nr}}(t^*) = L$.

c) The energy of the particles is

$$cp^0 = \frac{m_e c^2}{\sqrt{1 - u^2/c^2}} = c\sqrt{(m_e c)^2 + (qEt)^2}. \tag{3.674}$$

In the nonrelativistic limit, this gives

$$p^0 - m_e c \simeq \frac{1}{2}(qEt)^2, \tag{3.675}$$

which is the kinetic energy acquired by a nonrelativistic particle moving a time t in a constant electric field E.

1.132

a) Let S' be the particle's rest frame. We have $(A')^0 = \phi$ and $(A')^i = 0$, for $i = 1, 2, 3$, and thus, in this frame, we have

$$(E')^i(\boldsymbol{x}', t') = (F')^{i0} = (\partial')^i (A')^0 = -\partial'_i \phi = \frac{q(x')^i}{4\pi |\boldsymbol{x}'|^3} \quad \text{and} \quad (B')^i = \cdots = 0. \tag{3.676}$$

The transformation matrix from S' to S is

$$(\Lambda^\mu{}_\nu) = \begin{pmatrix} \gamma & -\beta\gamma & 0 & 0 \\ -\beta\gamma & \gamma & 0 & 0 \\ 0 & 0 & 1 & 0 \\ 0 & 0 & 0 & 1 \end{pmatrix}, \tag{3.677}$$

where $\beta = v/c$ and $\gamma = 1/\sqrt{1 - \beta^2}$, and thus, we obtain

$$F^{\mu\nu}(x) = \Lambda^\mu{}_\alpha \Lambda^\nu{}_\beta (F')^{\alpha\beta}(x'(x)), \tag{3.678}$$

where x is short for (x^λ) and $x'(x)$ means x' expressed in terms of the coordinates in S, i.e.,

$$(x')^0 = \gamma(x^0 + \beta x^1), \quad (x')^1 = \gamma(x^1 + \beta x^0), \quad (x')^2 = x^2, \quad (x')^3 = x^3. \tag{3.679}$$

We get (suppressing arguments for simplicity)

$$F^{\mu\nu} = (\Lambda^\mu{}_i \Lambda^\nu{}_0 - \Lambda^\mu{}_0 \Lambda^\nu{}_i)(E')^i, \tag{3.680}$$

and this leads to

$$E^j = F^{0j} = (\Lambda^0{}_i \Lambda^j{}_0 - \Lambda^0{}_0 \Lambda^j{}_i)(E')^i, \tag{3.681}$$

i.e., we have

$$(E^1, E^2, E^3) = ((E')^1, \gamma(E')^2, \gamma(E')^3) = \frac{q}{4\pi|\boldsymbol{x}'|^3}((x')^1, \gamma(x')^2, \gamma(x')^3)$$
$$= \frac{q\gamma}{4\pi\left[\gamma^2(x^1 + vt)^2 + (x^2)^2 + (x^3)^2\right]^{3/2}}((x^1 + vt), x^2, x^3). \tag{3.682}$$

A similar computation gives

$$(B^1, B^2, B^3) = (0, -\beta\gamma(E')^3, \beta\gamma(E')^2) = \frac{q\beta\gamma}{4\pi|\boldsymbol{x}'|^3}(0, -(x')^3, (x')^2)$$
$$= \frac{q\beta\gamma}{4\pi\left[\gamma^2(x^1 + vt)^2 + (x^2)^2 + (x^3)^2\right]^{3/2}}(0, -x^3, x^2). \tag{3.683}$$

b) $\boldsymbol{E} \cdot \boldsymbol{B}$ and $\boldsymbol{E}^2 - \boldsymbol{B}^2$ are Lorentz invariant, and thus, we should get $\boldsymbol{E} \cdot \boldsymbol{B} = \boldsymbol{E}' \cdot \boldsymbol{B}' = 0$ and

$$\boldsymbol{E}^2 - \boldsymbol{B}^2 = (\boldsymbol{E}')^2 = \frac{q^2}{16\pi^2|\boldsymbol{x}'|^4}. \tag{3.684}$$

Using the formulas above, it is straightforward to check that this is indeed the case.

1.133

a) The equation of motion follows from the Lorentz force law, i.e.,

$$\frac{dP^\mu}{d\tau} = \frac{q}{c^2} F^{\mu\nu} U_\nu, \tag{3.685}$$

with $P^\mu = m_0 U^\mu = m_0 c\, dx^\mu / d\tau$. The force is pure, and thus, we have

$$m_0 c \frac{d^2 x^\mu}{d\tau^2} = \frac{q}{c} F^\mu{}_\nu \frac{dx^\nu}{d\tau}, \tag{3.686}$$

which leads to

$$m_0 c \frac{d^2 x^1}{d\tau^2} = \frac{q}{c} F^1{}_2 \frac{dx^2}{d\tau} = q B^3 \frac{dx^2}{d\tau}, \tag{3.687}$$

since $F^1{}_2 = cB^3$ and $F^1{}_0 = F^1{}_1 = F^1{}_3 = 0$. Now, using $x^1 = x = R\cos\omega\tau$ and $x^2 = y = -R\sin\omega\tau$, we find that

$$m_0cR\frac{d^2}{d\tau^2}\cos\omega\tau = -m_0cR\omega^2\cos\omega\tau = -qB^3R\frac{d}{d\tau}\sin\omega\tau = -qB^3R\omega\cos\omega\tau, \tag{3.688}$$

which means that

$$\omega = \frac{qB^3}{m_0c}. \tag{3.689}$$

In order to calculate p_ip^i, we use $P^i = p^i = -p_i$ and $(B^3)^2 = (B^i)^2 = -B_iB^i$ as well as we set $c = 1$. Furthermore, we have

$$P^1 = m_0\frac{dx^1}{d\tau} = m_0\frac{d}{d\tau}R\cos\omega\tau = -m_0\omega R\sin\omega\tau, \tag{3.690}$$

$$P^2 = m_0\frac{dx^2}{d\tau} = -m_0\frac{d}{d\tau}R\sin\omega\tau = -m_0\omega R\cos\omega\tau. \tag{3.691}$$

Thus, using $\omega = qB^3/m_0$, we obtain

$$\begin{aligned} p_ip^i &= -(P^i)^2 = -\left[(P^1)^2 + (P^2)^2\right] = -m_0^2\omega^2R^2 \\ &= -m_0^2\left(\frac{qB^3}{m_0}\right)^2R^2 = -q^2R^2(B^3)^2 = q^2R^2B_iB^i, \end{aligned} \tag{3.692}$$

which is what we wanted to show.

b) Using conservation of 4-momentum, we have

$$p_\Sigma = p_\pi + p_X, \tag{3.693}$$

which implies that

$$\begin{aligned} M_X^2 = p_X^2 &= (p_\Sigma - p_\pi)^2 = p_\Sigma^2 + p_\pi^2 - 2p_\Sigma\cdot p_\pi \\ &= M_\Sigma^2 + M_\pi^2 - 2(p_\Sigma^0p_\pi^0 - |\mathbf{p}_\Sigma||\mathbf{p}_\pi|\cos\theta) \\ &= M_\Sigma^2 + M_\pi^2 - 2\left(\sqrt{M_\Sigma^2 + \mathbf{p}_\Sigma^2}\sqrt{M_\pi^2 + \mathbf{p}_\pi^2} - |\mathbf{p}_\Sigma||\mathbf{p}_\pi|\cos\theta\right), \end{aligned} \tag{3.694}$$

where M_X is the mass of the unknown uncharged particle X^0. Thus, solving for M_X, we obtain

$$M_X = \sqrt{M_\Sigma^2 + M_\pi^2 - 2\left(\sqrt{M_\Sigma^2 + \mathbf{p}_\Sigma^2}\sqrt{M_\pi^2 + \mathbf{p}_\pi^2} - |\mathbf{p}_\Sigma||\mathbf{p}_\pi|\cos\theta\right)}. \tag{3.695}$$

1.134

We start from the expression of the electromagnetic field tensor

$$F_{\mu\nu} = \partial_\mu A_\nu - \partial_\nu A_\mu = (k_\mu\varepsilon_\nu - k_\nu\varepsilon_\mu)\cos(k\cdot x). \tag{3.696}$$

We can now compute the invariants $I_1 = F_{\mu\nu}\tilde{F}^{\mu\nu} = \varepsilon^{\mu\nu\sigma\rho}F_{\mu\nu}F_{\sigma\rho} \propto \boldsymbol{E}\cdot\boldsymbol{B}$ and $I_2 = F_{\mu\nu}F^{\mu\nu} \propto \boldsymbol{E}^2 - \boldsymbol{B}^2$

$$\begin{aligned} I_1 &= \varepsilon^{\mu\nu\sigma\rho}(k_\mu\varepsilon_\nu - k_\nu\varepsilon_\mu)(k_\sigma\varepsilon_\rho - k_\rho\varepsilon_\sigma)\cos^2(k\cdot x) \\ &= 4\varepsilon^{\mu\nu\sigma\rho}k_\mu\varepsilon_\nu k_\sigma\varepsilon_\rho\cos^2(k\cdot x) \\ &= 0, \end{aligned} \tag{3.697}$$

where we have used the totally antisymmetric property of $\varepsilon^{\mu\nu\sigma\rho}$ and that $k_\mu k_\sigma$ is symmetric under the index change $\mu \leftrightarrow \sigma$ (we can also get the zero by using the symmetry of $\varepsilon_\nu\varepsilon_\rho$ under $\nu \leftrightarrow \rho$). For I_2, we obtain

$$\begin{aligned} I_2 &= (k_\mu\varepsilon_\nu - k_\nu\varepsilon_\mu)(k^\mu\varepsilon^\nu - k^\nu\varepsilon^\mu)\cos^2(k\cdot x) \\ &= 2(k^2\varepsilon^2 - (k\cdot\varepsilon)^2)\cos^2(k\cdot x). \end{aligned} \tag{3.698}$$

We now apply Maxwell's equation in vacuum, which state that

$$\partial_\mu F^{\mu\nu} = 0 \quad \Longrightarrow \quad -k_\mu(k^\mu\varepsilon^\nu - k^\nu\varepsilon^\mu)\sin(k\cdot x) = 0. \tag{3.699}$$

Multiplying this relation with ε^ν, we obtain

$$k^2\varepsilon^2 - (k\cdot\varepsilon)^2 = 0 \quad \Longrightarrow \quad I_2 = 0. \tag{3.700}$$

Thus, from $I_1 = 0$ we have $\boldsymbol{E}\cdot\boldsymbol{B} = 0$ and from $I_2 = 0$ we have $\boldsymbol{E}^2 = \boldsymbol{B}^2$. It follows that $\boldsymbol{E}$ and $\boldsymbol{B}$ are orthogonal with the same magnitude.

1.135

We start by noting that $F_{\mu\nu}F^{\mu\nu} \propto \boldsymbol{E}^2 - \boldsymbol{B}^2$ is a Lorentz invariant. Thus, if $|\boldsymbol{E}| = |\boldsymbol{B}|$ in one inertial frame, then $F_{\mu\nu}F^{\mu\nu} = 0$ in all inertial frames, and therefore, in the frame S',

$$\boldsymbol{E}'^2 - \boldsymbol{B}'^2 = 0 \quad \Longrightarrow \quad |\boldsymbol{E}'| = |\boldsymbol{B}'|. \tag{3.701}$$

Furthermore, the quantity $F_{\mu\nu}\tilde{F}^{\mu\nu} \propto \boldsymbol{E}\cdot\boldsymbol{B}$, where $\tilde{F}^{\mu\nu}$ is the dual field tensor, is also a Lorentz invariant. It follows that

$$\boldsymbol{E}\cdot\boldsymbol{B} = |\boldsymbol{E}|^2\cos\alpha = \boldsymbol{E}'\cdot\boldsymbol{B}' = |\boldsymbol{E}'|^2\cos\alpha', \tag{3.702}$$

where we have used that α (α') is the angle between $\boldsymbol{E}$ and $\boldsymbol{B}$ ($\boldsymbol{E}'$ and $\boldsymbol{B}'$) in S (S'). From this equality, it is straightforward to solve for $\cos\alpha'$:

$$\cos\alpha' = \frac{\boldsymbol{E}^2}{\boldsymbol{E}'^2}\cos\alpha. \tag{3.703}$$

1.136

The electromagnetic field tensor is given by

$$F_{\mu\nu} = \partial_\mu A_\nu - \partial_\nu A_\mu \quad \Longrightarrow \quad F_{0i} = -\partial_i A_0 = -(\nabla A_0)_i. \tag{3.704}$$

In the $(x, y, z) = (1, 0, 0)$, we have

$$-\nabla \frac{Q}{4\pi r} = \frac{Q}{4\pi r^2} e_r = \frac{Q}{4\pi} e_x. \tag{3.705}$$

It follows that

$$F = (F_{\mu\nu}) = \frac{Q}{4\pi} \begin{pmatrix} 0 & 1 & 0 & 0 \\ -1 & 0 & 0 & 0 \\ 0 & 0 & 0 & 0 \\ 0 & 0 & 0 & 0 \end{pmatrix}, \quad F' = (F^{\mu\nu}) = \frac{Q}{4\pi} \begin{pmatrix} 0 & -1 & 0 & 0 \\ 1 & 0 & 0 & 0 \\ 0 & 0 & 0 & 0 \\ 0 & 0 & 0 & 0 \end{pmatrix}. \tag{3.706}$$

The electromagnetic stress–energy tensor is given by

$$T^\mu_\nu = -\varepsilon_0 \left(F_{\nu\sigma} F^{\mu\sigma} - \frac{1}{4} \delta^\mu_\nu F_{\rho\sigma} F^{\rho\sigma} \right). \tag{3.707}$$

Computing the individual contributions, we have

$$(F_{\nu\sigma} F^{\mu\sigma}) = F F'^T = \frac{Q^2}{16\pi^2} \begin{pmatrix} -1 & 0 & 0 & 0 \\ 0 & -1 & 0 & 0 \\ 0 & 0 & 0 & 0 \\ 0 & 0 & 0 & 0 \end{pmatrix}, \tag{3.708}$$

and

$$F_{\rho\sigma} F^{\rho\sigma} = \operatorname{tr}(F F'^T) = \frac{Q^2}{16\pi^2} = -2\frac{Q^2}{16\pi^2}. \tag{3.709}$$

Thus, we have

$$(T^\mu_\nu) = \frac{Q\varepsilon_0}{32\pi^2} \begin{pmatrix} 1 & 0 & 0 & 0 \\ 0 & 1 & 0 & 0 \\ 0 & 0 & -1 & 0 \\ 0 & 0 & 0 & -1 \end{pmatrix}, \tag{3.710}$$

or equivalently

$$(T^{\mu\nu}) = \frac{Q\varepsilon_0}{32\pi^2} \begin{pmatrix} 1 & 0 & 0 & 0 \\ 0 & -1 & 0 & 0 \\ 0 & 0 & 1 & 0 \\ 0 & 0 & 0 & 1 \end{pmatrix}. \tag{3.711}$$

From here it is straightforward to obtain

$$T^\mu_\mu = 0. \tag{3.712}$$

However, we note that this is always the case as

$$T^\mu_\mu = -\varepsilon_0 \left(F_{\mu\sigma} F^{\mu\sigma} - \frac{1}{4} \delta^\mu_\mu F_{\sigma\rho} F^{\sigma\rho} \right) = -\varepsilon_0 F_{\mu\sigma} F^{\mu\sigma} (1 - 1) = 0, \tag{3.713}$$

as a direct result of $\delta^\mu_\mu = 4$.

1.137
Maxwell's equations state that

$$\partial_\sigma F^{\sigma\mu} = J^\mu, \quad \partial_\sigma \tilde{F}^{\sigma\mu} = 0. \tag{3.714}$$

The latter of these equations can be rewritten as

$$\partial^\sigma F^{\mu\nu} + \partial^\mu F^{\nu\sigma} + \partial^\nu F^{\sigma\mu} = 0. \tag{3.715}$$

It follows that

$$\partial^\sigma F^{\mu\nu} = \partial^\mu F^{\sigma\nu} - \partial^\nu F^{\sigma\mu}, \tag{3.716}$$

due to the antisymmetry of F. Using this, we find that

$$\Box F^{\mu\nu} = \partial_\sigma \partial^\sigma F^{\mu\nu} = \partial_\sigma (\partial^\mu F^{\sigma\nu} - \partial^\nu F^{\sigma\mu}). \tag{3.717}$$

We now change the order of the partial derivatives in each term, which leads to

$$\Box F^{\mu\nu} = \partial^\mu \partial_\sigma F^{\sigma\nu} - \partial^\nu \partial_\sigma F^{\sigma\mu} = \partial^\mu J^\nu - \partial^\nu J^\mu \equiv S^{\mu\nu}, \tag{3.718}$$

where we have used the first of Maxwell's equations in the second step. Thus, the components of the field tensor fulfill the wave equation with the source term $S^{\mu\nu} = \partial^\mu J^\nu - \partial^\nu J^\mu$. Note that S is antisymmetric as it must be due to the antisymmetry of F.

1.138
By imposing the Lorenz gauge $\partial_\mu A^\mu = 0$, we find that

$$k_\mu \varepsilon^\mu \cos(k \cdot x) = 0, \tag{3.719}$$

which is only fulfilled if k and ε are orthogonal, i.e., $k \cdot \varepsilon = 0$. For the electromagnetic field tensor, we find that

$$F_{\mu\nu} = \partial_\mu A_\nu - \partial_\nu A_\mu = (k_\mu \varepsilon_\nu - k_\nu \varepsilon_\mu) \cos(k \cdot x). \tag{3.720}$$

The requirement of fulfilling Maxwell's equations is thus given by

$$-k_\mu (k^\mu \varepsilon^\nu - k^\nu \varepsilon^\mu) \sin(k \cdot x) = 0. \tag{3.721}$$

Again, the sine function is nonzero, which results in

$$k^2 \varepsilon^\nu = k^\nu (k \cdot \varepsilon) = 0. \tag{3.722}$$

For nontrivial solutions, the amplitude $\varepsilon \neq 0$, resulting in $k^2 = 0$, i.e., the wavevector k is light-like.

We now find that

$$F_{\mu\sigma} F^{\nu\sigma} = (k_\mu \varepsilon_\sigma - k_\sigma \varepsilon_\mu)(k^\nu \varepsilon^\sigma - k^\sigma \varepsilon^\nu) \cos^2(k \cdot x) = \varepsilon^2 k_\mu k^\nu \cos^2(k \cdot x). \tag{3.723}$$

Contracting μ and ν, we also arrive at

$$F_{\rho\sigma}F^{\rho\sigma} = \varepsilon^2 k^2 \cos^2(k \cdot x) = 0. \tag{3.724}$$

Thus, the electromagnetic stress–energy tensor is therefore given by

$$T^{\nu}_{\mu} = -\varepsilon_0 \varepsilon^2 k^{\nu} k_{\mu} \cos^2(k \cdot x). \tag{3.725}$$

Note that the requirement $k \cdot \varepsilon = 0$ where k is lightlike necessarily implies that $\varepsilon^2 < 0$, i.e., ε is spacelike and the 00-component of T is positive.

1.139

The electromagnetic field tensor is given by

$$F_{\mu\nu} = \partial_\mu A_\nu - \partial_\nu A_\mu. \tag{3.726}$$

By defining the new 4-potential $\tilde{A}_\mu = A_\mu + \partial_\mu \varphi$, we find that the new electromagnetic field tensor is

$$\tilde{F}_{\mu\nu} = \partial_\mu \tilde{A}_\nu - \partial_\nu \tilde{A}_\mu = \partial_\mu A_\nu - \partial_\nu A_\mu + (\partial_\mu \partial_\nu - \partial_\nu \partial_\mu)\varphi = F_{\mu\nu} + (\partial_\mu \partial_\nu - \partial_\nu \partial_\mu)\varphi. \tag{3.727}$$

The additional term on the right-hand side disappears since the partial derivatives commute and we therefore end up with

$$\tilde{F}_{\mu\nu} = F_{\mu\nu}, \tag{3.728}$$

i.e, the electromagnetic field tensor is invariant under this transformation.

1.140

The quantity $F_{\mu\nu}\tilde{F}^{\mu\nu}$ is a Lorentz invariant and it therefore does not matter which inertial frame we compute it in. Furthermore, writing it out, we find that

$$F_{\mu\nu}\tilde{F}^{\mu\nu} \propto \boldsymbol{E} \cdot \boldsymbol{B}. \tag{3.729}$$

In S, the electric field $\boldsymbol{E}$ is nonzero, but the magnetic field $\boldsymbol{B}$ is not. It follows directly that

$$F_{\mu\nu}\tilde{F}^{\mu\nu} = 0, \tag{3.730}$$

which therefore also holds in S'.

1.141

In order for a frame where $\boldsymbol{E} = \boldsymbol{0}$ to exist, we must have $\boldsymbol{E} \cdot \boldsymbol{B} = 0$, since $\boldsymbol{E} \cdot \boldsymbol{B}$ is a Lorentz invariant. Furthermore, the invariant $\boldsymbol{E}^2 - \boldsymbol{B}^2$ requires $E^2 < B^2$. We find that

$$\boldsymbol{E} \cdot \boldsymbol{B} = \frac{q}{8\pi r^5} \boldsymbol{m} \cdot \boldsymbol{e}_r = 0, \tag{3.731}$$

implies that $\boldsymbol{m} \cdot \boldsymbol{e}_r = 0$. This is only possible in the plane perpendicular to the magnetic moment $\boldsymbol{m}$. Furthermore, we find that, in this plane,

$$\boldsymbol{E}^2 = \frac{q^2}{16\pi^2 r^4} < \boldsymbol{B}^2 = \frac{m^2}{16\pi r^6} \implies r < \frac{m}{q}, \tag{3.732}$$

where m is the magnitude of the magnetic moment $\boldsymbol{m}$. The set of points that fullfils the requirement is therefore the disk described by

$$\boldsymbol{m} \cdot \boldsymbol{e}_r = 0 \quad \text{and} \quad r < \frac{m}{q}. \tag{3.733}$$

1.142

The electric field is given by

$$\boldsymbol{E} = \frac{q}{4\pi}\left(\frac{\boldsymbol{r}_1}{r_1^3} + \frac{\boldsymbol{r}_2}{r_2^3}\right), \tag{3.734}$$

where we have introduced

$$\boldsymbol{r}_1 = \boldsymbol{x} - \frac{d}{2}\boldsymbol{e}_1 \quad \text{and} \quad \boldsymbol{r}_2 = \boldsymbol{x} + \frac{d}{2}\boldsymbol{e}_1. \tag{3.735}$$

This implies that

$$\boldsymbol{E}^2 = \frac{q^2}{16\pi^2}\left(\frac{1}{r_1^4} + \frac{1}{r_2^4} + \frac{2\boldsymbol{r}_1 \cdot \boldsymbol{r}_2}{r_1^3 r_2^3}\right). \tag{3.736}$$

Since there is no magnetic field, the components of the stress–energy tensor can now be identified as

$$T^{00} = \frac{1}{2}\boldsymbol{E}^2 = \frac{q^2}{32\pi^2}\left(\frac{1}{r_1^4} + \frac{1}{r_2^4} + \frac{2\boldsymbol{r}_1 \cdot \boldsymbol{r}_2}{r_1^3 r_2^3}\right), \tag{3.737}$$

$$\boldsymbol{g} = \boldsymbol{E} \times \boldsymbol{B} = 0, \tag{3.738}$$

$$\begin{aligned}\sigma_{ij} &= \frac{1}{2}\boldsymbol{E}^2\delta_{ij} - E_i E_j \\ &= T^{00}\delta_{ij} - \frac{q^2}{16\pi^2}\left(\frac{x_i - d\delta_{i1}/2}{r_1^3} + \frac{x_i + d\delta_{i1}/2}{r_2^3}\right) \\ &\quad \times \left(\frac{x_j - d\delta_{j1}/2}{r_1^3} + \frac{x_j + d\delta_{j1}/2}{r_2^3}\right).\end{aligned} \tag{3.739}$$

In particular, on the surface $x_1 = 0$, we can introduce polar coordinates ρ and ϕ in the x_2-x_3-plane, leading to $r_1 = r_2 = r = \sqrt{\rho^2 + d^2/4}$ and

$$\sigma_{ij} = \frac{\delta_{ij} q^2 \rho^2}{8\pi^2(\rho^2 + d^2/4)^3} - \frac{q^2 x_i x_j}{16\pi^2(\rho^2 + d^2/4)^3}. \tag{3.740}$$

Computing the force across the surface $x_1 = 0$ with normal $\boldsymbol{e}_1$, the second term in this expression will be proportional to $x_1 = 0$ and we find that

$$F_j = \int_{x_1=0} \sigma_{1j}\, dx_2 dx_3. \tag{3.741}$$

Computing the integral leads to

$$F_j = \frac{q^2\delta_{1j}}{4\pi}\int_0^\infty \frac{\rho^3\, d\rho}{(\rho^2 + d^2/4)^3} = \frac{q^2\delta_{1j}}{8\pi}\int_{d^2/4}^\infty \left(\frac{1}{t^2} - \frac{d^2}{4t^3}\right) dt = \frac{q^2}{4\pi d^2}\delta_{1j}. \tag{3.742}$$

It follows that the force on the electromagnetic field in the region $x_1 > 0$ from the electromagnetic field in the region $x_1 < 0$ is given by

$$\boldsymbol{F} = \frac{q^2 \boldsymbol{e}_1}{4\pi d^2}, \tag{3.743}$$

as expected. The time-component of the 4-force must be equal to zero as the energy current is given by $\boldsymbol{g} = \boldsymbol{0}$.

1.143

In the frame S, two electrons are moving in parallel with constant velocity v in the x-direction and a separation d orthogonal to the direction of motion, i.e., the y-direction. (In this problem, the z-direction can be neglected.) Consider the frame S' in which the electrons are at rest. In this frame, the separation is also d, since the motion of the electrons in the original frame S is orthogonal to the motion. See Figure 3.10 for the two frames S and S'.

In S', the force on the upper electron is given by

$$\boldsymbol{F}' = \frac{q^2}{4\pi d^2}\boldsymbol{e}_{y'}. \tag{3.744}$$

The 4-force on that electron is therefore

$$F' = \frac{q^2}{4\pi d^2}(0, 0, 1), \tag{3.745}$$

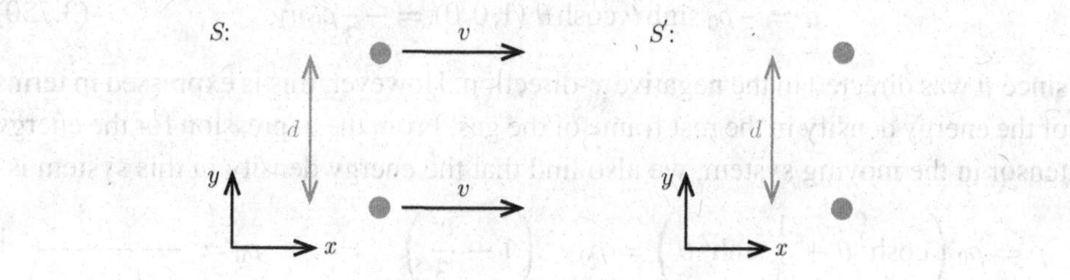

Figure 3.10 Two frames S and S' describing the two electrons.

where the z-component has been omitted. Lorentz transforming this 4-force to S, we find that

$$F = \frac{q^2}{4\pi d^2}(0,0,1), \tag{3.746}$$

since the only force component is in the y-direction.

1.144

In the rest frame of the gas, the energy tensor is given by

$$T = (T^{\mu\nu}) = \rho_0 \begin{pmatrix} 1 & 0 & 0 & 0 \\ 0 & 1/3 & 0 & 0 \\ 0 & 0 & 1/3 & 0 \\ 0 & 0 & 0 & 1/3 \end{pmatrix}, \tag{3.747}$$

since the equation of state for a relativistic gas gives $\rho_0 = 3p_0$ for the rest frame energy density and pressure. Without loss of generality, we can assume that the gas is moving in the negative x-direction. Thus, we can perform a Lorentz transformation from the gas rest frame to this frame which is given by

$$\Lambda = (\Lambda^\mu{}_\nu) = \begin{pmatrix} \cosh\theta & \sinh\theta & 0 & 0 \\ \sinh\theta & \cosh\theta & 0 & 0 \\ 0 & 0 & 1 & 0 \\ 0 & 0 & 0 & 1 \end{pmatrix}, \tag{3.748}$$

where $\tanh\theta = v$. By matrix multiplication, we can now obtain the components of $T^{\mu\nu}$ in the frame where the gas is moving

$$T' = (T'^{\mu\nu}) = \Lambda T \Lambda^T = \rho_0 \begin{pmatrix} \cosh^2\theta + \frac{1}{3}\sinh^2\theta & \frac{4}{3}\cosh\theta\sinh\theta & 0 & 0 \\ \frac{4}{3}\cosh\theta\sinh\theta & \sinh^2\theta + \frac{1}{3}\cosh^2\theta & 0 & 0 \\ 0 & 0 & \frac{1}{3} & 0 \\ 0 & 0 & 0 & \frac{1}{3} \end{pmatrix}. \tag{3.749}$$

Thus, the momentum density in the frame where the gas is moving is given by

$$\boldsymbol{g} = \frac{4}{3}\rho_0 \sinh\theta\cosh\theta\,(1,0,0) = -\frac{4}{3}\rho_0 \boldsymbol{v}\gamma^2, \tag{3.750}$$

since $\boldsymbol{v}$ was directed in the negative x-direction. However, this is expressed in terms of the energy density in the rest frame of the gas. From the expression for the energy tensor in the moving system, we also find that the energy density in this system is

$$\rho = \rho_0\left(\cosh^2\theta + \frac{1}{3}\sinh^2\theta\right) = \rho_0\gamma^2\left(1 + \frac{v^2}{3}\right) \quad \Longrightarrow \quad \rho_0 = \frac{\rho}{\gamma^2\left(1 + \frac{v^2}{3}\right)}. \tag{3.751}$$

After insertion, it follows that

$$g = -\frac{4}{3}\frac{\rho v}{1+\frac{v^2}{3}}. \tag{3.752}$$

For large velocities $v \to 1$, we obtain

$$g \to -\rho v \quad \Longrightarrow \quad |g| \to \rho, \tag{3.753}$$

which is in agreement with having an ensemble of photons traveling in the same direction (due to aberration).

1.145

The 4-force on the star cruiser is given by

$$F^\mu = \frac{dP^\mu}{d\tau}, \tag{3.754}$$

where P is the star cruiser 4-momentum and τ the star cruiser proper time. We can rewrite this as

$$\frac{dP^\mu}{d\tau} = \frac{dt}{d\tau}\frac{dP^\mu}{dt} = \gamma\frac{dP^\mu}{dt}. \tag{3.755}$$

The change in momentum during a time dt is given by the momentum stored in the gas that is absorbed by the cruiser. This can be obtained as

$$dP^\mu = T^{\mu 0}dV = \rho_0(1,0)^\mu Av\,dt = \rho_0 V^\mu_{\text{gas}} Av\,dt, \tag{3.756}$$

where we have used that $T^{\mu 0} = \rho_0(1,0)^\mu$ in the rest frame of the gas and that $dV = Av\,dt$. We have also used that $(1,0)^\mu$ is the 4-velocity of the gas rest frame V_{gas} in the gas rest frame. We thus obtain for the 4-force

$$F^\mu = \rho_0 V^\mu_{\text{gas}} Av. \tag{3.757}$$

Since V_{gas} is not parallel to V_{cruiser} as long as $v \neq 0$, the force is not heat-like. We also have

$$F \cdot V_{\text{cruiser}} = \rho_0 Av V_{\text{gas}} \cdot V_{\text{cruiser}} \neq 0, \tag{3.758}$$

which means that it is also not orthogonal to V_{cruiser} and thus not pure. Thus, it is neither pure nor heatlike (the cruiser is being both heated and accelerated by the force).

1.146

The transformation of the energy–momentum tensor is given by

$$(T') = \Lambda(T)\Lambda^T, \tag{3.759}$$

where Λ is the matrix

$$\Lambda = \gamma \begin{pmatrix} 1 & -v \\ -v & 1 \end{pmatrix}, \tag{3.760}$$

related to the Lorentz transformation with velocity v. We find that

$$(T') = \frac{1}{1-v^2} \begin{pmatrix} \rho_0 - v^2\sigma & v(\sigma - \rho_0) \\ v(\sigma - \rho_0) & v^2\rho_0 - \sigma \end{pmatrix}. \tag{3.761}$$

For the stress T^{11} to be equal to zero, we must fulfill

$$\sigma = v^2 \rho \quad \Rightarrow \quad v = \sqrt{\frac{\sigma}{\rho}} < 1. \tag{3.762}$$

Thus, *there exists a frame where the stress is zero.*

For the energy density T^{00}, we find that

$$T^{00} = \frac{\rho_0 - v^2\sigma}{1-v^2} \geq \rho_0 \frac{1-v^2}{1-v^2} = \rho_0, \tag{3.763}$$

with equality only when $v^2 = 0$. Thus, *there exists no frame where the energy density is smaller than* ρ_0.

1.147

The general form of the energy–momentum tensor of a perfect fluid is given by

$$T^{\mu\nu} = (\rho_0 + p)U^\mu U^\nu - p\eta^{\mu\nu}, \tag{3.764}$$

where U is the 4-velocity of the fluid. For our photon gas, the equation of state $p = \rho_0/3$ therefore leads to

$$T^{\mu\nu} = \rho_0 \left(\frac{4}{3} U^\mu U^\nu - \frac{1}{3} \eta^{\mu\nu} \right). \tag{3.765}$$

The energy density, momentum density, and stress tensor are therefore given by

$$\rho = T^{00} = \rho_0 \left(\frac{4}{3}\gamma^2 - \frac{1}{3} \right) = \rho_0 \gamma^2 \left(1 + \frac{v^2}{3} \right), \tag{3.766}$$

$$g^i = T^{0i} = -\frac{4}{3}\rho_0 \gamma^2 v^i, \tag{3.767}$$

$$\sigma^{ij} = \rho_0 \left(\frac{4}{3}\gamma^2 v^i v^j + \frac{1}{3}\delta^{ij} \right). \tag{3.768}$$

Clearly, the shear stress in a general frame is not going to be zero unless the 3-velocity $\boldsymbol{v}$ has at most one nonzero component. However, it is always possible to find such a frame by simple rotations. Note that the 4-velocity of the gas in the frame moving with velocity $\boldsymbol{v}$ in the gas rest frame is $-\boldsymbol{v}$.

1.148

The energy–momentum tensor of a perfect fluid is generally given by

$$T^{\mu\nu} = (\rho_0 + p)U^\mu U^\nu - p\eta^{\mu\nu}, \tag{3.769}$$

where U is the 4-velocity of the fluid. In an arbitrary frame, the energy density is given by the 00-component of the energy–momentum tensor and so

$$\rho = (\rho_0 + p)(U^0)^2 - p\eta^{00} = (\rho_0 + p)\gamma^2 - p. \tag{3.770}$$

The energy density divided by γ^2 is hence given by

$$\frac{\rho}{\gamma^2} = \rho_0 + p - \frac{p}{\gamma^2}. \tag{3.771}$$

Since $1 \leq \gamma < \infty$, we find that

$$\frac{\rho}{\gamma^2} < \rho_0 + p. \tag{3.772}$$

1.149

The stress–energy tensor of a perfect fluid is given by

$$T^{\mu\nu} = (\rho_0 + p)U^\mu U^\nu - p\eta^{\mu\nu}, \tag{3.773}$$

where U is the 4-velocity of the fluid. For an observer with 4-velocity V, we therefore find that

$$\rho = T^{\mu\nu} V_\mu V_\nu = (\rho_0 + p)(U \cdot V)^2 - pV^2 = \gamma^2 \rho_0 (1 + w) - w\rho_0, \tag{3.774}$$

where $\gamma = U \cdot V \geq 1$, since U and V are 4-velocities. The weak energy condition when $U = V$ is given by

$$\rho = \rho_0 \geq 0. \tag{3.775}$$

When $U \neq V$, we now obtain

$$\gamma^2(1 + w) - w \geq 0 \quad \Rightarrow \quad (\gamma^2 - 1)w \geq -\gamma^2 \quad w \geq \frac{-\gamma^2}{\gamma^2 - 1}, \tag{3.776}$$

since $\gamma^2 > 1$ in those situations. The right-hand side grows monotonously as γ^2 increases and the strongest limit is therefore obtained by letting $\gamma \to \infty$. Doing so, we obtain the condition

$$w \geq -1. \tag{3.777}$$

The weak energy condition therefore implies that

$$\rho_0 \geq 0 \quad \text{and} \quad w \geq -1. \tag{3.778}$$

1.150

a) Variation of the action

$$\mathscr{S} = \int_{s_0}^{s_1} \frac{1}{2} mc^2 \dot{x}_\mu \dot{x}^\mu \, ds = \int_{s_0}^{s_1} \mathscr{L} \, ds, \tag{3.779}$$

yields

$$\frac{d}{ds}\left(mc^2 \dot{x}^\mu\right) = c \frac{d}{ds} p^\mu = 0. \tag{3.780}$$

Integration gives $p^\mu = c^\mu$, where c^μ is a constant 4-vector.

b) Inserting the substitution $p \mapsto p + qA/c$ leads to the action

$$\begin{aligned} \mathscr{S}' &= \int_{s_0}^{s_1} \frac{1}{2m}\left(p^2 + 2\frac{pqA}{c} + \frac{q^2 A^2}{c^2}\right) ds \\ &= \int_{s_0}^{s_1} \left[\frac{1}{2} mc^2 \dot{x}^2 + q\dot{x}A + \mathcal{O}(q^2)\right] ds = \int_{s_0}^{s_1} \mathscr{L}' \, ds. \end{aligned} \tag{3.781}$$

Variation of this action, neglecting terms of order q^2, yields

$$\frac{\partial \mathscr{L}'}{\partial x^\mu} = q\dot{x}_\nu \frac{\partial}{\partial x^\mu} A^\nu, \quad \frac{\partial \mathscr{L}'}{\partial \dot{x}^\mu} = mc^2 \dot{x}_\mu + qA_\mu. \tag{3.782}$$

Thus, inserting this into the Euler–Lagrange variational equations, we find that

$$\frac{d}{ds}(mc^2 \dot{x}_\mu) + q\frac{d}{ds} A_\mu - q\dot{x}_\nu \frac{\partial}{\partial x^\mu} A^\nu = 0. \tag{3.783}$$

However, we have $\frac{d}{ds} A_\mu = \dot{x}^\nu \partial_\nu A_\mu$, from which we obtain

$$mc^2 \ddot{x}^\mu = q\dot{x}^\nu (\partial^\mu A_\nu - \partial_\nu A^\mu) = q\dot{x}^\nu F^\mu{}_\nu = q\dot{x}_\nu F^{\mu\nu}, \tag{3.784}$$

which is the desired result, i.e., the equations of the Lorentz force.

3.2 Solutions to Problems in General Relativity Theory

2.1

Using spherical coordinates θ and ϕ on the unit sphere, the function $f(x, y)$ restricted to $\mathbb{S}^2$ is given by

$$f = x^2 + y = \sin^2(\theta)\cos^2(\phi) + \sin(\theta)\sin(\phi), \tag{3.785}$$

which is a smooth function of the coordinates everywhere where the spherical coordinates are well defined, which is everywhere except the poles where $\theta = 0$ and π, respectively, as the coordinate system is singular in those points. In order to see that the function is smooth also at those points, we can pick another smooth coordinate system, such as basing a spherical coordinate system on the x-axis rather than the z-axis

$$x = \cos(\alpha), \quad y = \sin(\alpha)\cos(\beta), \quad z = \sin(\alpha)\sin(\beta). \tag{3.786}$$

In this coordinate system, we find that

$$f = \cos^2(\alpha) + \sin(\alpha)\cos(\beta), \tag{3.787}$$

which is again a smooth function of the coordinates.

2.2

The problem of defining tangent vectors at the poles is that the spherical coordinate system is not one-to-one at the poles, so a curve crossing a pole does not correspond to a continuous curve in the (θ, ϕ)-plane. The spherical coordinates are therefore not a good coordinate system for describing situations at the poles, instead we should use a different coordinate system. One possible coordinate system is a stereographic projection, which has no singularity at one pole. In this coordinate system, we can describe vectors at the poles without any problems, since it is one-to-one.

2.3

a) The Euclidean metric is $d\ell^2 = dx^2 + dy^2$. Differentiating the coordinate relations, we obtain

$$dx = e^t(ds + s\,dt), \quad dy = e^{-t}(ds - s\,dt), \tag{3.788}$$

leading to

$$\begin{aligned} d\ell^2 &= e^{2t}(ds^2 + 2s\,ds\,dt + s^2dt^2) + e^{-2t}(ds^2 - 2s\,ds\,dt + s^2dt^2) \\ &= 2\cosh(2t)ds^2 + 4s\sinh(2t)ds\,dt + 2s^2\cosh(2t)dt^2 = g_{ab}d\xi^a d\xi^b. \end{aligned} \tag{3.789}$$

The components of the metric tensor are therefore given by

$$g_{tt} = 2s^2\cosh(2t), \quad g_{ts} = g_{st} = 2s\sinh(2t), \quad g_{ss} = 2\cosh(2t). \tag{3.790}$$

The Christoffel symbols can be found through the geodesic equations. In Cartesian coordinates, they take the form $\ddot{x} = \ddot{y} = 0$, which leads to

$$\dot{x} = e^t(\dot{s} + s\dot{t}), \tag{3.791}$$

$$\ddot{x} = e^t(\ddot{s} + s\ddot{t} + 2\dot{s}\dot{t} + s\dot{t}^2) = 0, \tag{3.792}$$

$$\dot{y} = e^{-t}(\dot{s} - s\dot{t}), \tag{3.793}$$

$$\ddot{y} = e^{-t}(\ddot{s} - s\ddot{t} - 2\dot{s}\dot{t} + s\dot{t}^2) = 0. \tag{3.794}$$

Solving for $\ddot{s}$ and $\ddot{t}$, we find that

$$\ddot{s} + s\dot{t}^2 = 0, \quad \ddot{t} + 2\frac{\dot{s}\dot{t}}{s} = 0. \tag{3.795}$$

Comparing with the geodesic equations $\ddot{\xi}^a + \Gamma^a_{bc}\dot{\xi}^b\dot{\xi}^c = 0$ and keeping in mind that the Christoffel symbols are symmetric, the nonzero Christoffel symbols can be identified as

$$\Gamma^s_{tt} = s \quad \text{and} \quad \Gamma^t_{st} = \Gamma^t_{ts} = \frac{1}{s}. \tag{3.796}$$

The coordinates refer to the same point for all t when $s = 0$, they are singular at this point. We can choose to describe the part of Euclidean space, where $s > 0$ and $t \in \mathbb{R}$. Since e^t and e^{-t} are both larger than zero for all t and $\cosh(t) > 0$, this refers to the region $x, y > 0$.

Note that the geodesic equations can also be found by requiring the integral

$$L[\xi] = \int g_{ab}\dot{\xi}^a\dot{\xi}^b d\tau = 2\int [\cosh(2t)\dot{s}^2 + 2s\sinh(2t)\dot{s}\dot{t} + s^2\cosh(2t)\dot{t}^2]d\tau, \tag{3.797}$$

to be extremal (i.e., through the corresponding Euler–Lagrange equations). However, the computations become a bit lengthier.

b) As an alternative to the method presented in a), we can compute the tangent basis $\boldsymbol{E}_i = \partial\boldsymbol{x}/\partial x^i$ in order to find the metric. We obtain

$$\boldsymbol{E}_u = \frac{\partial\boldsymbol{x}}{\partial u} = \boldsymbol{e}_1, \quad \boldsymbol{E}_v = \frac{\partial\boldsymbol{x}}{\partial v} = 2v\boldsymbol{e}_2. \tag{3.798}$$

The components of the metric tensor are then generally given by $g_{ij} = \boldsymbol{E}_i \cdot \boldsymbol{E}_j$, and therefore,

$$g_{uu} = \boldsymbol{e}_1 \cdot \boldsymbol{e}_1 = 1, \quad g_{vv} = 2v\boldsymbol{e}_2 \cdot 2v\boldsymbol{e}_2 = 4v^2, \quad g_{uv} = g_{vu} = \boldsymbol{e}_1 \cdot 2v\boldsymbol{e}_2 = 0. \tag{3.799}$$

The Christoffel symbols in Euclidean space may be computed through $\partial_i\boldsymbol{E}_j = \Gamma^k_{ij}\boldsymbol{E}_k$. Since $\boldsymbol{E}_u$ is constant and $\boldsymbol{E}_v$ only depends on v, we find that the only nonzero derivative is

$$\partial_v\boldsymbol{E}_v = 2\boldsymbol{e}_2 = \frac{1}{v}\boldsymbol{E}_v. \tag{3.800}$$

It follows that the only nonzero Christoffel symbol is

$$\Gamma^v_{vv} = \frac{1}{v}. \tag{3.801}$$

Regarding where the coordinate system can be used, we note that, by definition, $y = v^2 \geq 0$. We also note that along $y = 0$, $\boldsymbol{E}_v = 0$ and therefore the coordinate system is singular along this line. The coordinate system is therefore only well defined for $y > 0$ and we can choose to use coordinates such that $v > 0$ in this regime.

2.4

We must show that if the condition

$$\left.\frac{d}{dt}x^i(\alpha(t))\right|_{t=t_0} = \left.\frac{d}{dt}x^i(\beta(t))\right|_{t=t_0} \quad \text{for } i = 1, 2, \ldots, n, \tag{3.802}$$

holds in a coordinate system y, then it also holds in another coordinate system x. Let x^i and y^i be two different local coordinates on the manifold M, i.e., x and y are smooth maps with smooth inverses. This means that the coordinates y^i may be written as functions $y^i(x)$ of the coordinates x^i, which are smooth with smooth inverses. Assume that $\alpha(t_0) = \beta(t_0)$, then $\alpha(t_0)$ can be described in both the coordinates of x and y and we have

$$\left.\frac{d}{dt}y^i(x(\alpha(t)))\right|_{t=t_0} = \left.\frac{\partial y^i}{\partial x^j}\frac{d}{dt}x^j(\alpha(t))\right|_{t=t_0}. \tag{3.803}$$

If the condition (3.802) holds in the coordinate system y, then using Eq. (3.803), we find that

$$\left.\frac{d}{dt}y^i(\alpha(t))\right|_{t=t_0} = \left.\frac{d}{dt}y^i(\beta(t))\right|_{t=t_0}$$

$$\Leftrightarrow \quad \left.\frac{\partial y^i}{\partial x^j}\frac{d}{dt}x^j(\alpha(t))\right|_{t=t_0} = \left.\frac{\partial y^i}{\partial x^k}\frac{d}{dt}x^k(\beta(t))\right|_{t=t_0}. \tag{3.804}$$

Since $y(x)$ is an invertible map, the matrix $A^i_j \equiv \partial y^i/\partial x^j$ is also invertible. Thus, multiplying Eq. (3.804) by A^{-1}, we obtain

$$\left.\frac{d}{dt}x^j(\alpha(t))\right|_{t=t_0} = \left.\frac{d}{dt}x^j(\beta(t))\right|_{t=t_0}, \tag{3.805}$$

which means that the condition in Eq. (3.802) is independent of the choice of coordinate system.

2.5

We show that if the system of first-order ordinary differential equations defining the parallel transport is fulfilled in a coordinate system, then it is also fulfilled in another coordinate system. First, we investigate the Christoffel symbols, and then, the parallel transport.

In a coordinate system, we have a set of basis vectors $\{\partial_i\}_{i=1}^n$, where the basis vectors depend on the coordinate x, which span T_xM for all x, so that all other vectors can be written as linear combinations of them. In particular, we have

$$\nabla_j \partial_i = \Gamma_{ij}^k \partial_k, \tag{3.806}$$

where the coefficients Γ_{ij}^k are the Christoffel symbols of the second kind. The derivative of a vector field $V = V^i \partial_i$ is then given by

$$\nabla_j V = \nabla_j (V^i \partial_i) = (\partial_j V^i + \Gamma_{jk}^i V^k)\partial_i \equiv (\nabla_j V^i)\partial_i. \tag{3.807}$$

It follows that the parallel transport equation is on the form

$$\dot{x}^k \nabla_k Y^i = \dot{x}^k \partial_k Y^i + \Gamma_{kj}^i \dot{x}^k Y^j = \dot{Y}^i + \Gamma_{kj}^i \dot{x}^k Y^j = 0. \tag{3.808}$$

Defining a new coordinate system y^a and referring to quantities in this coordinate system using primes, the transformation rules are given by

$$\dot{x}^k = \frac{dx^k}{ds} = \frac{dy^a}{ds}\frac{\partial x^k}{\partial y^a} = \dot{y}^a \frac{\partial x^k}{\partial y^a}, \tag{3.809}$$

$$Y^j = \frac{\partial x^j}{\partial y^a} Y'^a, \tag{3.810}$$

$$\Gamma'^a_{bc} = \Gamma^i_{jk} \frac{\partial y^a}{\partial x^i}\frac{\partial x^j}{\partial y^b}\frac{\partial x^k}{\partial y^c} + \frac{\partial^2 x^i}{\partial y^b \partial y^c}\frac{\partial y^a}{\partial x^i}. \tag{3.811}$$

Inserting these relations into the parallel transport equation results in

$$\begin{aligned} 0 &= \dot{y}^a \frac{\partial x^k}{\partial y^a}\frac{\partial}{\partial x^k}\left(\frac{\partial x^i}{\partial y^b} Y'^b\right) + \Gamma^i_{jk} \dot{y}^d \frac{\partial x^k}{\partial y^d}\frac{\partial x^j}{\partial y^e} Y'^e \\ &= \dot{y}^a \frac{\partial}{\partial y^a}\left(\frac{\partial x^i}{\partial y^b} Y'^b\right) + \Gamma^i_{jk} \dot{y}^d \frac{\partial x^j}{\partial y^d}\frac{\partial x^k}{\partial y^e} Y'^e \\ &= \frac{\partial x^i}{\partial y^b} \dot{y}^a \partial_a Y'^b + \frac{\partial^2 x^i}{\partial y^a \partial y^b} \dot{y}^a Y'^b + \Gamma^i_{jk} \frac{\partial x^j}{\partial y^a}\frac{\partial x^k}{\partial y^b} \dot{y}^a Y'^b. \end{aligned} \tag{3.812}$$

Multiplying both sides by $\partial y^c / \partial x^i$ results in

$$\begin{aligned} 0 &= \frac{\partial y^c}{\partial x^i}\left(\frac{\partial x^i}{\partial y^b} \dot{y}^a \partial_a Y'^b + \frac{\partial^2 x^i}{\partial y^a \partial y^b} \dot{y}^a Y'^b + \Gamma^i_{jk} \frac{\partial x^j}{\partial y^a}\frac{\partial x^k}{\partial y^b} \dot{y}^a Y'^b\right) \\ &= \dot{y}^a \partial_a Y'^c + \left(\Gamma^i_{jk} \frac{\partial x^j}{\partial y^a}\frac{\partial x^k}{\partial y^b}\frac{\partial y^c}{\partial x^i} + \frac{\partial y^c}{\partial x^i}\frac{\partial^2 x^i}{\partial y^a \partial y^b}\right) \dot{y}^a Y'^b \\ &= \dot{y}^a \partial_a Y'^c + \Gamma^c_{ab} \dot{y}^a Y'^b, \end{aligned} \tag{3.813}$$

which is the parallel transport equation in the y^a coordinates. Thus, the parallel transport equations in the x^i-coordinates imply the parallel transport equation in the y^a-coordinates, i.e., the parallel transport equations are preserved under coordinate transformations.

2.6
Consider the distance

$$\ell[\gamma] = \int_a^b \sqrt{\dot\theta(s)^2 + \sin^2\theta(s)\,\dot\phi(s)^2}\,ds \equiv \int_a^b \sqrt{\mathcal{L}(s)}\,ds, \tag{3.814}$$

where the Lagrangian is

$$\mathcal{L} = \dot\theta^2 + \sin^2\theta\,\dot\phi^2. \tag{3.815}$$

Using Euler–Lagrange equations for the coordinates θ and ϕ and reparametrizing the curve such that $d\mathcal{L}/ds = 0$, we find that

$$0 = \frac{\partial \mathcal{L}}{\partial \theta} - \frac{d}{ds}\frac{\partial \mathcal{L}}{\partial \dot\theta} = \sin 2\theta\,\dot\phi^2 - \frac{d}{ds}\left(2\dot\theta\right) = \sin 2\theta\,\dot\phi^2 - 2\ddot\theta, \tag{3.816}$$

$$0 = \frac{\partial \mathcal{L}}{\partial \phi} - \frac{d}{ds}\frac{\partial \mathcal{L}}{\partial \dot\phi} = 0 - \frac{d}{ds}\left(2\sin^2\theta\,\dot\phi\right) = -4\sin\theta\cos\theta\,\dot\theta\dot\phi - 2\sin^2\theta\,\ddot\phi, \tag{3.817}$$

which imply the two geodesic equations on $\mathbb{S}^2$ for θ and ϕ, namely

$$\ddot\theta - \frac{1}{2}\sin 2\theta\,\dot\phi^2 = 0, \quad \ddot\phi + 2\cot\theta\,\dot\theta\dot\phi = 0. \tag{3.818}$$

2.7
a) In spherical coordinates, the metric on the unit sphere $\mathbb{S}^2$ is given by

$$ds^2 = g_{ij}dx^i dx^j = d\Omega^2 = d\theta^2 + \sin^2\theta\,d\phi^2, \tag{3.819}$$

where

$$g = (g_{ij}) = \begin{pmatrix} g_{\theta\theta} & g_{\theta\phi} \\ g_{\phi\theta} & g_{\phi\phi} \end{pmatrix} = \begin{pmatrix} 1 & 0 \\ 0 & \sin^2\theta \end{pmatrix}$$

$$\Leftrightarrow \quad g^{-1} = (g^{ij}) = \begin{pmatrix} g^{\theta\theta} & g^{\theta\phi} \\ g^{\phi\theta} & g^{\phi\phi} \end{pmatrix} = \begin{pmatrix} 1 & 0 \\ 0 & \frac{1}{\sin^2\theta} \end{pmatrix}. \tag{3.820}$$

Using the formula for the Christoffel symbols in terms of the components of the metric tensor, i.e.,

$$\Gamma^i_{\ jk} = \frac{1}{2}g^{i\ell}\left(\partial_j g_{\ell k} + \partial_k g_{\ell j} - \partial_\ell g_{jk}\right), \tag{3.821}$$

we explicitly find that the eight Christoffel symbols on $\mathbb{S}^2$ are

$$\Gamma^{\theta}_{\theta\theta} = \frac{1}{2}g^{\theta\theta}\partial_\theta g_{\theta\theta} = 0, \qquad \Gamma^{\phi}_{\phi\phi} = \frac{1}{2}g^{\phi\phi}\partial_\phi g_{\phi\phi} = 0,$$

$$\Gamma^{\phi}_{\theta\theta} = -\frac{1}{2}g^{\phi\phi}\partial_\phi g_{\theta\theta} = 0, \quad \Gamma^{\theta}_{\phi\phi} = -\frac{1}{2}g^{\theta\theta}\partial_\theta g_{\phi\phi}$$

$$= -\frac{1}{2}\cdot 1\cdot 2\sin\theta\cos\theta = -\frac{1}{2}\sin 2\theta,$$

$$\Gamma^{\theta}_{\phi\theta} = \frac{1}{2}g^{\theta\theta}\partial_\phi g_{\theta\theta} = 0, \qquad \Gamma^{\phi}_{\theta\phi} = \frac{1}{2}g^{\phi\phi}\partial_\theta g_{\phi\phi} = \frac{1}{2}\cdot\frac{1}{\sin^2\theta}\cdot 2\sin\theta\cos\theta = \cot\theta,$$

$$\Gamma^{\theta}_{\theta\phi} = \frac{1}{2}g^{\theta\theta}\partial_\phi g_{\theta\theta} = 0, \qquad \Gamma^{\phi}_{\phi\theta} = \frac{1}{2}g^{\phi\phi}\partial_\theta g_{\phi\phi} = \frac{1}{2}\cdot\frac{1}{\sin^2\theta}\cdot 2\sin\theta\cos\theta = \cot\theta, \tag{3.822}$$

Thus, the three nonzero Christoffel symbols on $\mathbb{S}^2$ are

$$\Gamma^{\theta}_{\phi\phi} = -\frac{1}{2}\sin 2\theta, \quad \Gamma^{\phi}_{\theta\phi} = \cot\theta, \quad \Gamma^{\phi}_{\phi\theta} = \cot\theta. \tag{3.823}$$

b) Using the metric, we can write the Lagrangian as

$$\mathcal{L} = g_{ij}\dot{x}^i\dot{x}^j = \dot{\theta}^2 + \sin^2\theta\,\dot{\phi}^2. \tag{3.824}$$

Applying Euler–Lagrange equations to $\mathcal{L}$ for the θ and ϕ coordinates, we find that the geodesic equations are

$$0 = \frac{\partial\mathcal{L}}{\partial\theta} - \frac{d}{d\tau}\frac{\partial\mathcal{L}}{\partial\dot{\theta}} = \sin 2\theta\,\dot{\phi}^2 - \frac{d}{d\tau}\left(2\dot{\theta}\right) = \sin 2\theta\,\dot{\phi}^2 - 2\ddot{\theta}$$

$$\Rightarrow \quad \ddot{\theta} - \frac{1}{2}\sin 2\theta\,\dot{\phi}^2 = 0, \tag{3.825}$$

$$0 = \frac{\partial\mathcal{L}}{\partial\phi} - \frac{d}{d\tau}\frac{\partial\mathcal{L}}{\partial\dot{\phi}} = 0 - \frac{d}{d\tau}\left(2\sin^2\theta\,\dot{\phi}\right) = -4\sin\theta\cos\theta\,\dot{\theta}\dot{\phi} - 2\sin^2\theta\,\ddot{\phi}$$

$$\Rightarrow \quad \ddot{\phi} + 2\cot\theta\,\dot{\theta}\dot{\phi} = 0. \tag{3.826}$$

Thus, comparing the two geodesic equations with the general formula for the geodesic equations $\ddot{x}^i + \Gamma^i_{jk}\dot{x}^j\dot{x}^k = 0$, we obtain the three nonzero Christoffel symbols as

$$\Gamma^{\theta}_{\phi\phi} = -\frac{1}{2}\sin 2\theta, \quad \Gamma^{\phi}_{\theta\phi} = \cot\theta, \quad \Gamma^{\phi}_{\phi\theta} = \cot\theta. \tag{3.827}$$

Of course, the result is exactly the same as the solution to a).

2.8

We must show that the fact that two of the Christoffel symbols on the unit sphere $\mathbb{S}^2$ diverge in spherical coordinates θ and ϕ, i.e.,

$$\lim_{\theta\to 0,\pi} \Gamma^{\phi}_{\theta\phi} = \lim_{\theta\to 0,\pi} \Gamma^{\phi}_{\phi\theta} = \pm\infty, \tag{3.828}$$

is only due to our choice of coordinate system. If we can find another coordinate system, where the Christoffel symbols do not diverge as we approach the North or South Pole, we have shown that the singularities are only coordinate singularities. The problem with the spherical coordinates is that ϕ is undetermined for $\theta = 0$ and $\theta = \pi$, which means that ϕ can have any value at these two values of θ.

We can pick a coordinate system that is nonsingular at the poles by instead basing our spherical coordinate system on the x-axis, i.e., the embedding of the sphere in $\mathbb{R}^3$ would, instead, be given by

$$x = \cos(\alpha), \quad y = \sin(\alpha)\cos(\beta), \quad z = \sin(\alpha)\sin(\beta). \tag{3.829}$$

The poles where the (θ, φ) coordinates are singular are now represented by the new coordinates $(\alpha, \beta) = (\pi/2, \pm\pi/2)$ for $z = \pm 1$, respectively. The nonzero Christoffel symbols in the new coordinates are

$$\Gamma^{\alpha}_{\beta\beta} = -\frac{1}{2}\sin(2\alpha), \quad \Gamma^{\beta}_{\alpha\beta} = \Gamma^{\beta}_{\beta\alpha} = \cot(\alpha), \tag{3.830}$$

by the same computations that give the Christoffel symbols in regular spherical coordinates. Hence, at the points where the regular spherical coordinates lead to diverging Christoffel symbols, our new coordinate system results in

$$\Gamma^{\alpha}_{\beta\beta} = -\frac{1}{2}\sin(\pi) = 0, \quad \Gamma^{\beta}_{\alpha\beta} = \Gamma^{\beta}_{\beta\alpha} = \cot(\pi/2) = 0. \tag{3.831}$$

Thus, the singularities of the Christoffel symbols at the poles in spherical coordinates can be concluded to be coordinate singularities.

2.9

a) We note that if we assume θ to be constant, then the geodesic equations take the form

$$\sin(2\theta)\dot{\phi}^2 = 0, \quad \ddot{\phi} = 0. \tag{3.832}$$

Since θ is constant, we are only dealing with an actual curve is $\dot{\phi} \neq 0$, and therefore, we must have $\theta = \pi/2$ and $\ddot{\phi} = 0$ then implies $\phi = \alpha s + \beta$, which describes a *great circle*. Due to the symmetry of the sphere, any great circle will be a geodesic.

b) Using the general form for the set of first-order ordinary differential equations defining the parallel transport in Problem 2.5 and the Christoffel symbols in Problem 2.8 (or the solution to Problem 2.7) as well as that $(\theta, \phi) = (\alpha s + \beta, \phi_0)$, we find that

$$\dot{X}^{\theta} = 0, \tag{3.833}$$

$$\dot{X}^{\phi} + \cot(\theta)\,\dot{\theta} X^{\phi} = \dot{X}^{\phi} + \alpha\cot(\alpha s + \beta) = 0. \tag{3.834}$$

This set of equations has the solution

$$X^\theta = \text{const.}, \quad X^\phi = \text{const.} \cdot [\sin(\alpha s + \beta)]^{-1} = \text{const.} \cdot (\sin\theta)^{-1}. \tag{3.835}$$

We now assume $X = (1,1)$ at $(\theta,\phi) = (\pi/4, 0)$ as the initial condition for the vector field. We find that

$$X^\theta = 1, \quad X^\phi(\theta) = 1 \cdot \sin\tfrac{\pi}{4} \cdot (\sin\theta)^{-1} = \frac{1}{\sqrt{2}}(\sin\theta)^{-1}. \tag{3.836}$$

Then, parallel transporting this to the point $(\theta,\phi) = \left(\frac{\pi}{2}, 0\right)$ yields

$$X^\theta = 1, \quad X^\phi\left(\tfrac{\pi}{2}\right) = \frac{1}{\sqrt{2}}\left(\sin\tfrac{\pi}{2}\right)^{-1} = \frac{1}{\sqrt{2}}. \tag{3.837}$$

Thus, the parallel-transported vector is determined to be

$$X(\pi/2, 0) = \left(1, \frac{1}{\sqrt{2}}\right). \tag{3.838}$$

2.10

We use cylindrical coordinates and determine the path $z(\varphi)$ that minimizes

$$L[z] = \int_0^{\pi/2} \sqrt{a^2 z(\varphi)^2 + (1+a^2) z'(\varphi)^2}\, d\varphi, \tag{3.839}$$

and goes through the start and end points. The solution is

$$z(\varphi) = -h \frac{\cos\frac{a\pi/4}{\sqrt{1+a^2}}}{\cos\frac{a(\varphi-\pi/4)}{\sqrt{1+a^2}}}. \tag{3.840}$$

2.11

Short solution: The first section of the path is part of a great circle, i.e., a geodesic, on the sphere (here: Earth), and therefore, the angle between the parallel transported vector and the tangent to the circle is constant.

From the metric $ds^2 = R^2\left(d\theta^2 + \sin^2\theta\, d\phi^2\right)$ on a sphere, one obtains the Christoffel symbols $\Gamma^\theta_{\phi\phi} = -\frac{1}{2}\sin 2\theta$, $\Gamma^\phi_{\theta\phi} = \Gamma^\phi_{\phi\theta} = \cot\theta$, and all other Γ's are equal to zero; here $0 \le \theta \le \pi$ and $0 \le \phi < 2\pi$. For the second part of the journey, we can choose the angle ϕ as the path parameter. Thus, the equations for parallel transport of a vector $X = (X^\theta, X^\phi)$ are

$$\begin{cases} \dot{X}^\theta - \frac{1}{2}\sin(2\theta) X^\phi = 0 \\ \dot{X}^\phi + \cot(\theta) X^\theta = 0 \end{cases}, \tag{3.841}$$

or

$$\begin{cases} \dot{u} - \cos(\theta) v = 0 \\ \dot{v} + \cos(\theta) u = 0 \end{cases}, \tag{3.842}$$

where $u = X^\theta$ and $v = \sin\theta X^\phi$. Now, $\theta = 45^\circ = \text{const.}$ and $-60^\circ \leq \phi \leq -30^\circ$, so the solution is given by

$$\begin{pmatrix} u \\ v \end{pmatrix} = \begin{pmatrix} \cos[\alpha(\phi - \phi_0)] & \sin[\alpha(\phi - \phi_0)] \\ -\sin[\alpha(\phi - \phi_0)] & \cos[\alpha(\phi - \phi_0)] \end{pmatrix} \begin{pmatrix} u_0 \\ v_0 \end{pmatrix}, \tag{3.843}$$

with $\alpha \equiv \cos\theta = \frac{1}{\sqrt{2}}$ and $\phi_0 = -30^\circ$. Thus, the vector (u, v) is rotated by the angle

$$\alpha(-60^\circ - \phi_0) = \frac{1}{\sqrt{2}}[-60^\circ - (-30^\circ)] = -\frac{1}{\sqrt{2}} \cdot 30^\circ = -\frac{\pi}{6\sqrt{2}}, \tag{3.844}$$

and the final direction is: $45^\circ - \frac{30^\circ}{\sqrt{2}} \approx 23.8^\circ$ $\left(\frac{3-\sqrt{2}}{12}\pi \approx 0.415\right)$.
Note the sign: Compass directions are taken clockwise.

Detailed solution: In order to find how the vector is parallel transported, we must solve the parallel transport equations, which are given by

$$\begin{cases} \dot{X}^\theta - \sin(\theta)\cos(\theta) X^\phi \dot{\phi} = 0 \\ \dot{X}^\phi + \cot(\theta)(X^\phi \dot{\theta} + X^\theta \dot{\phi}) = 0 \end{cases}. \tag{3.845}$$

First, we choose our coordinate system such that Cape Verde has the coordinates $\theta = \frac{\pi}{2}$ and $\phi = 0$. Then, we have

1. Cape Verde: $\theta = \frac{\pi}{2}, \phi = 0$,
2. Azores: $\theta = \frac{\pi}{4}, \phi = 0$,
3. Nova Scotia: $\theta = \frac{\pi}{4}, \phi = -\frac{\pi}{6}$.

For the trip from 1 to 2, we set $\theta = \frac{\pi}{2} - \tau$ and $\phi = 0$ when $0 \leq \tau \leq \frac{\pi}{4}$. The parallel transport equations then become

$$\begin{cases} \dot{X}^\theta = 0 \\ \dot{X}^\phi - \cot(\theta) X^\phi = 0 \end{cases}, \tag{3.846}$$

with initial conditions

$$\begin{cases} X^\theta(0) = \frac{1}{\sqrt{2}} \\ X^\phi(0) = \frac{1}{\sqrt{2}} \end{cases}. \tag{3.847}$$

The first equation can be easily solved giving $X^\theta\left(\frac{\pi}{4}\right) = X^\theta(0) = \frac{1}{\sqrt{2}}$, whereas the second equation can be integrated and we find that

$$\begin{aligned} \int_0^{\frac{\pi}{4}} \frac{\dot{X}^\phi}{X^\phi} d\tau &= \ln X^\phi\left(\tfrac{\pi}{4}\right) - \ln X^\phi(0) = \ln \frac{X^\phi\left(\frac{\pi}{4}\right)}{X^\phi(0)} \\ &= \int_0^{\frac{\pi}{4}} \cot\left(\tfrac{\pi}{2} - \tau\right) d\tau = \left[-\ln\left|\sin\left(\tfrac{\pi}{2} - \tau\right)\right|\right]_0^{\frac{\pi}{4}} \\ &= -\ln\left(\frac{1/\sqrt{2}}{1}\right) = \ln\sqrt{2}, \end{aligned} \tag{3.848}$$

which yields $X^\phi\left(\frac{\pi}{4}\right) = \sqrt{2}X^\phi(0) = 1$. Thus, we obtain

$$X^\theta\left(\tfrac{\pi}{4}\right) = \frac{1}{\sqrt{2}}, \quad X^\phi\left(\tfrac{\pi}{4}\right) = 1. \tag{3.849}$$

Note that the vector still has the same length as

$$g_{\mu\nu}X^\mu X^\nu\big|_{\tau=0} = \left(\frac{1}{\sqrt{2}}\right)^2 + \sin^2\frac{\pi}{2}\cdot\left(\frac{1}{\sqrt{2}}\right)^2 = 1, \tag{3.850}$$

$$g_{\mu\nu}X^\mu X^\nu\big|_{\tau=\frac{\pi}{4}} = \left(\frac{1}{\sqrt{2}}\right)^2 + \sin^2\frac{\pi}{4}\cdot 1^2 = 1. \tag{3.851}$$

For the trip from 2 to 3, we set $\theta = \frac{\pi}{4}$ and $\phi = -\tau$ when $0 \leq \tau \leq \frac{\pi}{6}$. The parallel transport equations then become

$$\begin{cases} \dot{X}^\theta + \frac{1}{2}X^\phi = 0 \\ \dot{X}^\phi - X^\phi = 0 \end{cases}, \tag{3.852}$$

with the initial conditions

$$\begin{cases} X^\theta(0) = \frac{1}{\sqrt{2}} \\ X^\phi(0) = 1 \end{cases}. \tag{3.853}$$

Now, we can write the two equations as

$$\begin{pmatrix} \dot{X}^\theta \\ \dot{X}^\phi \end{pmatrix} = \begin{pmatrix} 0 & \frac{1}{2} \\ -1 & 0 \end{pmatrix}\begin{pmatrix} X^\theta \\ X^\phi \end{pmatrix} \equiv A\begin{pmatrix} X^\theta \\ X^\phi \end{pmatrix}. \tag{3.854}$$

There are many ways to solve these equations, e.g.,

- Differentiate the equations once more to obtain uncoupled second-order equations, solve them, and check consistency with the first-order equations.
- Diagonalize the matrix A and solve the uncoupled first-order equations, and then transform back to the original basis to find the solution.
- Compute the matrix exponential of the matrix A to find the solution.

Here, we will use the third method, but the choice of method depends on the problem. We have

$$A = \begin{pmatrix} 0 & \frac{1}{2} \\ -1 & 0 \end{pmatrix}, \tag{3.855}$$

so the exponential becomes

$$\exp(A\tau) = \begin{pmatrix} \cos\left(\frac{1}{\sqrt{2}}\tau\right) & \frac{1}{\sqrt{2}}\sin\left(\frac{1}{\sqrt{2}}\tau\right) \\ -\sqrt{2}\sin\left(\frac{1}{\sqrt{2}}\tau\right) & \cos\left(\frac{1}{\sqrt{2}}\tau\right) \end{pmatrix}, \tag{3.856}$$

and we find that

$$\begin{pmatrix} X^\theta\left(\frac{\pi}{6}\right) \\ X^\phi\left(\frac{\pi}{6}\right) \end{pmatrix} = \exp\left(\tfrac{\pi}{6}A\right)\begin{pmatrix} X^\theta(0) \\ X^\phi(0) \end{pmatrix} = \begin{pmatrix} \cos\left(\frac{\pi}{6\sqrt{2}}\right) & \frac{1}{\sqrt{2}}\sin\left(\frac{\pi}{6\sqrt{2}}\right) \\ -\sqrt{2}\sin\left(\frac{\pi}{6\sqrt{2}}\right) & \cos\left(\frac{\pi}{6\sqrt{2}}\right) \end{pmatrix}\begin{pmatrix} \frac{1}{\sqrt{2}} \\ 1 \end{pmatrix}. \tag{3.857}$$

Thus, we obtain

$$X^\theta\left(\tfrac{\pi}{6}\right) = \frac{1}{\sqrt{2}}\left[\cos\left(\frac{\pi}{6\sqrt{2}}\right) + \sin\left(\frac{\pi}{6\sqrt{2}}\right)\right], \tag{3.858}$$

$$X^\phi\left(\tfrac{\pi}{6}\right) = -\sin\left(\frac{\pi}{6\sqrt{2}}\right) + \cos\left(\frac{\pi}{6\sqrt{2}}\right). \tag{3.859}$$

At 3, i.e., at Nova Scotia, the unit vector pointing 45° north-east is $\mathbf{w} = (w^\theta, w^\phi) = \left(\frac{1}{\sqrt{2}}, 1\right)$. Therefore, the inner product of $\mathbf{v}$ and $\mathbf{w}$ is

$$\begin{aligned} g_{\mu\nu}v^\mu w^\nu &= 1 \cdot \frac{1}{\sqrt{2}} \cdot \frac{1}{\sqrt{2}}\left[\cos\left(\frac{\pi}{6\sqrt{2}}\right) + \sin\left(\frac{\pi}{6\sqrt{2}}\right)\right] \\ &\quad + \sin^2\frac{\pi}{4} \cdot 1 \cdot \left[-\sin\left(\frac{\pi}{6\sqrt{2}}\right) + \cos\left(\frac{\pi}{6\sqrt{2}}\right)\right] = \cos\left(\frac{\pi}{6\sqrt{2}}\right), \end{aligned} \tag{3.860}$$

which means that the transported vector has deviated an angle $\frac{\pi}{6\sqrt{2}} \simeq 21.2°$ during the parallel transport. Thus, its final direction is $\frac{\pi}{4} - \frac{\pi}{6\sqrt{2}} \simeq 23.8°$.

2.12

In a parallel transport all angles and lengths are preserved; on the other hand, the tangent vectors of a geodesic are parallel transported by definition. Let v_A be the unit tangent vector to the great circle AN, at the point A, pointing toward the North Pole N. After parallel transport to the position N, it becomes a unit tangent vector to the curve AN at N. This vector v_N forms an angle $\pi - \theta$ with the tangent vector to the curve NB at N. After parallel transport to the position B, it becomes a vector v_B that forms an angle $(\pi - \theta) - \frac{\pi}{2} = \frac{\pi}{2} - \theta$ with the equator. Thus, the final vector v'_A at A forms an angle $\frac{\pi}{2} - \theta$ with the equator, i.e., an angle $\frac{\pi}{2} - \left(\frac{\pi}{2} - \theta\right) = \theta$ with the vector v_A. However, the area of ANB is θR^2. The parallel transports described here are depicted in Figure 3.11.

2.13

The line element in $\mathbb{R}^3$ is $ds^2 = dx^2 + dy^2 + dz^2$. With the standard coordinates on the unit sphere given by

$$x = \sin(\theta)\cos(\phi), \quad y = \sin(\theta)\sin(\phi), \quad z = \cos(\theta), \tag{3.861}$$

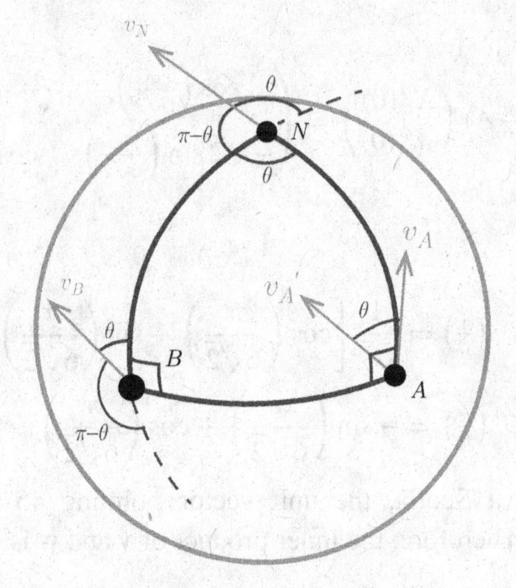

Figure 3.11 Illustration of the parallel transports in the solution to Problem 2.12.

we find that the induced line element on the sphere becomes

$$ds^2 = d\theta^2 + \sin^2(\theta)d\phi^2. \tag{3.862}$$

Thus, the nonzero components of the metric tensor on the sphere are $g_{\theta\theta} = 1$ and $g_{\phi\phi} = \sin^2(\theta)$.

We compute the inner product of the vectors $u = (u^\theta, u^\phi) = (1, 2)$ and $v = (v^\theta, v^\phi) = (2, -1)$ (in the θ- and ϕ-coordinates) to be

$$\begin{aligned} g(u, v) = g_{ij}u^i v^j &= g_{\theta\theta}u^\theta v^\theta + g_{\phi\phi}u^\phi v^\phi \\ &= 1 \cdot 1 \cdot 2 + \sin^2\theta \cdot 2 \cdot (-1) = 2(1 - \sin^2\theta), \end{aligned} \tag{3.863}$$

at the point given by the coordinates (θ, ϕ).

2.14

Since there are only two independent coordinates, i.e., θ and ϕ, all nonzero components of R are in fact determined by the single independent component

$$R_{\theta\phi\theta\phi} = g_{\theta\theta}R^\theta{}_{\phi\theta\phi} = R^\theta{}_{\phi\theta\phi}, \tag{3.864}$$

where the last equality uses $g_{\theta\theta} = 1$. From the definition of the Riemann curvature tensor, we find that

$$\begin{aligned} R^i{}_{\phi\theta\phi}\partial_i &= \nabla_\theta\nabla_\phi\partial_\phi - \nabla_\phi\nabla_\theta\partial_\phi = \nabla_\theta(\Gamma^\theta_{\phi\phi}\partial_\theta) - \nabla_\phi(\Gamma^\phi_{\theta\phi}\partial_\phi) \\ &= (\partial_\theta\Gamma^\theta_{\phi\phi})\partial_\theta - \Gamma^\phi_{\theta\phi}\Gamma^\theta_{\phi\phi}\partial_\theta = -\frac{1}{2}[\partial_\theta \sin(2\theta)]\partial_\theta + \frac{1}{2}\cot(\theta)\sin(2\theta)\partial_\theta \end{aligned}$$

$$= -\cos(2\theta)\partial_\theta + \frac{\cos(\theta)\sin(\theta)\cos(\theta)}{\sin(\theta)}\partial_\theta$$
$$= [-\cos^2(\theta) + \sin^2(\theta) + \cos^2(\theta)]\partial_\theta = \sin^2(\theta)\partial_\theta. \quad (3.865)$$

We conclude that $R_{\theta\phi\theta\phi} = \sin^2(\theta)$. From this, all of the nonzero components of the Riemann curvature tensor may be determined as

$$R^\theta{}_{\phi\theta\phi} = -R^\theta{}_{\phi\phi\theta} = g^{\theta\theta} R_{\theta\phi\theta\phi} = \sin^2(\theta), \quad R^\phi{}_{\theta\phi\theta} = -R^\phi{}_{\theta\theta\phi} = g^{\phi\phi} R_{\phi\theta\phi\theta} = 1. \quad (3.866)$$

2.15

a) Inserting the vector fields

$$X = x\frac{\partial}{\partial y} - y\frac{\partial}{\partial x} \quad \text{and} \quad Y = x\frac{\partial}{\partial x} + y\frac{\partial}{\partial y}, \quad (3.867)$$

into the definition of the commutator $[X,Y]$, one obtains

$$[X,Y] = XY - YX$$
$$= \left(x\frac{\partial}{\partial y} - y\frac{\partial}{\partial x}\right)\left(x\frac{\partial}{\partial x} + y\frac{\partial}{\partial y}\right) - \left(x\frac{\partial}{\partial x} + y\frac{\partial}{\partial y}\right)\left(x\frac{\partial}{\partial y} - y\frac{\partial}{\partial x}\right)$$
$$= \left(x^2\frac{\partial^2}{\partial y\partial x} + x\frac{\partial}{\partial y} + xy\frac{\partial^2}{\partial y^2} - y\frac{\partial}{\partial x} - yx\frac{\partial^2}{\partial x^2} - y^2\frac{\partial^2}{\partial x\partial y}\right)$$
$$- \left(x\frac{\partial}{\partial y} + x^2\frac{\partial^2}{\partial x\partial y} - xy\frac{\partial^2}{\partial x^2} + yx\frac{\partial^2}{\partial y^2} - y\frac{\partial}{\partial x} - y^2\frac{\partial^2}{\partial y\partial x}\right)$$
$$= \left\{\frac{\partial^2}{\partial x\partial y} = \frac{\partial^2}{\partial y\partial x}\right\} = 0, \quad (3.868)$$

i.e., $[X,Y] = 0$.

Note that in polar coordinates the vector fields can be written as

$$X = x\frac{\partial}{\partial y} - y\frac{\partial}{\partial x} = \frac{\partial}{\partial \phi} \quad \text{and} \quad Y = x\frac{\partial}{\partial x} + y\frac{\partial}{\partial y} = r\frac{\partial}{\partial r}, \quad (3.869)$$

which means that

$$[X,Y] = \left[\frac{\partial}{\partial\phi}, r\frac{\partial}{\partial r}\right] = \frac{\partial}{\partial\phi}\left(r\frac{\partial}{\partial r}\right) - r\frac{\partial}{\partial r}\frac{\partial}{\partial\phi} = r\frac{\partial}{\partial\phi}\frac{\partial}{\partial r} - r\frac{\partial}{\partial\phi}\frac{\partial}{\partial r} = 0. \quad (3.870)$$

b) Using the definition of the torsion, namely

$$T(X,Y) = \nabla_X Y - \nabla_Y X - [X,Y], \quad (3.871)$$

together with the facts that the torsion is zero, $\nabla_Y X = X$, and $[X,Y] = 0$, one obtains

$$0 = \nabla_X Y - X - 0, \quad (3.872)$$

i.e., $\nabla_X Y = X$.

Now, using the definition of the curvature, namely

$$R(X,Y)Z = [\nabla_X, \nabla_Y]\,Z - \nabla_{[X,Y]}Z, \tag{3.873}$$

we obtain

$$\begin{aligned} R(X,Y)X &= \nabla_X(\nabla_Y X) - \nabla_Y(\nabla_X X) - \nabla_{[X,Y]}X \\ &= \nabla_X X - \nabla_Y(-Y) = -Y + Y = 0, \end{aligned} \tag{3.874}$$

$$R(X,Y)Y = \nabla_X(\nabla_Y Y) - \nabla_Y(\nabla_X Y) - \nabla_{[X,Y]}Y = \nabla_X Y - \nabla_Y X = X - X = 0, \tag{3.875}$$

which implies that $R(X,Y) = 0$, since $\{X,Y\}$ is a basis of vector fields at all points $(x,y) \neq (0,0)$.

2.16

In the XY-basis, the Christoffel symbols are given by the 2×2-matrices

$$\Gamma^{\bullet}_{X\bullet} = \begin{pmatrix} 0 & 1 \\ 0 & 1 \end{pmatrix} \quad \text{and} \quad \Gamma^{\bullet}_{Y\bullet} = \begin{pmatrix} 1 & 0 \\ -1 & 0 \end{pmatrix}. \tag{3.876}$$

Thus, for the Riemann curvature tensor, we have

$$R^{\bullet}{}_{\bullet XY} = X \cdot \Gamma^{\bullet}_{Y\bullet} - Y \cdot \Gamma^{\bullet}_{X\bullet} + [\Gamma_X, \Gamma_Y] = \begin{pmatrix} -1 & -1 \\ -1 & 1 \end{pmatrix}. \tag{3.877}$$

Note that $R^{\bullet}{}_{\bullet YX} = -R^{\bullet}{}_{\bullet XY}$ and $R^{\bullet}{}_{\bullet XX} = R^{\bullet}{}_{\bullet YY} = 0$. Furthermore, we can express the vector fields as

$$\begin{pmatrix} X \\ Y \end{pmatrix} = \begin{pmatrix} a_{11} & a_{12} \\ a_{21} & a_{22} \end{pmatrix} \begin{pmatrix} \partial_1 \\ \partial_2 \end{pmatrix} = \begin{pmatrix} x^2 & -x^1 \\ x^1 & x^2 \end{pmatrix} \begin{pmatrix} \partial_1 \\ \partial_2 \end{pmatrix}, \tag{3.878}$$

$$\begin{pmatrix} \partial_1 \\ \partial_2 \end{pmatrix} = \begin{pmatrix} b_{11} & b_{12} \\ b_{21} & b_{22} \end{pmatrix} \begin{pmatrix} X \\ Y \end{pmatrix} = \frac{1}{r^2} \begin{pmatrix} x^2 & x^1 \\ -x^1 & x^2 \end{pmatrix} \begin{pmatrix} X \\ Y \end{pmatrix}, \tag{3.879}$$

where $r^2 = (x^1)^2 + (x^2)^2$.

Let us denote the Riemann curvature tensor in the XY-system as $\tilde{R}^{\ell}{}_{ijk}$, where $i,j,k,\ell = 1,2$ corresponds to XY. Then, in the x^1x^2-coordinates, we have

$$R^{\ell}{}_{ijk} = a_{\alpha\ell} b_{i\beta} b_{j\gamma} b_{k\omega} \tilde{R}^{\alpha}{}_{\beta\gamma\omega}. \tag{3.880}$$

Inserting the matrices a and b above as well as the components of $\tilde{R}$ into this formula for $R = (R^{\ell}{}_{ijk})$, we obtain

$$R^1{}_{112} = \frac{1}{r^4}\left[(x^1)^2 - 2x^1x^2 - (x^2)^2\right]. \tag{3.881}$$

Note that $R^1{}_{121} = -R^1{}_{112}$ and always $R^1{}_{1ii} = 0$ (no summation).

2.17
The commutation relations of the vector fields are given by

$$\left[L_i, L_j\right] = \epsilon_{ijk} L_k. \tag{3.882}$$

Now, by definition

$$\nabla_i L_j = \Gamma^k_{ij} L_k. \tag{3.883}$$

On the other hand, the torsion is zero, i.e.,

$$T_{ij} = \nabla_i L_j - \nabla_j L_i - \left[L_i, L_j\right] \equiv 0. \tag{3.884}$$

Inserting Eqs. (3.883) and (3.882) into Eq. (3.884) yields

$$\Gamma^k_{ij} = \Gamma^k_{ji} + \epsilon_{ijk}. \tag{3.885}$$

The orthonormal metric is given by $g(L_i, L_j) = \delta_{ij}$, which implies that

$$L_i g(L_j, L_k) = 0 = g(\nabla_i L_j, L_k) + g(L_j, \nabla_i L_k). \tag{3.886}$$

Using Eq. (3.886), we obtain

$$\Gamma^k_{ij} = -\Gamma^j_{ik}. \tag{3.887}$$

Working out the symmetries gives the relations

$$\Gamma^k_{ij} = -\Gamma^j_{ik} = -\Gamma^j_{ki} - \epsilon_{ikj} = \Gamma^i_{kj} - \epsilon_{ikj} = \Gamma^i_{jk} = -\Gamma^k_{ji} = -\Gamma^k_{ij} - \epsilon_{jik}. \tag{3.888}$$

From this, we obtain

$$2\Gamma^k_{ij} = -\epsilon_{jik}, \tag{3.889}$$

and finally, we find that

$$\Gamma^k_{ij} = \frac{1}{2}\epsilon_{ijk}. \tag{3.890}$$

Define the matrix $\Gamma_i \equiv \left(\Gamma^k_{ij}\right)$. Then, we have the commutation relations

$$[\Gamma_1, \Gamma_2] = \frac{1}{2}\Gamma_3, \tag{3.891}$$

etc. From this, we obtain the Riemann curvature tensor R as

$$R(L_1, L_2) = [\nabla_1, \nabla_2] - \nabla_{[L_1, L_2]} = [\Gamma_1, \Gamma_2] - \Gamma_3 = -\frac{1}{2}\Gamma_3, \tag{3.892}$$

etc.

2.18

a) We have that

$$\nabla_1 e_1 = \nabla_x \frac{\partial}{\partial x} = (x+y)e_1 - \frac{\partial}{\partial y} = (x+y)e_1 - e_2 - (x+y)e_1 = -e_2, \tag{3.893}$$

$$\begin{aligned}\nabla_1 e_2 &= \nabla_x \left[-(x+y)\frac{\partial}{\partial x} + \frac{\partial}{\partial y}\right] \\ &= -\frac{\partial}{\partial x} - (x+y)\left[(x+y)\frac{\partial}{\partial x} - \frac{\partial}{\partial y}\right] + [2+(x+y)^2]\frac{\partial}{\partial x} - (x+y)\frac{\partial}{\partial y} \\ &= \frac{\partial}{\partial x} = e_1, \end{aligned}\tag{3.894}$$

$$\begin{aligned}\nabla_2 e_1 &= [-(x+y)\nabla_x + \nabla_y]\frac{\partial}{\partial x} \\ &= -(x+y)\left[(x+y)\frac{\partial}{\partial x} - \frac{\partial}{\partial y}\right] + (x+y)(x+y+1)\frac{\partial}{\partial x} - (x+y+1)\frac{\partial}{\partial y} \\ &= (x+y)\frac{\partial}{\partial x} - \frac{\partial}{\partial y} = -e_2, \end{aligned}\tag{3.895}$$

$$\nabla_2 e_2 = [-(x+y)\nabla_x + \nabla_y]\left[-(x+y)\frac{\partial}{\partial x} + \frac{\partial}{\partial y}\right] = \{\ldots\} = \frac{\partial}{\partial x} = e_1. \tag{3.896}$$

We have used that $\nabla_{\alpha X + \beta Y} Z = \alpha \nabla_X Z + \beta \nabla_Y Z$ for all vector fields X, Y, and Z and smooth functions α, β, and $\nabla_X \alpha Y = (X \cdot \alpha) Y + \alpha \nabla_X Y$.

b) Now, ∇_1 and ∇_2 are represented by *antisymmetric* matrices

$$\begin{pmatrix} 0 & 1 \\ -1 & 0 \end{pmatrix},$$

in the orthogonal bases $\{e_1, e_2\}$. It follows that the metric g_{ij} is *compatible* with the connection ∇, i.e.,

$$\begin{aligned} \nabla_i g_{jk} &= 0 \quad \text{in this basis or} \\ e_i \cdot g_{jk} &= 0 = g(\nabla_i e_j, e_k) + g(e_j, \nabla_i e_k). \end{aligned}\tag{3.897}$$

Therefore, angles and lengths are preserved in parallel transport. Thus, the final angle is $\pi/3$.

2.19

The two-dimensional surface is given by the parametrization

$$x = r\cos\phi, \quad y = r\sin\phi, \quad z = \frac{2}{3}r^{3/2}. \tag{3.898}$$

a) Using the differentials of this parametrization

$$dx = \frac{\partial x}{\partial r}\,dr + \frac{\partial x}{\partial \phi}\,d\phi = \cos\phi\,dr - r\sin\phi\,d\phi, \tag{3.899}$$

$$dy = \frac{\partial y}{\partial r}\,dr + \frac{\partial y}{\partial \phi}\,d\phi = \sin\phi\,dr + r\cos\phi\,d\phi, \tag{3.900}$$

$$dz = \frac{\partial z}{\partial r}\,dr + \frac{\partial z}{\partial \phi}\,d\phi = \frac{2}{3}\cdot\frac{3}{2}r^{1/2}\,dr + 0\,d\phi = \sqrt{r}\,dr, \tag{3.901}$$

we find that the flat three-dimensional Euclidean metric induces the two-dimensional metric on the surface such that

$$\begin{aligned}
ds^2 &= dx^2 + dy^2 + dz^2 \\
&= (\cos\phi\,dr - r\sin\phi\,d\phi)^2 + (\sin\phi\,dr + r\cos\phi\,d\phi)^2 + \left(\sqrt{r}\,dr\right)^2 \\
&= \cos^2\phi\,dr^2 + r^2\sin^2\phi\,d\phi^2 - 2r\sin\phi\cos\phi\,drd\phi \\
&\quad + \sin^2\phi\,dr^2 + r^2\cos^2\phi\,d\phi^2 + 2r\sin\phi\cos\phi\,drd\phi \\
&\quad + r\,dr^2 \\
&= dr^2 + r^2\,d\phi^2 + r\,dr = (1+r)\,dr^2 + r^2 d\phi^2 \equiv g_{ij}\,dx^i dx^j.
\end{aligned} \tag{3.902}$$

Thus, we obtain the induced metric on the surface as

$$g = (g_{ij}) = \begin{pmatrix} g_{rr} & g_{r\phi} \\ g_{\phi r} & g_{\phi\phi} \end{pmatrix} = \begin{pmatrix} 1+r & 0 \\ 0 & r^2 \end{pmatrix} = \operatorname{diag}(1+r, r^2). \tag{3.903}$$

b) In order to investigate if the three different curves are geodesics, we consider the Lagrangian

$$\mathcal{L} = \sqrt{g_{ij}\dot{x}^i\dot{x}^j} = \left[(1+r)\dot{r}^2 + r^2\dot{\phi}^2\right]^{1/2}, \tag{3.904}$$

where a dot indicates differentiation with respect to the curve parameter λ. Inserting this Lagrangian into the Euler–Lagrange equations for the two coordinates r and ϕ, we find the two geodesic equations

$$\begin{gathered}
\frac{\partial\mathcal{L}}{\partial r} - \frac{d}{d\lambda}\frac{\partial\mathcal{L}}{\partial\dot{r}} = 0, \quad \frac{\partial\mathcal{L}}{\partial r} = \frac{1}{2\mathcal{L}}\left(\dot{r}^2 + 2r\dot{\phi}^2\right), \quad \frac{\partial\mathcal{L}}{\partial\dot{r}} = \frac{1}{\mathcal{L}}(1+r)\dot{r} \\
\Rightarrow \ddot{r} + \frac{1}{2(1+r)}\dot{r}^2 - \frac{r}{1+r}\dot{\phi}^2 - \frac{\dot{r}}{2\mathcal{L}^2}\left[\dot{r}^3 + 2(1+r)\dot{r}\ddot{r} + 2r\dot{r}\dot{\phi}^2 + 2r^2\dot{\phi}\ddot{\phi}\right] = 0,
\end{gathered} \tag{3.905}$$

$$\frac{\partial\mathcal{L}}{\partial\phi} - \frac{d}{d\lambda}\frac{\partial\mathcal{L}}{\partial\dot{\phi}} = 0, \quad \frac{\partial\mathcal{L}}{\partial\phi} = 0, \quad \frac{\partial\mathcal{L}}{\partial\dot{\phi}} = \frac{1}{\mathcal{L}}r^2\dot{\phi} \quad \Rightarrow \quad \frac{d}{d\lambda}\left(\frac{1}{\mathcal{L}}r^2\dot{\phi}\right) = 0. \tag{3.906}$$

First, for Ant #1, we have $r = \lambda$ and $\phi = 0$, which lead to $\dot{r} = 1$, $\ddot{r} = 0$, $\dot{\phi} = 0$, and $\ddot{\phi} = 0$. Therefore, the geodesic equation for the r-coordinate becomes

$$\frac{1}{2(1+\lambda)} - \frac{1}{2\mathcal{L}^2} = 0. \tag{3.907}$$

Now, inserting $r = \lambda$, $\dot{r} = 1$, and $\dot{\phi} = 0$ into the Lagrangian, we find that $\mathcal{L} = \sqrt{1+\lambda}$, which implies that the geodesic equation for the r-coordinate is immediately satisfied. The geodesic equation for the ϕ-coordinate is trivially satisfied, since $\dot{\phi} = 0$. Thus, Ant #1 follows a geodesic.

Second, for Ant #2, we have $r = \lambda^{2/3} - 1$ and $\phi = \pi/2$, which lead to $\dot{r} = 2\lambda^{-1/3}/3$, $\ddot{r} = -2\lambda^{-4/3}/9$, $\dot{\phi} = 0$, and $\ddot{\phi} = 0$. Therefore, the geodesic equation for the r-coordinate becomes

$$\begin{aligned} &-\frac{2}{9}\lambda^{-4/3} + \frac{1}{2}\lambda^{-2/3} \cdot \frac{4}{9}\lambda^{-2/3} \\ &-\frac{1}{3}\lambda^{-1/3} \cdot \frac{1}{\mathcal{L}^2}\left[\frac{8}{27}\lambda^{-1} + 2\lambda^{2/3} \cdot \frac{2}{3}\lambda^{-1/3} \cdot \left(-\frac{2}{9}\lambda^{-4/3}\right)\right] = 0, \end{aligned} \tag{3.908}$$

which turns out to be trivially satisfied for all $\lambda > 1$. Again, the geodesic equation for the ϕ-coordinate is trivially satisfied, since also $\dot{\phi} = 0$. Thus, Ant #2 follows a geodesic.

Note that the curves for Ants #1 and #2 are indeed similar. In principle, the two curves just differ by different values of the ϕ-coordinate, but they do have different parametrizations for the r-coordinate.

Finally, for Ant #3, we have $r = \lambda^{1/2}$ and $\phi = \ln \lambda$ with $\lambda > 0$, which lead to $\dot{r} = \lambda^{-1/2}/2$ and $\dot{\phi} = \lambda^{-1}$. Therefore, we find that

$$\mathcal{L} = \frac{1}{2}\left(5\lambda^{-1} + \lambda^{-1/2}\right)^{1/2}, \quad \frac{\partial \mathcal{L}}{\partial \dot{\phi}} = \frac{\lambda \cdot \lambda^{-1}}{\mathcal{L}} = \frac{1}{\mathcal{L}} = 2\left(5\lambda^{-1} + \lambda^{-1/2}\right)^{-1/2}, \tag{3.909}$$

so that the left-hand side of the geodesic equation for the ϕ-coordinate becomes

$$\frac{d}{d\lambda}\frac{\partial \mathcal{L}}{\partial \dot{\phi}} = \frac{5\lambda^{-2} + \frac{1}{2}\lambda^{-3/2}}{\left(5\lambda^{-1} + \lambda^{-1/2}\right)^{3/2}} \neq 0, \tag{3.910}$$

which means that the geodesic equations are not satisfied. Thus, Ant #3 does not follow a geodesic. In conclusion, Ants #1 and #2 are walking on geodesics, whereas Ant #3 is not.

2.20

Since $z^2 = r^2 - a^2$, we get $z\,dz = r\,dr$, i.e., $dz = \pm r\,dr/\sqrt{r^2 - a^2}$ and the induced metric is therefore given by

$$dz^2 = dr^2 + r^2 d\varphi^2 + dz^2 = \left(1 + \frac{r^2}{r^2 - a^2}\right) dr^2 + r^2 d\varphi^2. \tag{3.911}$$

Thus, we can compute the geodesic equation using Euler–Lagrange equations for the following Lagrangian

$$\mathcal{L} = \frac{1}{2}\left(\frac{2r^2 - a^2}{r^2 - a^2}\dot{r}^2 + r^2\dot{\varphi}^2\right). \tag{3.912}$$

We get

$$\frac{d}{d\tau}\frac{\partial \mathcal{L}}{\partial \dot{r}} - \frac{\partial \mathcal{L}}{\partial r} = \frac{d}{d\tau}\left(\frac{2r^2 - a^2}{r^2 - a^2}\dot{r}\right) + \frac{a^2 r}{(r^2 - a^2)^2}\dot{r}^2 - r\dot{\varphi}^2 = 0, \tag{3.913}$$

$$\frac{d}{d\tau}\frac{\partial \mathcal{L}}{\partial \dot{\varphi}} - \frac{\partial \mathcal{L}}{\partial \varphi} = \frac{d}{d\tau}(r^2\dot{\varphi}) = 0, \tag{3.914}$$

or equivalently, we have

$$\ddot{r} - \frac{a^2 r}{(r^2 - a^2)(2r^2 - a^2)}\dot{r}^2 - \frac{(r^2 - a^2)r}{2r^2 - a^2}\dot{\varphi}^2 = 0, \tag{3.915}$$

$$\ddot{\varphi} + \frac{2}{r}\dot{r}\dot{\varphi} = 0. \tag{3.916}$$

Comparing with the general form of the geodesic equation $\ddot{x}^\mu + \Gamma^\mu_{\alpha\beta}\dot{x}^\alpha\dot{x}^\beta = 0$, we obtain the following nonzero Christoffel symbols

$$\Gamma^r_{rr} = -\frac{a^2 r}{(r^2 - a^2)(2r^2 - a^2)}, \quad \Gamma^r_{\varphi\varphi} = -\frac{(r^2 - a^2)r}{2r^2 - a^2}, \quad \Gamma^\varphi_{r\varphi} = \Gamma^\varphi_{\varphi r} = \frac{1}{r}. \tag{3.917}$$

2.21

a) In order to find the metric tensor on the given surface, we first parametrize this surface. We set $c = 1$, so the equation for the surface becomes $t^2 - x^2 - y^2 = -K^2$. We choose the parametrization to be $t = K\sinh(\theta)$, $x = K\cosh(\theta)\cos(\varphi)$, and $y = K\cosh(\theta)\sin(\varphi)$. Next, we set $X = (t, x, y)$. Thus, the tangent vectors of the manifold are

$$\frac{\partial X}{\partial \theta} = K(\cosh(\theta), \sinh(\theta)\cos(\varphi), \sinh(\theta)\sin(\varphi)), \tag{3.918}$$

$$\frac{\partial X}{\partial \varphi} = K\cosh(\theta)(0, -\sin(\varphi), \cos(\varphi)). \tag{3.919}$$

Now, the metric tensor of the embedded surface ($g_{\mu\nu}$) is given by

$$g_{\mu\nu} = \eta_{ij}\frac{\partial X^i}{\partial x^\mu}\frac{\partial X^j}{\partial x^\nu}, \tag{3.920}$$

where (η_{ij}) is the metric tensor of the embedding space, i.e., the three-dimensional Minkowski metric $(\eta_{ij}) = \mathrm{diag}(+1, -1, -1)$. We find that

$$g_{\theta\theta} = \frac{\partial X}{\partial \theta} \cdot \frac{\partial X}{\partial \theta} = K^2(\cosh^2(\theta) - \sinh^2(\theta)\cos^2(\varphi) - \sinh^2(\theta)\sin^2(\varphi)) = K^2, \tag{3.921}$$

$$g_{\theta\varphi} = \frac{\partial X}{\partial \theta} \cdot \frac{\partial X}{\partial \varphi} = K^2 \sinh(\theta)\cosh(\theta)(0 + \cos(\varphi)\sin(\varphi) - \sin(\varphi)\cos(\varphi)) = 0, \tag{3.922}$$

$$g_{\varphi\theta} = g_{\theta\varphi} = 0, \tag{3.923}$$

$$g_{\varphi\varphi} = \frac{\partial X}{\partial \varphi} \cdot \frac{\partial X}{\partial \varphi} = K^2 \cosh^2(\theta)(-\sin^2(\varphi) - \cos^2(\varphi)) = -K^2\cosh^2(\theta), \tag{3.924}$$

which imply that the metric tensor of the embedded surface is

$$(g_{\mu\nu}) = \begin{pmatrix} K^2 & 0 \\ 0 & -K^2\cosh^2(\theta) \end{pmatrix}. \tag{3.925}$$

Therefore, we obtain

$$ds^2 = g_{\mu\nu}dx^\mu dx^\nu = K^2(d\theta^2 - \cosh^2(\theta)\, d\varphi^2), \tag{3.926}$$

which means that the metric on the given surface is not positive definite but indefinite and has signature $+-$.

b) The tangent vector is proportional to $(1, 0, 0)$. Solving $(t, x, y) = (0, 0, K) = p$, we find that $\theta = 0$ and $\varphi = 0$, and using

$$\begin{aligned} K(1,0,0) = \dot\theta(0) \left.\frac{\partial X}{\partial \theta}\right|_p + \dot\varphi(0) \left.\frac{\partial X}{\partial \varphi}\right|_p &= \dot\theta(0) K(1,0,0) + \dot\varphi(0) K(0,1,0) \\ &= K(\dot\theta(0), \dot\varphi(0), 0), \end{aligned} \tag{3.927}$$

we obtain $\dot\theta(0) = 1$ and $\dot\varphi(0) = 0$. The Euler–Lagrange equations of the Lagrangian

$$\mathcal{L} = g_{\mu\nu}\dot x^\mu \dot x^\nu = K^2(\dot\theta^2 - \cosh^2(\theta)\,\dot\varphi^2), \tag{3.928}$$

are the geodesic equations. We find the equations

$$\begin{aligned} &\frac{\partial \mathcal{L}}{\partial \theta} - \frac{d}{d\tau}\frac{\partial \mathcal{L}}{\partial \dot\theta} = K^2\left[-2\cosh(\theta)\sinh(\theta)\,\dot\varphi^2 - \frac{d}{d\tau}(2\dot\theta)\right] = 0 \\ &\Rightarrow \quad \ddot\theta + \cosh(\theta)\sinh(\theta)\,\dot\varphi^2 = 0, \end{aligned} \tag{3.929}$$

$$\frac{\partial \mathcal{L}}{\partial \varphi} - \frac{d}{d\tau}\frac{\partial \mathcal{L}}{\partial \dot\varphi} = 0 - \frac{d}{d\tau}\left(-2K^2\cosh^2(\theta)\,\dot\varphi\right) = 0 \quad \Rightarrow \quad \ddot\varphi + 2\tanh(\theta)\,\dot\theta\dot\varphi = 0, \tag{3.930}$$

which mean that the geodesic equations are

$$\ddot{\theta} + \cosh(\theta)\sinh(\theta)\,\dot{\varphi}^2 = 0, \quad \ddot{\varphi} + 2\tanh(\theta)\,\dot{\theta}\dot{\varphi} = 0, \tag{3.931}$$

with the initial conditions

$$\theta(0) = 0, \quad \varphi(0) = 0, \quad \dot{\theta}(0) = 1, \quad \dot{\varphi}(0) = 0. \tag{3.932}$$

When we derived the Euler–Lagrange equation for φ, we found that $\cosh^2(\theta)\,\dot{\varphi} = I = \text{const}$. Inserting the initial conditions, we find that $I = 0$. Since $\cosh(\theta) \neq 0$, this implies that $\dot{\varphi}(\tau) = 0$ for all τ, which means that $\varphi(\tau) = 0$, since $\varphi(0) = 0$. Therefore, the Euler–Lagrange equation for θ becomes $\ddot{\theta} = 0$, which has the solution $\theta(\tau) = \dot{\theta}(0)\tau + \theta(0) = \tau$, since $\theta(0) = 0$ and $\dot{\theta}(0) = 1$. Thus, the geodesic equations have the solution

$$\theta(\tau) = \tau, \quad \varphi(\tau) = 0. \tag{3.933}$$

One can easily check that this is a solution to the geodesic equations.

2.22

Introduce $(x^1, x^2) = R(\cos\theta, \sin\theta)$ and $(x^3, x^4) = r(\cos\phi, \sin\phi)$. We then have $R^2 - r^2 = 1$ on S, and thus, we can write $R = \sqrt{1 + r^2}$. The pseudo-Riemannian metric in $\mathbb{R}^4$ then induces the pseudo-Riemannian metric on S such that

$$ds^2 = dR^2 + R^2 d\theta^2 - dr^2 - r^2 d\phi^2 = \left(\frac{rdr}{\sqrt{1+r^2}}\right)^2 + R^2 d\theta^2 - dr^2 - r^2 d\phi^2$$
$$= (1 + r^2)d\theta^2 - \frac{1}{1+r^2}dr^2 - r^2 d\phi^2, \tag{3.934}$$

which has one positive sign and two negative signs. Using Euler–Lagrange variational equations for the Lagrangian $\mathcal{L} = (1 + r^2)\dot{\theta}^2 - \frac{1}{1+r^2}\dot{r}^2 - r^2\dot{\phi}^2$, the geodesic equations for θ, ϕ, and r become

$$\frac{d}{ds}(R^2\dot{\theta}) = 0, \tag{3.935}$$

$$\frac{d}{ds}(r^2\dot{\phi}) = 0, \tag{3.936}$$

$$\frac{d}{ds}\left(\frac{2\dot{r}}{1+r^2}\right) + \frac{\partial \mathcal{L}}{\partial r} = 0, \tag{3.937}$$

respectively. Thus, the first two equations give the pair of constants of motion, i.e., $k_\theta \equiv (1 + r^2)\dot{\theta}$ and $k_\phi \equiv r^2\dot{\phi}$.

2.23

In local coordinates, the differential equation describing the flow lines are given by

$$\dot{x}^i = X^i. \tag{3.938}$$

By differentiating this relation once with respect to the curve parameter, we obtain

$$\ddot{x}^i = \frac{d}{ds}X^i = \frac{dx^j}{ds}\partial_j X^i = \dot{x}^j \partial_j X^i = X^j \partial_j X^i. \tag{3.939}$$

For the flow lines to be geodesics, they must fulfill the geodesic equations

$$\ddot{x}^i + \Gamma^i_{\ jk}\dot{x}^j\dot{x}^k = 0. \tag{3.940}$$

If inserting our expressions for $\ddot{x}^i$ and $\dot{x}^i$ in terms of the vector field components X^i into this equation, then we obtain

$$X^j(\partial_j X^i + \Gamma^i_{\ jk}X^k) = 0, \tag{3.941}$$

which is a necessary and sufficient condition for the flow lines to be geodesics. (This can be rewritten as

$$\nabla_X X = 0, \tag{3.942}$$

without any reference to local coordinates, i.e., the directional derivative of the vector X in the direction of X vanishes.)

2.24

The metric induced by the given embedding into $\mathbb{R}^4$ is given by

$$ds^2 = \sum_i dy_i^2 = [d\cos(r)]^2 + \left\{d\left[\frac{\sin(r)}{r}\boldsymbol{x}\right]\right\}^2, \tag{3.943}$$

where y_i are the coordinates in $\mathbb{R}^4$. Computing this metric in terms of the coordinates r, θ, and ϕ, we obtain

$$ds^2 = dr^2 + \sin^2(r)[d\theta^2 + \sin^2(\theta)d\phi^2]. \tag{3.944}$$

By using the Euler–Lagrange equations for $\mathcal{L} = g_{\mu\nu}\dot{x}^\mu\dot{x}^\nu$, we find the geodesic equations

$$\ddot{r} = \frac{1}{2}\sin(2r)\left[\dot{\theta}^2 + \sin^2(\theta)\dot{\phi}^2\right], \tag{3.945}$$

$$\frac{d}{ds}\sin^2(r)\dot{\theta} = \frac{1}{2}\sin^2(r)\sin(2\theta)\dot{\phi}^2, \tag{3.946}$$

$$\frac{d}{ds}\sin^2(r)\sin^2(\theta)\dot{\phi} = 0. \tag{3.947}$$

The given one-parameter subgroups are given by the coordinates $r(t) = at$, $\theta(t) = \theta_0$, and $\phi(t) = \phi_0$, and thus, $\dot{r} = a$, $\ddot{r} = 0$, $\dot{\theta} = 0$, and $\dot{\phi} = 0$. By insertion, we observe that the geodesic equations are fulfilled.

2.25

a) Using the definition of the covariant derivative $D_\mu A_\nu = \partial_\mu A_\nu - \Gamma^\lambda_{\mu\nu} A_\lambda$ and the fact that the metric tensor is covariantly constant, i.e., $D_\mu g_{\alpha\nu} = 0$ [the given condition (i)], we have

$$D_\mu g_{\alpha\nu} = \partial_\mu g_{\alpha\nu} - \Gamma^\rho_{\mu\alpha} g_{\rho\nu} - \Gamma^\rho_{\mu\nu} g_{\alpha\rho} = 0. \tag{3.948}$$

Furthermore, permuting the indices μ, α, and ν cyclically, we obtain

$$D_\nu g_{\mu\alpha} = \partial_\nu g_{\mu\alpha} - \Gamma^\rho_{\nu\mu} g_{\rho\alpha} - \Gamma^\rho_{\nu\alpha} g_{\mu\rho} = 0, \tag{3.949}$$

$$-D_\alpha g_{\nu\mu} = -\partial_\alpha g_{\nu\mu} + \Gamma^\rho_{\alpha\nu} g_{\rho\mu} + \Gamma^\rho_{\alpha\mu} g_{\nu\rho} = 0. \tag{3.950}$$

Adding the three equations and using that the Christoffel symbols are symmetric with respect to the two lower indices, i.e., $\Gamma^\rho_{\mu\nu} = \Gamma^\rho_{\nu\mu}$ [the given condition (ii)], we find that

$$\partial_\mu g_{\alpha\nu} + \partial_\nu g_{\mu\alpha} - \partial_\alpha g_{\nu\mu} - 2\Gamma^\rho_{\mu\nu} g_{\alpha\rho} = 0, \tag{3.951}$$

and contracting with $g^{\lambda\alpha}$, we finally obtain

$$\Gamma^\lambda_{\mu\nu} = \frac{1}{2} g^{\lambda\alpha} \left(\partial_\mu g_{\alpha\nu} + \partial_\nu g_{\alpha\mu} - \partial_\alpha g_{\mu\nu} \right), \tag{3.952}$$

which is the fundamental theorem in Riemannian geometry.

b) We have

$$T_0{}^0 = 0, \quad T_0{}^1 = -1, \quad T_1{}^0 = 1, \quad T_1{}^1 = 0, \tag{3.953}$$

since $D_\mu V^\nu = \partial_\mu V^\nu$ in Minkowski coordinates. Furthermore, we have

$$T'_0{}^1 = \frac{\partial x^\mu}{\partial x'^0} \frac{\partial x'^1}{\partial x^\nu} T_\mu{}^\nu = \frac{\partial x^1}{\partial x'^0} \frac{\partial x'^1}{\partial x^0} - \frac{\partial x^0}{\partial x'^0} \frac{\partial x'^1}{\partial x^1} = \frac{\partial x}{\partial \lambda} \frac{\partial a}{\partial t} - \frac{\partial t}{\partial \lambda} \frac{\partial a}{\partial x}. \tag{3.954}$$

Now, since $t = a \sinh \lambda$ and $x = a \cosh \lambda$, the inverse coordinate transformation is given by $\lambda = \operatorname{artanh}(t/x)$ and $a = \sqrt{x^2 - t^2}$, which leads to

$$\frac{\partial x}{\partial \lambda} = a \sinh \lambda, \quad \frac{\partial a}{\partial t} = -\frac{t}{\sqrt{x^2 - t^2}} = -\sinh \lambda,$$

$$\frac{\partial t}{\partial \lambda} = a \cosh \lambda, \quad \frac{\partial a}{\partial x} = \frac{x}{\sqrt{x^2 - t^2}} = \cosh \lambda. \tag{3.955}$$

Thus, we obtain

$$T'_0{}^1 = a \sinh \lambda \cdot (-\sinh \lambda) - a \cosh \lambda \cdot \cosh \lambda$$
$$= -a(\sinh^2 \lambda + \cosh^2 \lambda) = -a \cosh 2\lambda. \tag{3.956}$$

2.26

a) The general formula is given by

$$S^{\mu\nu} \mapsto S'^{\mu\nu} = \frac{\partial x'^{\mu}}{\partial x^{\alpha}}\frac{\partial x'^{\nu}}{\partial x^{\beta}}S^{\alpha\beta}. \tag{3.957}$$

b) The covariant derivative $D_\mu S^{\mu\nu}$ in terms of partial derivatives and Christoffel symbols is given by

$$D_\mu S^{\mu\nu} = \partial_\mu S^{\mu\nu} + \Gamma^{\mu}_{\mu\alpha}S^{\alpha\nu} + \Gamma^{\nu}_{\mu\alpha}S^{\mu\alpha}. \tag{3.958}$$

c) (i) Using the general formula in a), we have

$$\begin{aligned} S'^{\mu\nu} &= \frac{\partial x'^{\mu}}{\partial x^{\alpha}}\frac{\partial x'^{\nu}}{\partial x^{\beta}}S^{\alpha\beta} = \frac{\partial x'^{\mu}}{\partial x^{1}}\frac{\partial x'^{\nu}}{\partial x^{2}}S^{12} + \frac{\partial x'^{\mu}}{\partial x^{2}}\frac{\partial x'^{\nu}}{\partial x^{1}}S^{21} \\ &= \left(\frac{\partial x'^{\mu}}{\partial x^{1}}\frac{\partial x'^{\nu}}{\partial x^{2}} - \frac{\partial x'^{\mu}}{\partial x^{2}}\frac{\partial x'^{\nu}}{\partial x^{1}}\right)S^{12}. \end{aligned} \tag{3.959}$$

Thus, we obtain

$$\begin{aligned} S'^{12} &= \left(\frac{\partial x'^{1}}{\partial x^{1}}\frac{\partial x'^{2}}{\partial x^{2}} - \frac{\partial x'^{1}}{\partial x^{2}}\frac{\partial x'^{2}}{\partial x^{1}}\right)S^{12} = \left(\frac{x}{r}\frac{x}{r^2} - \frac{y}{r}\frac{-y}{r^2}\right)2xy \\ &= \frac{1}{r^2}(x^2+y^2)2r^2\cos\varphi\sin\varphi = r\sin 2\varphi = -S'^{21}, \end{aligned} \tag{3.960}$$

$$S'^{11} = S'^{22} = 0. \tag{3.961}$$

(ii) In this case, we have

$$D_\mu S^{\mu\nu} = \partial_\mu S^{\mu\nu} = \partial_1 S^{1\nu} + \partial_2 S^{2\nu}, \tag{3.962}$$

which implies that $D_\mu S^{\mu 1} = \partial_2 S^{21} = -2x$ and $D_\mu S^{\mu 2} = \partial_1 S^{12} = 2y$ and can be formally written as $(D_\mu S^{\mu\nu}) = (-2x, 2y)$.

2.27

a) We can compute the geodesic equations from the Lagrangian

$$\mathcal{L} = \frac{1}{2}\left[(r^2-1)\dot{t}^2 - \frac{1}{r^2-1}\dot{r}^2\right], \tag{3.963}$$

where $\dot{t} = dt/d\tau$, $\dot{r} = dr/d\tau$, and τ is the proper time. Using Euler–Lagrange equations for this Lagrangian

$$\frac{d}{d\tau}\frac{\partial\mathcal{L}}{\partial\dot{t}} - \frac{\partial\mathcal{L}}{\partial t} = 0 \quad\Rightarrow\quad \frac{d}{d\tau}\left[(r^2-1)\dot{t}\right] = (r^2-1)\ddot{t} + 2r\dot{r}\dot{t} = 0, \tag{3.964}$$

$$\frac{d}{d\tau}\frac{\partial\mathcal{L}}{\partial\dot{r}} - \frac{\partial\mathcal{L}}{\partial r} = 0 \quad\Rightarrow\quad \frac{d}{d\tau}\left(-\frac{\dot{r}}{r^2-1}\right) - r\dot{t}^2 - \frac{r\dot{r}^2}{(r^2-1)^2} = 0, \tag{3.965}$$

we obtain the geodesic equations

$$\ddot{t} + \frac{2r}{r^2 - 1}\dot{t}\dot{r} = 0, \tag{3.966}$$

$$\ddot{r} + r(r^2 - 1)\dot{t}^2 - \frac{r}{r^2 - 1}\dot{r}^2 = 0. \tag{3.967}$$

Comparing the two geodesic equations with the general form of the geodesic equations

$$\ddot{x}^\mu + \Gamma^\mu_{\nu\lambda}\dot{x}^\nu\dot{x}^\lambda = 0, \tag{3.968}$$

we find the following nonzero Christoffels symbols

$$\Gamma^0_{01} = \Gamma^0_{10} = \frac{r}{r^2 - 1}, \quad \Gamma^1_{00} = r(r^2 - 1), \quad \Gamma^1_{11} = -\frac{r}{r^2 - 1}. \tag{3.969}$$

b) Note that $S_{\mu\nu} = (a/v)g_{\mu\nu}$, where $g_{\mu\nu}$ is the metric tensor with the components $g_{00} = v(r^2 - 1)$, $g_{11} = -v/(r^2 - 1)$, and $g_{01} = g_{10} = 0$. Thus, we have

$$g^{00} = \frac{1}{v(r^2 - 1)}, \quad g^{11} = -\frac{r^2 - 1}{v}, \quad g^{01} = g^{10} = 0, \tag{3.970}$$

and therefore, we obtain $S^{\mu\nu} = (a/v)g^{\mu\nu}$. Moreover, $D_\lambda g^{\mu\nu} = 0$ implies that

$$D_\mu S^{\mu\nu} = 0. \tag{3.971}$$

c) Using the given coordinate transformations, we obtain

$$d\theta = a\,dt, \quad dr = \sinh(\eta)d\eta, \tag{3.972}$$

and thus, we have the metric

$$\begin{aligned} ds^2 &= v\left[(\cosh(\eta)^2 - 1)d\theta^2/a^2 - \frac{1}{\cosh(\eta)^2 - 1}\sinh(\eta)^2 d\eta^2\right] \\ &= \frac{v}{a^2}\sinh^2\eta d\theta^2 - v d\eta^2. \end{aligned} \tag{3.973}$$

This gives

$$g'_{00} = \frac{v}{a^2}\sinh^2\eta, \quad g'_{11} = -v, \quad g'_{01} = g'_{10} = 0, \tag{3.974}$$

and $S'_{\mu\nu} = (a/v)g'_{\mu\nu}$.

2.28

a) Using the Rindler coordinate system, we have

$$dt = a\cosh(\lambda)\,d\lambda + \sinh(\lambda)\,da, \quad dx = a\sinh(\lambda)\,d\lambda + \cosh(\lambda)\,da, \tag{3.975}$$

and thus, we obtain the line element in Rindler coordinates λ and a as

$$ds^2 = dt^2 - dx^2 = a^2 d\lambda^2 - da^2, \tag{3.976}$$

which means that $g_{00} = g_{\lambda\lambda} = a^2$, $g_{11} = g_{aa} = -1$, and $g_{01} = g_{10} = 0$. The nonzero Christoffel symbols can be computed using the Lagrangian given by

$$\mathcal{L} = \frac{1}{2}\left(a^2\dot{\lambda}^2 - \dot{a}^2\right). \tag{3.977}$$

Using Euler–Lagrange equations for this Lagrangian yields

$$\ddot{\lambda} + \frac{2}{a}\dot{\lambda}\dot{a} = 0, \quad \ddot{a} + a\dot{\lambda}^2 = 0, \tag{3.978}$$

which can be compared with the general form of the geodesic equations $\ddot{x}^\mu + \Gamma^\mu_{\nu\rho}\dot{x}^\nu\dot{x}^\rho = 0$ to identify the Christoffel symbols. Thus, we find the following nonzero Christoffel symbols in Rindler coordinates λ and a

$$\Gamma^\lambda_{\lambda a} = \Gamma^\lambda_{a\lambda} = \frac{1}{a}, \quad \Gamma^a_{\lambda\lambda} = a. \tag{3.979}$$

b) The divergence of a vector field V^μ is given by

$$D_\mu V^\mu = \frac{1}{\sqrt{-\bar{g}}}\partial_\mu\left(\sqrt{-\bar{g}}V^\mu\right) = \partial_\mu V^\mu + \frac{1}{\sqrt{-\bar{g}}}\partial_\mu\left(\sqrt{-\bar{g}}\right)V^\mu, \tag{3.980}$$

whereas the Laplacian of a scalar field Φ is

$$\begin{aligned} D_\mu\partial^\mu\Phi &= \frac{1}{\sqrt{-\bar{g}}}\partial_\mu\left(\sqrt{-\bar{g}}g^{\mu\nu}\partial_\nu\Phi\right) = \frac{1}{\sqrt{-\bar{g}}}\partial_\mu\left(\sqrt{-\bar{g}}\partial^\mu\Phi\right) \\ &= \partial_\mu\partial^\mu\Phi + \frac{1}{\sqrt{-\bar{g}}}\left(\partial_\mu\sqrt{-\bar{g}}\right)\partial^\mu\Phi. \end{aligned} \tag{3.981}$$

In Rindler coordinates, it holds that $\bar{g} \equiv \det g = a^2\cdot(-1) - 0^2 = -a^2$, which means that $\sqrt{-\bar{g}} = a$. Furthermore, it holds that $g^{00} = g^{\lambda\lambda} = 1/a^2$ and $g^{11} = g^{aa} = -1$. Thus, in Rindler coordinates, we obtain

$$D_\mu V^\mu = \partial_\lambda V^\lambda + \partial_a V^a + \frac{1}{a}\left[(\partial_\lambda a)V^\lambda + (\partial_a a)V^a\right] = \partial_\lambda V^\lambda + \partial_a V^a + \frac{1}{a}V^a, \tag{3.982}$$

$$\begin{aligned} D_\mu\partial^\mu\Phi &= \partial_\lambda\partial^\lambda\Phi + \partial_a\partial^a\Phi + \frac{1}{a}\partial^a\Phi = \partial_\lambda g^{\lambda\lambda}\partial_\lambda\Phi + \partial_a g^{aa}\partial_a\Phi + \frac{1}{a}g^{aa}\partial_a\Phi \\ &= \left(\frac{1}{a^2}\partial_\lambda^2 - \partial_a^2 - \frac{1}{a}\partial_a\right)\Phi. \end{aligned} \tag{3.983}$$

c) In general, the transformation rule for a tensor of rank two from $T^\mu{}_\nu(t,x)$ to $T'^\mu{}_\nu(\lambda,a)$ is given by

$$T'^\mu{}_\nu(\lambda,a) = \frac{\partial x'^\mu}{\partial x^\alpha}\frac{\partial x^\beta}{\partial x'^\nu}T^\alpha{}_\beta(t(\lambda,a),x(\lambda,a)). \tag{3.984}$$

In particular, for the $T'^0{}_0$ component, we have

$$T'^0{}_0 = \frac{\partial x'^0}{\partial x^\alpha}\frac{\partial x^\beta}{\partial x'^0}T^\alpha{}_\beta = \frac{\partial x'^0}{\partial x^0}\frac{\partial x^0}{\partial x'^0}T^0{}_0 + \frac{\partial x'^0}{\partial x^1}\frac{\partial x^0}{\partial x'^0}T^1{}_0 + \frac{\partial x'^0}{\partial x^0}\frac{\partial x^1}{\partial x'^0}T^0{}_1 + \frac{\partial x'^0}{\partial x^1}\frac{\partial x^1}{\partial x'^0}T^1{}_1. \tag{3.985}$$

Now, the components of the given tensor (in the coordinates t and x) are

$$T^0{}_0 = -T^1{}_1 = x^2 - t^2, \quad T^1{}_0 = T^0{}_1 = 0, \tag{3.986}$$

which means that

$$T'^0{}_0 = \frac{\partial x'^0}{\partial x^0}\frac{\partial x^0}{\partial x'^0}T^0{}_0 + \frac{\partial x'^0}{\partial x^1}\frac{\partial x^1}{\partial x'^0}T^1{}_1 = \left(\frac{\partial \lambda}{\partial t}\frac{\partial t}{\partial \lambda} - \frac{\partial \lambda}{\partial x}\frac{\partial x}{\partial \lambda}\right)(x^2 - t^2). \tag{3.987}$$

Then, since $t = a\sinh\lambda$ and $x = a\cosh\lambda$ in Rindler coordinates, we have $x^2 - t^2 = a^2$. Furthermore, the inverse coordinate transformation is given by $\lambda = \operatorname{artanh}(t/x)$ and $a = \sqrt{x^2 - t^2}$, which leads to

$$\frac{\partial \lambda}{\partial t} = \frac{1}{a}\cosh\lambda, \quad \frac{\partial t}{\partial \lambda} = a\cosh\lambda, \quad \frac{\partial \lambda}{\partial x} = -\frac{1}{a}\sinh\lambda, \quad \frac{\partial x}{\partial \lambda} = a\sinh\lambda. \tag{3.988}$$

Thus, we obtain the $T'^0{}_0$ component of the tensor in Rindler coordinates λ and a as

$$\begin{aligned} T'^0{}_0 &= \left[\frac{1}{a}\cosh\lambda \cdot a\cosh\lambda - \left(-\frac{1}{a}\sinh\lambda\right)\cdot a\sinh\lambda\right]a^2 \\ &= a^2(\cosh^2\lambda + \sinh^2\lambda) = a^2\cosh 2\lambda. \end{aligned} \tag{3.989}$$

2.29

Using the given metric $ds^2 = dt^2 - dr^2 - r^2 d\phi^2$, we find that

$$g_{tt} = g^{tt} = 1, \quad g_{rr} = g^{rr} = -1, \quad g_{\phi\phi} = \frac{1}{g^{\phi\phi}} = -r^2, \tag{3.990}$$

which only give rise to one nonzero ordinary derivative, i.e., $\partial_r g_{\phi\phi} = -2r$. Using the general expression for the Christoffel symbols, we obtain the following nonzero Christoffel symbols

$$\Gamma^\phi_{r\phi} = \Gamma^\phi_{\phi r} = \frac{1}{r}, \quad \Gamma^r_{\phi\phi} = -r. \tag{3.991}$$

Using the definition of the covariant derivative $\nabla_\mu V_\nu = \partial_\mu V_\nu - \Gamma^\lambda_{\mu\nu} V_\lambda$, we obtain the components of the covariant derivative

$$\nabla_t V_\nu = \partial_t V_\nu, \tag{3.992}$$

$$\nabla_r V_t = \partial_r V_t, \tag{3.993}$$

$$\nabla_r V_r = \partial_r V_r, \tag{3.994}$$

$$\nabla_r V_\phi = \partial_r V_\phi - \Gamma^\phi_{r\phi} V_\phi = \partial_r V_\phi - \frac{1}{r} V_\phi, \tag{3.995}$$

$$\nabla_\phi V_t = \partial_\phi V_t, \tag{3.996}$$

$$\nabla_\phi V_r = \partial_\phi V_r - \Gamma^\phi_{\phi r} V_\phi = \partial_\phi V_r - \frac{1}{r} V_\phi, \tag{3.997}$$

$$\nabla_\phi V_\phi = \partial_\phi V_\phi - \Gamma^r_{\phi\phi} V_r = \partial_\phi V_\phi + r V_r, \tag{3.998}$$

which imply the divergence

$$\nabla_\mu V^\mu = \nabla_t V^t + \nabla_r V^r + \nabla_\phi V^\phi = \partial_t V^t + \partial_r V^r + \Gamma^r_{r\nu} V^\nu + \partial_\phi V^\phi + \Gamma^\phi_{\phi\nu} V^\nu$$
$$= \partial_\mu V^\mu + \Gamma^\phi_{\phi r} V^r = \partial_\mu V^\mu + \frac{1}{r} V^r, \tag{3.999}$$

where we used that $\partial_t V^t + \partial_r V^r + \partial_\phi V^\phi = \partial_\mu V^\mu$, $\Gamma^r_{r\nu} = 0$ and $\Gamma^\phi_{\phi t} = \Gamma^\phi_{\phi\phi} = 0$.

2.30

a) First, let us write the tangent vector as $T^\mu = \frac{dx^\mu}{d\lambda}$ and the vector field as $V^\nu = \frac{dx^\nu}{d\lambda}$, then we insert this into the given condition, so that

$$0 = T^\mu \nabla_\mu V^\nu = \frac{dx^\mu}{d\lambda} \nabla_\mu \frac{dx^\nu}{d\lambda} = \frac{dx^\mu}{d\lambda}\left(\partial_\mu \frac{dx^\nu}{d\lambda} + \Gamma^\nu_{\mu\alpha} \frac{dx^\alpha}{d\lambda}\right)$$
$$= \frac{d^2 x^\nu}{d\lambda^2} + \Gamma^\nu_{\mu\alpha} \frac{dx^\mu}{d\lambda} \frac{dx^\alpha}{d\lambda}, \tag{3.1000}$$

where we used $\nabla_\mu V^\nu = \partial_\mu V^\nu + \Gamma^\nu_{\mu\alpha} V^\alpha$ and the chain rule for differentiation, i.e., $\frac{d}{d\lambda} = \frac{dx^\mu}{d\lambda} \frac{\partial}{\partial x^\mu}$. Finally, if we set the parameter λ equal to the proper time τ, i.e., $\lambda = \tau$, we obtain the geodesic equation $\ddot{x}^\nu + \Gamma^\nu_{\mu\alpha} \dot{x}^\mu \dot{x}^\alpha = 0$.

b) Using the generalized condition $T^\mu \nabla_\mu T^\nu = \alpha T^\nu$ and $T^\mu = \frac{dx^\mu}{d\lambda(T)}$, we find that

$$\frac{dx^\mu}{d\lambda(\tau)} \nabla_\mu \frac{dx^\nu}{d\lambda(\tau)} = \alpha \frac{dx^\nu}{d\lambda(\tau)} \quad \Rightarrow \quad \frac{d\tau}{d\lambda} \frac{dx^\mu}{d\tau} \nabla_\mu \left(\frac{d\tau}{d\lambda} \frac{dx^\nu}{d\tau}\right) = \alpha \frac{d\tau}{d\lambda} \frac{dx^\nu}{d\tau}, \tag{3.1001}$$

where we again used the chain rule of differentiation. Now, removing the common factor $\frac{d\tau}{d\lambda}$ and using Leibniz' rule of differentiation, we obtain

$$\frac{dx^\mu}{d\tau}\left[\frac{dx^\nu}{d\tau} \partial_\mu \frac{d\tau}{d\lambda} + \frac{d\tau}{d\lambda} \nabla_\mu \left(\frac{dx^\nu}{d\tau}\right)\right] = \alpha \frac{dx^\nu}{d\tau}, \tag{3.1002}$$

which means that

$$\frac{dx^\mu}{d\tau}\nabla_\mu\left(\frac{dx^\nu}{d\tau}\right)=0, \tag{3.1003}$$

is equivalent to

$$\frac{d}{d\tau}\frac{d\tau}{d\lambda}=\alpha, \tag{3.1004}$$

which may be solved. Adding the extra term to the calculation in a) gives

$$\ddot{x}^\mu+\Gamma^\mu_{\nu\lambda}\dot{x}^\nu\dot{x}^\lambda=\alpha\dot{x}^\mu. \tag{3.1005}$$

2.31

a) The definition of the angle θ between the vectors A^μ and B^μ is given by

$$\cos\theta=\frac{A\cdot B}{\sqrt{A^2B^2}}. \tag{3.1006}$$

If we transform the metric according to $g_{\mu\nu}\mapsto f(x)g_{\mu\nu}$, then the inner product changes accordingly as

$$A\cdot B=g_{\mu\nu}A^\mu B^\nu\mapsto f(x)g_{\mu\nu}A^\mu B^\nu. \tag{3.1007}$$

Thus, it follows that

$$\cos\theta\mapsto\frac{f(x)g_{\mu\nu}A^\mu B^\nu}{\sqrt{f(x)^2g_{\mu\nu}A^\mu A^\nu g_{\sigma\rho}B^\sigma B^\rho}}=\frac{g_{\mu\nu}A^\mu B^\nu}{\sqrt{g_{\mu\nu}A^\mu A^\nu g_{\sigma\rho}B^\sigma B^\rho}}. \tag{3.1008}$$

Thus, the angle does not change under this transformation.

b) For a null curve, the tangent vector given by $V^\mu=dx^\mu/d\tau$ fulfills $g_{\mu\nu}V^\mu V^\nu=0$. If we change the metric according to the conformal transformation, we instead obtain

$$V^2=f(x)g_{\mu\nu}V^\mu V^\nu=f(x)\cdot 0=0. \tag{3.1009}$$

Thus, null curves are still null curves after a conformal transformation.

2.32

The metric tensor in $\mathbb{R}^3$ is given by the line element

$$ds^2=dx^2+dy^2+dz^2. \tag{3.1010}$$

In order to compute the induced metric on the sphere from the given embedding, we therefore compute dx, dy, and dz

$$dx=R[\cos(\theta)\cos(\varphi)d\theta-\sin(\theta)\sin(\varphi)d\varphi], \tag{3.1011}$$

$$dy = R[\cos(\theta)\sin(\varphi)d\theta + \sin(\theta)\cos(\varphi)d\varphi], \tag{3.1012}$$

$$dz = -\alpha R\sin(\theta)d\theta. \tag{3.1013}$$

Squaring and summing leads to

$$ds^2 = R^2\{[\cos^2(\theta) + \alpha^2\sin^2(\theta)]d\theta^2 + \sin^2(\theta)d\varphi^2\}. \tag{3.1014}$$

The metric components can thus be identified as

$$g_{\theta\theta} = R^2[\cos^2(\theta) + \alpha^2\sin^2(\theta)], \quad g_{\varphi\varphi} = R^2\sin^2(\theta), \quad g_{\theta\varphi} = g_{\varphi\theta} = 0. \tag{3.1015}$$

In order to identify the Christoffel symbols, we find the geodesic equations by variation of

$$\mathcal{S} = \frac{1}{2}\int g_{ij}\dot{x}^i\dot{x}^j dt = \frac{R^2}{2}\int \{[\cos^2(\theta) + \alpha^2\sin^2(\theta)]\dot{\theta}^2 + \sin^2(\theta)\dot{\varphi}^2\}dt. \tag{3.1016}$$

The variation yields the Euler–Lagrange equations

$$\begin{aligned}\frac{d}{dt}\{[\cos^2(\theta) + \alpha^2\sin^2(\theta)]\dot{\theta}\} &= \ddot{\theta}[\cos^2(\theta) + \alpha^2\sin^2(\theta)] + (\alpha^2 - 1)\sin(2\theta)\dot{\theta}^2 \\ &= \frac{1}{2}(\alpha^2 - 1)\sin(2\theta)\dot{\theta}^2 + \frac{1}{2}\sin(2\theta)\dot{\varphi}^2,\end{aligned} \tag{3.1017}$$

$$\frac{d}{dt}[\sin^2(\theta)\dot{\varphi}] = \sin^2(\theta)\ddot{\varphi} + \sin(2\theta)\dot{\varphi}\dot{\theta} = 0. \tag{3.1018}$$

Dividing by the multipliers of $\ddot{\theta}$ and $\ddot{\varphi}$, respectively, leads to the geodesic equations

$$\ddot{\theta} + \frac{1}{2}\frac{(\alpha^2 - 1)\sin(2\theta)}{\cos^2(\theta) + \alpha^2\sin^2(\theta)}\dot{\theta}^2 - \frac{\sin(2\theta)}{2[\cos^2(\theta) + \alpha^2\sin^2(\theta)]}\dot{\varphi}^2 = 0, \tag{3.1019}$$

$$\ddot{\varphi} + 2\cot(\theta)\dot{\varphi}\dot{\theta} = 0. \tag{3.1020}$$

Identifying these equations with the geodesic equations, assuming that the connection is torsion free, we find that the nonzero Christoffel symbols are given by

$$\Gamma^\theta_{\theta\theta} = \frac{(\alpha^2 - 1)\sin(2\theta)}{\cos^2(\theta) + \alpha^2\sin^2(\theta)} \to 0, \tag{3.1021}$$

$$\Gamma^\theta_{\varphi\varphi} = -\frac{\sin(2\theta)}{2[\cos^2(\theta) + \alpha^2\sin^2(\theta)]} \to -\frac{\sin(2\theta)}{2}, \tag{3.1022}$$

$$\Gamma^\varphi_{\theta\varphi} = \Gamma^\varphi_{\varphi\theta} = \cot(\theta) \to \cot(\theta), \tag{3.1023}$$

where the limits shown correspond to the limit as $\alpha \to 1$, where we recover the regular embedding of the sphere of radius R in $\mathbb{R}^3$.

2.33

a) Using the definition of the Levi-Civita connection in local coordinates, i.e.,

$$\Gamma^{\lambda}_{\mu\nu} = \frac{1}{2} g^{\lambda\omega} \left(\partial_\mu g_{\nu\omega} + \partial_\nu g_{\mu\omega} - \partial_\omega g_{\mu\nu}\right), \tag{3.1024}$$

with $\lambda = \mu$, one obtains

$$\begin{aligned}
\Gamma^{\mu}_{\mu\nu} &= \frac{1}{2} g^{\mu\omega} \left(\partial_\mu g_{\nu\omega} + \partial_\nu g_{\mu\omega} - \partial_\omega g_{\mu\nu}\right) = \frac{1}{2}\partial^{\omega} g_{\nu\omega} + \frac{1}{2} g^{\mu\omega}\partial_\nu g_{\mu\omega} - \frac{1}{2}\partial^{\mu} g_{\mu\nu} \\
&= \{g_{\mu\nu} \text{ symmetric}\} = \frac{1}{2} g^{\omega\mu}\partial_\nu g_{\mu\omega} = \frac{1}{2}\operatorname{tr}\left(g^{-1}\partial_\nu g\right) = \frac{1}{2}\operatorname{tr}\partial_\nu \ln g \\
&= \frac{1}{2}\partial_\nu \operatorname{tr}\ln g = \frac{1}{2}\partial_\nu \ln\det g = \frac{1}{2}(\det g)^{-1}\partial_\nu \det g = \frac{1}{2}\overline{g}^{-1}\partial_\nu \overline{g}.
\end{aligned} \tag{3.1025}$$

b) Let us investigate how the Christoffel symbols $\Gamma^{\lambda}_{\mu\nu} = \Gamma^{\lambda}_{\mu\nu}(x)$ change under a general coordinate transformation $y = y(x)$. Let us denote by ∇_μ the *covariant derivative* such that $\nabla_\mu = \nabla_{\frac{\partial}{\partial x^\mu}} = \nabla_{\partial_\mu}$. We then have

$$\nabla_\mu \partial_\nu = \Gamma^{\lambda}_{\mu\nu}\partial_\lambda. \tag{3.1026}$$

Denoting $\partial'_\mu = \frac{\partial}{\partial y^\mu}$ and using $\nabla'_\mu = \frac{\partial x^\alpha}{\partial y^\mu}\nabla_\alpha$ (since $\nabla_{fX} Y = f\nabla_X Y$ for any vector fields X and Y and any smooth real valued function f), we find that

$$\begin{aligned}
\nabla'_\mu \partial'_\nu &= \Gamma'^{\lambda}_{\mu\nu}\partial'_\lambda = \frac{\partial x^\alpha}{\partial y^\mu}\nabla_\alpha\left(\frac{\partial x^\beta}{\partial y^\nu}\partial_\beta\right) = \frac{\partial x^\alpha}{\partial y^\mu}\left[\frac{\partial x^\beta}{\partial y^\nu}\nabla_\alpha\partial_\beta + \partial_\alpha\left(\frac{\partial x^\beta}{\partial y^\nu}\right)\partial_\beta\right] \\
&= \frac{\partial x^\alpha}{\partial y^\mu}\frac{\partial x^\beta}{\partial y^\nu}\Gamma^{\gamma}_{\alpha\beta}\partial_\gamma + \frac{\partial^2 x^\beta}{\partial y^\mu \partial y^\nu}\partial_\beta \\
&= \frac{\partial x^\alpha}{\partial y^\mu}\frac{\partial x^\beta}{\partial y^\nu}\frac{\partial y^\lambda}{\partial x^\gamma}\Gamma^{\gamma}_{\alpha\beta}\partial'_\lambda + \frac{\partial^2 x^\beta}{\partial y^\mu \partial y^\nu}\frac{\partial y^\lambda}{\partial x^\beta}\partial'_\lambda.
\end{aligned} \tag{3.1027}$$

Identifying terms, we obtain

$$\Gamma'^{\lambda}_{\mu\nu}(y) = \frac{\partial x^\alpha}{\partial y^\mu}\frac{\partial x^\beta}{\partial y^\nu}\frac{\partial y^\lambda}{\partial x^\gamma}\Gamma^{\gamma}_{\alpha\beta}(x) + \frac{\partial y^\lambda}{\partial x^\beta}\frac{\partial^2 x^\beta}{\partial y^\mu \partial y^\nu}, \tag{3.1028}$$

which is the transformation rule for the Christoffel symbols with respect to general coordinate transformations. Note that in linear coordinate transformations, the inhomogeneous term containing second-order derivatives vanishes and the Christoffel symbols transform like components of a third-rank tensor.

c) Consider a path $t \mapsto x(t)$. It holds that

$$\frac{d}{dt}g(Y,Y) = \frac{d}{dt}\left[g_{\mu\nu}(x(t))Y^\mu Y^\nu\right] = \dot{x}^\lambda(\partial_\lambda g_{\mu\nu})Y^\mu Y^\nu + g_{\mu\nu}\dot{Y}^\mu Y^\nu + g_{\mu\nu}Y^\mu \dot{Y}^\nu$$
$$= \{g_{\mu\nu} \text{ symmetric}\} = \dot{x}^\lambda(\partial_\lambda g_{\mu\nu})Y^\mu Y^\nu + 2g_{\mu\nu}\dot{Y}^\mu Y^\nu$$
$$= \dot{x}^\lambda(\partial_\lambda g_{\omega\nu})Y^\omega Y^\nu - 2g_{\mu\nu}\Gamma^\mu_{\lambda\omega}\dot{x}^\lambda Y^\omega Y^\nu$$
$$= \dot{x}^\lambda Y^\omega Y^\nu(\partial_\lambda g_{\omega\nu} - 2g_{\mu\nu}\Gamma^\mu_{\lambda\omega}), \tag{3.1029}$$

where the formula for parallel transport, i.e., $\dot{Y}^\mu + \Gamma^\mu_{\lambda\omega}\dot{x}^\lambda Y^\omega = 0$, has been used. Then, inserting the expression for the Levi-Civita connection (or the Christoffel symbols) $\Gamma^\mu_{\lambda\omega}$ in terms of the metric into the last parenthesis, one obtains

$$\partial_\lambda g_{\omega\nu} - 2g_{\mu\nu}\Gamma^\mu_{\lambda\omega} = \partial_\lambda g_{\omega\nu} - 2g_{\mu\nu}\cdot\frac{1}{2}g^{\mu\rho}\left(\partial_\lambda g_{\omega\rho} + \partial_\omega g_{\lambda\rho} - \partial_\rho g_{\lambda\omega}\right)$$
$$= \partial_\lambda g_{\omega\nu} - (\partial_\lambda g_{\omega\nu} + \partial_\omega g_{\lambda\nu} - \partial_\nu g_{\lambda\omega}) = \partial_\nu g_{\lambda\omega} - \partial_\omega g_{\lambda\nu}, \tag{3.1030}$$

which is antisymmetric in ν and ω. Finally, since the contraction between an antisymmetric tensor and a symmetric tensor vanishes, it follows that

$$\frac{d}{dt}g(Y,Y) = \dot{x}^\lambda Y^\omega Y^\nu\left(\partial_\nu g_{\lambda\omega} - \partial_\omega g_{\lambda\nu}\right) = \dot{x}^\lambda\left(Y^\omega Y^\nu\partial_\nu g_{\lambda\omega} - Y^\omega Y^\nu\partial_\omega g_{\lambda\nu}\right)$$
$$= \dot{x}^\lambda\left(Y^\omega Y^\nu\partial_\nu g_{\lambda\omega} - Y^\nu Y^\omega\partial_\nu g_{\lambda\omega}\right) = \{Y^\nu Y^\omega = Y^\omega Y^\nu\}$$
$$= \dot{x}^\lambda\left(Y^\omega Y^\nu\partial_\nu g_{\lambda\omega} - Y^\omega Y^\nu\partial_\nu g_{\lambda\omega}\right) = 0. \tag{3.1031}$$

Thus, $g(Y,Y)$ is constant.

2.34
For

$$W^\nu = U^\mu\nabla_\mu V^\nu, \tag{3.1032}$$

we can show that

$$W_\nu = g_{\nu\sigma}W^\sigma = g_{\nu\sigma}U^\mu\nabla_\mu V^\sigma = U^\mu\nabla_\mu g_{\nu\sigma}V^\sigma = U^\mu\nabla_\mu V_\nu, \tag{3.1033}$$

where we have used the fact that $\nabla_\nu g_{\mu\rho} = 0$, which follows from $\nabla_\mu A_{\nu\sigma} = \partial_\mu A_{\nu\sigma} - \Gamma^\rho_{\mu\nu}A_{\rho\sigma} - \Gamma^\rho_{\mu\sigma}A_{\nu\rho}$ and the form of the Christoffel symbols in terms of the metric tensor.

2.35

a) Using the given orthonormal basis as well as the nonzero Christoffel symbols in spherical coordinates, we find that

$$\nabla_\theta e_1 = \nabla_\theta \partial_\theta = 0, \tag{3.1034}$$

$$\nabla_\theta e_2 = \nabla_\theta \left(\frac{1}{\sin\theta} \partial_\phi \right) = -\frac{\cos\theta}{\sin^2\theta} \partial_\phi + \frac{1}{\sin\theta} \nabla_\theta \partial_\phi = -\frac{\cos\theta}{\sin^2\theta} \partial_\phi + \frac{1}{\sin\theta} \Gamma^\phi_{\theta\phi} \partial_\phi$$
$$= -\frac{\cos\theta}{\sin^2\theta} \partial_\phi + \frac{1}{\sin\theta} \cot\theta \, \partial_\phi = 0, \tag{3.1035}$$

$$\nabla_\phi e_1 = \nabla_\phi \partial_\theta = \Gamma^\phi_{\phi\theta} \partial_\phi = \cot\theta \, \partial_\phi = \cos\theta \cdot e_2, \tag{3.1036}$$

$$\nabla_\phi e_2 = \nabla_\phi \left(\frac{1}{\sin\theta} \partial_\phi \right) = \frac{1}{\sin\theta} \nabla_\phi \partial_\phi = \frac{1}{\sin\theta} \Gamma^\theta_{\phi\phi} \partial_\phi$$
$$= \frac{1}{\sin\theta} \left(-\frac{1}{2} \sin 2\theta \right) \partial_\phi = -\cos\theta \cdot e_1, \tag{3.1037}$$

which leads to the Christoffel symbols (in the given orthonormal basis) $\Gamma^i_{\theta j} = 0$ and

$$\Gamma^\bullet_{\phi\bullet} = \cos\theta \begin{pmatrix} 0 & -1 \\ 1 & 0 \end{pmatrix}. \tag{3.1038}$$

b) The equations of motion for the parallel transport of the new coordinates u^i are

$$\dot{u}^i(s) + \Gamma^i_{\phi j} \dot{\phi}(s) u^j(s) + \Gamma^i_{\theta j} \dot{\theta}(s) u^j(s) = 0. \tag{3.1039}$$

Since $\Gamma^i_{\theta j} = 0$, we obtain

$$\begin{pmatrix} \dot{u}^1(s) \\ \dot{u}^2(s) \end{pmatrix} = -\dot{\phi}(s) \cos\theta \begin{pmatrix} 0 & -1 \\ 1 & 0 \end{pmatrix} \begin{pmatrix} u^1 \\ u^2 \end{pmatrix}. \tag{3.1040}$$

The solution to this equation is $u(t) = e^{A(t)} u(0)$, where

$$A(t) = -\int_0^t \cos\theta(s) \dot{\phi}(s) \, ds \begin{pmatrix} 0 & -1 \\ 1 & 0 \end{pmatrix}. \tag{3.1041}$$

Now, we have

$$\int_0^T \cos\theta(s) \dot{\phi}(s) \, ds = \int_0^T [\cos\theta(s) \cdot \dot{\phi}(s) + 0 \cdot \dot{\theta}(s)] \, ds, \tag{3.1042}$$

where the integration is over a closed loop γ enclosing the surface area S. Using Green's formula, we can write this as

$$\int_0^T \cos\theta(s)\dot{\phi}(s)\,ds = \int_S \left(\partial_\theta \cdot \cos\theta - \partial_\phi \cdot 0\right)\,d\phi\,d\theta$$
$$= -\int_S \sin\theta\,d\phi\,d\theta = -\text{area}(S) = -\Omega. \tag{3.1043}$$

Thus, we find that $u' = Ru$, where

$$R = \exp\left(\Omega\begin{pmatrix}0 & -1\\ 1 & 0\end{pmatrix}\right) = \begin{pmatrix}\cos\Omega & -\sin\Omega\\ \sin\Omega & \cos\Omega\end{pmatrix}. \tag{3.1044}$$

2.36

The spherical coordinates are given by

$$x = r\sin(\theta)\cos(\phi), \quad y = r\sin(\theta)\sin(\phi), \quad z = r\cos(\theta), \tag{3.1045}$$

From these relations, we can deduce that the line element in spherical coordinates is given by

$$ds^2 = dr^2 + r^2[d\theta^2 + \sin^2(\theta)d\phi^2], \tag{3.1046}$$

which, in turn, implies that

$$\mathcal{L} = g_{ij}\dot{x}^i\dot{x}^j = \dot{r}^2 + r^2[\dot{\theta}^2 + \sin^2(\theta)d\phi^2]. \tag{3.1047}$$

The geodesic equations are given by extremizing the integral $\int \mathcal{L}\,ds$. From the Euler–Lagrange equations, we obtain

$$\ddot{r} - r[\dot{\theta}^2 + \sin^2(\theta)\dot{\phi}^2] = 0, \tag{3.1048}$$

$$\ddot{\theta} + \frac{2}{r}\dot{\theta}\dot{r} - \frac{1}{2}\sin(2\theta)\dot{\phi}^2 = 0, \tag{3.1049}$$

$$\ddot{\phi} + \frac{2}{r}\dot{\phi}\dot{r} + 2\cot(\theta)\dot{\theta}\dot{\phi} = 0. \tag{3.1050}$$

Comparing with the geodesic equations $\dot{x}^i\nabla_i\dot{x}^j = 0$, we get

$$\Gamma^r_{\theta\theta} = -r, \quad \Gamma^r_{\phi\phi} = -r\sin^2(\theta), \quad \Gamma^\theta_{\theta r} = \Gamma^\theta_{r\theta} = \Gamma^\phi_{\phi r} = \Gamma^\phi_{r\phi} = \frac{1}{r},$$
$$\Gamma^\theta_{\phi\phi} = -\frac{1}{2}\sin(2\theta), \quad \Gamma^\phi_{\theta\phi} = \Gamma^\phi_{\phi\theta} = \cot(\theta), \tag{3.1051}$$

as the nonvanishing Christoffel symbols.

2.37

a) We can express θ and ϕ in terms of x and y as

$$\theta = 2\arctan\left(\frac{r}{2R}\right), \quad \phi = \arctan\left(\frac{y}{x}\right), \tag{3.1052}$$

where $r^2 = x^2 + y^2$. We can now use the relations

$$d\theta = \frac{\partial\theta}{\partial x}dx + \frac{\partial\theta}{\partial y}dy, \quad d\phi = \frac{\partial\phi}{\partial x}dx + \frac{\partial\phi}{\partial y}dy, \tag{3.1053}$$

to obtain the following expressions for $d\theta^2$ and $d\phi^2$:

$$d\theta^2 = \frac{1}{r^2R^2}\frac{1}{\left(1+\frac{r^2}{4R^2}\right)^2}\left(x^2dx^2 + y^2dy^2 + 2xydxdy\right), \tag{3.1054}$$

$$d\phi^2 = \frac{1}{r^4}\left(y^2dx^2 + x^2dy^2 - 2xydxdy\right). \tag{3.1055}$$

We can also express $\sin\theta$ in terms of r as

$$\sin\theta = 2\cos\frac{\theta}{2}\sin\frac{\theta}{2} = \frac{2\tan\frac{\theta}{2}}{1+\tan^2\frac{\theta}{2}} = \frac{4Rr}{4R^2+r^2}. \tag{3.1056}$$

Inserting into the original metric, we obtain

$$ds^2 = \left(\frac{1}{1+\frac{r^2}{4R^2}}\right)^2 (dx^2 + dy^2). \tag{3.1057}$$

b) The Christoffel symbols can be computed from

$$\Gamma^\mu_{\nu\lambda} = \frac{1}{2}g^{\mu\omega}(\partial_\nu g_{\lambda\omega} + \partial_\lambda g_{\nu\omega} - \partial_\omega g_{\nu\lambda}). \tag{3.1058}$$

Using the fact that this is symmetric under $\nu \leftrightarrow \lambda$ as well as that the metric is symmetric under $x \leftrightarrow y$, we only need to compute three of the Christoffel symbols:

$$\Gamma^x_{xx} = -\frac{x}{2(R^2 + r^2/4)}, \tag{3.1059}$$

$$\Gamma^x_{xy} = -\frac{y}{2(R^2 + r^2/4)}, \tag{3.1060}$$

$$\Gamma^x_{yy} = -\Gamma^x_{xx}. \tag{3.1061}$$

c) The stereographic projection can be visualized according to Figure 3.12.

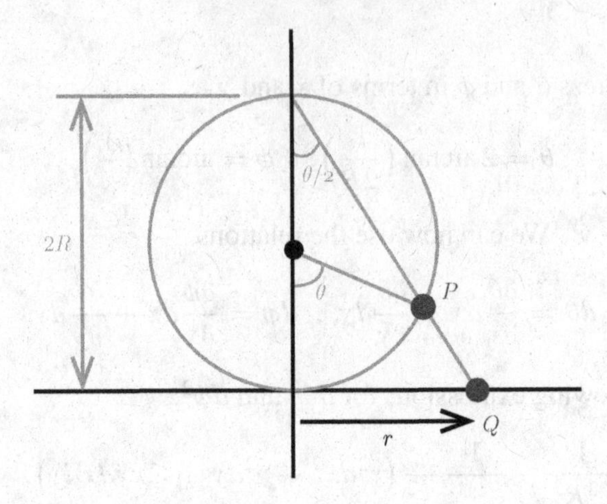

Figure 3.12 An illustration of the stereographic projection. The point P on the sphere is represented by the point Q in the plane.

2.38

We have a two-dimensional hyperbolic subspace $x^2+y^2-t^2 = 1, z = 0$, embedded into a four-dimensional Minkowski space.

a) The surface can be parametrized by introducing the parameters r and ϕ according to

$$x = r\cos\phi, \tag{3.1062}$$

$$y = r\sin\phi, \tag{3.1063}$$

$$t^2 = r^2 - 1. \tag{3.1064}$$

b) The metric is given by

$$ds^2 = dt^2 - dx^2 - dy^2, \tag{3.1065}$$

where

$$2rdr = 2tdt, \tag{3.1066}$$

$$dx = \frac{\partial x}{\partial r}dr + \frac{\partial x}{\partial \phi}d\phi, \tag{3.1067}$$

$$dy = \frac{\partial y}{\partial r}dr + \frac{\partial y}{\partial \phi}d\phi, \tag{3.1068}$$

and thus, the metric can be written as

$$ds^2 = dt^2 - dx^2 - dy^2 = dt^2 - \left(\frac{\partial x}{\partial r}dr + \frac{\partial x}{\partial \phi}d\phi\right)^2 - \left(\frac{\partial y}{\partial r}dr + \frac{\partial y}{\partial \phi}d\phi\right)^2$$
$$= dt^2 - (\cos\phi dr - r\sin\phi d\phi)^2 - (\sin\phi dr + r\cos\phi d\phi)^2$$
$$= dt^2 - dr^2 - r^2 d\phi^2$$
$$= dt^2 - \frac{t^2}{1+t^2}dt^2 - r^2 d\phi^2 = \frac{1}{1+t^2}dt^2 - (1+t^2)d\phi^2. \tag{3.1069}$$

c) The Christoffel symbols can be determined from

$$\Gamma^\rho_{\mu\nu} = \frac{1}{2}g^{\rho\lambda}(\partial_\mu g_{\nu\lambda} + \partial_\nu g_{\mu\lambda} - \partial_\lambda g_{\mu\nu}). \tag{3.1070}$$

The nonzero Christoffel symbols are given by

$$\Gamma^t_{tt} = -\frac{t}{1+t^2}, \tag{3.1071}$$
$$\Gamma^t_{\phi\phi} = t(1+t^2), \tag{3.1072}$$
$$\Gamma^\phi_{t\phi} = \Gamma^\phi_{\phi t} = \frac{t}{1+t^2}. \tag{3.1073}$$

2.39

a) Inserting $X = \partial_\mu$, $Y = \partial_\nu$, and $Z = \partial_\lambda$ into the definition of the Riemann curvature tensor, one finds that

$$R(\partial_\mu, \partial_\nu)\partial_\lambda = R^\omega{}_{\lambda\mu\nu}\partial_\omega = \left[\nabla_\mu, \nabla_\nu\right]\partial_\lambda - \nabla_{[\partial_\mu,\partial_\nu]}\partial_\lambda = \left[\nabla_\mu, \nabla_\nu\right]\partial_\lambda$$
$$= \nabla_\mu(\Gamma^\rho_{\nu\lambda}\partial_\rho) - \nabla_\nu(\Gamma^\rho_{\mu\lambda}\partial_\rho)$$
$$= (\partial_\mu\Gamma^\rho_{\nu\lambda})\partial_\rho + \Gamma^\rho_{\nu\lambda}\Gamma^\omega_{\mu\rho}\partial_\omega - (\partial_\nu\Gamma^\rho_{\mu\lambda})\partial_\rho - \Gamma^\rho_{\mu\lambda}\Gamma^\omega_{\nu\rho}\partial_\omega, \tag{3.1074}$$

which means that

$$R^\omega{}_{\lambda\mu\nu} = \partial_\mu\Gamma^\omega_{\nu\lambda} - \partial_\nu\Gamma^\omega_{\mu\lambda} + \Gamma^\omega_{\mu\rho}\Gamma^\rho_{\nu\lambda} - \Gamma^\omega_{\nu\rho}\Gamma^\rho_{\mu\lambda}. \tag{3.1075}$$

When $T = 0$, one has $\Gamma^\lambda_{\mu\nu} = \Gamma^\lambda_{\nu\mu}$. Thus, one obtains

$$\begin{aligned} R^\omega{}_{\mu\nu\lambda} + R^\omega{}_{\nu\lambda\mu} + R^\omega{}_{\lambda\mu\nu} &= \partial_\nu\Gamma^\omega_{\lambda\mu} - \partial_\lambda\Gamma^\omega_{\nu\mu} + \partial_\lambda\Gamma^\omega_{\mu\nu} - \partial_\mu\Gamma^\omega_{\lambda\nu} + \partial_\mu\Gamma^\omega_{\nu\lambda} - \partial_\nu\Gamma^\omega_{\mu\lambda} \\ &\quad + \Gamma^\omega_{\nu\rho}\Gamma^\rho_{\lambda\mu} - \Gamma^\omega_{\lambda\rho}\Gamma^\rho_{\nu\mu} + \Gamma^\omega_{\lambda\rho}\Gamma^\rho_{\mu\nu} - \Gamma^\omega_{\mu\rho}\Gamma^\rho_{\lambda\nu} \\ &\quad + \Gamma^\omega_{\mu\rho}\Gamma^\rho_{\nu\lambda} - \Gamma^\omega_{\nu\rho}\Gamma^\rho_{\mu\lambda} \\ &= 0. \end{aligned} \tag{3.1076}$$

b) The second Bianchi identity $R_{\alpha\beta\mu\nu;\lambda} + R_{\alpha\beta\nu\lambda;\mu} + R_{\alpha\beta\lambda\mu;\nu} = 0$ can be written in matrix notation as

$$\partial_\lambda R^{\bullet}{}_{\bullet\mu\nu} + [\Gamma^{\bullet}_{\lambda\bullet}, R^{\bullet}{}_{\bullet\mu\nu}] + \partial_\mu R^{\bullet}{}_{\bullet\nu\lambda} + [\Gamma^{\bullet}_{\mu\bullet}, R^{\bullet}{}_{\bullet\nu\lambda}] + \partial_\nu R^{\bullet}{}_{\bullet\lambda\mu} + [\Gamma^{\bullet}_{\nu\bullet}, R^{\bullet}{}_{\bullet\lambda\mu}] = 0, \tag{3.1077}$$

which follows from the definition of the Riemann curvature tensor

$$R^{\omega}{}_{\lambda\mu\nu} = \partial_\mu \Gamma^{\omega}_{\nu\lambda} - \partial_\nu \Gamma^{\omega}_{\mu\lambda} + \Gamma^{\omega}_{\mu\rho}\Gamma^{\rho}_{\nu\lambda} - \Gamma^{\omega}_{\nu\rho}\Gamma^{\rho}_{\mu\lambda}, \tag{3.1078}$$

in matrix notation, i.e.,

$$R^{\bullet}{}_{\bullet\mu\nu} = \partial_\mu \Gamma^{\bullet}_{\nu\bullet} - \partial_\nu \Gamma^{\bullet}_{\mu\bullet} + [\Gamma^{\bullet}_{\mu\bullet}, \Gamma^{\bullet}_{\nu\bullet}] = [\partial_\mu + \Gamma^{\bullet}_{\mu\bullet}, \partial_\nu + \Gamma^{\bullet}_{\nu\bullet}], \tag{3.1079}$$

and the Jacobi identity for matrices (and linear operators), i.e.,

$$[X,[Y,Z]] + [Y,[Z,X]] + [Z,[X,Y]] = 0. \tag{3.1080}$$

Explicitly, using $X = \nabla_\lambda$, $Y = \nabla_\mu$, and $Z = \nabla_\nu$ with $\nabla_\lambda = \partial_\lambda + \Gamma^{\bullet}_{\lambda\bullet}$, we find that

$$\begin{aligned} [X,[Y,Z]] &= [\nabla_\lambda, [\nabla_\mu, \nabla_\nu]] = [\nabla_\lambda, [\partial_\mu + \Gamma^{\bullet}_{\mu\bullet}, \partial_\nu + \Gamma^{\bullet}_{\nu\bullet}]] \\ &= [\nabla_\lambda, R^{\bullet}{}_{\bullet\mu\nu}] = [\partial_\lambda + \Gamma^{\bullet}_{\lambda\bullet}, R^{\bullet}{}_{\bullet\mu\nu}] \\ &= [\partial_\lambda, R^{\bullet}{}_{\bullet\mu\nu}] + [\Gamma^{\bullet}_{\lambda\bullet}, R^{\bullet}{}_{\bullet\mu\nu}] = \partial_\lambda R^{\bullet}{}_{\bullet\mu\nu} + [\Gamma^{\bullet}_{\lambda\bullet}, R^{\bullet}{}_{\bullet\mu\nu}]. \end{aligned} \tag{3.1081}$$

Similarly, cyclicly permuting the indices in the above equation, we have

$$[Y,[Z,X]] = \partial_\mu R^{\bullet}{}_{\bullet\nu\lambda} + [\Gamma^{\bullet}_{\mu\bullet}, R^{\bullet}{}_{\bullet\nu\lambda}], \quad [Z,[X,Y]] = \partial_\nu R^{\bullet}{}_{\bullet\lambda\mu} + [\Gamma^{\bullet}_{\nu\bullet}, R^{\bullet}{}_{\bullet\lambda\mu}]. \tag{3.1082}$$

Inserting the results for $[X,[Y,Z]$, $[Y,[Z,X]]$, and $[Z,[X,Y]]$ into the Jacobi identity, we obtain the second Bianchi identity

$$\partial_\lambda R^{\bullet}{}_{\bullet\mu\nu} + [\Gamma^{\bullet}_{\lambda\bullet}, R^{\bullet}{}_{\bullet\mu\nu}] + \partial_\mu R^{\bullet}{}_{\bullet\nu\lambda} + [\Gamma^{\bullet}_{\mu\bullet}, R^{\bullet}{}_{\bullet\nu\lambda}] + \partial_\nu R^{\bullet}{}_{\bullet\lambda\mu} + [\Gamma^{\bullet}_{\nu\bullet}, R^{\bullet}{}_{\bullet\lambda\mu}] = 0. \tag{3.1083}$$

Now, use the second Bianchi identity to show that the covariant derivative of the energy–momentum tensor $T^{\mu\nu}$ in Einstein's equations

$$G^{\mu\nu} = \frac{8\pi G}{c^4} T^{\mu\nu}, \tag{3.1084}$$

vanishes.

In the second Bianchi identity

$$R_{\alpha\beta\mu\nu;\lambda} + R_{\alpha\beta\nu\lambda;\mu} + R_{\alpha\beta\lambda\mu;\nu} = 0, \tag{3.1085}$$

contracting the α and μ indices with the inverse metric tensor $g^{\alpha\mu}$, we obtain

$$g^{\alpha\mu}(R_{\alpha\beta\mu\nu;\lambda} + R_{\alpha\beta\nu\lambda;\mu} + R_{\alpha\beta\lambda\mu;\nu}) = 0. \tag{3.1086}$$

By the definition of the Ricci tensor $R_{\beta\nu} = R^{\mu}{}_{\beta\mu\nu} = g^{\alpha\mu} R_{\alpha\beta\mu\nu}$ and the fact that $R_{\alpha\beta\lambda\mu} = -R_{\alpha\beta\mu\lambda}$, this can be written as

$$R_{\beta\nu;\lambda} + R^{\mu}{}_{\beta\nu\lambda;\mu} - R_{\beta\lambda;\nu} = 0, \tag{3.1087}$$

where we have taken into account that the covariant derivative of $g^{\alpha\mu}$ vanishes, implying that multiplication with components of the metric tensor commutes with covariant differentiation; in particular, index raising or lowering commutes with covariant derivatives. Contracting once again with $g^{\beta\nu}$ and using the definition of the Ricci scalar $R = g^{\mu\nu} R_{\mu\nu}$, we find that

$$g^{\beta\nu}(R_{\beta\nu;\lambda} + R^{\mu}{}_{\beta\nu\lambda;\mu} - R_{\beta\lambda;\nu}) = R_{;\lambda} + g^{\beta\nu} R^{\mu}{}_{\beta\nu\lambda;\mu} - R^{\nu}{}_{\lambda;\nu} = 0. \tag{3.1088}$$

Now, using the facts that $R_{\alpha\beta\nu\lambda} = -R_{\beta\alpha\nu\lambda} = -R_{\alpha\beta\lambda\nu} = R_{\nu\lambda\alpha\beta}$, we have

$$\begin{aligned} g^{\beta\nu} R^{\mu}{}_{\beta\nu\lambda;\mu} &= g^{\beta\nu} g^{\alpha\mu} R_{\alpha\beta\nu\lambda;\mu} = -g^{\beta\nu} g^{\alpha\mu} R_{\beta\alpha\nu\lambda;\mu} \\ &= -g^{\mu\alpha} R^{\nu}{}_{\alpha\nu\lambda;\mu} = -g^{\mu\alpha} R_{\alpha\lambda;\mu} = -R^{\mu}{}_{\lambda;\mu}. \end{aligned} \tag{3.1089}$$

Inserting this into the second term, we obtain

$$R_{;\lambda} - R^{\mu}{}_{\lambda;\mu} - R^{\nu}{}_{\lambda;\nu} = R_{;\lambda} - 2R^{\mu}{}_{\lambda;\mu} = 0. \tag{3.1090}$$

Note that $R_{;\mu} = \partial_{\mu} R$, since R is a scalar. An equivalent form of the previous equation is

$$\left(2R^{\mu}{}_{\lambda} - \delta^{\mu}_{\lambda} R\right)_{;\mu} = 0. \tag{3.1091}$$

Finally, raising the index λ by applying $g^{\lambda\nu}$ and dividing by a factor of two lead to

$$\left(R^{\mu\nu} - \frac{1}{2} R g^{\mu\nu}\right)_{;\mu} = 0, \tag{3.1092}$$

which implies that $\nabla_{\mu} G^{\mu\nu} = G^{\mu\nu}{}_{;\mu} = 0$, since the Einstein tensor is defined as $G^{\mu\nu} = R^{\mu\nu} - \frac{1}{2} R g^{\mu\nu}$. Thus, we have now shown that the covariant divergence of the Einstein tensor $G^{\mu\nu}$ vanishes, and therefore, it immediately follows that $\nabla_{\mu} T^{\mu\nu} = 0$, i.e., the covariant divergence of the energy–momentum tensor $T^{\mu\nu}$ vanishes, since $T^{\mu\nu}$ is directly proportional to $G^{\mu\nu}$ via Einstein's equations $G^{\mu\nu} = aT^{\mu\nu}$, where $a = \frac{8\pi G}{c^4}$ is a constant.

Note that, for a flat spacetime in Minkowski coordinates, it holds that $\nabla_{\mu} T^{\mu\nu} = \partial_{\mu} T^{\mu\nu}$. Furthermore, the Christoffel symbols, the Riemann curvature tensor, the Ricci tensor, and the Ricci scalar are all zero since they are all either first-order or second-order derivatives of the metric, which are constant in Minkowski coordinates, and therefore, it implies that the Einstein tensor is also zero, i.e., $G^{\mu\nu} = 0$.

Thus, since $T^{\mu\nu} \propto G^{\mu\nu}$, it means that $T^{\mu\nu} = 0$, and therefore, the covariant divergence of the energy–momentum tensor $T^{\mu\nu}$ for a flat spacetime trivially vanishes, since $\nabla_\mu 0 = 0$.

2.40

The *curvature* is related to the *parallel transport* in the following way. Consider a very small parallelogram with corners at x, $x + \delta x$, $x + \delta x + \delta y$, and $x + \delta y$. Furthermore, consider the parallel transport equations

$$\dot{Y}^i(s) + \Gamma^i_{kj}(x(s))\dot{x}^k(s)Y^j(s) = 0, \quad i = 1, 2, \ldots, n, \tag{3.1093}$$

where $Y(s)$ is an unknown vector field along the curve $x(s)$. According to this system of equations, a tangent vector Y at x when parallel transported to the point $x + \delta x$ becomes approximately (in given local coordinates)

$$Y^i(x + \delta x) = Y^i(x) - \Gamma^i_{jk}(x)Y^k(x)\delta x^j. \tag{3.1094}$$

At the next point $x + \delta x + \delta y$, we obtain

$$\begin{aligned} Y^i(x + \delta x + \delta y) &= Y^i(x) - \Gamma^i_{jk}(x)Y^k(x)\delta x^j \\ &\quad - \Gamma^i_{jk}(x + \delta x)[Y^k(x) - \Gamma^k_{\ell m}(x)Y^m(x)\delta x^\ell]\delta y^j \\ &= Y^i(x) - \Gamma^i_{jk}(x)Y^k(x)\delta x^j - \Gamma^i_{jk}(x)Y^k(x)\delta y^j \\ &\quad - \partial_m\Gamma^i_{jk}(x)\delta x^m \delta y^j Y^k(x) + \Gamma^i_{jk}(x)\Gamma^k_{\ell m}(x)Y^m(x)\delta x^\ell \delta y^j. \end{aligned} \tag{3.1095}$$

In the last step, we have neglected the terms, which are of third order in the coordinate differentials. In the same way, we can compute the parallel transport of Y from x to $x + \delta y$ and further to $x + \delta y + \delta x$. The parallel transport around the parallelogram is then obtained as a combination of the right-hand side of the above formula and the latter transport (note the direction of motion); the result is

$$\delta Y^i = R^i{}_{k\ell j}(x)Y^k(x)\delta x^\ell \delta y^j = \frac{1}{2}R^i{}_{k\ell j}(x)Y^k(x)(\delta x^\ell \delta y^j - \delta x^j \delta y^\ell), \tag{3.1096}$$

where

$$R^i{}_{k\ell j} = \partial_\ell \Gamma^i{}_{jk} - \partial_j \Gamma^i_{\ell k} + \Gamma^i_{\ell m}\Gamma^m_{jk} - \Gamma^i_{jm}\Gamma^m_{\ell k}, \tag{3.1097}$$

is the Riemann curvature tensor (sometimes just called the curvature). Thus, the parallel transport around the small parallelogram is proportional to the curvature at x.

2.41

The Cartesian coordinates (x, y, z) can be parametrized in cylindrical coordinates as $(x, y, z) = (R\cos\phi, R\sin\phi, z)$. The metric in $\mathbb{R}^3$ is then given by

$$ds^2 = dR^2 + R^2 d\phi^2 + dz^2, \tag{3.1098}$$

which, where (R, ϕ, z) are the cylindrical coordinates, means that the metric tensor is $\tilde{g} = (\tilde{g}_{\mu\nu}) = \operatorname{diag}(1, R^2, 1)$ in the cylindrical coordinates (R, ϕ, z). On the paraboloid, it holds that

$$z = x^2 + y^2 = R^2, \tag{3.1099}$$

which gives $dz = 2RdR$. Inserting this into the metric, one obtains

$$ds^2 = (4R^2 + 1)dR^2 + R^2 d\phi^2, \tag{3.1100}$$

and the corresponding metric tensor is $g = (g_{\mu\nu}) = \operatorname{diag}(4R^2 + 1, R^2)$. The Christoffel symbols are now calculated directly from the metric to be

$$\Gamma^R_{RR} = \frac{4R}{4R^2 + 1}, \tag{3.1101}$$

$$\Gamma^\phi_{R\phi} = \Gamma^\phi_{\phi R} = \frac{1}{R}, \tag{3.1102}$$

$$\Gamma^R_{\phi\phi} = -\frac{R}{4R^2 + 1}, \tag{3.1103}$$

$$\Gamma^R_{R\phi} = \Gamma^R_{\phi R} = \Gamma^\phi_{RR} = \Gamma^\phi_{\phi\phi} = 0. \tag{3.1104}$$

Using the Christoffel symbols, define the 2×2 matrices

$$\Gamma_R \equiv \Gamma^\bullet_{R\bullet} = \begin{pmatrix} \frac{4R}{4R^2+1} & 0 \\ 0 & \frac{1}{R} \end{pmatrix} \quad \text{and} \quad \Gamma_\phi \equiv \Gamma^\bullet_{\phi\bullet} = \begin{pmatrix} 0 & -\frac{R}{4R^2+1} \\ \frac{1}{R} & 0 \end{pmatrix}. \tag{3.1105}$$

Then, the four components of the Riemann curvature tensor, $R^\bullet{}_{\bullet R\phi} = -R^\bullet{}_{\bullet\phi R}$, are given by the 2×2 matrix

$$R^\bullet{}_{\bullet R\phi} = \partial_R \Gamma_\phi - \partial_\phi \Gamma_R + \left[\Gamma_R, \Gamma_\phi\right] = \begin{pmatrix} 0 & \frac{4R^2}{(4R^2+1)^2} \\ -\frac{4}{4R^2+1} & 0 \end{pmatrix}. \tag{3.1106}$$

2.42

a) First, we need the Christoffel symbols. There are basically two ways in which we can obtain the Christoffel symbols, either immediately using the equation

$$\ddot{x}^\beta + \Gamma^\beta_{\mu\nu}\dot{x}^\mu \dot{x}^\nu = 0, \tag{3.1107}$$

or derive the Euler–Lagrange equations and compare them with the above equation. Let us do the latter. The Euler–Lagrange equation with respect to r becomes

$$4r^3\dot{\phi}^2 - 2r\dot{r}^2 - \frac{d}{d\tau}(r^2 2\dot{r}) = 4r^3\dot{\phi}^2 - 2r\dot{r} - 2r^2\ddot{r} = 0 \;\Rightarrow\; \ddot{r} + \dot{r}^2/r - 2r\dot{\phi}^2 = 0, \tag{3.1108}$$

and with respect to ϕ

$$\frac{d}{d\tau}(2r^4\dot{\phi}) = 0 \quad \Rightarrow \quad \ddot{\phi} + 4\dot{r}\dot{\phi}/r = 0. \tag{3.1109}$$

Reading off the Christoffel symbols gives

$$\Gamma^r_{rr} = \frac{1}{r}, \quad \Gamma^r_{\phi\phi} = -2r, \quad \Gamma^\phi_{\phi r} = \frac{2}{r}. \tag{3.1110}$$

Second, the Christoffel symbols must be used in the formula for the Riemann tensor:

$$\begin{aligned} R^r{}_{\phi r\phi} &= \partial_r \Gamma^r_{\phi\phi} - \partial_\phi \Gamma^r_{r\phi} + \Gamma^r_{r\alpha}\Gamma^\alpha_{\phi\phi} - \Gamma^r_{\phi\alpha}\Gamma^\alpha_{r\phi} \\ &= \partial_r(-2r) + \frac{1}{r}(-2r) - (-2r)\frac{2}{r} = -2 - 2 + 4 = 0. \end{aligned} \tag{3.1111}$$

b) The area is given by

$$A = \int \sqrt{\det(g)}\, dr d\phi = \{\det(g) = r^6\} = \int \sqrt{r^6}\, dr d\phi = \int r^3\, dr d\phi = \frac{\pi r^4}{2}. \tag{3.1112}$$

The circumference (which comes from integrating the metric with dr being constant) is

$$ds = \sqrt{r^4}\, d\phi = r^2\, d\phi \quad \Rightarrow \quad C = s = \int_0^{2\pi} r^2\, d\phi = 2\pi r^2. \tag{3.1113}$$

Thus, in this two-dimensional metric, the relation between area and circumference of a circle is $C^2 = 8\pi A$, and not $C^2 = 4\pi A$ as in the case of Euclidean space.

2.43

By the metric compatibility of the Levi-Civita connection, it holds that

$$V \cdot g(U, W) = g(\nabla_V U, W) + g(U, \nabla_V W), \tag{3.1114}$$

where V, U, and W are vector fields. In particular, if U and W are equal to X, Y, or Z, then

$$g(\nabla_V U, W) + g(U, \nabla_V W) = 0, \tag{3.1115}$$

since $g(U, W)$ is constant in this case.

First, let $U = W = X$, then we have

$$0 = 2g(\nabla_V X, X) = 2g(\Gamma^X_{VX}X + \Gamma^Y_{VX}Y + \Gamma^Z_{VX}Z, X) = -2\Gamma^X_{VX} \Rightarrow \Gamma^X_{VX} = 0, \tag{3.1116}$$

where $V = X, Y, Z$. In the same way, exchanging X for Y and Z, we find that $\Gamma^Y_{VY} = \Gamma^Z_{VZ} = 0$. Next, let $V = U = X$ and $W = Y, Z$, then we obtain

$$0 = g(\nabla_X X, W) + g(X, \nabla_X W) = g(\nabla_X X, W) = -\Gamma^W_{XX} \Rightarrow \Gamma^W_{XX} = 0, \tag{3.1117}$$

where the second term is zero, since $\nabla_X W = \nabla_W X - [W,X]$, $\Gamma^X_{WX} = 0$, and $g(X,[W,X]) = 0$. In a similar fashion, $\Gamma^Y_{WY} = 0$ and $\Gamma^Z_{WZ} = 0$.

From the above follows that $\nabla_X Y = aZ$, $\nabla_Y Z = bX$, and $\nabla_Z X = cY$, where a, b, and c are functions. Since the torsion on M is zero, the following relations hold

$$\nabla_Y X = \nabla_X Y - [X,Y] = (a+1)Z, \tag{3.1118}$$

$$\nabla_Z Y = \nabla_Y Z - [Y,Z] = (b-1)X, \tag{3.1119}$$

$$\nabla_X Z = \nabla_Z X - [Z,X] = (c-1)Y. \tag{3.1120}$$

This gives a linear system of equations for a, b, and c as follows

$$0 = X \cdot g(Y,Z) = g(\nabla_X Y, Z) + g(Y, \nabla_X Z) = a - c + 1, \tag{3.1121}$$

$$0 = Y \cdot g(Z,X) = g(\nabla_Y Z, X) + g(Z, \nabla_Y X) = a - b + 1, \tag{3.1122}$$

$$0 = Z \cdot g(X,Y) = g(\nabla_Z X, Y) + g(X, \nabla_Z Y) = 1 - c - b. \tag{3.1123}$$

Solving this system of equations gives $-a = b = c = 1/2$.

Since the affine connection is known, it is straightforward to compute all independent components of the Riemann curvature tensor, i.e., $R(V,U)W = [\nabla_V, \nabla_U]W - \nabla_{[V,U]}W$. The result of this computation is

$$R(X,Y)Z = R(Y,Z)X = R(Z,X)Y = 0, \tag{3.1124}$$

$$R(X,Y)Y = R(Z,X)Z = -X/4, \tag{3.1125}$$

$$R(X,Y)X = R(Y,Z)Z = Y/4, \tag{3.1126}$$

$$R(Y,Z)Y = -R(Z,X)X = Z/4. \tag{3.1127}$$

2.44

a) Using the definition of the Riemann curvature tensor constructed in terms of a metric and that the first and third indices are equal, we obtain the Ricci tensor as

$$R_{\mu\nu} = R^{\lambda}{}_{\mu\lambda\nu} = \partial_\lambda \Gamma^\lambda_{\nu\mu} - \partial_\nu \Gamma^\lambda_{\lambda\mu} + \Gamma^\lambda_{\lambda\omega}\Gamma^\omega_{\nu\mu} - \Gamma^\lambda_{\nu\omega}\Gamma^\omega_{\lambda\mu}. \tag{3.1128}$$

The fact that the Christoffel symbols are symmetric in the two lower indices, i.e., $\Gamma^\lambda_{\mu\nu} = \Gamma^\lambda_{\nu\mu}$, implies that the first and third terms are symmetric. Furthermore, the fourth term is symmetric due to interchange of summation indices $\lambda \leftrightarrow \omega$ and again using the symmetry property of the Christoffel symbols. Therefore, it remains to show that the second term is symmetric. However, using the fact that [see Problem 2.33 a)]

$$\Gamma^\lambda_{\lambda\mu} = \frac{1}{2}(\det g)^{-1}\,\partial_\mu(\det g), \tag{3.1129}$$

which implies that $\partial_\nu \Gamma^\lambda_{\lambda\mu} = \partial_\mu \Gamma^\lambda_{\lambda\nu}$. Thus, the Ricci tensor is symmetric, i.e., $R_{\mu\nu} = R_{\nu\mu}$.

In addition, the Einstein tensor

$$G_{\mu\nu} = R_{\mu\nu} - \frac{1}{2} g_{\mu\nu} R, \tag{3.1130}$$

is obviously symmetric, since the metric $g_{\mu\nu}$ and the Ricci tensor $R_{\mu\nu}$ are symmetric and R is the Ricci scalar.

b) The Ricci tensor can be written as

$$R_{\mu\nu} = R^{\lambda}{}_{\mu\lambda\nu} = g^{\lambda\alpha} R_{\alpha\mu\lambda\nu}. \tag{3.1131}$$

Using the (anti)symmetries

$$R_{\alpha\mu\lambda\nu} = -R_{\mu\alpha\lambda\nu} = -R_{\alpha\mu\nu\lambda} = R_{\mu\alpha\nu\lambda}, \tag{3.1132}$$

we obtain

$$R_{00} = g^{11} R_{1010}, \quad R_{01} = -g^{01} R_{1010}, \quad R_{10} = -g^{10} R_{1010}, \quad R_{11} = g^{00} R_{1010}. \tag{3.1133}$$

This implies that the Ricci tensor is given by

$$R_{\mu\nu} = \frac{1}{\overline{g}} g_{\mu\nu} R_{1010}, \quad \text{where } \overline{g} = \det g = \det(g_{\mu\nu}). \tag{3.1134}$$

The Ricci scalar can now be calculated to be

$$R = g^{\mu\nu} R_{\mu\nu} = \frac{2}{\overline{g}} R_{1010}. \tag{3.1135}$$

Inserting the metric, the Ricci tensor, and the Ricci scalar in the sought-after tensor, we find that

$$R_{\mu\nu} - k g_{\mu\nu} R = \frac{1}{\overline{g}} g_{\mu\nu} R_{1010} - k g_{\mu\nu} \frac{2}{\overline{g}} R_{1010} = \frac{R_{1010}}{\overline{g}} (1 - 2k)\, g_{\mu\nu}. \tag{3.1136}$$

Thus, since $\overline{g} \neq 0$ and in general $R_{1010} \neq 0$, the tensor $R_{\mu\nu} - k g_{\mu\nu} R$ vanishes when $k = \frac{1}{2}$, i.e.,

$$R_{\mu\nu} - \frac{1}{2} g_{\mu\nu} R = 0. \tag{3.1137}$$

c) As found in b), $G_{\mu\nu} = R_{\mu\nu} - g_{\mu\nu} R/2 = 0$ in this spacetime. The Einstein field equations $G_{\mu\nu} = 0$ in vacuum are therefore satisfied.

2.45

a) Using Euler–Lagrange equations for the t and y coordinates with the Lagrangian

$$\mathcal{L} = g_{\mu\nu} \dot{x}^{\mu} \dot{x}^{\nu} = \frac{1}{y^2} \left(\dot{t}^2 - \dot{y}^2 \right), \tag{3.1138}$$

we find that

$$\frac{\partial \mathcal{L}}{\partial t} - \frac{d}{d\tau}\frac{\partial \mathcal{L}}{\partial \dot{t}} = 0, \quad \frac{\partial \mathcal{L}}{\partial t} = 0, \quad \frac{\partial \mathcal{L}}{\partial \dot{t}} = \frac{2}{y^2}\dot{t} \quad \Rightarrow \quad \frac{d}{d\tau}\frac{\dot{t}}{y^2} = 0, \tag{3.1139}$$

$$\frac{\partial \mathcal{L}}{\partial y} - \frac{d}{d\tau}\frac{\partial \mathcal{L}}{\partial \dot{y}} = 0, \quad \frac{\partial \mathcal{L}}{\partial y} = -\frac{2}{y^3}\left(\dot{t}^2 - \dot{y}^2\right), \quad \frac{\partial \mathcal{L}}{\partial \dot{y}} = -\frac{2}{y^2}\dot{y}$$

$$\Rightarrow \quad -\frac{2}{y^3}\left(\dot{t}^2 - \dot{y}^2\right) + \frac{d}{d\tau}\frac{2\dot{y}}{y^2} = 0, \tag{3.1140}$$

which lead to the geodesic equations

$$\ddot{t} - \frac{2}{y}\dot{t}\dot{y} = 0, \quad \ddot{y} - \frac{1}{y}\dot{t}^2 - \frac{1}{y}\dot{y}^2 = 0. \tag{3.1141}$$

Thus, using the two geodesic equations for the t and y coordinates, we identify the nonzero Christoffel symbols as

$$\Gamma^t_{\ ty} = \Gamma^t_{\ yt} = \Gamma^y_{\ tt} = \Gamma^y_{\ yy} = -\frac{1}{y}. \tag{3.1142}$$

b) The Christoffel symbols can be written in matrix form as

$$\Gamma^{\bullet}_{\ t\bullet} = \begin{pmatrix} \Gamma^t_{\ tt} & \Gamma^t_{\ ty} \\ \Gamma^y_{\ tt} & \Gamma^y_{\ ty} \end{pmatrix} = -\frac{1}{y}\begin{pmatrix} 0 & 1 \\ 1 & 0 \end{pmatrix}, \quad \Gamma^{\bullet}_{\ y\bullet} = \begin{pmatrix} \Gamma^t_{\ yt} & \Gamma^t_{\ yy} \\ \Gamma^y_{\ yt} & \Gamma^y_{\ yy} \end{pmatrix} = -\frac{1}{y}\begin{pmatrix} 1 & 0 \\ 0 & 1 \end{pmatrix}. \tag{3.1143}$$

Now, the Riemann curvature tensor in matrix form is given by

$$R^{\bullet}_{\ \bullet\mu\nu} = \partial_\mu \Gamma^{\bullet}_{\ \nu\bullet} - \partial_\nu \Gamma^{\bullet}_{\ \mu\bullet} + \Gamma^{\bullet}_{\ \mu\bullet}\Gamma^{\bullet}_{\ \nu\bullet} - \Gamma^{\bullet}_{\ \nu\bullet}\Gamma^{\bullet}_{\ \mu\bullet} = \partial_\mu \Gamma^{\bullet}_{\ \nu\bullet} - \partial_\nu \Gamma^{\bullet}_{\ \mu\bullet} = -R^{\bullet}_{\ \bullet\nu\mu}. \tag{3.1144}$$

Note that $\Gamma^{\bullet}_{\ \mu\bullet}\Gamma^{\bullet}_{\ \nu\bullet} - \Gamma^{\bullet}_{\ \nu\bullet}\Gamma^{\bullet}_{\ \mu\bullet} = 0$, since $\Gamma^{\bullet}_{\ y\bullet} \propto \mathbb{1}_2$. Therefore, we have

$$R^{\bullet}_{\ \bullet ty} = \partial_t \Gamma^{\bullet}_{\ y\bullet} - \partial_y \Gamma^{\bullet}_{\ t\bullet} = -\partial_y \Gamma^{\bullet}_{\ t\bullet} = -\partial_y \begin{pmatrix} 0 & -\frac{1}{y} \\ -\frac{1}{y} & 0 \end{pmatrix} = \begin{pmatrix} 0 & -\frac{1}{y^2} \\ -\frac{1}{y^2} & 0 \end{pmatrix} = -R^{\bullet}_{\ \bullet yt}. \tag{3.1145}$$

Note that $\partial_t \Gamma^{\bullet}_{\ y\bullet} = 0$, since $\Gamma^{\bullet}_{\ y\bullet}$ is independent of t. Thus, we obtain the nonzero components of the Riemann curvature tensor as

$$R^t_{\ yty} = R^y_{\ tty} = -\frac{1}{y^2} = -R^t_{\ yyt} = -R^y_{\ tyt}. \tag{3.1146}$$

Then, the Ricci tensor is given by $R_{\mu\nu} = R^{\alpha}_{\ \mu\alpha\nu} = R^t_{\ \mu t\nu} + R^y_{\ \mu y\nu}$, and we find that its components are $R_{tt} = R^t_{\ ttt} + R^y_{\ tyt} = 1/y^2$, $R_{ty} = R^t_{\ tty} + R^y_{\ tyy} = 0$, $R_{yt} = R^t_{\ ytt} + R^y_{\ yyt} = 0$, and $R_{yy} = R^t_{\ yty} + R^y_{\ yyy} = -1/y^2$. Thus, we obtain $R_{tt} = -R_{yy} = 1/y^2$ and $R_{ty} = R_{yt} = 0$. Finally, the Ricci scalar is given by a contraction of the inverse metric tensor and the Ricci tensor as

$$R = R^\mu_\mu = g^{\mu\nu} R_{\mu\nu} = g^{tt} R_{tt} + g^{yy} R_{yy} = y^2 \cdot \frac{1}{y^2} + (-y^2) \cdot \left(-\frac{1}{y^2}\right) = 1 + 1 = 2. \tag{3.1147}$$

2.46

Using Euler–Lagrange equations for the ρ and ϕ coordinates with the Lagrangian

$$\mathcal{L} = g_{ij}\dot{x}^i\dot{x}^j = \dot{\rho}^2 + (a^2 + \rho^2)\dot{\phi}^2, \tag{3.1148}$$

we find that

$$\frac{\partial \mathcal{L}}{\partial \rho} - \frac{d}{d\tau}\frac{\partial \mathcal{L}}{\partial \dot{\rho}} = 0, \quad \frac{\partial \mathcal{L}}{\partial \rho} = 2\rho\,\dot{\phi}^2, \quad \frac{\partial \mathcal{L}}{\partial \dot{\rho}} = 2\dot{\rho} \quad \Rightarrow \quad 2\rho\,\dot{\phi}^2 - 2\ddot{\rho} = 0, \tag{3.1149}$$

$$\frac{\partial \mathcal{L}}{\partial \phi} - \frac{d}{d\tau}\frac{\partial \mathcal{L}}{\partial \dot{\phi}} = 0, \quad \frac{\partial \mathcal{L}}{\partial \phi} = 0, \quad \frac{\partial \mathcal{L}}{\partial \dot{\phi}} = 2(a^2 + \rho^2)\,\dot{\phi}$$
$$\Rightarrow \quad 0 - 4\rho\,\dot{\rho}\dot{\phi} - 2(a^2 + \rho^2)\,\ddot{\phi} = 0, \tag{3.1150}$$

which lead to the geodesic equations

$$\ddot{\rho} - \rho\,\dot{\phi}^2 = 0, \quad \ddot{\phi} + \frac{2\rho}{a^2 + \rho^2}\,\dot{\rho}\dot{\phi} = 0. \tag{3.1151}$$

Now, using the general formula for the geodesic equations $\ddot{x}^i + \Gamma^i_{jk}\dot{x}^j\dot{x}^k = 0$, we identify the nonzero Christoffel symbols as

$$\Gamma^\rho_{\phi\phi} = -\rho, \quad \Gamma^\phi_{\rho\phi} = \Gamma^\phi_{\phi\rho} = \frac{\rho}{a^2 + \rho^2}, \tag{3.1152}$$

which can be written in matrix form as

$$\Gamma^\bullet_{\rho\bullet} = \begin{pmatrix} \Gamma^\rho_{\rho\rho} & \Gamma^\rho_{\rho\phi} \\ \Gamma^\phi_{\rho\rho} & \Gamma^\phi_{\rho\phi} \end{pmatrix} = \begin{pmatrix} 0 & 0 \\ 0 & \frac{\rho}{a^2+\rho^2} \end{pmatrix}, \quad \Gamma^\bullet_{\phi\bullet} = \begin{pmatrix} \Gamma^\rho_{\phi\rho} & \Gamma^\rho_{\phi\phi} \\ \Gamma^\phi_{\phi\rho} & \Gamma^\phi_{\phi\phi} \end{pmatrix} = \begin{pmatrix} 0 & -\rho \\ \frac{\rho}{a^2+\rho^2} & 0 \end{pmatrix}. \tag{3.1153}$$

Then, the Riemann curvature tensor in matrix form is given by

$$\begin{aligned} R^\bullet{}_{\bullet\rho\phi} &= \partial_\rho \Gamma^\bullet_{\phi\bullet} - \partial_\phi \Gamma^\bullet_{\rho\bullet} + \Gamma^\bullet_{\rho\bullet}\Gamma^\bullet_{\phi\bullet} - \Gamma^\bullet_{\phi\bullet}\Gamma^\bullet_{\rho\bullet} \\ &= \partial_\rho \begin{pmatrix} 0 & -\rho \\ \frac{\rho}{a^2+\rho^2} & 0 \end{pmatrix} - \partial_\phi \begin{pmatrix} 0 & 0 \\ 0 & \frac{\rho}{a^2+\rho^2} \end{pmatrix} + \begin{pmatrix} 0 & 0 \\ 0 & \frac{\rho}{a^2+\rho^2} \end{pmatrix} \begin{pmatrix} 0 & -\rho \\ \frac{\rho}{a^2+\rho^2} & 0 \end{pmatrix} \\ &\quad - \begin{pmatrix} 0 & -\rho \\ \frac{\rho}{a^2+\rho^2} & 0 \end{pmatrix} \begin{pmatrix} 0 & 0 \\ 0 & \frac{\rho}{a^2+\rho^2} \end{pmatrix} \\ &= \begin{pmatrix} 0 & -1 \\ \frac{a^2-\rho^2}{(a^2+\rho^2)^2} & 0 \end{pmatrix} + \begin{pmatrix} 0 & 0 \\ \frac{\rho^2}{(a^2+\rho^2)^2} & 0 \end{pmatrix} - \begin{pmatrix} 0 & -\frac{\rho^2}{a^2+\rho^2} \\ 0 & 0 \end{pmatrix} \\ &= \begin{pmatrix} 0 & -\frac{a^2}{a^2+\rho^2} \\ \frac{a^2}{(a^2+\rho^2)^2} & 0 \end{pmatrix} = -R^\bullet{}_{\bullet\phi\rho}, \end{aligned} \tag{3.1154}$$

which leads to the nonzero components of the Riemann curvature tensor, i.e.,

$$R^{\rho}{}_{\phi\rho\phi} = -\frac{a^2}{a^2+\rho^2} = -R^{\rho}{}_{\phi\phi\rho}, \quad R^{\phi}{}_{\rho\rho\phi} = \frac{a^2}{(a^2+\rho^2)^2} = -R^{\phi}{}_{\rho\phi\rho}. \tag{3.1155}$$

Next, the Ricci tensor is given by $R_{ij} = R^{k}{}_{ikj} = R^{\rho}{}_{i\rho j} + R^{\phi}{}_{i\phi j}$, and explicitly, the components are

$$R_{\rho\rho} = R^{\rho}{}_{\rho\rho\rho} + R^{\phi}{}_{\rho\phi\rho} = 0 + \left[-\frac{a^2}{(a^2+\rho^2)^2}\right] = -\frac{a^2}{(a^2+\rho^2)^2}, \tag{3.1156}$$

$$R_{\rho\phi} = R^{\rho}{}_{\rho\rho\phi} + R^{\phi}{}_{\rho\phi\phi} = 0 + 0 = 0, \tag{3.1157}$$

$$R_{\phi\rho} = R^{\rho}{}_{\phi\rho\rho} + R^{\phi}{}_{\phi\phi\rho} = 0 + 0 = 0, \tag{3.1158}$$

$$R_{\phi\phi} = R^{\rho}{}_{\phi\rho\phi} + R^{\phi}{}_{\phi\phi\phi} = -\frac{a^2}{a^2+\rho^2} + 0 = -\frac{a^2}{a^2+\rho^2}. \tag{3.1159}$$

Finally, the Ricci scalar is given by the contraction of the inverse metric tensor and the Ricci tensor, i.e.,

$$\begin{aligned} R &= g^{ij}R_{ij} = g^{\rho\rho}R_{\rho\rho} + g^{\phi\phi}R_{\phi\phi} \\ &= 1 \cdot \left[-\frac{a^2}{(a^2+\rho^2)^2}\right] + \frac{1}{a^2+\rho^2}\left(-\frac{a^2}{a^2+\rho^2}\right) = -\frac{2a^2}{(a^2+\rho^2)^2}. \end{aligned} \tag{3.1160}$$

2.47

From the solution to Problem 2.28 for the Rindler coordinates λ and a, we have the Christoffel symbols

$$\Gamma^{\lambda}_{\lambda a} = \Gamma^{\lambda}_{a\lambda} = \frac{1}{a}, \quad \Gamma^{a}_{\lambda\lambda} = a. \tag{3.1161}$$

Therefore, we can write the Christoffel symbols in matrix form as

$$\Gamma^{\bullet}_{\lambda\bullet} = \begin{pmatrix} 0 & 1/a \\ a & 0 \end{pmatrix}, \quad \Gamma^{\bullet}_{a\bullet} = \begin{pmatrix} 1/a & 0 \\ 0 & 0 \end{pmatrix}. \tag{3.1162}$$

Now, the Riemann curvature tensor in matrix form is given by

$$R^{\bullet}{}_{\bullet\mu\nu} = \partial_\mu \Gamma^{\bullet}_{\nu\bullet} - \partial_\nu \Gamma^{\bullet}_{\mu\bullet} + \Gamma^{\bullet}_{\mu\bullet}\Gamma^{\bullet}_{\nu\bullet} - \Gamma^{\bullet}_{\nu\bullet}\Gamma^{\bullet}_{\mu\bullet} = -R^{\bullet}{}_{\bullet\nu\mu}, \tag{3.1163}$$

which implies the potentially nonzero components of the Riemann curvature tensor

$$R^{\bullet}{}_{\bullet\lambda a} = \partial_\lambda \Gamma^{\bullet}_{a\bullet} - \partial_a \Gamma^{\bullet}_{\lambda\bullet} + \Gamma^{\bullet}_{\lambda\bullet}\Gamma^{\bullet}_{a\bullet} - \Gamma^{\bullet}_{a\bullet}\Gamma^{\bullet}_{\lambda\bullet} = -\partial_a \Gamma^{\bullet}_{\lambda\bullet} + \Gamma^{\bullet}_{\lambda\bullet}\Gamma^{\bullet}_{a\bullet} - \Gamma^{\bullet}_{a\bullet}\Gamma^{\bullet}_{\lambda\bullet}, \tag{3.1164}$$

where $\partial_\lambda \Gamma^{\bullet}_{a\bullet} = 0$, since $\Gamma^{\bullet}_{a\bullet}$ is independent of λ. Therefore, inserting $\Gamma^{\bullet}_{\lambda\bullet}$ and $\Gamma^{\bullet}_{a\bullet}$ into $R^{\bullet}{}_{\bullet\lambda a}$, we find that

$$R^{\bullet}{}_{\bullet\lambda a} = -\partial_a \begin{pmatrix} 0 & 1/a \\ a & 0 \end{pmatrix} + \begin{pmatrix} 0 & 1/a \\ a & 0 \end{pmatrix}\begin{pmatrix} 1/a & 0 \\ 0 & 0 \end{pmatrix} - \begin{pmatrix} 1/a & 0 \\ 0 & 0 \end{pmatrix}\begin{pmatrix} 0 & 1/a \\ a & 0 \end{pmatrix}$$
$$= \begin{pmatrix} 0 & 1/a^2 \\ -1 & 0 \end{pmatrix} + \begin{pmatrix} 0 & 0 \\ 1 & 0 \end{pmatrix} - \begin{pmatrix} 0 & 1/a^2 \\ 0 & 0 \end{pmatrix} = \begin{pmatrix} 0 & 0 \\ 0 & 0 \end{pmatrix}. \tag{3.1165}$$

Thus, all components of the Riemann curvature tensor for the Rindler space are zero, and hence, the Riemann curvature tensor is zero. It follows trivially that both the Ricci tensor and the Ricci scalar are also zero.

2.48

Given the coordinate system provided by the problem, we obtain $dt = \cosh(\lambda)d\lambda$, $dx = \sinh(\lambda)\cos(\varphi)d\lambda - \cosh(\lambda)\sin(\varphi)d\varphi$, $dy = \sinh(\lambda)\sin(\varphi)d\lambda + \cosh(\lambda)\cos(\varphi)d\varphi$, and the induced metric

$$ds^2 = d\lambda^2 - \cosh^2\lambda d\varphi^2. \tag{3.1166}$$

We compute the Christoffel symbols from the geodesic equations, using $\mathcal{L} = \frac{1}{2}\left(\dot\lambda^2 - \cosh^2\lambda\dot\varphi^2\right)$, which gives

$$\ddot\lambda + \cosh(\lambda)\sinh(\lambda)\dot\varphi^2 = 0, \tag{3.1167}$$
$$\ddot\varphi + 2\tanh(\lambda)\dot\lambda\dot\varphi = 0, \tag{3.1168}$$

and thus, the nonzero components of the Christoffel symbols are

$$\Gamma^{\lambda}_{\varphi\varphi} = \cosh(\lambda)\sinh(\lambda), \quad \Gamma^{\varphi}_{\lambda\varphi} = \Gamma^{\varphi}_{\varphi\lambda} = \tanh(\lambda). \tag{3.1169}$$

The computation of the Riemann tensor is most easily performed using the following matrix notation

$$R^{\bullet}{}_{\bullet\mu\nu} = \left[\partial_\mu + \Gamma^{\bullet}_{\mu\bullet}, \partial_\nu + \Gamma^{\bullet}_{\nu\bullet}\right], \tag{3.1170}$$

where $\Gamma^{\bullet}_{\mu\bullet}$ stands for the matrix with elements $\Gamma^{\alpha}_{\mu\beta}$, etc., and the matrix product is used. We get the matrices

$$\Gamma^{\bullet}_{\lambda\bullet} = \begin{pmatrix} 0 & 0 \\ 0 & \tanh(\lambda) \end{pmatrix}, \quad \Gamma^{\bullet}_{\varphi\bullet} = \begin{pmatrix} 0 & \cosh(\lambda)\sinh(\lambda) \\ \tanh(\lambda) & 0 \end{pmatrix}, \tag{3.1171}$$

and by straightforward computations, we find

$$R^{\bullet}{}_{\bullet\lambda\varphi} = -R^{\bullet}{}_{\bullet\varphi\lambda} = \begin{pmatrix} 0 & \sinh^2\lambda \\ 1 & 0 \end{pmatrix}. \tag{3.1172}$$

Therefore, the nonzero components of the Riemann tensor are

$$R^{\lambda}{}_{\varphi\lambda\varphi} = -R^{\lambda}{}_{\varphi\varphi\lambda} = \cosh^2\lambda, \quad R^{\varphi}{}_{\lambda\lambda\varphi} = -R^{\varphi}{}_{\lambda\varphi\lambda} = 1. \tag{3.1173}$$

Thus, we obtain the following nonzero components of the Ricci tensor

$$R_{\lambda\lambda} = R^{\varphi}{}_{\lambda\varphi\lambda} = -1, \quad R_{\varphi\varphi} = R^{\lambda}{}_{\varphi\lambda\varphi} = \cosh^2\lambda. \tag{3.1174}$$

Since $ds^2 = d\lambda^2 - \cosh^2\lambda d\varphi^2$ implies that the nonzero components of the metric tensor are as follows

$$g_{\lambda\lambda} = 1, \quad g_{\varphi\varphi} = -\cosh^2\lambda, \tag{3.1175}$$

we observe that

$$R_{\mu\nu} = -g_{\mu\nu}, \tag{3.1176}$$

i.e., $\Lambda = 1$.

2.49

We can use the results in the solution to Problem 2.27 to compute

$$R^{0}{}_{101} = \partial_0\Gamma^0_{11} - \partial_1\Gamma^0_{01} + \Gamma^0_{0\mu}\Gamma^\mu_{11} - \Gamma^0_{1\mu}\Gamma^\mu_{01}. \tag{3.1177}$$

Keeping only the nonzero terms, we find that

$$\begin{aligned} R^{0}{}_{101} &= -\partial_1\Gamma^0_{01} + \Gamma^0_{01}\Gamma^1_{11} - \Gamma^0_{10}\Gamma^0_{01} \\ &= -\frac{\partial}{\partial r}\frac{r}{r^2-1} + \frac{r}{r^2-1}\left(-\frac{r}{r^2-1}\right) - \left(\frac{r}{r^2-1}\right)^2 \\ &= \frac{r^2+1}{(r^2-1)^2} - \frac{2r^2}{(r^2-1)^2} = -\frac{r^2-1}{(r^2-1)^2} = -\frac{1}{r^2-1}. \end{aligned} \tag{3.1178}$$

Thus, we have the following nonzero components of the Riemann curvature tensor

$$R_{0101} = g_{00}R^{0}{}_{101} = v(r^2-1)\left(-\frac{1}{r^2-1}\right) = -v, \tag{3.1179}$$

$$R_{0101} = R_{1010} = -R_{1001} = -R_{0110}, \tag{3.1180}$$

where we used the symmetry properties of the Riemann curvature tensor. Furthermore, we obtain the following nonzero components of the Ricci tensor

$$R_{00} = g^{11}R_{1010} = -\frac{r^2-1}{v}(-v) = r^2 - 1, \tag{3.1181}$$

$$R_{11} = g^{00}R_{0101} = \frac{1}{v(r^2-1)}(-v) = -\frac{1}{r^2-1}. \tag{3.1182}$$

Thus, we have that $R_{\mu\nu} = \frac{1}{v}g_{\mu\nu}$.

2.50

a) Letting $r^2 = t^2 + u^2$, the surface is described by $r^2 - x^2 = \alpha^2$. This is a hyperbola that may be parametrized as

$$r = \alpha \cosh\theta, \quad x = \alpha \sinh\theta. \tag{3.1183}$$

Furthermore, the relation $r^2 = t^2 + u^2$ describes a circle that we may parametrize as

$$t = r\cos\varphi = \alpha\cosh\theta\cos\varphi, \quad u = r\sin\varphi = \alpha\cosh\theta\sin\varphi. \tag{3.1184}$$

Thus, any point on the surface may be specified using the coordinates θ and φ. This leads to

$$dt = \alpha(\sinh\theta\cos\varphi d\theta - \cosh\theta\sin\varphi d\varphi), \tag{3.1185}$$

$$du = \alpha(\sinh\theta\sin\varphi d\theta + \cosh\theta\cos\varphi d\varphi), \tag{3.1186}$$

$$dx = \alpha\cosh\theta d\theta. \tag{3.1187}$$

Inserted into the given line element $ds^2 = dt^2 + du^2 - dx^2$, we obtain

$$ds^2 = \alpha^2(\cosh^2\theta d\varphi^2 - d\theta^2). \tag{3.1188}$$

b) The geodesic equations may be derived from extremizing

$$\mathscr{S} = \frac{1}{2}\int g_{\mu\nu}\dot{\chi}^\mu\dot{\chi}^\nu\, ds = \frac{\alpha^2}{2}\int(\cosh^2\theta\dot{\varphi}^2 - \dot{\theta}^2)\, ds. \tag{3.1189}$$

Requiring $\delta\mathscr{S} = 0$ yields

$$\sinh\theta\cosh\theta\dot{\varphi}^2 + \ddot{\theta} = 0, \tag{3.1190}$$

$$\frac{d}{dt}(\cosh^2\theta\dot{\varphi}) = \ddot{\varphi}\cosh^2\theta + 2\sinh\theta\cosh\theta\dot{\varphi}\dot{\theta} = 0, \tag{3.1191}$$

which simplifies to

$$\ddot{\theta} + \frac{\sinh 2\theta}{2}\dot{\varphi}^2 = 0, \quad \ddot{\varphi} + 2\tanh\theta\dot{\varphi}\dot{\theta} = 0. \tag{3.1192}$$

Comparison and identification with the geodesic equation $\ddot{\chi}^a + \Gamma^a_{bc}\dot{\chi}^b\dot{\chi}^c = 0$ and requiring the connection to be torsion free, results in the following nonzero Christoffel symbols

$$\Gamma^\theta_{\varphi\varphi} = \sinh\theta\cosh\theta = \frac{\sinh 2\theta}{2}, \quad \Gamma^\varphi_{\theta\varphi} = \Gamma^\varphi_{\varphi\theta} = \tanh\theta. \tag{3.1193}$$

c) In order to compute the Ricci scalar, we first compute the one independent entry of the curvature tensor $R_{\theta\varphi\theta\varphi}$. We have by definition

$$R^a{}_{\varphi\theta\varphi}\partial_a = \nabla_\theta\nabla_\varphi\partial_\varphi - \nabla_\varphi\nabla_\theta\partial_\varphi = \nabla_\theta(\Gamma^b_{\varphi\varphi}\partial_b) - \nabla_\varphi(\Gamma^b_{\theta\varphi}\partial_b)$$
$$= \nabla_\theta\left(\frac{\sinh 2\theta}{2}\,\partial_\theta\right) - \nabla_\varphi\left(\tanh\theta\,\partial_\varphi\right)$$
$$= \cosh 2\theta\,\partial_\theta - \tanh\theta\,\Gamma^b_{\varphi\varphi}\partial_b = \cosh 2\theta\,\partial_\theta - \tanh\theta\,\sinh\theta\cosh\theta\,\partial_\theta$$
$$= (\cosh^2\theta + \sinh^2\theta - \sinh^2\theta)\,\partial_\theta = \cosh^2\theta\,\partial_\theta, \tag{3.1194}$$

which implies that

$$R^\theta{}_{\varphi\theta\varphi} = \cosh^2\theta. \tag{3.1195}$$

Thus, we have

$$R_{\theta\varphi\theta\varphi} = g_{\theta a}R^a{}_{\varphi\theta\varphi} = g_{\theta\theta}R^\theta{}_{\varphi\theta\varphi} = -\alpha^2\cosh^2\theta. \tag{3.1196}$$

For the Ricci scalar, we find

$$R = g^{ab}R^c{}_{acb} = g^{ab}g^{cd}R_{dacb} = 2g^{\theta\theta}g^{\varphi\varphi}R_{\theta\varphi\theta\varphi} = \frac{-2}{\alpha^4\cosh^2\theta}(-\alpha^2\cosh^2\theta) = \frac{2}{\alpha^2}. \tag{3.1197}$$

2.51

a) Given the spherically symmetric metric, the metric tensor can be written as

$$g = \mathrm{diag}(g_{tt}, g_{rr}, g_{\theta\theta}, g_{\phi\phi}) = \mathrm{diag}(e^\nu, -e^\rho, -r^2, -r^2\sin^2\theta), \tag{3.1198}$$

with the inverse metric tensor being

$$g^{-1} = \mathrm{diag}(g^{tt}, g^{rr}, g^{\theta\theta}, g^{\phi\phi}) = \mathrm{diag}\left(e^{-\nu}, -e^{-\rho}, -\frac{1}{r^2}, -\frac{1}{r^2\sin^2\theta}\right). \tag{3.1199}$$

Using Euler–Lagrange equations for the t, r, θ, and ϕ coordinates with the Lagrangian

$$\mathcal{L} = e^\nu\dot{t}^2 - e^\rho\dot{r}^2 - r^2\dot{\theta}^2 - r^2\sin^2\theta\,\dot{\phi}^2, \tag{3.1200}$$

we find that the geodesic equations are

$$\ddot{t} + \frac{1}{2}\mathring{\nu}\,\dot{t}^2 + \nu'\,\dot{t}\dot{r} + \frac{1}{2}e^{\rho-\nu}\mathring{\rho}\,\dot{r}^2 = 0, \tag{3.1201}$$

$$\ddot{r} + \frac{1}{2}e^{\nu-\rho}\nu'\,\dot{t}^2 + \mathring{\rho}\,\dot{t}\dot{r} + \frac{1}{2}\rho'\,\dot{r}^2 - re^{-\rho}\,\dot{\theta}^2 - r\sin^2\theta\,e^{-\rho}\,\dot{\phi}^2 = 0, \tag{3.1202}$$

$$\ddot{\theta} + \frac{2}{r}\dot{r}\dot{\theta} - \sin\theta\cos\theta\,\dot{\phi}^2 = 0, \tag{3.1203}$$

$$\ddot{\phi} + \frac{2}{r}\dot{r}\dot{\phi} + 2\cot\theta\,\dot{\theta}\dot{\phi} = 0, \tag{3.1204}$$

where the dot, the circle, and the prime denote differentiations with respect to the path parameter s, time t, and radial coordinate r, respectively. Comparing the four geodesic equations with the general formula for the geodesic equations $\ddot{x}^\alpha + \Gamma^\alpha_{\beta\gamma}\dot{x}^\beta\dot{x}^\gamma = 0$, we identify the 17 nonzero Christoffel symbols as

$$
\begin{aligned}
&\Gamma^t_{tt} = \frac{1}{2}\mathring{\nu}, \quad \Gamma^t_{tr} = \Gamma^t_{rt} = \frac{1}{2}\nu', \quad \Gamma^t_{rr} = \frac{1}{2}e^{\rho-\nu}\mathring{\rho},\\
&\Gamma^r_{tt} = \frac{1}{2}e^{\nu-\rho}\nu', \quad \Gamma^r_{tr} = \Gamma^r_{rt} = \frac{1}{2}\mathring{\rho}, \quad \Gamma^r_{rr} = \frac{1}{2}\rho',\\
&\Gamma^r_{\theta\theta} = -re^{-\rho}, \quad \Gamma^r_{\phi\phi} = -r\sin^2\theta e^{-\rho},\\
&\Gamma^\theta_{r\theta} = \Gamma^\theta_{\theta r} = \frac{1}{r}, \quad \Gamma^\theta_{\phi\phi} = -\sin\theta\cos\theta,\\
&\Gamma^\phi_{r\phi} = \Gamma^\phi_{\phi r} = \frac{1}{r}, \quad \Gamma^\phi_{\theta\phi} = \Gamma^\phi_{\phi\theta} = \cot\theta.
\end{aligned}
\tag{3.1205}
$$

Expressing the Christoffel symbols in matrix form, the Riemann curvature tensor in matrix form is given by

$$
R^{\bullet}{}_{\bullet\alpha\beta} = \partial_\alpha\Gamma^\bullet_{\beta\bullet} - \partial_\beta\Gamma^\bullet_{\alpha\bullet} + \Gamma^\bullet_{\alpha\bullet}\Gamma^\bullet_{\beta\bullet} - \Gamma^\bullet_{\beta\bullet}\Gamma^\bullet_{\alpha\bullet} = -R^{\bullet}{}_{\bullet\beta\alpha}, \tag{3.1206}
$$

which leads to the nonzero components of the Riemann curvature tensor, i.e.,

$$
\begin{aligned}
&R^t{}_{rtr} = e^{\rho-\nu}\left(\frac{1}{2}\overset{\circ\circ}{\rho} + \frac{1}{4}\mathring{\rho}^2 - \frac{1}{4}\mathring{\nu}\mathring{\rho}\right) - \frac{1}{2}\nu'' - \frac{1}{4}\nu'^2 + \frac{1}{4}\nu'\rho',\\
&R^r{}_{ttr} = \frac{1}{2}\overset{\circ\circ}{\rho} + \frac{1}{4}\mathring{\rho}^2 - \frac{1}{4}\mathring{\nu}\mathring{\rho} + e^{\nu-\rho}\left(-\frac{1}{2}\nu'' - \frac{1}{4}\nu'^2 + \frac{1}{4}\nu'\rho'\right),\\
&R^t{}_{\theta t\theta} = -\frac{1}{2}\nu' re^{-\rho}, \quad R^r{}_{\theta t\theta} = \frac{1}{2}\mathring{\rho}re^{-\rho}, \quad R^\theta{}_{tt\theta} = -\frac{1}{2r}e^{\nu-\rho}\nu', \quad R^\theta{}_{rt\theta} = -\frac{1}{2r}\mathring{\rho},\\
&R^t{}_{\phi t\phi} = -\frac{1}{2}\nu' r\sin^2\theta e^{-\rho}, \quad R^r{}_{\phi t\phi} = \frac{1}{2}\mathring{\rho}r\sin^2\theta e^{-\rho},\\
&R^\phi{}_{tt\phi} = -\frac{1}{2r}e^{\nu-\rho}\nu', \quad R^\phi{}_{rt\phi} = -\frac{1}{2r}\mathring{\rho},\\
&R^t{}_{\theta r\theta} = -\frac{1}{2}\mathring{\rho}re^{-\nu}, \quad R^r{}_{\theta r\theta} = \frac{1}{2}\rho' re^{-\rho}, \quad R^\theta{}_{tr\theta} = -\frac{1}{2r}\mathring{\rho}, \quad R^\theta{}_{rr\theta} = -\frac{1}{2r}\rho',\\
&R^t{}_{\phi r\phi} = -\frac{1}{2}\mathring{\rho}r\sin^2\theta e^{-\nu}, \quad R^r{}_{\phi r\phi} = \frac{1}{2}\rho' r\sin^2\theta e^{-\rho},\\
&R^\phi{}_{tr\phi} = -\frac{1}{2r}\mathring{\rho}, \quad R^\phi{}_{rr\phi} = -\frac{1}{2r}\rho',\\
&R^\theta{}_{\phi\theta\phi} = \sin^2\theta\left(1 - e^{-\rho}\right), \quad R^\phi{}_{\theta\theta\phi} = -1 + e^{-\rho}.
\end{aligned}
\tag{3.1207}
$$

The Ricci tensor is given by $R_{\alpha\beta} = R^{\gamma}{}_{\alpha\gamma\beta}$, and explicitly, the nonzero components are

$$R_{tt} = R^{t}{}_{ttt} + R^{r}{}_{trt} + R^{\theta}{}_{t\theta t} + R^{\phi}{}_{t\phi t}$$
$$= e^{\nu-\rho}\left(\frac{1}{2}\nu'' + \frac{1}{4}\nu'^2 - \frac{1}{4}\nu'\rho' + \frac{1}{r}\nu'\right) - \left(\frac{1}{2}\overset{\circ\circ}{\rho} + \frac{1}{4}\overset{\circ}{\rho}^2 - \frac{1}{4}\overset{\circ}{\nu}\overset{\circ}{\rho}\right),$$

$$R_{tr} = R^{t}{}_{ttr} + R^{r}{}_{trr} + R^{\theta}{}_{t\theta r} + R^{\phi}{}_{t\phi r} = \frac{1}{r}\overset{\circ}{\rho} = R_{rt},$$

$$R_{rr} = R^{t}{}_{rtr} + R^{r}{}_{rrr} + R^{\theta}{}_{r\theta r} + R^{\phi}{}_{r\phi r}$$
$$= -\left(\frac{1}{2}\nu'' + \frac{1}{4}\nu'^2 - \frac{1}{4}\nu'\rho' - \frac{1}{r}\rho'\right) + e^{\rho-\nu}\left(\frac{1}{2}\overset{\circ\circ}{\rho} + \frac{1}{4}\overset{\circ}{\rho}^2 - \frac{1}{4}\overset{\circ}{\nu}\overset{\circ}{\rho}\right),$$

$$R_{\theta\theta} = R^{t}{}_{\theta t\theta} + R^{r}{}_{\theta r\theta} + R^{\theta}{}_{\theta\theta\theta} + R^{\phi}{}_{\theta\phi\theta} = -\left[1 + \frac{1}{2}r\left(\nu' - \rho'\right)\right]e^{-\rho} + 1,$$

$$R_{\phi\phi} = R^{t}{}_{\phi t\phi} + R^{r}{}_{\phi r\phi} + R^{\theta}{}_{\phi\theta\phi} + R^{\phi}{}_{\phi\phi\phi} = \sin^2\theta\, R_{\theta\theta}. \tag{3.1208}$$

Finally, the Ricci scalar is given by the contraction of the inverse metric tensor and the Ricci tensor, i.e.,

$$R = g^{\alpha\beta}R_{\alpha\beta} = g^{tt}R_{tt} + g^{rr}R_{rr} + g^{\theta\theta}R_{\theta\theta} + g^{\phi\phi}R_{\phi\phi}$$
$$= \left[\nu'' + \frac{1}{2}\nu'^2 - \frac{1}{2}\nu'\rho' + \frac{2}{r}\nu' + \frac{1}{r}\left(\nu' - \rho'\right)\right]e^{-\rho}$$
$$- \left(\overset{\circ\circ}{\rho} + \frac{1}{2}\overset{\circ}{\rho}^2 - \frac{1}{2}\overset{\circ}{\nu}\overset{\circ}{\rho}\right)e^{-\nu} + \frac{2}{r^2}\left(e^{-\rho} - 1\right). \tag{3.1209}$$

b) Einstein's equations $G_{\alpha\beta} = 0$ are

$$R_{\alpha\beta} - \frac{1}{2}Rg_{\alpha\beta} = 0. \tag{3.1210}$$

Due to Birkhoff's theorem, which we will not show in this problem (see Problem 2.52), the arbitrary functions $\nu = \nu(t,r)$ and $\rho = \rho(t,r)$ must be independent of time t, i.e., $\nu = \nu(r)$ and $\rho = \rho(r)$, which imply that $\overset{\circ}{\nu} = \overset{\circ\circ}{\nu} = 0$ and $\overset{\circ}{\rho} = \overset{\circ\circ}{\rho} = 0$. Therefore, the nonzero components of the Ricci tensor and the Ricci scalar simplify to

$$R_{tt} = e^{\nu-\rho}\left(\frac{1}{2}\nu'' + \frac{1}{4}\nu'^2 - \frac{1}{4}\nu'\rho' + \frac{1}{r}\nu'\right),$$
$$R_{rr} = -\left(\frac{1}{2}\nu'' + \frac{1}{4}\nu'^2 - \frac{1}{4}\nu'\rho' - \frac{1}{r}\rho'\right),$$

$$R_{\theta\theta} = -\left[1 + \frac{1}{2} r \left(\nu' - \rho'\right)\right] e^{-\rho} + 1,$$

$$R_{\phi\phi} = \sin^2\theta \, R_{\theta\theta},$$

$$R = \left[\nu'' + \frac{1}{2}\nu'^2 - \frac{1}{2}\nu'\rho' + \frac{2}{r}\nu' + \frac{1}{r}\left(\nu' - \rho'\right)\right] e^{-\rho} + \frac{2}{r^2}\left(e^{-\rho} - 1\right). \tag{3.1211}$$

Using the two equations

$$G_{tt} = R_{tt} - \frac{1}{2} R g_{tt} = 0, \tag{3.1212}$$

$$G_{rr} = R_{rr} - \frac{1}{2} R g_{rr} = 0, \tag{3.1213}$$

we immediately obtain that

$$\nu' - \rho' + \frac{2}{r}\left(1 - e^{\rho}\right) = 0, \quad \nu' = -\rho', \tag{3.1214}$$

and combining these two expressions, we find the equation

$$\rho' - \frac{1}{r}\left(1 - e^{\rho}\right) = 0. \tag{3.1215}$$

Performing the substitution $\lambda = e^{-\rho}$, which implies that $\lambda' = -\rho' e^{-\rho}$, we instead have the ordinary first-order differential equation

$$\lambda' + \frac{1}{r}\lambda = \frac{1}{r} \quad \Longrightarrow \quad (\lambda r)' = 1, \tag{3.1216}$$

which has the solution $\lambda r = r - r_*$ or, equivalently, $\lambda = 1 - r_*/r$, where r_* is an integration constant. We therefore find

$$\lambda(r) = 1 - \frac{r_*}{r} = e^{-\rho} = -\frac{1}{g_{rr}}. \tag{3.1217}$$

Finally, using $\nu' = -\rho'$ and $e^{-\rho} = 1 - \frac{r_*}{r}$, we find that

$$g_{tt}(r) = e^{\nu} = 1 - \frac{r_*}{r}, \quad g_{rr}(r) = -e^{\rho} = -\left(1 - \frac{r_*}{r}\right)^{-1}. \tag{3.1218}$$

Thus, we obtain the solution to Einstein's equations $G_{\alpha\beta} = 0$, namely

$$ds^2 = g_{\alpha\beta} dx^\alpha dx^\beta = \left(1 - \frac{r_*}{r}\right) dt^2 - \left(1 - \frac{r_*}{r}\right)^{-1} dr^2 - r^2\left(d\theta^2 + \sin^2\theta d\phi^2\right), \tag{3.1219}$$

which is the *Schwarzschild metric* and known as the Schwarzschild solution to the free Einstein field equations, i.e., $G_{\alpha\beta} = R_{\alpha\beta} - \frac{1}{2} R g_{\alpha\beta} = 0$.

2.52

Assume that Einstein's equations in empty space hold, i.e., $G_{\alpha\beta} = 0$, where $G_{\alpha\beta}$ is the Einstein tensor. In general, using the definition of the Einstein tensor, we have

$$G_{\alpha\beta} \equiv R_{\alpha\beta} - \frac{1}{2} R g_{\alpha\beta} = 0, \tag{3.1220}$$

where $g_{\alpha\beta}$ is a spherically symmetric metric such that (see Problem 2.51 and its solution)

$$ds^2 = g_{\alpha\beta} dx^\alpha dx^\beta = e^\nu dt^2 - e^\rho dr^2 - r^2 \left(d\theta^2 + \sin^2\theta d\phi^2\right), \tag{3.1221}$$

and $R_{\alpha\beta}$ and R are the Ricci tensor and the Ricci scalar, respectively. In particular, we have

$$G_{tr} = R_{tr} - \frac{1}{2} R g_{tr} = 0. \tag{3.1222}$$

However, $g_{tr} = 0$, which implies that $R_{tr} = 0$. Using the results of Problem 2.51, we have that $R_{tr} = \overset{\circ}{\rho}/r$, where the circle denotes differentiation with respect to the coordinate time t. Therefore, it must hold that $\overset{\circ}{\rho} = 0$ (which leads to $\overset{\circ\circ}{\rho} = 0$), which means that ρ has no time dependence. Since ρ and, hence also $\overset{\circ}{\rho}$, have no time dependence, Einstein's equation (see, again, the solution to Problem 2.51)

$$G_{\theta\theta} = R_{\theta\theta} - \frac{1}{2} R g_{\theta\theta} = -\left[1 + \frac{1}{2} r \left(\nu' - \rho'\right)\right] e^{-\rho} + 1$$
$$- \frac{1}{2} \left\{ \left[\nu'' + \frac{1}{2} \nu'^2 - \frac{1}{2} \nu' \rho' + \frac{2}{r} \nu' + \frac{1}{r} \left(\nu' - \rho'\right) \right] e^{-\rho} + \frac{2}{r^2} \left(e^{-\rho} - 1\right) \right\} = 0, \tag{3.1223}$$

implies that ν' has also no time dependence, since the entire equation has no time dependence. Therefore, the statement

$$\nu' \equiv \frac{\partial \nu}{\partial r} = f(r), \tag{3.1224}$$

means that ν must depend on t and r separately such that

$$\nu = \nu(r) + g(t). \tag{3.1225}$$

The appearance of $\nu(r)$ and $g(t)$ in ds^2 has a form such that a possible time-dependence $g(t)$ can be absorbed in a new time variable t' such that

$$e^\nu dt^2 = e^{\nu(r) + g(t)} dt^2 = e^{\nu(r)} e^{g(t)} dt^2 = e^{\nu(r)} dt'^2. \tag{3.1226}$$

In terms of the coordinates t', r, θ, and ϕ, the metric functions $g_{t't'} = e^{\nu(r)}$, $g_{rr} = -e^{\rho(r)}$, $g_{\theta\theta} = -r^2$, and $g_{\phi\phi} = -r^2 \sin^2\theta$ are time independent, i.e., a spherically symmetric solution to $G_{\alpha\beta} = 0$ must be static, and thus, it holds that $\nu = \nu(r)$ and $\rho = \rho(r)$ such that $\mathring{\nu} = 0$ and $\mathring{\rho} = 0$. This completes the proof of Birkhoff's theorem.

2.53

For a Levi-Civita connection, we have $\nabla g^{\mu\nu} \equiv 0$. Therefore, we find that

$$\nabla_\mu T^{\mu\nu} = \epsilon_0(\nabla_\mu F^\mu{}_\lambda)F^{\lambda\nu} + \epsilon_0 F^\mu{}_\lambda \nabla_\mu F^{\lambda\nu} + \frac{\epsilon_0}{2} g^{\mu\nu} F_{\lambda\omega}\nabla_\mu F^{\lambda\omega}. \tag{3.1227}$$

The first term is equal to $\epsilon_0 J_\lambda F^{\lambda\nu}$ by the first set of Maxwell's equations, i.e., $\nabla_\mu F^\mu{}_\lambda = J_\lambda$. Using the second set of Maxwell's equations, i.e., $\partial^\mu F^{\nu\lambda} + \partial^\nu F^{\lambda\mu} + \partial^\lambda F^{\mu\nu} = 0$ (note that here one can use ∂ instead of ∇ by the antisymmetry of $F^{\mu\nu}$ and by $\Gamma^\lambda_{\mu\nu} = \Gamma^\lambda_{\nu\mu}$), we obtain

$$\begin{aligned}
\frac{1}{\epsilon_0}\nabla_\mu T^{\mu\nu} &= J_\lambda F^{\lambda\nu} + F_{\mu\lambda}\nabla^\mu F^{\lambda\nu} + \frac{1}{2}F_{\lambda\omega}\nabla^\nu F^{\lambda\omega} \\
&= J_\lambda F^{\lambda\nu} + F_{\mu\lambda}\nabla^\mu F^{\lambda\nu} - \frac{1}{2}(F_{\lambda\omega}\nabla^\lambda F^{\omega\nu} + F_{\lambda\omega}\nabla^\omega F^{\nu\lambda}) \\
&= \{F_{\lambda\omega}\nabla^\omega F^{\nu\lambda} = \left[F^{\mu\nu} = -F^{\nu\mu}\right] = (-F_{\omega\lambda})\nabla^\omega(-F^{\lambda\nu}) \\
&= F_{\omega\lambda}\nabla^\omega F^{\lambda\nu} = [\omega \leftrightarrow \lambda] = F_{\lambda\omega}\nabla^\lambda F^{\omega\nu}\} \\
&= J_\lambda F^{\lambda\nu} + F_{\mu\lambda}\nabla^\mu F^{\lambda\nu} - F_{\lambda\omega}\nabla^\lambda F^{\omega\nu} = J_\lambda F^{\lambda\nu},
\end{aligned} \tag{3.1228}$$

i.e.,

$$\nabla_\mu T^{\mu\nu} = \epsilon_0 J_\mu F^{\mu\nu}. \tag{3.1229}$$

2.54

We know that the correct formula for half of Maxwell's equations in general relativity is given by

$$\nabla_\alpha F_{\beta\gamma} + \nabla_\beta F_{\gamma\alpha} + \nabla_\gamma F_{\alpha\beta} = 0, \tag{3.1230}$$

since this is a tensor equation, which in a local inertial frame equals the corresponding formula in special relativity. It remains to be shown that it is sufficient to use partial derivatives even in the case of general relativity.

Now, the covariant derivative of a second-order tensor is given by

$$(\nabla_\alpha F)_{\beta\gamma} = \partial_\alpha F_{\beta\gamma} - \Gamma^\kappa_{\alpha\beta}F_{\kappa\gamma} - \Gamma^\kappa_{\alpha\gamma}F_{\beta\kappa}. \tag{3.1231}$$

Therefore, we have $\nabla_\alpha F_{\beta\gamma} + \nabla_\beta F_{\gamma\alpha} + \nabla_\gamma F_{\alpha\beta} = \partial_\alpha F_{\beta\gamma} + \partial_\beta F_{\gamma\alpha} + \partial_\gamma F_{\alpha\beta}$ if

$$\Gamma^\kappa_{\alpha\beta}F_{\kappa\gamma} + \Gamma^\kappa_{\alpha\gamma}F_{\beta\kappa} + \Gamma^\kappa_{\beta\gamma}F_{\kappa\alpha} + \Gamma^\kappa_{\beta\alpha}F_{\gamma\kappa} + \Gamma^\kappa_{\gamma\alpha}F_{\kappa\beta} + \Gamma^\kappa_{\gamma\beta}F_{\alpha\kappa} = 0, \tag{3.1232}$$

which is true, since the terms cancel pairwise (due to $F_{\alpha\beta}$ being antisymmetric and $\Gamma^{\alpha}_{\beta\gamma} = \Gamma^{\alpha}_{\gamma\beta}$ being symmetric in the two lower indices). Thus, we have shown that half of Maxwell's equations can be written as

$$\partial_\alpha F_{\beta\gamma} + \partial_\beta F_{\gamma\alpha} + \partial_\gamma F_{\alpha\beta} = 0, \tag{3.1233}$$

even in general relativity.

2.55

For any vector field X, we have the covariant divergence

$$\nabla_\mu X^\mu = \partial_\mu X^\mu + \Gamma^{\mu}_{\mu\nu} X^\nu. \tag{3.1234}$$

Using the definition of the Christoffel symbols in terms of the metric, we obtain for $\Gamma^{\mu}_{\mu\nu}$ the following

$$\Gamma^{\mu}_{\mu\nu} = \frac{1}{2} g^{\mu\lambda}(\partial_\nu g_{\lambda\mu}) = \frac{1}{2}\,\mathrm{tr}\left(g^{-1}\partial_\nu g\right) = \frac{1}{2}\,\mathrm{tr}\left[(-g)^{-1}\partial_\nu(-g)\right] = \frac{1}{2}\partial_\nu\,\mathrm{tr}\ln(-g), \tag{3.1235}$$

which, using the matrix identity $\mathrm{tr}\ln A = \ln\det A$, can be rewritten in terms of $\bar{g} \equiv \det(-g) = -\det g$ as

$$\Gamma^{\mu}_{\mu\nu} = \frac{1}{2}\partial_\nu \ln\det(-g) = \frac{1}{2}\partial_\nu \ln\bar{g} = \frac{1}{2\bar{g}}\partial_\nu \bar{g}. \tag{3.1236}$$

Inserting this result for the Christoffel symbol $\Gamma^{\mu}_{\mu\nu}$ into the equation for the covariant derivative yields

$$\begin{aligned}\nabla_\mu X^\mu &= \partial_\mu X^\mu + \frac{1}{2\bar{g}}(\partial_\nu \bar{g})X^\nu = \partial_\mu X^\mu + \frac{1}{2\bar{g}}(\partial_\mu \bar{g})X^\mu \\ &= \frac{1}{\sqrt{\bar{g}}}\partial_\mu\left(\sqrt{\bar{g}}X^\mu\right) = \bar{g}^{-\frac{1}{2}}\partial_\mu(\bar{g}^{\frac{1}{2}}X^\mu).\end{aligned} \tag{3.1237}$$

Thus, from this follows that $\nabla_\mu j^\mu = 0$ can be written as $\bar{g}^{-\frac{1}{2}}\partial_\mu(\bar{g}^{\frac{1}{2}} j^\mu) = 0$.

Now, we wish to prove that $\nabla_\mu j^\mu = 0$ is compatible with the generally covariant form of Maxwell's equations, i.e., we wish to show that $a = \nabla_\nu\nabla_\mu F^{\mu\nu} = 0$. First, expanding the covariant derivative ∇_μ and expressing a in terms of Christoffel symbols, we obtain

$$a = \nabla_\nu(\partial_\mu F^{\mu\nu} + \Gamma^{\mu}_{\mu\omega}F^{\omega\nu} + \Gamma^{\nu}_{\mu\omega}F^{\mu\omega}), \tag{3.1238}$$

where the last term vanishes due to $F^{\mu\omega} = -F^{\omega\mu}$ and $\Gamma^{\nu}_{\mu\omega} = \Gamma^{\nu}_{\omega\mu}$. Second, expanding the covariant derivative ∇_ν, we find that

$$
\begin{aligned}
a &= \partial_\nu \partial_\mu F^{\mu\nu} + \Gamma^\nu_{\nu\lambda} \partial_\mu F^{\mu\lambda} + \partial_\nu (\Gamma^\mu_{\mu\omega} F^{\omega\nu}) + \Gamma^\nu_{\nu\lambda} \Gamma^\mu_{\mu\omega} F^{\omega\lambda} \\
&= \Gamma^\nu_{\nu\lambda} \partial_\mu F^{\mu\lambda} + \Gamma^\mu_{\mu\omega} \partial_\nu F^{\omega\nu} + F^{\omega\nu} \partial_\nu \Gamma^\mu_{\mu\omega} = \left(\Gamma^\nu_{\nu\lambda} \partial_\mu + \Gamma^\nu_{\nu\mu} \partial_\lambda\right) F^{\mu\lambda} + F^{\omega\nu} \partial_\nu \Gamma^\mu_{\mu\omega} \\
&= F^{\omega\nu} \partial_\nu \Gamma^\mu_{\mu\omega} = \frac{1}{2} F^{\omega\nu} \partial_\nu \partial_\omega \ln \bar{g} = 0,
\end{aligned}
\tag{3.1239}
$$

where we have used several times the fact that an antisymmetric tensor (such as F) that is contracted with a symmetric tensor vanishes. Thus, we have proven that $\nabla_\nu \nabla_\mu F^{\mu\nu} = 0$.

2.56

a) To show that the connection is the Levi-Civita connection, we need to show that it is metric compatible and torsion free. This is a simple matter of checking the relations

$$
X_i \cdot g(X_j, X_k) = g(\nabla_{X_i} X_j, X_k) + g(X_j, \nabla_{X_i} X_k), \tag{3.1240}
$$

and

$$
T(X_i, X_j) = \nabla_{X_i} X_j - \nabla_{X_j} X_i - [X_i, X_j] = 0, \tag{3.1241}
$$

cf., Problem 2.43.

b) The curvature tensor can be calculated in the same way as in Problem 2.43, this leads to the following nonzero components of the curvature tensor:

$$
\begin{aligned}
&{R^1}_{221} = {R^1}_{010} = {R^2}_{112} = {R^2}_{020} = {R^0}_{220} = {R^0}_{110} \\
&\quad = -{R^1}_{212} = -{R^1}_{001} = -{R^2}_{121} = -{R^2}_{002} = -{R^0}_{202} = -{R^0}_{101} = \frac{1}{4}.
\end{aligned}
\tag{3.1242}
$$

Computing the Ricci tensor from its definition $R_{ij} = {R^k}_{ikj}$, we find that $R_{ij} = g_{ij}/2$. From this follows that the Ricci scalar is given by $R = g^{ij} R_{ij} = \delta^i_i/2 = 3/2$, and thus,

$$
G_{ij} = R_{ij} - \frac{1}{2} g_{ij} R = \frac{1}{2} g_{ij} \left(1 - \frac{3}{2}\right) = -\frac{1}{4} g_{ij}. \tag{3.1243}
$$

By the Einstein equations, we have

$$
T_{ij} = \frac{1}{8\pi} G_{ij} = -\frac{1}{32\pi} g_{ij}, \tag{3.1244}
$$

where we have put $G = c = 1$.

2.57

Introducing coordinates χ^a and having the particle worldline parametrized by $\chi^a = f^a(\tau)$, the action can be written in the form

$$\mathscr{S}_M = M \int \delta^{(4)}(\chi - f)\sqrt{g_{ab}\dot{f}^a \dot{f}^b}\, d^4\chi d\tau, \tag{3.1245}$$

where $\delta^{(4)}(\chi - f)$ is a four-dimensional δ-function in the coordinates and g_{ab} is a function of the coordinates χ. Assuming coordinates such that χ^a can be used to parametrize the particle worldline (e.g., a coordinate time), we can use the δ-function $\delta(\chi^0 - f^0(\tau))$ to perform the τ-integral:

$$\mathscr{S}_M = M \int \delta^{(3)}(\boldsymbol{\chi} - \boldsymbol{f}(\tau))\frac{1}{\dot{f}^0}\sqrt{g_{ab}\dot{f}^a \dot{f}^b}\, d^4\chi, \tag{3.1246}$$

where any reference to τ should be seen as an implicit function of χ^0 given by the inverse of $\chi^0 = f^0(\tau)$. Varying this action with respect to g_{ab} leads to

$$T^{ab} = \frac{2}{\sqrt{-g}}\frac{\delta \mathscr{S}_M}{\delta g_{ab}} = \frac{M\dot{f}^a \dot{f}^b \delta^{(3)}(\boldsymbol{\chi} - \boldsymbol{f})}{\sqrt{-g}\,\dot{f}^0\sqrt{g_{cd}\dot{f}^c \dot{f}^d}} = \frac{\dot{f}^a \dot{f}^b \delta^{(3)}(\boldsymbol{\chi} - \boldsymbol{f})}{\sqrt{-g}\,\dot{f}^0} M, \tag{3.1247}$$

where the last step assumes that the 4-velocity of the particle is properly normalized.

For the case of standard coordinates on Minkowski space, $\chi^0 = t$, $\chi^i = x^i$, we find that $\sqrt{-g} = 1$, and therefore,

$$T^{00} = \delta^{(3)}(\boldsymbol{\chi} - \boldsymbol{f}(t))\dot{t}M = M\gamma\delta^{(3)}(\boldsymbol{\chi} - \boldsymbol{f}(t)) = E\delta^{(3)}(\boldsymbol{\chi} - \boldsymbol{f}(t)), \tag{3.1248}$$

$$T^{0i} = \delta^{(3)}(\boldsymbol{\chi} - \boldsymbol{f}(t))M\dot{x}^i = M\gamma v^i \delta^{(3)}(\boldsymbol{\chi} - \boldsymbol{f}(t)) = p^i \delta^{(3)}(\boldsymbol{\chi} - \boldsymbol{f}(t)). \tag{3.1249}$$

This is reasonable. The energy density is the energy of the particle located at $\boldsymbol{x} = \boldsymbol{f}(t)$ and the momentum density is the momentum of the particle located at $\boldsymbol{x} = \boldsymbol{f}(t)$.

2.58

The action is given by

$$\begin{aligned}\mathscr{S}_{\text{EM}} &= \int \left[-\frac{1}{4}(\partial_\mu A_\nu - \partial_\nu A_\mu)g^{\mu\sigma}g^{\nu\rho}(\partial_\sigma A_\rho - \partial_\rho A_\sigma) + J^\mu A_\mu\right]\sqrt{g}\, d^4x \\ &= \int \mathcal{L}\, d^4x. \end{aligned} \tag{3.1250}$$

Varying the action with respect to A_λ, we find

$$\frac{\partial \mathcal{L}}{\partial A_\lambda} = \sqrt{g}J^\lambda,$$

$$\frac{\partial \mathcal{L}}{\partial(\partial_\kappa A_\lambda)} = -\frac{1}{2}(\delta^\kappa_\mu \delta^\lambda_\nu - \delta^\lambda_\mu \delta^\kappa_\nu) g^{\mu\sigma} g^{\nu\rho} (\partial_\sigma A_\rho - \partial_\rho A_\sigma)\sqrt{g}$$
$$= -\frac{1}{2}(F^{\kappa\lambda} - F^{\lambda\kappa})\sqrt{g} = F^{\lambda\kappa}\sqrt{g}. \quad (3.1251)$$

The Euler–Lagrange equations are therefore

$$\frac{\partial \mathcal{L}}{\partial A_\lambda} - \partial_\kappa\left(\frac{\partial \mathcal{L}}{\partial(\partial_\kappa A_\lambda)}\right) = \sqrt{g} J^\lambda - \partial_\kappa(F^{\lambda\kappa}\sqrt{g}) = 0, \quad (3.1252)$$

or in other words, it holds that

$$\frac{1}{\sqrt{g}}\partial_\kappa(F^{\lambda\kappa}\sqrt{g}) = J^\lambda. \quad (3.1253)$$

Note that, for an antisymmetric tensor, $\frac{1}{\sqrt{g}}\partial_\kappa(F^{\lambda\kappa}\sqrt{g}) = \nabla_\kappa F^{\lambda\kappa}$ and so this is also possible to express as

$$\nabla_\kappa F^{\lambda\kappa} = J^\lambda. \quad (3.1254)$$

2.59

With the given Lagrangian, the action is given by

$$\mathscr{S} = \mathscr{S}_{\text{EM}} + \int \mathcal{L}_\phi \sqrt{g}\, d^4x = \mathscr{S}_{\text{EM}} + \int \frac{1}{2}\left[g^{\mu\nu}(\partial_\mu\phi)(\partial_\nu\phi) - m^2\phi^2\right]\sqrt{g}\, d^4x, \quad (3.1255)$$

where the Einstein–Hilbert action $\mathscr{S}_{\text{EM}}$ does not depend on ϕ. Varying with respect to ϕ then leads to $\mathscr{S}_{\text{EM}} = 0$, and therefore, $\delta \int \mathcal{L}_\phi \sqrt{g}\, d^4x \equiv \delta \int L\, d^4x = 0$. From the Euler–Lagrange equations for ϕ with $L = \mathcal{L}_\phi\sqrt{g}$, we find the equation of motion for ϕ is

$$\frac{\partial L}{\partial \phi} - \partial_\mu \frac{\partial L}{\partial(\partial_\mu \phi)} = 0. \quad (3.1256)$$

We have

$$\frac{\partial L}{\partial \phi} = \sqrt{g}\frac{\partial \mathcal{L}_\phi}{\partial \phi} = -\sqrt{g} m^2 \phi, \quad (3.1257)$$

$$\frac{\partial L}{\partial(\partial_\mu\phi)} = \sqrt{g}\frac{\partial \mathcal{L}_\phi}{\partial(\partial_\mu\phi)} = \sqrt{g} g^{\mu\nu}\partial_\nu\phi, \quad (3.1258)$$

and the equation of motion is therefore

$$\frac{1}{\sqrt{g}}\partial_\mu\left(\sqrt{g} g^{\mu\nu}\partial_\nu\phi\right) + m^2\phi = 0. \quad (3.1259)$$

Note that this can also be expressed in terms of the connection ∇ as

$$\Box\phi + m^2\phi = 0, \quad (3.1260)$$

where $\Box \equiv g^{\mu\nu}\nabla_\mu\nabla_\nu$. In Minkowski spacetime, this reduces to the Klein-Gordon equation

$$\partial_\mu\partial^\mu\phi + m^2\phi = 0. \tag{3.1261}$$

2.60

To find the stress–energy tensor, we vary the action $\mathscr{S} = \int \mathcal{L}\sqrt{-g}\, d^4x$ with respect to the metric inverse. We find that

$$\delta\mathscr{S} = \int (\delta\mathcal{L}\sqrt{-g} + \mathcal{L}\delta\sqrt{-g})\, d^4x. \tag{3.1262}$$

Furthermore, we have

$$\delta\mathcal{L} = \frac{1}{2}(\partial_\mu\phi)(\partial_\nu\phi)\delta g^{\mu\nu}. \tag{3.1263}$$

Thus, we find that

$$\delta\mathscr{S} = \int \left\{\frac{1}{2}(\partial_\mu\phi)(\partial_\nu\phi) - \frac{1}{2}\left[\frac{1}{2}g^{\sigma\rho}(\partial_\sigma\phi)(\partial_\rho\phi) - V(\phi)\right] g_{\mu\nu}\right\} \delta g^{\mu\nu}\sqrt{-g}\, d^4x. \tag{3.1264}$$

It follows that

$$T_{\mu\nu} = \frac{2}{\sqrt{-g}}\frac{\delta\mathscr{S}}{\delta g^{\mu\nu}} = (\partial_\mu\phi)(\partial_\nu\phi) - \frac{1}{2}g_{\mu\nu}g^{\sigma\rho}(\partial_\sigma\phi)(\partial_\rho\phi) + V(\phi)g_{\mu\nu}. \tag{3.1265}$$

For the case where the stress–energy tensor is dominated by $V(\phi)$, we find that

$$T_{\mu\nu} \simeq V(\phi)g_{\mu\nu}, \tag{3.1266}$$

i.e., the stress–energy tensor is proportional to the metric. Note that this would lead to the equation-of-state parameter $w = p/\rho = -1$.

2.61

If the solution would be a vacuum solution, it would satisfy $G_{\mu\nu} = 8\pi G T_{\mu\nu} = 0$. Our aim is therefore to show that $G_{\mu\nu} \neq 0$, or equivalently $R_{\mu\nu} \neq 0$. We note that our spacetime is a spatially flat Robertson–Walker spacetime with $a = \cosh(Ht)$. For the general spatially flat Robertson–Walker spacetime, we can find the Christoffel symbols by extremizing

$$\mathcal{S} = \frac{1}{2}\int g_{\mu\nu}x^\mu x^\nu d\tau. \tag{3.1267}$$

The Euler–Lagrange equations for $\delta\mathcal{S} = 0$ take the form

$$\ddot{t} = -aa'\dot{\boldsymbol{x}}^2, \quad \ddot{x}^i + 2\frac{a'}{a}\dot{t}\dot{x}^i = 0, \tag{3.1268}$$

from which we can identify the nonzero Christoffel symbols

$$\Gamma^t_{ij} = aa'\delta_{ij} \quad \text{and} \quad \Gamma^i_{jt} = \frac{a'}{a}\delta_{ij}. \tag{3.1269}$$

For the general spatially flat Robertson–Walker metric, we therefore find that

$$R_{tt} = -\frac{3a''}{a} \quad \text{and} \quad R_{ij} = \delta_{ij}(aa'' + 2a'^2). \tag{3.1270}$$

It follows that the Ricci scalar is given by

$$R = g^{tt}R_{tt} + g^{ij}R_{ij} = 6\frac{a'^2}{a}. \tag{3.1271}$$

For our case, $a = \cosh(Ht)$ and $a' = H\sinh(Ht)$ and therefore

$$R = 6H^2\tanh^2(Ht). \tag{3.1272}$$

From Einstein's equations follows that

$$g^{\mu\nu}G_{\mu\nu} = -R = 8\pi G g^{\mu\nu}T_{\mu\nu}. \tag{3.1273}$$

Since our R is nonzero in our case, the vacuum solution $T_{\mu\nu} = 0$ is therefore not allowed as it would lead to $R = 0$. The spacetime is therefore not a vacuum solution to Einstein's equations.

2.62

The action corresponding to the electromagnetic field is

$$\mathcal{S}_{\text{EM}} = -\frac{1}{4}\int F_{\mu\nu}F_{\sigma\rho}g^{\mu\sigma}g^{\nu\rho}\sqrt{|\bar{g}|}d^4x. \tag{3.1274}$$

Varying this action and setting the variation to zero results in

$$\delta\mathcal{S}_{\text{EM}} = -\frac{1}{2}\int\left(g^{\mu\sigma}F_{\mu\nu}F_{\sigma\rho}\delta g^{\nu\rho} + g^{\mu\sigma}g^{\nu\rho}F_{\mu\nu}\delta F_{\sigma\rho}\right)\sqrt{|\bar{g}|}d^4x - \int\mathcal{L}\delta\sqrt{|\bar{g}|}d^4x. \tag{3.1275}$$

The variation of $\sqrt{|\bar{g}|}$ can be found by considering the relation $\delta^\mu_\nu = g^{\mu\lambda}g_{\lambda\nu}$ from which we find

$$0 = \delta\delta^\mu_\nu = g_{\lambda\nu}\delta g^{\mu\lambda} + g^{\mu\lambda}\delta g_{\lambda\nu}. \tag{3.1276}$$

This relation takes the matrix form $g\delta g^{-1} = -g^{-1}\delta g$. From the matrix identity $\text{tr}(\ln(A)) = \ln(\det(A))$ now follows

$$\text{tr}(g^{-1}\delta g) = \delta\ln(|\bar{g}|) = \frac{\delta|\bar{g}|}{|\bar{g}|}, \tag{3.1277}$$

or, in components, we have

$$\delta|\bar{g}| = |\bar{g}|g^{\mu\nu}\delta g_{\mu\nu} = -|\bar{g}|g_{\mu\nu}\delta g^{\mu\nu}. \tag{3.1278}$$

This means that

$$\delta\sqrt{|\bar{g}|} = -\frac{1}{2}\sqrt{|\bar{g}|}g_{\mu\nu}\delta g^{\mu\nu}. \tag{3.1279}$$

At the same time, the field $F_{\mu\nu} = \partial_\mu A_\nu - \partial_\nu A_\mu$ is independent of $g^{\mu\nu}$ and therefore does not vary when the metric is varied. It follows that

$$\frac{\delta S_{\text{EM}}}{\delta g^{\nu\rho}} = -\frac{1}{2}\left(F_{\mu\nu}F^\mu{}_\rho - \frac{1}{4}g_{\nu\rho}F^{\alpha\beta}F_{\alpha\beta}\right)\sqrt{|\bar{g}|}, \tag{3.1280}$$

and therefore, we find that the stress–energy tensor of the electromagnetic field is

$$T_{\nu\rho} = -F_{\mu\nu}F^\mu{}_\rho + \frac{1}{4}g_{\nu\rho}F^{\alpha\beta}F_{\alpha\beta}. \tag{3.1281}$$

2.63

Since $x(s)$ is a geodesic, it holds that

$$\frac{dx^i}{ds}\nabla_i\left(\frac{dx^j}{ds}\right) = 0. \tag{3.1282}$$

Therefore, we find that

$$\begin{aligned}
\dot{W} \equiv \frac{d}{ds}W &= \frac{dx^i}{ds}\nabla_i W = \frac{dx^i}{ds}\nabla_i\left(X_j\frac{dx^j}{ds}\right) \\
&= \frac{dx^i}{ds}\left(\nabla_i X_j\right)\frac{dx^j}{ds} + \frac{dx^i}{ds}\left(\nabla_i\frac{dx^j}{ds}\right)X_j \\
&= \frac{dx^i}{ds}\left(\nabla_i X_j\right)\frac{dx^j}{ds} + 0 = \left(\nabla_i X_j\right)\frac{dx^i}{ds}\frac{dx^j}{ds} = 0,
\end{aligned} \tag{3.1283}$$

since $\nabla_i X_j$ is antisymmetric and $\frac{dx^i}{ds}\frac{dx^j}{ds}$ is symmetric. Thus, $\dot{W} = 0$ implies that $W = X_i\frac{dx^i}{ds}$ is a conserved quantity.

2.64

a) By multiplying the coordinate-based expression with $A^\mu B^\nu$, where A^μ and B^ν are arbitrary vectors, we obtain

$$A^\mu B^\nu X^\lambda \partial_\lambda g_{\mu\nu} = -g_{\lambda\nu}A^\mu B^\nu \partial_\mu X^\lambda - g_{\mu\lambda}A^\mu B^\nu \partial_\nu X^\lambda. \tag{3.1284}$$

The left-hand side of this equation can be written as

$$\begin{aligned}
A^\mu B^\nu X^\lambda \partial_\lambda g_{\mu\nu} &= X^\lambda \partial_\lambda A^\mu B^\nu g_{\mu\nu} - g_{\mu\nu}A^\mu X^\lambda \partial_\lambda B^\nu - g_{\mu\nu}B^\nu X^\lambda \partial_\lambda A^\mu \\
&= X \cdot g(A, B) - g_{\mu\nu}A^\mu X^\lambda \partial_\lambda B^\nu - g_{\mu\nu}B^\nu X^\lambda \partial_\lambda A^\mu.
\end{aligned} \tag{3.1285}$$

Moving the last two terms of this expression to the right-hand side of the first equation, the right-hand side simplifies to $g([X,A],B) + g(A,[X,B])$ and we have the condition

$$X \cdot g(A,B) = g([X,A],B) + g(A,[X,B]). \tag{3.1286}$$

In addition, since A and B are arbitrary, we do not lose any information in the multiplication. The wanted coordinate-free version of the condition follows.

b) The result follows from the given equations with $X^\mu = 1$ if $\mu = \lambda$ and $X^\mu = 0$ otherwise. The left-hand side takes the form $\partial_\lambda g_{\mu\nu}$ and the right-hand side vanishes since X^μ are constant.

c) In this case, the coordinate-free condition should be used. The left-hand side vanishes and the right-hand side vanishes if any of the vector fields are equal, this follows from the form of the commutators (the commutator of two vector fields is orthogonal to both fields according to the definition). The remaining cases can be checked explicitly.

2.65

a) The flow of $K = y\partial_x - x\partial_y$ should satisfy the differential equations

$$\dot{x} = K^x = y, \quad \dot{y} = K^y = -x. \tag{3.1287}$$

Solving this system of differential equations with $x(0) = x_0$ and $y(0) = y_0$ leads to

$$x = x_0\cos(s) + y_0\sin(s), \quad y = -x_0\sin(s) + y_0\cos(s), \tag{3.1288}$$

corresponding to a rotation of the coordinates. Since the metric in Euclidean space is given by

$$g_{xx} = g_{yy} = 1, \quad g_{xy} = g_{yx} = 0, \tag{3.1289}$$

we particularly find that

$$\mathcal{L}_K g_{xx} = g_{xi}\partial_x K^i + g_{ix}\partial_x K^i = 2\partial_x K^x = 2\partial_x y = 0, \tag{3.1290}$$

$$\mathcal{L}_K g_{xy} = g_{xi}\partial_y K^i + g_{iy}\partial_x K^i = \partial_y K^x + \partial_x K^y = \partial_y y + \partial_x(-x) = 1 - 1 = 0, \tag{3.1291}$$

$$\mathcal{L}_K g_{yy} = 2g_{yi}\partial_y K^i = 2\partial_y(-x) = 0. \tag{3.1292}$$

The vector field K is therefore a Killing vector field in two-dimensional Euclidean space.

b) By letting $x \mapsto t$ and $y \mapsto x$ in a), we find that the flow is given by

$$t = t_0\cos(s) + x_0\sin(s), \quad x = -t_0\sin(s) + x_0\cos(s). \tag{3.1293}$$

However, the Minkowski metric is given by

$$g_{tt} = -g_{xx} = 1, \quad g_{tx} = g_{xt} = 0, \tag{3.1294}$$

and therefore, in particular, we obtain

$$L_K g_{tx} = g_{t\mu}\partial_x K^\mu + g_{\mu x}\partial_t K^\mu = \partial_x K^t - \partial_t K^x = \partial_x x - \partial_t(-t) = 1 + 1 = 2 \neq 0. \tag{3.1295}$$

The field K is therefore not a Killing vector field in two-dimensional Minkowski space.

2.66

a) Using the given coordinates r and φ, the remaining coordinate z of the embedding is given by

$$z = \alpha(x^2 + y^2) = \alpha r^2. \tag{3.1296}$$

Differentiating the coordinate functions in $\mathbb{R}^3$ leads to

$$dx = \cos\varphi\, dr - r\sin\varphi\, d\varphi, \tag{3.1297}$$

$$dy = \sin\varphi\, dr + r\cos\varphi\, d\varphi, \tag{3.1298}$$

$$dz = 2\alpha r\, dr. \tag{3.1299}$$

Inserting this into the Euclidean metric on $\mathbb{R}^3$, we find that

$$\begin{aligned} ds^2 &= dx^2 + dy^2 + dz^2 \\ &= (\cos\varphi\, dr - r\sin\varphi\, d\varphi)^2 + (\sin\varphi\, dr + r\cos\varphi\, d\varphi)^2 + 4\alpha^2 r^2 dr^2 \\ &= (1 + 4\alpha^2 r^2)dr^2 + r^2 d\varphi^2. \end{aligned} \tag{3.1300}$$

The nonzero components of the metric are therefore

$$g_{rr} = 1 + 4\alpha^2 r^2, \quad g_{\varphi\varphi} = r^2. \tag{3.1301}$$

b) The Killing equation in terms of the metric and partial derivatives takes the form

$$\mathcal{L}_K g_{ab} = K^c\partial_c g_{ab} + g_{ac}\partial_b K^c + g_{cb}\partial_a K^c. \tag{3.1302}$$

For $K = \partial_\varphi$, we have $K^r = 0$, $K^\varphi = 1$, and therefore,

$$\mathcal{L}_K g_{ab} = 1 \cdot \partial_\varphi g_{ab} + g_{a\varphi}\partial_b 1 + g_{\varphi b}\partial_a 1 = \partial_\varphi g_{ab} = 0, \tag{3.1303}$$

where the last step is taking into account that the metric components do not depend on φ. $K = \partial_\varphi$ is therefore a Killing vector field. The corresponding conserved quantity along a geodesic is given by

$$C = g(\dot{\gamma}, K) = g_{ab}\dot{x}^a K^b = g_{\varphi\varphi}\dot{\varphi} = r^2\dot{\varphi}. \tag{3.1304}$$

2.67

a) The geodesic equations are found by finding the stationary paths of

$$\mathscr{S} = \frac{1}{2}\int g_{ab}\dot{x}^a\dot{x}^b\,d\tau = \int \frac{1}{2}\left[(R+\rho\sin\varphi)^2\dot{\theta}^2 + \rho^2\dot{\varphi}^2\right]d\tau \equiv \int \mathcal{L}\,d\tau. \tag{3.1305}$$

The Euler–Lagrange equations are

$$\begin{aligned}\frac{\partial\mathcal{L}}{\partial\theta} - \frac{d}{d\tau}\frac{\partial\mathcal{L}}{\partial\dot{\theta}} &= -\frac{d}{d\tau}\left[(R+\rho\sin\varphi)^2\dot{\theta}\right]\\ &= -(R+\rho\sin\varphi)^2\ddot{\theta} - 2\rho\cos\varphi(R+\rho\sin\varphi)\dot{\varphi}\dot{\theta} = 0\\ \Rightarrow\quad &\ddot{\theta} + \frac{2\rho\cos\varphi}{R+\rho\sin\varphi}\dot{\varphi}\dot{\theta} = 0,\end{aligned} \tag{3.1306}$$

$$\begin{aligned}\frac{\partial\mathcal{L}}{\partial\varphi} - \frac{d}{d\tau}\frac{\partial\mathcal{L}}{\partial\dot{\varphi}} &= \rho\cos\varphi(R+\rho\sin\varphi)\dot{\theta}^2 - \rho^2\ddot{\varphi} = 0\\ \Rightarrow\quad &\ddot{\varphi} - \frac{\cos\varphi(R+\rho\sin\varphi)}{\rho}\dot{\theta}^2 = 0.\end{aligned} \tag{3.1307}$$

From this and the symmetry of the Levi-Civita connection, we can identify the nonzero Christoffel symbols to be

$$\Gamma^\theta_{\theta\varphi} = \Gamma^\theta_{\varphi\theta} = \frac{\rho\cos\varphi}{R+\rho\sin\varphi}, \quad \Gamma^\varphi_{\theta\theta} = -\frac{\cos\varphi}{\rho}(R+\rho\sin\varphi). \tag{3.1308}$$

b) Since the metric does not depend explicitly on θ, the coordinate basis vector field ∂_θ is a Killing vector. The corresponding conserved quantity along a geodesic is given by

$$g(\partial_\theta, \dot{\gamma}) = g_{\theta a}\dot{x}^a = g_{\theta\theta}\dot{\theta} = (R+\rho\sin\varphi)^2\dot{\theta} = \text{const}. \tag{3.1309}$$

c) With all of the connection coefficients being equal to zero, the components of $\tilde{\nabla}_a g_{bc}$ are given by

$$\tilde{\nabla}_a g_{bc} = \partial_a g_{bc} - \tilde{\Gamma}^d_{ab} g_{dc} - \tilde{\Gamma}^d_{ac} g_{bd} = \partial_a g_{bc}. \tag{3.1310}$$

As $g_{\theta\varphi} = g_{\varphi\theta} = 0$ and $g_{\varphi\varphi}$ is constant, the only possibly nonzero components are

$$\tilde{\nabla}_\varphi g_{\theta\theta} = \partial_\varphi g_{\theta\theta} = 2(R+\rho\sin\varphi)\rho\cos\varphi \quad\text{and}\quad \tilde{\nabla}_\theta g_{\theta\theta} = \partial_\theta g_{\theta\theta} = 0. \tag{3.1311}$$

All other components are zero.

2.68

a) The metric tensor components can be found through the Euclidean line element

$$ds^2 = dx^2 + dy^2 + dz^2, \tag{3.1312}$$

by insertion of the expressions for the coordinates in $\mathbb{R}^3$ as defined in the embedding. This leads to

$$ds^2 = d\rho^2 \left[1 + \sin^2\left(\frac{\rho}{R_0}\right)\right] + \rho^2 d\varphi^2 = g_{ab}dx^a dx^b. \tag{3.1313}$$

Identification of the metric components yields

$$g_{\rho\rho} = 1 + \sin^2\left(\frac{\rho}{R_0}\right), \quad g_{\varphi\varphi} = \rho^2, \quad g_{\rho\varphi} = g_{\varphi\rho} = 0. \tag{3.1314}$$

To find the geodesic equations from which we will be able to identify the Christoffel symbols, we extremize

$$\mathcal{S} = \int \left\{\dot{\rho}^2 \left[1 + \sin^2\left(\frac{\rho}{R_0}\right)\right] + \rho^2\dot{\varphi}^2\right\} d\tau, \tag{3.1315}$$

leading to the Euler–Lagrange equations

$$\ddot{\rho} + \left[1 + \sin^2\left(\frac{\rho}{R_0}\right)\right]^{-1} \left[\sin\left(\frac{2\rho}{R_0}\right)\frac{\dot{\rho}^2}{R_0} - \rho\dot{\varphi}^2\right] = 0, \tag{3.1316}$$

$$\ddot{\varphi} + \frac{2}{\rho}\dot{\varphi}\dot{\rho} = 0. \tag{3.1317}$$

The nonzero Christoffel symbols can therefore be identified as

$$\Gamma^\rho_{\rho\rho} = \left[R_0 + R_0 \sin^2\left(\frac{\rho}{R_0}\right)\right]^{-1} \sin\left(\frac{2\rho}{R_0}\right), \tag{3.1318}$$

$$\Gamma^\rho_{\varphi\varphi} = -\rho\left[1 + \sin^2\left(\frac{\rho}{R_0}\right)\right]^{-1}, \tag{3.1319}$$

$$\Gamma^\varphi_{\rho\varphi} = \Gamma^\varphi_{\varphi\rho} = \frac{1}{\rho}. \tag{3.1320}$$

b) For $K = \partial_\rho$, the flow equations are

$$\dot{\rho} = K^\rho = 1, \quad \dot{\varphi} = K^\varphi = 0, \tag{3.1321}$$

with the solution $\rho = \rho_0 + \tau$, $\varphi = \varphi_0$. Similarly, for $Q = \partial_\varphi$, we find the flow equations

$$\dot{\rho} = Q^\rho = 0, \quad \dot{\varphi} = Q^\varphi = 1, \tag{3.1322}$$

with the solution $\rho = \rho_0$, $\varphi = \varphi_0 + \tau$, describing a rotation.

To determine if K or Q are Killing vector fields, we compute $\mathcal{L}_K g$ and $\mathcal{L}_Q g$, respectively. We find that

$$\mathcal{L}_K g_{ab} = \partial_\rho g_{ab}. \tag{3.1323}$$

In particular, $\mathcal{L}_K g_{\varphi\varphi} \neq 0$ as $g_{\varphi\varphi}$ depends on ρ. The field K is therefore not a Killing vector field. However, for Q, we obtain

$$\mathcal{L}_Q g_{ab} = \partial_\varphi g_{ab} = 0, \tag{3.1324}$$

since none of the metric components depend on φ. It follows that Q is a Killing vector field and the corresponding rotations are symmetries of the embedded surface.

2.69

a) In the xy-plane, we introduce polar coordinates according to

$$x = r\cos\varphi, \quad y = r\sin\varphi. \tag{3.1325}$$

Inserting this into the dS_2 relation yields

$$t^2 - r^2 = -r_0^2. \tag{3.1326}$$

This hyperbola can be parametrized according to

$$r = r_0\cosh\tau, \quad t = r_0\sinh\tau \tag{3.1327}$$

by the single parameter τ. We therefore have coordinates τ and φ for dS_2 where

$$t = r_0\sinh\tau, \quad x = r_0\cosh\tau\cos\varphi, \quad y = r_0\cosh\tau\sin\varphi. \tag{3.1328}$$

b) The induced metric can be found through the Minkowski line element

$$ds^2 = dt^2 - dx^2 - dy^2. \tag{3.1329}$$

With the parametrization from a), we find

$$dt = r_0\cosh\tau d\tau, \tag{3.1330}$$

$$dx = r_0[\sinh\tau\cos\varphi d\tau - \cosh\tau\sin\varphi d\varphi], \tag{3.1331}$$

$$du = r_0[\sinh\tau\sin\varphi d\tau + \cosh\tau\cos\varphi d\varphi], \tag{3.1332}$$

leading to

$$ds^2 = r_0^2\cosh^2\tau d\tau^2 - r_0^2\sinh^2\tau d\tau^2 - r_0^2\cosh^2\tau d\varphi^2 = r_0^2[d\tau^2 - \cosh^2\tau d\varphi^2]. \tag{3.1333}$$

From this we can identify the metric components

$$g_{\tau\tau} = r_0^2, \quad g_{\varphi\varphi} = -r_0^2\cosh^2\tau, \quad g_{\varphi\tau} = g_{\tau\varphi} = 0. \tag{3.1334}$$

c) The metric components do not depend on the coordinate φ. This means that the coordinate vector field ∂_φ is a Killing vector field.

A second Killing field can be found by noting that Lorentz boosts in 1+2-dimensional Minkowski space preserves dS_2. A boost with rapidity θ in the x-direction is given by

$$t \to t\cosh\theta - x\sinh\theta, \quad x \to x\cosh\theta - t\sinh\theta, \quad y \to y. \tag{3.1335}$$

Denoting differentiation with respect to θ by $\cdot$, we find, at $\theta = 0$,

$$\dot{t} = -x = -r_0\cosh\tau\cos\varphi = r_0\dot{\tau}\cosh\tau, \tag{3.1336}$$

$$\dot{y} = 0 = r_0[\dot{\tau}\sinh\tau\sin\varphi + \dot{\varphi}\cosh\tau\cos\varphi], \tag{3.1337}$$

leading to

$$\dot{\tau} = -\cos\varphi, \tag{3.1338}$$

$$\dot{\varphi} = -\dot{\tau}\tanh\tau\tan\varphi = \tanh\tau\sin\varphi. \tag{3.1339}$$

The corresponding generating vector field on dS_2 is therefore

$$K = -\cos\varphi\partial_\tau + \sin\varphi\tanh\tau\partial_\varphi. \tag{3.1340}$$

We can check explicitly that this is indeed a Killing vector field by taking the Lie derivative of the metric:

$$\mathcal{L}_K g_{ab} = K^c\partial_c g_{ab} + g_{ac}\partial_b K^c + g_{cb}\partial_a K^c. \tag{3.1341}$$

Explicitly, we have

$$\mathcal{L}_K g_{\tau\tau} = 2g_{\tau\tau}\partial_\tau K^\tau = 2r_0^2\partial_\tau(-\cos\varphi) = 0, \tag{3.1342}$$

$$\begin{aligned}\mathcal{L}_K g_{\varphi\varphi} &= \cos\varphi\partial_t(r_0^2\cosh^2\tau) + 2g_{\varphi\varphi}\partial_\varphi(\sin\varphi\tanh\tau)\\ &= \cos\varphi\cdot 2r_0^2\sinh\tau\cosh\tau - 2r_0^2\cosh^2\tau\cos\varphi\tanh\tau = 0,\end{aligned} \tag{3.1343}$$

$$\begin{aligned}\mathcal{L}_K g_{\varphi\tau} &= 0 + g_{\varphi\varphi}\partial_\tau K^\varphi + g_{\tau\tau}\partial_\varphi K^\tau\\ &= -r_0^2\cosh^2\tau\partial_\tau(\sin\varphi\tanh\tau) + r_0^2\partial_\varphi(-\cos\varphi) = 0.\end{aligned} \tag{3.1344}$$

It follows that K is a Killing vector field on dS_2.

Note that a third Killing vector field can also be found by considering Lorentz boosts in the y-direction in 1+2-dimensional Minkowski space. Thus, dS_2 is a maximally symmetric space.

2.70

Introduce the quantity

$$F = \left(1 - \frac{\alpha}{r}\right)\dot{x}_0^2 - \left(1 - \frac{\alpha}{r}\right)^{-1}\dot{r}^2 - r^2\dot{\phi}^2, \tag{3.1345}$$

where dot is the derivative with respect to the curve parameter s. Now, the geodesic distance along the curve is a functional of F, i.e.,

$$L = \int F^{1/2}\, ds. \tag{3.1346}$$

Using Euler–Lagrange variational equations

$$\frac{d}{ds}\left(\frac{\partial F}{\partial \dot{x}}\right) - \frac{\partial F}{\partial x} = 0, \tag{3.1347}$$

for each of the functions $x = x^0, r, \phi$, we obtain

$$\ddot{x}^0 + \left(1 - \frac{\alpha}{r}\right)^{-1} \frac{\alpha}{r^2}\dot{x}^0\dot{r} = 0, \tag{3.1348}$$

$$\ddot{r} + \frac{1}{2}\left(1 - \frac{\alpha}{r}\right)\frac{\alpha}{r^2}(\dot{x}^0)^2 - \frac{1}{2}\left(1 - \frac{\alpha}{r}\right)^{-1}\frac{\alpha}{r^2}\dot{r}^2 - \left(1 - \frac{\alpha}{r}\right) r\dot{\phi}^2 = 0, \tag{3.1349}$$

$$\ddot{\phi} + \frac{2}{r}\dot{r}\dot{\phi} = 0, \tag{3.1350}$$

which are the geodesic equations of motion for a test particle in the Schwarzschild metric when $\theta = \pi/2$.

2.71

The Schwarzschild metric in spherical coordinates is given by

$$ds^2 = \left(1 - \frac{2GM}{c^2 r}\right)(dx^0)^2 - \left(1 - \frac{2GM}{c^2 r}\right)^{-1} dr^2 - r^2 d\Omega^2, \tag{3.1351}$$

where $d\Omega^2 = d\theta^2 + \sin^2\theta d\phi^2$. Now, we use

$$\boldsymbol{x} = (x^1, x^2, x^3) = r(\sin\theta\cos\phi, \sin\theta\sin\phi, \cos\theta), \tag{3.1352}$$

so that

$$d\boldsymbol{x}^2 = (dx^1)^2 + (dx^2)^2 + (dx^3)^2 = dr^2 + r^2 d\Omega^2, \tag{3.1353}$$

where $d\boldsymbol{x} = (dx^1, dx^2, dx^3)$. Using $r^2 = (x^1)^2 + (x^2)^2 + (x^3)^2 = \boldsymbol{x}^2$, we obtain $rdr = \boldsymbol{x} \cdot d\boldsymbol{x}$ and so we have

$$dr^2 = \frac{(\boldsymbol{x} \cdot d\boldsymbol{x})^2}{r^2} = \frac{(x^1dx^1 + x^2dx^2 + x^3dx^3)^2}{r^2}, \tag{3.1354}$$

$$r^2 d\Omega^2 = d\boldsymbol{x}^2 - dr^2 = d\boldsymbol{x}^2 - \frac{(\boldsymbol{x} \cdot d\boldsymbol{x})^2}{r^2}. \tag{3.1355}$$

Thus, the Schwarzschild metric in the new coordinates becomes

$$ds^2 = \left(1 - \frac{2GM}{c^2 r}\right)(dx^0)^2 - \left(1 - \frac{2GM}{c^2 r}\right)^{-1} \frac{(\boldsymbol{x}\cdot d\boldsymbol{x})^2}{r^2} - \left[d\boldsymbol{x}^2 - \frac{(\boldsymbol{x}\cdot d\boldsymbol{x})^2}{r^2}\right]$$
$$= \left(1 - \frac{2GM}{c^2 r}\right)(dx^0)^2 - d\boldsymbol{x}^2 - \left[\left(1 - \frac{2GM}{c^2 r}\right)^{-1} - 1\right]\frac{(\boldsymbol{x}\cdot d\boldsymbol{x})^2}{r^2}, \tag{3.1356}$$

where $r = \sqrt{(x^1)^2 + (x^2)^2 + (x^3)^2}$.

2.72

a) Circular motion in the equatorial plane, i.e., $r = r_0 = \text{const.}$ and $\theta = \frac{\pi}{2}$, implies that $\dot{r} = 0$ and $\ddot{r} = 0$ as well as $\dot{\theta} = 0$ (and $\ddot{\theta} = 0$). Inserting these conditions into the geodesic equation for the r-coordinate, see the solution to 2.70, gives a constraint

$$\frac{1}{2}\frac{r_* c^2}{r_0^2}\left(1 - \frac{r_*}{r_0}\right)\dot{t}^2 - r_0\left(1 - \frac{r_*}{r_0}\right)\dot{\phi}^2 = 0, \tag{3.1357}$$

which leads to

$$\dot{t}^2 = \frac{2r_0^3}{r_* c^2}\dot{\phi}^2. \tag{3.1358}$$

This implies that

$$\left(\frac{dt}{d\phi}\right)^2 = \frac{2r_0^3}{r_* c^2} \quad \Rightarrow \quad dt = \sqrt{\frac{2r_0^3}{r_* c^2}}\, d\phi, \tag{3.1359}$$

where it was assumed that $\frac{dt}{d\phi} > 0$. Integrating this over one period (in universal time), i.e., Δt, one finds that

$$\Delta t = \int_{t_1=0}^{t_2=\Delta t} dt = \int_{\phi_1=0}^{\phi_2=2\pi} \sqrt{\frac{2r_0^3}{r_* c^2}}\, d\phi = \sqrt{\frac{2r_0^3}{r_* c^2}} \int_{\phi_1=0}^{\phi_2=2\pi} d\phi$$
$$= \sqrt{\frac{2r_0^3}{r_* c^2}} \cdot 2\pi = 2\pi\sqrt{\frac{2r_0^3}{r_* c^2}}. \tag{3.1360}$$

Finally, reinserting $r_* = \frac{2GM}{c^2}$ in the above formula for Δt, one obtains

$$\Delta t = 2\pi\sqrt{\frac{r_0^3}{GM}}, \tag{3.1361}$$

which is Kepler's third law.

b) For circular motion in the equatorial plane, i.e., $r = r_0 = \text{const.}$, $\dot{r} = 0$, $\theta = \frac{\pi}{2}$, and $\dot{\theta} = 0$, the Euler–Lagrange equations arising from the Lagrangian

$$\mathscr{L}_0 = \frac{1}{2}\left[\left(1 - \frac{r_*}{r_0}\right)c^2\dot{t}^2 - r_0^2\dot{\phi}^2\right], \tag{3.1362}$$

are simply $\ddot{t} = 0$ and $\ddot{\phi} = 0$. Thus, $\dot{t} \equiv \alpha$ and $\dot{\phi} \equiv \beta$ are constants of motion and $\frac{dt}{d\phi} = \frac{\alpha}{\beta}$. Using the constraint in a), we find that

$$\frac{\alpha^2}{\beta^2} = \left(\frac{dt}{d\phi}\right)^2 = \frac{2r_0^3}{r_* c^2}. \tag{3.1363}$$

Furthermore, using s as the curve parameter ($s = c\tau$, where τ is the proper time), we have the conservation law

$$\frac{1}{2} = \mathscr{L}_0 = \frac{1}{2}\left[\left(1 - \frac{r_*}{r_0}\right)c^2\alpha^2 - r_0^2\beta^2\right], \tag{3.1364}$$

since $ds^2 = g_{\mu\nu}dx^\mu dx^\nu$ and $\mathscr{L} = \frac{1}{2}g_{\mu\nu}\dot{x}^\mu\dot{x}^\nu$.

Solving for α in the two equations

$$\frac{\alpha^2}{\beta^2} = \frac{2r_0^3}{r_* c^2}, \quad 1 = \left(1 - \frac{r_*}{r_0}\right)c^2\alpha^2 - r_0^2\beta^2, \tag{3.1365}$$

we find that

$$\alpha = \frac{1}{c\sqrt{1 - \frac{3r_*}{2r_0}}} = \frac{dt}{ds} = \frac{1}{c}\frac{dt}{d\tau} \quad \Rightarrow \quad d\tau = \sqrt{1 - \frac{3r_*}{2r_0}}\, dt, \tag{3.1366}$$

where it is again assumed that $\frac{dt}{d\phi} > 0$. Thus, the proper time period $\Delta\tau$ is given by

$$\Delta\tau = \int_0^{\Delta t}\sqrt{1 - \frac{3r_*}{2r_0}}\, dt = \sqrt{1 - \frac{3r_*}{2r_0}}\,\Delta t. \tag{3.1367}$$

Finally, inserting $r_* = \frac{2GM}{c^2}$ and Δt found in a) into the above formula for $\Delta\tau$, we obtain

$$\Delta\tau = 2\pi\sqrt{\frac{2r_0^3}{r_* c^2}}\sqrt{1 - \frac{3r_*}{2r_0}} = \frac{2\pi r_0}{c}\sqrt{\frac{r_0 c^2}{GM} - 3}. \tag{3.1368}$$

2.73

The geodesic equations follow from the Euler–Lagrange equations with the Lagrangian

$$\mathcal{L} = g_{\mu\nu}\dot{x}^\mu\dot{x}^\nu = \left(1 - \frac{r_*}{r}\right)(\dot{x}^0)^2 - \left(1 - \frac{r_*}{r}\right)^{-1}\dot{r}^2 - r^2(\dot{\theta}^2 + \sin^2\theta\,\dot{\phi}^2). \tag{3.1369}$$

For radially freely falling bodies, it holds that $\dot{\theta} = 0$ and $\dot{\phi} = 0$, which means that the Lagrangian reduces to

$$\mathcal{L} = \left(1 - \frac{r_*}{r}\right)(\dot{x}^0)^2 - \left(1 - \frac{r_*}{r}\right)^{-1} \dot{r}^2. \tag{3.1370}$$

Now, the geodesic equations for x^0 and r coordinates are

$$\frac{\partial \mathcal{L}}{\partial x^0} - \frac{d}{d\tau}\frac{\partial \mathcal{L}}{\partial \dot{x}^0} = -2\frac{d}{d\tau}\left[\left(1 - \frac{r_*}{r}\right)\dot{x}^0\right] = 0, \tag{3.1371}$$

$$\frac{\partial \mathcal{L}}{\partial r} - \frac{d}{d\tau}\frac{\partial \mathcal{L}}{\partial \dot{r}} = \frac{r_*}{r^2}(\dot{x}^0)^2 - \left(1 - \frac{r_*}{r}\right)^{-2}\frac{r_*}{r^2}\dot{r}^2 + 2\frac{d}{d\tau}\left[\left(1 - \frac{r_*}{r}\right)^{-1}\dot{r}\right] = 0. \tag{3.1372}$$

The geodesic equation for the x^0-coordinate leads to

$$\left(1 - \frac{r_*}{r}\right)\dot{x}^0 = E = \text{const.} \tag{3.1373}$$

In addition, we can find more information about the geodesics by using that

$$\frac{d\mathcal{L}}{d\tau} = 0. \tag{3.1374}$$

Thus, we have that

$$\mathcal{L} = \left(1 - \frac{r_*}{r}\right)(\dot{x}^0)^2 - \left(1 - \frac{r_*}{r}\right)^{-1} \dot{r}^2 = -\epsilon = \text{const.} \tag{3.1375}$$

Combining the two expressions defining the quantities E and ϵ, we find that

$$\dot{r}^2 = \epsilon\left(1 - \frac{r_*}{r}\right) + E^2. \tag{3.1376}$$

In order to find $r(\tau)$, we solve for $\dot{r}$, separate the variables τ and r, and integrate both variables, i.e.,

$$\frac{dr}{d\tau} = -\sqrt{\epsilon\left(1 - \frac{r_*}{r}\right) + E^2} \quad \Rightarrow \quad d\tau = -\frac{dr}{\sqrt{\epsilon\left(1 - \frac{r_*}{r}\right) + E^2}}$$

$$\Rightarrow \quad \tau = \int_0^\tau d\tau' = -\int \frac{dr}{\sqrt{\epsilon\left(1 - \frac{r_*}{r}\right) + E^2}}, \tag{3.1377}$$

which describes the spacetime curves. Note that we use a minus sign for the square root, since the bodies are radially freely falling *toward* $r = 0$, i.e., $\dot{r} < 0$.

Usually, when solving equations of this type, one tries to find a solution $r = r(\tau)$, but one can also find a solution $\tau = \tau(r)$, since $r(\tau)$ is one-to-one. In principle, one can then find x^0 as

$$x^0(\tau) = \int_0^\tau \left[1 - \frac{r_*}{r(\tau')}\right]^{-1} E\, d\tau'. \tag{3.1378}$$

2.74

By using the Euler–Lagrange equations in the same manner as in Problem 2.70, the equations of motion are given by

$$\frac{d}{ds}\left(1-\frac{r_*}{r}\right)\dot{t}=0 \quad\Rightarrow\quad \left(1-\frac{r_*}{r}\right)\dot{t}=E, \tag{3.1379}$$

$$\frac{d}{ds}r^2\dot{\phi}=0 \quad\Rightarrow\quad r^2\dot{\phi}=h, \tag{3.1380}$$

$$\frac{r_*}{r^2}\dot{t}^2-2r\dot{\phi}^2=\frac{r_*}{r^2}\left(1-\frac{r_*}{r}\right)^{-2}\dot{r}^2+2\frac{d}{ds}\left[\left(1-\frac{r_*}{r}\right)^{-1}\dot{r}\right], \tag{3.1381}$$

where E and h are constants.

For circular orbits, $r=r_0$ is constant. From the third of the equations of motion, we then obtain

$$\frac{r_*}{r_0^2}\dot{t}^2=2r_0\dot{\phi}^2 \quad\Rightarrow\quad \Delta t=\Delta\phi\sqrt{\frac{2r_0^3}{r_*}}. \tag{3.1382}$$

By inserting this into the line element ds^2, we obtain

$$\Delta s^2=2r_0^2\left(\frac{r_0}{r_*}-\frac{3}{2}\right)\Delta\phi^2. \tag{3.1383}$$

Since $\Delta s^2\geq 0$ for any physically viable worldline, we obtain $r_0\geq 3r_*/2$. Thus, there are no circular orbits inside the $r=3r_*/2$.

2.75

Geodesics for massive test objects in the Schwarzschild metric satisfy the equation of motion

$$E^2-\frac{\dot{r}^2}{2}=\frac{1}{2}\left(1+\frac{L^2}{r^2}\right)\left(1-\frac{r_*}{r}\right)=V(r), \tag{3.1384}$$

where $L=r^2\dot{\varphi}$ is the angular momentum. Approximately circular orbits have an energy close to the minimum of $V(r)$. To find this minimum, we define $U(x)=V\left(\frac{1}{x}\right)$ and set $U'(x_0)=0$:

$$U'(x_0)=x_0L^2(1-r_*x_0)-\frac{r_*}{2}(1+L^2x_0^2)=0, \tag{3.1385}$$

with the solution

$$x_0=\frac{1}{3r_*}\left(1-\sqrt{1-\frac{3r_*^2}{L^2}}\right), \tag{3.1386}$$

where the solution with a plus sign in front of the square root is unstable. For $r_0 = \frac{1}{x_0} \gg r_*$, we require that $L^2 \gg r_*^2$. The second derivative of the potential is

$$V''(r) = \frac{d^2}{dr^2}U(x) = \frac{d}{dr}\left(\frac{dx}{dr}U'(x)\right) = \left(\frac{dx}{dr}\right)^2 U''(x) + \frac{d^2x}{dr^2}U'(x). \quad (3.1387)$$

At $r_0 = \frac{1}{x_0}$, we therefore obtain

$$V''(r_0) = \left(\frac{dx}{dr}\right)^2 U''(x) + 0 = x_0^4 U''(x_0) = \frac{U''(x_0)}{r_0^4}. \quad (3.1388)$$

We find that

$$U''(x_0) = -3r_* L^2 x_0 + L^2 = L^2 - L^2\left(1 - \sqrt{1 - \frac{3r_*^2}{L^2}}\right) = L^2\sqrt{1 - \frac{3r_*^2}{L^2}}. \quad (3.1389)$$

Thus, for small $\rho = r - r_0$, we find that

$$V(r) \simeq V_0 + \frac{1}{2}\frac{U''(x_0)}{r_0^4}\rho^2. \quad (3.1390)$$

The equation of motion for ρ becomes

$$\ddot{\rho} + \frac{U''(x_0)}{r_0^4}\rho = \ddot{\rho} + \omega^2\rho = 0. \quad (3.1391)$$

The proper time period for small radial oscillations is therefore

$$T_\rho = \frac{2\pi}{\omega} = \frac{2\pi r_0^2}{\sqrt{U''(x_0)}} = \frac{2\pi}{x_0^2\sqrt{U''(x_0)}} = \frac{9\pi r_*^2}{2L}\left[1 + \frac{9r_*^2}{4L^2} + \mathcal{O}\left(\frac{r_*}{L}\right)^4\right]. \quad (3.1392)$$

To find the orbital period, we consider

$$L = (r_0 + \rho)^2\dot{\varphi} \simeq r_0^2\dot{\varphi} \quad \Rightarrow \quad \frac{1}{\dot{\varphi}} = \frac{r_0^2}{L}, \quad (3.1393)$$

assuming that $\rho \ll r_0$. The orbital period is the proper time taken to complete a full turn, i.e.,

$$T_\varphi = \int d\tau = \int \frac{d\varphi}{\dot{\varphi}} = 2\pi\frac{r_0^2}{L} = \frac{2\pi}{x_0^2 L}. \quad (3.1394)$$

Inserting the expression for x_0, we find that

$$T_\varphi = \frac{9\pi r_*^2}{2L}\left[1 + \frac{3r_*^2}{2L^2} + \mathcal{O}\left(\frac{r_*}{L}\right)^4\right]. \quad (3.1395)$$

Thus, finally, the ratio between the periods is

$$\frac{T_\rho}{T_\varphi} = \frac{1+\frac{9r_*^2}{4L^2}+\mathcal{O}\left(\frac{r_*}{L}\right)^4}{1+\frac{3r_*^2}{2L^2}+\mathcal{O}\left(\frac{r_*}{L}\right)^4} = 1+\left(\frac{9}{4}-\frac{3}{2}\right)\frac{r_*^2}{L^2}+\mathcal{O}\left(\frac{r_*}{L}\right)^4 = 1+\frac{3r_*^2}{4L^2}+\mathcal{O}\left(\frac{r_*}{L}\right)^4. \tag{3.1396}$$

For $r_0 \gg r_*$, we also have

$$\frac{1}{r_0} = \left(1-\sqrt{1-\frac{3r_*^2}{L^2}}\right)\frac{1}{3r_*} \simeq \frac{1}{3r_*}\frac{3r_*^2}{2L^2} = \frac{r_*}{2L^2}. \tag{3.1397}$$

Inserting this into the above leads to

$$\frac{T_\rho}{T_\varphi} \simeq 1+\frac{3r_*}{2r_0}. \tag{3.1398}$$

2.76

For the past-null geodesics to originate at infinity, it is necessary that there exists a classical turning point. The effective potential for a light signal is given by

$$V(r) = \frac{1}{2}\frac{L^2}{r^2}\left(1-\frac{R}{r}\right). \tag{3.1399}$$

For $r \to \infty$, we also have $V(r) \to 0$, and consequently, we find that

$$E = \frac{\dot{r}^2}{2} = \frac{1}{2}, \tag{3.1400}$$

when using a parametrization such that $\dot{r} = 1$ at $r \to \infty$. For a classical turning point r_* to exist, it must hold that the maximum of $V(r) \geq E$ for some r. The extreme points of V satisfy (with $x = 1/r$)

$$V'(r) = \frac{dx}{dr}\frac{d}{dx}V\left(\frac{1}{x}\right) = \frac{dx}{dr}\frac{1}{2}L^2(2-3xR)x = 0. \tag{3.1401}$$

For $r < \infty$, this is solved by $r = \frac{3R}{2}$, where the potential obtains a maximum. We find that the condition for a classical turning point to exist is therefore

$$E = \frac{1}{2} \leq V\left(\frac{3R}{2}\right) = \frac{1}{2}\frac{4L^2}{9R^2}\left(1-\frac{2}{3}\right) = \frac{1}{2}\frac{4}{27}\frac{L^2}{R^2} \quad\Rightarrow\quad L^2 \leq \frac{27}{4}R^2. \tag{3.1402}$$

The smallest impact parameter with a classical turning point is therefore (see Figure 3.13)

$$b^2 = \frac{27}{4}R^2, \tag{3.1403}$$

giving an optical size of

$$4\pi b^2 = 27\pi R^2. \tag{3.1404}$$

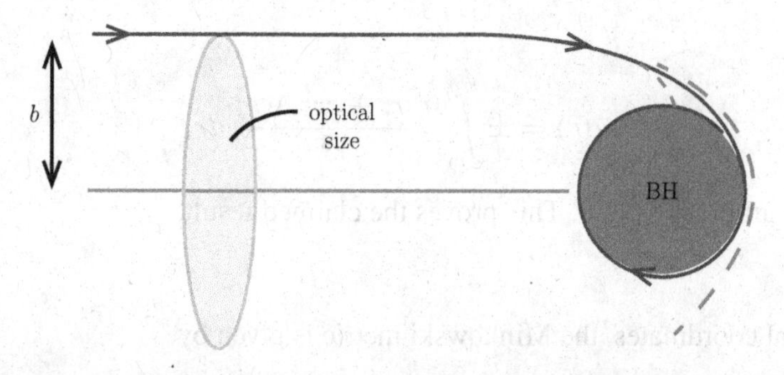

Figure 3.13 Illustration of the optical size of a Schwarzschild black hole ("BH"), where b is the minimal impact parameter of the optical size $4\pi b^2$.

In other words, note that the optical size of the black hole is significantly larger than $4\pi R^2$.

2.77
Using the given ansatz, we have

$$dZ_1 = \cosh\frac{t}{2r_*} f(r)\,dt + 2r_* \sinh\frac{t}{2r_*} f'(r)\,dr, \tag{3.1405}$$

$$dZ_2 = \sinh\frac{t}{2r_*} f(r)\,dt + 2r_* \cosh\frac{t}{2r_*} f'(r)\,dr, \tag{3.1406}$$

$$dZ_3 = g'(r)\,dr, \tag{3.1407}$$

$$dZ_4 = \sin\theta\cos\varphi\,dr + r\cos\theta\cos\varphi\,d\theta - r\sin\theta\sin\varphi\,d\varphi, \tag{3.1408}$$

$$dZ_5 = \sin\theta\sin\varphi\,dr + r\cos\theta\sin\varphi\,d\theta + r\sin\theta\cos\varphi\,d\varphi, \tag{3.1409}$$

$$dZ_6 = \cos\theta\,dr - r\sin\theta\,d\theta. \tag{3.1410}$$

Therefore, we find that

$$dZ_1^2 - dZ_2^2 - dZ_3^2 = f(r)^2 dt^2 - \left[(2r_*)^2 f'(r)^2 + g'(r)^2\right] dr^2, \tag{3.1411}$$

$$-dZ_4^2 - dZ_5^2 - dZ_6^2 = -dr^2 - r^2\left[d\theta^2 + \sin(\theta)^2 d\varphi^2\right], \tag{3.1412}$$

and thus, the result claimed is equivalent to the following conditions

$$f(r)^2 = 1 - \frac{r_*}{r}, \quad -(2r_*)^2 f'(r)^2 - g'(r)^2 - 1 = -\left(1 - \frac{r_*}{r}\right)^{-1}. \tag{3.1413}$$

After some computations, we obtain

$$f(r) = \sqrt{1 - \frac{r_*}{r}}, \quad g'(r) = \frac{r^2 r_* + r r_*^2 + r_*^3}{r^3}, \tag{3.1414}$$

i.e.,

$$g(r) = \pm \int_{r_0}^{r} \frac{r^2 r_* + r r_*^2 + r_*^3}{r^3} \, dr, \tag{3.1415}$$

with some arbitrary $r_0 > 0$. This proves the claimed result.

2.78

In spherical coordinates, the Minkowski metric is given by

$$ds^2 = (dx^0)^2 - dr^2 - r^2(d\theta^2 + \sin^2\theta d\phi^2). \tag{3.1416}$$

Using the restriction to the three-dimensional hyperboloid M_3, i.e., $(x^0)^2 - r^2 = -a^2$, we find that

$$r = \sqrt{(x^0)^2 + a^2} \quad \Rightarrow \quad dr = \frac{x^0 dx^0}{\sqrt{(x^0)^2 + a^2}}. \tag{3.1417}$$

Now, inserting the restriction to M_3 into the Minkowski metric induces the curved metric on M_3, i.e.,

$$ds^2 = \frac{a^2}{(x^0)^2 + a^2}(dx^0)^2 - \left[(x^0)^2 + a^2\right](d\theta^2 + \sin^2\theta d\phi^2). \tag{3.1418}$$

In the case of lightlike geodesics with $d\phi = 0$, since ϕ is assumed to be constant, we have

$$0 = ds^2 = \frac{a^2}{(x^0)^2 + a^2}(dx^0)^2 - \left[(x^0)^2 + a^2\right] d\theta^2. \tag{3.1419}$$

Now, we can define the Lagrangian as

$$\mathcal{L} = \frac{a^2}{(x^0)^2 + a^2}(\dot{x}^0)^2 - \left[(x^0)^2 + a^2\right]\dot{\theta}^2. \tag{3.1420}$$

Using the Euler–Lagrange equation for the θ-coordinate, we obtain

$$\frac{d}{ds}\frac{\partial L}{\partial \dot{\theta}} - \frac{\partial L}{\partial \theta} = -2\frac{d}{ds}\left\{\left[(x^0)^2 + a^2\right]\dot{\theta}\right\} = 0, \tag{3.1421}$$

which defines a constant of motion that can be immediately integrated to give $\dot{\theta} = A/[(x^0)^2 + a^2]$, where A is some integration constant (in fact, the constant of motion). Furthermore, the condition $ds^2 = 0$ implies that

$$\frac{a^2}{(x^0)^2 + a^2}(\dot{x}^0)^2 - \left[(x^0)^2 + a^2\right]\dot{\theta}^2 = 0. \tag{3.1422}$$

Hence, combining the two conditions on $\dot{x}^0$ and $\dot{\theta}$ yields

$$\dot{x}^0 = \frac{(x^0)^2 + a^2}{a}\dot{\theta} = \frac{A}{a} \quad \Rightarrow \quad x^0 = \frac{A}{a}s + \alpha, \tag{3.1423}$$

where α is an integration constant. Thus, inserting the solution for x^0 into the Euler–Lagrange equation for the θ-coordinate, we obtain the lightlike geodesics with $d\phi = 0$ as

$$\dot{\theta} = \frac{A}{(x^0)^2 + a^2} = \frac{A}{(As/a + \alpha)^2 + a^2} \quad \Rightarrow \quad \theta = \arctan\left(\frac{A}{a^2}s + \frac{\alpha}{a}\right) + \beta, \tag{3.1424}$$

where β is an integration constant.

2.79

Introduce the Lagrangian

$$\mathscr{L} = g_{\mu\nu}\dot{x}^\mu\dot{x}^\nu = (\dot{x}^0)^2 - \dot{r}^2 - r^2\dot{\phi}^2 = \left\{(x^0)^2 - (x^1)^2 - (x^2)^2 = -1\right\}$$
$$= \frac{r^2}{r^2-1}\dot{r}^2 - \dot{r}^2 - r^2\dot{\phi}^2 = \frac{1}{r^2-1}\dot{r}^2 - r^2\dot{\phi}^2, \tag{3.1425}$$

where it holds that $x^0 = \sqrt{r^2-1}$. The Euler–Lagrange variational equations, i.e., $\frac{d}{ds}\left(\frac{\partial \mathscr{L}}{\partial \dot{x}^\mu}\right) - \frac{\partial \mathscr{L}}{\partial x^\mu} = 0$, then become

$$\frac{d}{ds}\left(\frac{\dot{r}}{r^2-1}\right) + r\dot{\phi}^2 + \frac{r\dot{r}^2}{(r^2-1)^2} = 0 \quad \text{for } r, \tag{3.1426}$$

$$\frac{d}{ds}(r^2\dot{\phi}) = 0 \quad \text{for } \phi. \tag{3.1427}$$

Furthermore, along a lightlike curve $\mathscr{L} = 0$, and thus, we have

$$\frac{1}{r^2-1}\dot{r}^2 - r^2\dot{\phi}^2 = 0. \tag{3.1428}$$

From the Euler–Lagrange variational equation for ϕ, we find that $r^2\dot{\phi} = A = \text{const.}$ Inserting this into the lightlike condition, we obtain

$$\frac{1}{r^2-1}\dot{r}^2 - \frac{A^2}{r^2} = 0 \quad \Rightarrow \quad \frac{dr}{ds} = \dot{r} = \frac{A}{r}\sqrt{r^2-1} \quad \Rightarrow \quad \frac{d}{ds}\sqrt{r^2-1} = A, \tag{3.1429}$$

so that $\sqrt{r^2-1} = As + s_0$ or $r = \sqrt{1 + (As + s_0)^2}$. From $r^2\dot{\phi} = A$, we find that

$$\frac{d\phi}{ds} = \dot{\phi} = \frac{A}{1 + (As + s_0)^2}, \tag{3.1430}$$

which implies that

$$\phi = \phi_0 + \arctan(As + s_0) \equiv \phi_0 + \arctan\tilde{s}. \tag{3.1431}$$

Now, $x^0 = \sqrt{r^2 - 1} = As + s_0 = \tilde{s} = \tan(\phi - \phi_0)$, i.e., $x^0 = \tan(\phi - \phi_0)$, and finally, $\Delta\phi \equiv \phi - \phi_0 = \frac{\pi}{2}$, which corresponds to $\Delta x^0 = \tan \Delta\phi \to \infty$. Thus, it takes an infinite global time difference Δx^0 for a light signal to travel from the point $\phi_0 = 0$ to the point $\phi = \frac{\pi}{2}$ on the surface.

2.80

The Schwarzschild metric (for $\theta = \pi/2$) is given by

$$ds^2 = \alpha(dx^0)^2 - \alpha^{-1}dr^2 - r^2 d\phi^2, \tag{3.1432}$$

where $\alpha \equiv \alpha(r) = 1 - \frac{2GM}{r}$. The geodesic equations are derived by varying $\mathcal{L} = g_{\mu\nu}\dot{x}^\mu \dot{x}^\nu$, and therefore, using Euler–Lagrange equations for the x^0 and ϕ coordinates, we obtain

$$\ddot{x}^0 + \frac{1}{\alpha}\frac{d\alpha}{dr}\dot{x}^0\dot{r} = 0, \tag{3.1433}$$

$$\ddot{\phi} + \frac{2}{r}\dot{r}\dot{\phi} = 0, \tag{3.1434}$$

which are the geodesic equations $\ddot{x}^\lambda + \Gamma^\lambda_{\mu\nu}\dot{x}^\mu\dot{x}^\nu = 0$ (see also Problem 2.70). In this case, we actually have the simpler geodesic equations, namely

$$\alpha\dot{x}^0 = E, \tag{3.1435}$$

$$r^2\dot{\phi} = h, \tag{3.1436}$$

where E and h are constants. For a lightlike geodesic ($\mathcal{L} = 0$) in the plane $\theta = \frac{\pi}{2}$, we have

$$0 = g_{00}(\dot{x}^0)^2 + g_{rr}\dot{r}^2 + g_{\phi\phi}\dot{\phi}^2, \tag{3.1437}$$

where $g_{00} = \alpha$, $g_{rr} = -\frac{1}{\alpha}$, and $g_{\phi\phi} = -r^2$. Now, the geodesic equation for x^0 implies that $\dot{x}^0 = \frac{E}{\alpha}$, where E is a constant of motion, whereas the geodesic equation for ϕ implies that $\dot{\phi} = \frac{h}{r^2}$, where h is another constant of motion. Inserting the two constants of motion into the condition for a lightlike geodesic yields

$$0 = g_{00}\frac{E^2}{\alpha^2} + g_{rr}\dot{r}^2 + g_{\phi\phi}\frac{h^2}{r^4}. \tag{3.1438}$$

Solving the above equation for $\dot{r}$ gives

$$\dot{r} = \sqrt{-\frac{g_{00}}{g_{rr}}\frac{E^2}{\alpha^2} - \frac{g_{\phi\phi}}{g_{rr}}\frac{h^2}{r^4}} = \sqrt{E^2 - \alpha\frac{h^2}{r^2}}. \tag{3.1439}$$

Thus, using the fact that $\dot{r} = \frac{dr}{ds} = f(r)$ and $\alpha = 1 - \frac{2GM}{r}$, we obtain the answer

$$\frac{dr}{ds} = f(r) = \sqrt{E^2 - \left(1 - \frac{2GM}{r}\right)\frac{h^2}{r^2}}, \tag{3.1440}$$

where E and h are constants of motion, which is the sought-after differential equation for $r(s)$ when restricted to the plane $\theta = \frac{\pi}{2}$.

2.81

Consider the metric

$$ds^2 = g_{\mu\nu}dx^\mu dx^\nu = c^2dt^2 - S(t)^2(dx^2 + dy^2 + dz^2), \tag{3.1441}$$

where $S(t)$ is an increasing function of time t with $S(0) = 0$. To find the geodesic equations of motion, we vary the action

$$S = \int \mathcal{L}\, ds, \tag{3.1442}$$

where the Lagrangian is given by

$$\mathcal{L} = g_{\mu\nu}\dot{x}^\mu\dot{x}^\nu = c^2\dot{t}^2 - S(t)^2(\dot{x}^2 + \dot{y}^2 + \dot{z}^2). \tag{3.1443}$$

The Euler–Lagrange equations are

$$\frac{d}{ds}\frac{\partial \mathcal{L}}{\partial \dot{x}^\mu} - \frac{\partial \mathcal{L}}{\partial x^\mu} = 0, \tag{3.1444}$$

which give the following geodesic equations

$$\ddot{t} + \frac{1}{c^2}S(t)S'(t)(\dot{x}^2 + \dot{y}^2 + \dot{z}^2) = 0, \tag{3.1445}$$

$$\ddot{x} + 2S(t)^{-1}S'(t)\dot{t}\dot{x} = 0, \tag{3.1446}$$

$$\ddot{y} + 2S(t)^{-1}S'(t)\dot{t}\dot{y} = 0, \tag{3.1447}$$

$$\ddot{z} + 2S(t)^{-1}S'(t)\dot{t}\dot{z} = 0. \tag{3.1448}$$

Now, for lightlike geodesics, we have $ds = 0$. Note that we may always rotate a coordinate system such that the motion is directed along one spatial coordinate only (say x) and the motion is taking place in the positive direction. Then, inserting $ds = 0$ and the given condition $S(t) = t/t_0$ for $t_0 > 0$ into the metric, we find that

$$0 = c^2dt^2 - \left(\frac{t}{t_0}\right)^2 dx^2 \Rightarrow c\,dt = +\frac{t}{t_0}\,dx \Rightarrow dx = \frac{ct_0}{t}\,dt. \tag{3.1449}$$

Integrating this equation yields

$$x = ct_0 \ln(t/\tau), \tag{3.1450}$$

where τ is some constant that can be determined from the initial conditions. Thus, the lightlike geodesics are given by

$$\boldsymbol{x} = ct_0 \ln(t/\tau)\boldsymbol{e}_r + \boldsymbol{k}, \tag{3.1451}$$

where $\boldsymbol{e}_r$ is the initial direction of motion and $\boldsymbol{k}$ is a constant.

Finally, for the event (or spacetime point) $p = (ct_0, ct_0, 0, 0)$, the points on the future light cone are given by

$$\boldsymbol{x} = ct_0 \ln(t/t_0)\boldsymbol{e}_r + \boldsymbol{k}, \tag{3.1452}$$

where $\boldsymbol{k} = (ct_0, 0, 0)$. Therefore, the set of events $J^+(p)$ that are causally connected to p are the points inside this light cone, i.e.,

$$J^+(p) = \{(ct, x, y, z) \in \mathbb{R}^4 : (x - ct_0)^2 + y^2 + z^2 \leq [ct_0 \ln(t/t_0)]^2\}. \tag{3.1453}$$

2.82

From the setup of the problem (see Figure 3.14), we have (at the equator $\theta = \pi/2$):

$$\dot{\theta} = 1, \quad \dot{\varphi} = 0, \qquad \text{for the geodesics,} \tag{3.1454}$$

$$X^\theta = 0, \quad X^\varphi = \delta, \qquad \text{for the separation.} \tag{3.1455}$$

It follows that

$$A^\theta = R^\theta{}_{\theta\theta\varphi}\delta, \quad A^\varphi = R^\varphi{}_{\theta\theta\varphi}\delta. \tag{3.1456}$$

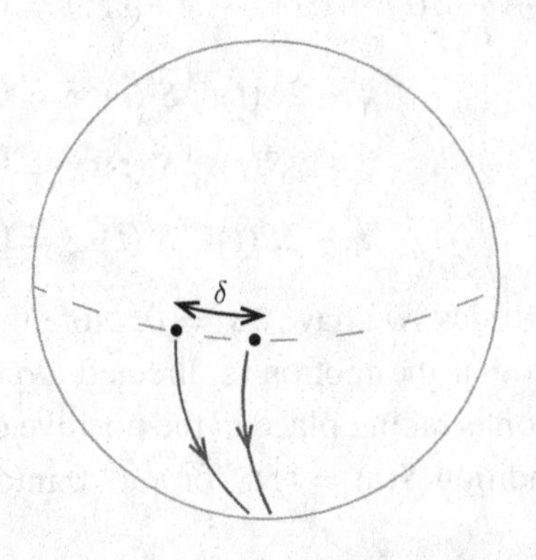

Figure 3.14 Illustration of the setup of the problem. On the unit sphere $\mathbb{S}^2$, two geodesics are separated by a small distance δ and both are orthogonal to the equator $\theta = \pi/2$.

Noting that $R^{\theta}{}_{\theta\theta\varphi} = g^{\theta\theta} R_{\theta\theta\theta\varphi} = 0$, it thus follows that $A^{\theta} = 0$. For $R^{\varphi}{}_{\theta\theta\varphi}$, we find that

$$R^a{}_{\theta\theta\varphi}\partial_a = \nabla_\theta\nabla_\varphi\partial_\theta - \nabla_\varphi\nabla_\theta\partial_\theta = \nabla_\theta(\Gamma^b_{\varphi\theta}\partial_b) - \nabla_\varphi(\Gamma^b_{\theta\theta}\partial_b) = \nabla_\theta(\Gamma^b_{\varphi\theta}\partial_b) - 0$$
$$= \nabla_\theta\left(\frac{\cos\theta}{\sin\theta}\partial_\varphi\right) = \left(-\frac{\sin\theta}{\sin\theta} - \frac{\cos^2\theta}{\sin^2\theta}\right)\partial_\varphi + \cot\theta\,\Gamma^b_{\theta\varphi}\partial_b$$
$$= \left(-1 - \cot^2\theta + \cot^2\theta\right)\partial_\varphi = -\partial_\varphi, \tag{3.1457}$$

which implies that $R^{\varphi}{}_{\theta\theta\varphi} = -1$. We therefore conclude that $A^{\varphi} = -\delta$.

2.83

a) The trajectories $x^\mu(\tau)$ of freely falling particles or photons are the geodesics of the given metric and can be obtained from the variational principle

$$\delta\int\frac{1}{2}\dot{s}^2\,d\tau = \delta\int\frac{1}{2}\left[\dot{t}^2 - e^{2t/R}(\dot{x}^2+\dot{y}^2+\dot{z}^2)\right]d\tau = 0, \tag{3.1458}$$

with the dot indicating differentiation with respect to the parameter τ. The Euler–Lagrange equations for this variational problem are

$$\ddot{t} = -\frac{1}{R}e^{2t/R}(\dot{x}^2+\dot{y}^2+\dot{z}^2), \tag{3.1459}$$

$$\frac{d}{d\tau}(-e^{2t/R}\dot{x}^i) = 0, \tag{3.1460}$$

where $i = 1,2,3$ and $(x^1,x^2,x^3) = (x,y,z)$. The latter Euler–Lagrange equations can be integrated and give

$$\dot{x}^i = c^i e^{-2t/R}, \tag{3.1461}$$

for certain integration constants c^i, which are fixed by the initial conditions. Thus, we find

$$\frac{dx^i}{dx^j} = \frac{c^i}{c^j} = \text{const.} \tag{3.1462}$$

for all $i \neq j$, which can be solved in the following simple manner, e.g.,

$$x = a + c^1 z/c^3, \quad y = b + c^2 z/c^3, \tag{3.1463}$$

with further integration constants a and b. This proves that all geodesics for the given metric are straight lines.

b) The photon moves on a lightlike trajectory along the x-axis, i.e., $ds = dy = dz = 0$. This gives

$$dt = -e^{t/R} dx, \tag{3.1464}$$

where we assume the minus sign, since we want $\dot{t} > 0$ and $\dot{x} < 0$ at $t = 0$. This implies

$$\int_0^{t(x)} e^{-t/R}\, dt = -\int_X^0 dx, \tag{3.1465}$$

which yields

$$-R(e^{-t/R} - 1) = X, \tag{3.1466}$$

and thus, we obtain

$$t = -R \log(1 - X/R). \tag{3.1467}$$

2.84

a) A suitable coordinate system for the problem is given by using spherical coordinates in the subspace spanned by the spatial directions of the five-dimensional Minkowski space, i.e.,

$$x^1 = r \sin(\chi) \sin(\theta) \cos(\phi), \tag{3.1468}$$

$$x^2 = r \sin(\chi) \sin(\theta) \sin(\phi), \tag{3.1469}$$

$$x^3 = r \sin(\chi) \cos(\theta), \tag{3.1470}$$

$$x^4 = r \cos(\chi). \tag{3.1471}$$

In these coordinates, dS_4 is the restriction of the five-dimensional Minkowski metric to the hyperboloid $t^2 - r^2 = -T_0^2$. This condition can be parametrized as

$$r = T_0 \cosh(\tau/T_0), \quad t = T_0 \sinh(\tau/T_0). \tag{3.1472}$$

In the coordinates $(\tau, \chi, \theta, \phi)$, the line element becomes

$$ds^2 = d\tau^2 - T_0^2 \cosh^2(\tau/T_0) d\Omega^2, \tag{3.1473}$$

where $d\Omega^2 = d\chi^2 + \sin^2(\chi)[d\theta^2 + \sin^2(\theta) d\phi^2]$ is the standard line element on the three-dimensional sphere.

b) The geodesic equations are given by varying the integral $S = \int \mathcal{L}\, ds$, where

$$\mathcal{L} = g_{ij}\dot{x}^i \dot{x}^j = \dot{\tau}^2 - T_0^2 \cosh^2(\tau/T_0)\{\dot{\chi}^2 + \sin^2(\chi)[\dot{\theta}^2 + \sin^2(\theta)\dot{\phi}^2]\}. \tag{3.1474}$$

From the Euler–Lagrange equation, we deduce

$$\ddot{\tau} + \frac{T_0}{2} \sinh(2\tau/T_0)\{\dot{\chi}^2 + \sin^2(\chi)[\dot{\theta}^2 + \sin^2(\theta)\dot{\phi}^2]\} = 0, \tag{3.1475}$$

$$\ddot{\chi} - \frac{1}{2}\sin(2\chi)\left[\dot{\theta}^2 + \sin^2(\theta)\dot{\phi}^2\right] + \frac{2}{T_0}\tanh(\tau/T_0)\dot{\tau}\dot{\chi} = 0, \tag{3.1476}$$

$$\ddot{\theta} + 2\cot(\chi)\dot{\chi}\dot{\theta} - \frac{1}{2}\sin(2\theta)\dot{\phi}^2 + \frac{2}{T_0}\tanh(\tau/T_0)\dot{\tau}\dot{\theta} = 0, \tag{3.1477}$$

$$\ddot{\phi} + 2\cot(\chi)\dot{\chi}\dot{\phi} + 2\cot(\theta)\dot{\theta}\dot{\phi} + \frac{2}{T_0}\tanh(\tau/T_0)\dot{\tau}\dot{\phi} = 0. \tag{3.1478}$$

c) Comparing with the Robertson–Walker metric, we obtain

$$S(\tau) = T_0\cosh(\tau/T_0), \tag{3.1479}$$

i.e., a universe which is contracting for $\tau < 0$ and expanding for $\tau > 0$.

If we compute the Ricci tensor and Ricci scalar for dS_4, then we obtain

$$R_{\mu\nu} = \frac{3}{T_0^2}g_{\mu\nu}, \quad R = \frac{12}{T_0^2}. \tag{3.1480}$$

Thus, the Einstein tensor is given by

$$G_{\mu\nu} = R_{\mu\nu} - \frac{1}{2}g_{\mu\nu}R = -\frac{3}{T_0^2}g_{\mu\nu} = 8\pi G T_{\mu\nu}. \tag{3.1481}$$

This gives $T_{\mu\nu} = -\Lambda g_{\mu\nu}/(8\pi G)$, which is typical for so-called *dark energy* (Λ is also know as *cosmological constant*, in dS_4 we have a positive cosmological constant $\Lambda = 3/T_0^2$).

2.85

a) Let the Lagrangian be

$$\mathcal{L} = \frac{1}{2}\left[(r^2-1)\dot{t}^2 - \frac{1}{r^2-1}\dot{r}^2\right], \tag{3.1482}$$

see Problem 2.27. The trajectory of a light ray then obeys $\mathcal{L} = 0$, i.e., $(r^2-1)^2\dot{t}^2 = \dot{r}^2$. Taking the square root, we can solve this equation by separation and find that

$$\frac{dr}{r^2-1} = \frac{1}{2}\left(\frac{1}{r-1} - \frac{1}{r+1}\right) = \pm dt. \tag{3.1483}$$

Integrating, we obtain

$$\frac{1}{2}\left(\ln\frac{r-1}{r_0-1} - \ln\frac{r+1}{r_0+1}\right) = \pm t, \tag{3.1484}$$

where we integrated the left-hand side from r_0 to $r = r(t)$ and the right-hand side from t_0 to t so that $r(0) = r_0$. Thus, we can write this as

$$\frac{r-1}{r+1} = \frac{r_0-1}{r_0+1}\exp(\pm 2t), \tag{3.1485}$$

and after some computations, we find

$$r(t) = \frac{r_0 + 1 + (r_0 - 1)\exp(\pm 2t)}{r_0 + 1 - (r_0 - 1)\exp(\pm 2t)}. \tag{3.1486}$$

Note that the different signs $\pm$ correspond to the two possible directions of the light ray.

b) The trajectory of a particle can also be obtained from the Lagrangian $\mathcal{L}$. Using $\partial\mathcal{L}/\partial t = 0$, we find a conservation law

$$\frac{d}{d\tau}\frac{\partial\mathcal{L}}{\partial\dot{t}} = 0, \tag{3.1487}$$

i.e.,

$$\frac{\partial\mathcal{L}}{\partial\dot{t}} = (r^2 - 1)\dot{t} = c_1 \tag{3.1488}$$

for some constant c_1. To compute c_1, we set $t = 0$ for which it holds that $r = r_0$ and $\dot{t} = \gamma = 1$, since the velocity is zero. Thus, we find $c_1 = r_0^2 - 1$, and therefore, we have

$$\dot{t} = \frac{r_0^2 - 1}{r^2 - 1}. \tag{3.1489}$$

A second conservation law is $\mathcal{L} = c_2$ for some other constant c_2. Inserting again $t = 0$, we find $c_2 = (r_0^2 - 1)/2$, since $\dot{t} = 1$, $\dot{r} = 0$, and $r = r_0$ at $t = 0$. Combining the first and the second conservation laws and multiplying with a factor of two, we obtain

$$(r^2 - 1)\dot{t}^2 - \frac{1}{r^2 - 1}\dot{r}^2 = \frac{(r_0^2 - 1)^2 - \dot{r}^2}{r^2 - 1} = r_0^2 - 1 \quad \Rightarrow \quad \dot{r}^2 = (r_0^2 - 1)(r_0^2 - r^2). \tag{3.1490}$$

Separating this equation, we obtain

$$d\tau = -\frac{dr}{\sqrt{(r_0^2 - 1)(r_0^2 - r^2)}}, \tag{3.1491}$$

where we choose the sign of the square root so that $\dot{r} < 0$. Integrating the left-hand side from 0 to τ and the right-hand side from r_0 to r_1 yields the proper time

$$\tau = -\int_{r_0}^{r_1} \frac{dr}{\sqrt{(r_0^2 - 1)(r_0^2 - r^2)}} = \frac{1}{\sqrt{r_0^2 - 1}}\left(\frac{\pi}{2} - \arctan\frac{r_1}{\sqrt{r_0^2 - r_1^2}}\right). \tag{3.1492}$$

To compute the coordinate time, we use the first conservation law and the result for $d\tau$, i.e.,

$$dt = \frac{r_0^2 - 1}{r^2 - 1} d\tau = -\frac{r_0^2 - 1}{r^2 - 1} \frac{dr}{\sqrt{(r_0^2 - 1)(r_0^2 - r^2)}} = -\sqrt{r_0^2 - 1} \frac{dr}{(r^2 - 1)\sqrt{r_0^2 - r^2}}. \tag{3.1493}$$

Thus, integrating, we have

$$t = -\sqrt{r_0^2 - 1} \int_{r_0}^{r_1} \frac{dr}{(r^2 - 1)\sqrt{r_0^2 - r^2}}. \tag{3.1494}$$

When $r_1 \to 1$, the integral diverges as $t \sim -\ln(r_1 - 1)/2$. The point $r = 1$ is a singularity similar to the singularity of the Schwarzschild metric at the event horizon: Our computation above shows that the proper time for a particle falling toward $r = 1$ is finite, whereas the coordinate time for that is infinite. Thus, this suggests that $r = 1$ is a coordinate singularity.

2.86

a) Using the given coordinate transformations, we have

$$dU = \cos(\alpha)dt - t\sin(\alpha)d\alpha, \tag{3.1495}$$

$$dV = \sin(\alpha)dt + t\cos(\alpha)d\alpha, \tag{3.1496}$$

$$dX = \sin(\theta)\cos(\varphi)dr + r\cos(\theta)\cos(\varphi)d\theta - r\sin(\theta)\sin(\varphi)d\varphi, \tag{3.1497}$$

$$dY = \sin(\theta)\sin(\varphi)dr + r\cos(\theta)\sin(\varphi)d\theta + r\sin(\theta)\cos(\varphi)d\varphi, \tag{3.1498}$$

$$dZ = \cos(\theta)dr - r\sin(\theta)d\theta. \tag{3.1499}$$

Thus, the metric of AdS_4 in the coordinates α, λ, θ, and φ is given by

$$ds^2 = dt^2 + t^2 d\alpha^2 - dr^2 - r^2 d\Omega^2 = \cosh(\lambda)^2 d\alpha^2 - d\lambda^2 - \sinh(\lambda)^2 d\Omega^2, \tag{3.1500}$$

where $d\Omega^2 = d\theta^2 + \sin(\theta)^2 d\varphi^2$.

b) To find the trajectories, we can solve (since $\dot{\theta} = \dot{\varphi} = 0$)

$$\mathcal{L} = \frac{1}{2}\left[\cosh(\lambda)^2 \dot{\alpha}^2 - \dot{\lambda}^2\right] = 0. \tag{3.1501}$$

This gives

$$\frac{d\lambda}{\cosh(\lambda)} = \pm d\alpha \quad \Rightarrow \quad 2\arctan\tanh\frac{\lambda}{2} = \pm(\alpha - \alpha_0)$$

$$\Rightarrow \quad \lambda = \pm 2\,\text{artanh}\left(\tan\frac{\alpha - \alpha_0}{2}\right). \tag{3.1502}$$

c) The Euler–Lagrange equations following from the variational problem defined in the hints are

$$\ddot{X}^a = 2\lambda X^a, \quad X^a X_a - 1 = 0. \tag{3.1503}$$

Contracting the first equation $\ddot{X}^a = 2\lambda X^a$ with X_a and using the second equation $X^a X_a = 1$, we obtain

$$2\lambda = \ddot{X}^a X_a = \frac{d}{d\tau}(\dot{X}^a X_a) - \dot{X}^a \dot{X}_a, \tag{3.1504}$$

but $\dot{X}^a X_a = 0$, which follows from differentiating $X^a X_a = 1$ with respect to τ, and thus, we have

$$2\lambda = -\dot{X}^a \dot{X}_a. \tag{3.1505}$$

To prove that this is a constant of motion, as claimed in the hints, we compute

$$\frac{d}{d\tau}(\dot{X}^a \dot{X}_a) = 2\ddot{X}^a \dot{X}_a = 4\lambda X^a \dot{X}_a = 0. \tag{3.1506}$$

Thus, we can set $\dot{X}^a \dot{X}_a = 0$, which implies that $\ddot{X}^a = 0$.

2.87

a) We determine the Christoffel symbols by deriving the geodesic equation with $\mathcal{L} = g_{\mu\nu}\dot{x}^\mu \dot{x}^\nu = \dot{t}^2 - a(t)^2 \dot{x}^2$. We obtain

$$\frac{d}{d\tau}\frac{\partial \mathcal{L}}{\partial \dot{t}} - \frac{\partial \mathcal{L}}{\partial t} = 2\ddot{t} + 2aa'\dot{x}^2 = 0 \quad \Leftrightarrow \quad \ddot{t} + aa'\dot{x}^2 = 0, \tag{3.1507}$$

$$\frac{d}{d\tau}\frac{\partial \mathcal{L}}{\partial \dot{x}} - \frac{\partial \mathcal{L}}{\partial x} = \frac{d}{d\tau}(2a^2\dot{x}) = 2a^2\ddot{x} + 4aa'\dot{t}\dot{x} = 0 \quad \Leftrightarrow \quad \ddot{x} + 2\frac{a'}{a}\dot{t}\dot{x} = 0 \tag{3.1508}$$

with $a'(t) = da(t)/dt$. Comparing with the general form of the geodesic equations $\ddot{x}^\lambda + \Gamma^\lambda_{\mu\nu}\dot{x}^\mu \dot{x}^\nu = 0$, we can read off the following nonzero Christoffel symbols

$$\Gamma^t_{xx} = aa', \quad \Gamma^x_{tx} = \Gamma^x_{xt} = \frac{a'}{a}. \tag{3.1509}$$

Now, we compute

$$\begin{aligned} R^t{}_{xtx} &= \partial_t \Gamma^t_{xx} - \partial_x \Gamma^t_{tx} + \Gamma^t_{t\alpha}\Gamma^\alpha_{xx} - \Gamma^t_{x\alpha}\Gamma^\alpha_{tx} = \partial_t \Gamma^t_{xx} - \Gamma^t_{xx}\Gamma^x_{tx} \\ &= \partial_t(aa') - aa'\frac{a'}{a} = (a')^2 + aa'' - (a')^2 = aa''. \end{aligned} \tag{3.1510}$$

Furthermore, we have $g_{tt} = 1$, $g_{xx} = -a^2$, and $g_{tx} = g_{xt} = 0$, which imply that

$$g^{tt} = 1, \quad g^{xx} = -\frac{1}{a^2}, \quad g^{tx} = g^{xt} = 0. \tag{3.1511}$$

Thus, we find the nonzero components of the Riemann curvature tensor

$$R_{txtx} = g_{tt} R^{t}{}_{xtx} = aa'' = R_{xtxt} = -R_{txxt} = -R_{xttx}, \tag{3.1512}$$

where the last identities follow from general symmetry properties of the Riemann curvature tensor. The other components of the Riemann curvature tensor $R_{\mu\nu\alpha\beta}$ are zero due to general symmetry properties. Thus, we obtain the components of the Ricci tensor $R_{\mu\nu}$ as

$$R_{tt} = g^{\alpha\beta} R_{t\alpha t\beta} = g^{xx} R_{txtx} = -\frac{1}{a^2} aa'' = -\frac{a''}{a}, \tag{3.1513}$$

$$R_{xx} = g^{\alpha\beta} R_{x\alpha x\beta} = g^{tt} R_{xtxt} = 1 \cdot aa'' = aa'', \tag{3.1514}$$

$$R_{tx} = g^{\alpha\beta} R_{t\alpha x\beta} = 0, \tag{3.1515}$$

$$R_{xt} = g^{\alpha\beta} R_{x\alpha t\beta} = 0, \tag{3.1516}$$

which can be summarized as $R_{\mu\nu} = -g_{\mu\nu} a''/a$.

Note that the Ricci scalar $R \equiv g^{\mu\nu} R_{\mu\nu}$ is given by

$$R = g^{tt} R_{tt} + g^{xx} R_{xx} = 1 \cdot \left(-\frac{a''}{a}\right) + \left(-\frac{1}{a^2}\right) aa'' = -\frac{2a''}{a}, \tag{3.1517}$$

and thus, the Einstein tensor $G_{\mu\nu} \equiv R_{\mu\nu} - \frac{1}{2} R g_{\mu\nu}$ vanishes, i.e.,

$$G_{\mu\nu} = -g_{\mu\nu} \frac{a''}{a} - \frac{1}{2} \left(-\frac{2a''}{a}\right) g_{\mu\nu} = 0. \tag{3.1518}$$

b) We can find the trajectory of a light ray from $\mathcal{L} = \dot{t}^2 - a(t)^2 \dot{x}^2 = 0$, where $\dot{t} = dt/d\tau$, $\dot{x} = dx/d\tau$, and τ being the proper time, i.e.,

$$\left(\frac{dx}{dt}\right)^2 = \frac{1}{a^2} \quad \Rightarrow \quad \frac{dx(t)}{dt} = \frac{1}{a(t)} = A^2 + B^2 t^2, \tag{3.1519}$$

where we choose the plus sign of the square root, since it should be assumed that $dx/dt > 0$ at $t = 0$. Thus, integrating, we obtain the trajectory of the light ray as

$$x(t) = x_0 + A^2 t + \frac{B^2}{3} t^3, \tag{3.1520}$$

where $x_0 = x(0)$.

2.88

The metric tensor for the three-dimensional Robertson–Walker spacetime can be written as

$$g = (g_{\mu\nu}) = \begin{pmatrix} g_{tt} & g_{tr} & g_{t\phi} \\ g_{rt} & g_{rr} & g_{r\phi} \\ g_{\phi t} & g_{\phi r} & g_{\phi\phi} \end{pmatrix} = \begin{pmatrix} 1 & 0 & 0 \\ 0 & -\frac{a^2}{1-kr^2} & 0 \\ 0 & 0 & -a^2r^2 \end{pmatrix}. \tag{3.1521}$$

Inserting the Lagrangian

$$\mathcal{L} = g_{\mu\nu}\dot{x}^\mu \dot{x}^\nu = \dot{t}^2 - \frac{a^2}{1-kr^2}\dot{r}^2 - a^2r^2\dot{\phi}^2, \tag{3.1522}$$

into Euler–Lagrange equations, we find that

$$\frac{\partial \mathcal{L}}{\partial t} - \frac{d}{d\tau}\frac{\partial \mathcal{L}}{\partial \dot{t}} = 0, \quad \frac{\partial \mathcal{L}}{\partial t} = -\frac{2aa'}{1-kr^2}\dot{r}^2 - 2aa'r^2\,\dot{\phi}^2, \quad \frac{\partial \mathcal{L}}{\partial \dot{t}} = 2\dot{t},$$

$$\Rightarrow \quad -\frac{2aa'}{1-kr^2}\dot{r}^2 - 2aa'r^2\,\dot{\phi}^2 - 2\ddot{t} = 0, \tag{3.1523}$$

$$\frac{\partial \mathcal{L}}{\partial r} - \frac{d}{d\tau}\frac{\partial \mathcal{L}}{\partial \dot{r}} = 0, \quad \frac{\partial \mathcal{L}}{\partial r} = -\frac{2ka^2r}{(1-kr^2)^2}\dot{r}^2 - 2a^2r\,\dot{\phi}^2, \quad \frac{\partial \mathcal{L}}{\partial \dot{r}} = -\frac{2a^2}{1-kr^2}\dot{r},$$

$$\Rightarrow \quad \frac{2ka^2r}{(1-kr^2)^2}\dot{r}^2 - 2a^2r\,\dot{\phi}^2 + \frac{4a}{1-kr^2}\dot{a}\dot{r} + \frac{2a^2}{1-kr^2}\ddot{r} = 0, \tag{3.1524}$$

$$\frac{\partial \mathcal{L}}{\partial \phi} - \frac{d}{d\tau}\frac{\partial \mathcal{L}}{\partial \dot{\phi}} = 0, \quad \frac{\partial \mathcal{L}}{\partial \phi} = 0, \quad \frac{\partial \mathcal{L}}{\partial \dot{\phi}} = -2a^2r^2\,\dot{\phi},$$

$$\Rightarrow \quad 4ar^2\,\dot{a}\dot{\phi} + 4a^2r\,\dot{r}\dot{\phi} + 2a^2r^2\,\ddot{\phi} = 0. \tag{3.1525}$$

Now, using that

$$\dot{a} = \frac{da}{d\tau} = \frac{da}{dt}\frac{dt}{d\tau} = a'\,\dot{t}, \tag{3.1526}$$

we obtain the geodesic equations as

$$\ddot{t} + \frac{aa'}{1-kr^2}\dot{r}^2 + aa'r^2\,\dot{\phi}^2 = 0, \tag{3.1527}$$

$$\ddot{r} + \frac{2a'}{a}\dot{t}\dot{r} + \frac{kr}{1-kr^2}\dot{r}^2 - (1-kr^2)r\,\dot{\phi}^2 = 0, \tag{3.1528}$$

$$\ddot{\phi} + \frac{2a'}{a}\dot{t}\dot{\phi} + \frac{2}{r}\dot{r}\dot{\phi} = 0. \tag{3.1529}$$

Finally, using the general formula for the geodesic equations $\ddot{x}^\mu + \Gamma^\mu_{\nu\lambda}\dot{x}^\nu\dot{x}^\lambda = 0$, we identify the nonzero Christoffel symbols as

$$\Gamma^t_{rr} = \frac{aa'}{1-kr^2}, \tag{3.1530}$$

$$\Gamma^t_{\phi\phi} = aa'r^2, \tag{3.1531}$$

$$\Gamma^r_{tr} = \Gamma^r_{rt} = \frac{a'}{a}, \tag{3.1532}$$

$$\Gamma^r_{rr} = \frac{kr}{1-kr^2}, \tag{3.1533}$$

$$\Gamma^r_{\phi\phi} = -(1-kr^2)r, \tag{3.1534}$$

$$\Gamma^\phi_{t\phi} = \Gamma^\phi_{\phi t} = \frac{a'}{a}, r \tag{3.1535}$$

$$\Gamma^\phi_{r\phi} = \Gamma^\phi_{\phi r} = \frac{1}{r}. \tag{3.1536}$$

2.89

a) We compute the Euler–Lagrange equation for

$$\mathcal{L} = \frac{1}{2} g_{\mu\nu}(x(s))\dot{x}^\mu(s)\dot{x}^\nu(s), \tag{3.1537}$$

since we know that they are equivalent to

$$\ddot{x}^\mu + \Gamma^\mu_{\alpha\beta}\dot{x}^\alpha\dot{x}^\beta = 0, \tag{3.1538}$$

and thus allow us to read off the nonzero Christoffel symbols in a simple manner.

Thus, we obtain

$$\mathcal{L} = \frac{1}{2}\left\{\dot{t}^2 - e^{-2t/a}\left[(1+r^2/a^2)^{-1}\dot{r}^2 + r^2\sin^2\theta\dot{\varphi}^2 + r^2\dot{\theta}^2\right]\right\} \tag{3.1539}$$

and the Euler–Lagrange equations

$$\frac{d}{ds}\frac{\partial\mathcal{L}}{\partial\dot{t}} - \frac{\partial\mathcal{L}}{\partial t} = \ddot{t} - \frac{1}{a}e^{-2t/a}\left[(1+r^2/a^2)^{-1}\dot{r}^2 + r^2\sin^2\theta\dot{\varphi}^2 + r^2\dot{\theta}^2\right] = 0, \tag{3.1540}$$

$$\begin{aligned}\frac{d}{ds}\frac{\partial\mathcal{L}}{\partial\dot{r}} - \frac{\partial\mathcal{L}}{\partial r} &= -\frac{d}{ds}\left[e^{-2t/a}(1+r^2/a^2)^{-1}\dot{r}\right] \\ &\quad + e^{-2t/a}\left[-(1+r^2/a^2)^{-2}\frac{r}{a^2}\dot{r}^2 + r\sin^2\theta\dot{\varphi}^2 + r\dot{\theta}^2\right] \\ &= -\frac{e^{-2t/a}}{1+r^2/a^2}\left[\ddot{r} - \frac{2}{a}\dot{t}\dot{r} - \frac{r}{a^2+r^2}\dot{r}^2\right. \\ &\quad \left. - (1+r^2/a^2)(r\sin^2\theta\dot{\varphi}^2 + r\dot{\theta}^2)\right] = 0,\end{aligned} \tag{3.1541}$$

$$\begin{aligned}\frac{d}{ds}\frac{\partial\mathcal{L}}{\partial\dot{\theta}} - \frac{\partial\mathcal{L}}{\partial\theta} &= -\frac{d}{ds}\left(e^{-2t/a}r^2\dot{\theta}\right) + e^{-2t/a}r^2\sin\theta\cos\theta\dot{\varphi}^2 \\ &= -r^2\left(\ddot{\theta} - \frac{2}{a}\dot{t}\dot{\theta} + \frac{2}{r}\dot{r}\dot{\theta} - \sin\theta\cos\theta\dot{\varphi}^2\right) = 0,\end{aligned} \tag{3.1542}$$

$$\frac{d}{ds}\frac{\partial \mathcal{L}}{\partial \dot\varphi} - \frac{\partial \mathcal{L}}{\partial \varphi} = -\frac{d}{ds}\left(e^{-2t/a} r^2 \sin^2\theta \dot\varphi\right)$$

$$= -e^{-2t/a} r^2 \sin^2\theta \left(\ddot\varphi - \frac{2}{a}\dot\varphi \dot t + \frac{2}{r}\dot r \dot\varphi + 2\cot\theta \dot\theta \dot\varphi\right) = 0 \quad (3.1543)$$

from which we find the nonzero Christoffel symbols $\Gamma^\mu_{\alpha\beta} = \Gamma^\mu_{\beta\alpha}$ such that

$$\Gamma^t_{rr} = -\frac{1}{a}e^{-2t/a}(1+r^2/a^2)^{-1}, \quad \Gamma^t_{\theta\theta} = -\frac{1}{a}e^{-2t/a}r^2, \quad \Gamma^t_{\varphi\varphi} = -\frac{1}{a}e^{-2t/a}r^2\sin^2\theta,$$

$$\Gamma^r_{rt} = -\frac{1}{a}, \quad \Gamma^r_{rr} = -\frac{r}{a^2+r^2},$$

$$\Gamma^r_{\theta\theta} = -(1+r^2/a^2)r, \quad \Gamma^r_{\varphi\varphi} = -(1+r^2/a^2)r\sin^2\theta,$$

$$\Gamma^\theta_{t\theta} = -\frac{1}{a}, \quad \Gamma^\theta_{r\theta} = \frac{1}{r}, \quad \Gamma^\theta_{\varphi\varphi} = -\sin\theta\cos\theta,$$

$$\Gamma^\varphi_{t\varphi} = -\frac{1}{a}, \quad \Gamma^\varphi_{r\varphi} = \frac{1}{r}, \quad \Gamma^\varphi_{\theta\varphi} = \cot\theta. \quad (3.1544)$$

We now compute the covariant derivative $\nabla_\mu A^\nu$ of the vector field A^ν

$$\nabla_\mu A^\nu = \partial_\mu A^\nu + \Gamma^\nu_{\mu\alpha}A^\alpha, \quad (3.1545)$$

i.e., in component form, we obtain

$$\nabla_t A^t = \partial_t A^t + \Gamma^t_{tt}A^t + \Gamma^t_{tr}A^r = \frac{1}{a}, \quad (3.1546)$$

$$\nabla_r A^t = \partial_r A^t + \Gamma^t_{rt}A^t + \Gamma^t_{rr}A^r = -\frac{r}{a^2}e^{-2t/a}(1+r^2/a^2)^{-1}, \quad (3.1547)$$

$$\nabla_t A^r = \partial_t A^r + \Gamma^r_{tt}A^t + \Gamma^r_{tr}A^r = -\frac{r}{a^2}, \quad (3.1548)$$

$$\nabla_r A^r = \partial_r A^r + \Gamma^r_{rt}A^t + \Gamma^r_{rr}A^r = \frac{1}{a} - \frac{t}{a^2} - \frac{1}{a}\frac{r^2/a^2}{1+r^2/a^2}. \quad (3.1549)$$

b) We have the conservation law

$$g_{\mu\nu}(x(s))\dot x^\mu(s)\dot x^\nu(s) = \dot t^2 - e^{-2t/a}(1+r^2/a^2)^{-1}\dot r^2 = 0, \quad (3.1550)$$

since the tangent should be a lightlike vector. From this, we conclude that

$$\frac{dr}{dt} = \frac{\dot r}{\dot t} = \sqrt{1+r^2/a^2}\, e^{t/a}, \quad (3.1551)$$

with $r(0) = a$. This can be solved by separation

$$\int_a^{r(t)} \frac{dr}{\sqrt{1+r^2/a^2}} = \int_0^t e^{t/a}\, dt, \quad (3.1552)$$

i.e.,

$$a[\operatorname{arsinh}(r/a) - \operatorname{arsinh}(1)] = a(e^{t/a} - 1). \tag{3.1553}$$

We thus can obtain the following explicit trajectory of the light pulse

$$r(t) = a \sinh(e^{t/a} + \operatorname{arsinh}(1) - 1). \tag{3.1554}$$

2.90

For $k = 1$ and $d\Omega = 0$, we can write the Robertson–Walker metric as

$$ds^2 = c^2 dt^2 - S(t)^2 d\chi^2 = \left[c^2 \dot{t}^2 - S(t)^2 \dot{\chi}^2\right] d\tau^2, \tag{3.1555}$$

where $\chi = \arcsin r$ and dot is differentiation with respect to the parameter τ. Thus, we have the metric condition $c^2\dot{t}^2 - S(t)^2\dot{\chi}^2 = \beta = \text{const.}$ for an affinely parametrized geodesic.

a) The Euler–Lagrange equations yield the geodesic equations for t and χ, i.e.,

$$\ddot{t} + \frac{1}{c^2} S(t) S'(t) \dot{\chi}^2 = 0, \tag{3.1556}$$

$$\frac{d}{d\tau}\left[S(t)^2 \dot{\chi}\right] = 0. \tag{3.1557}$$

See also the solution to Problem 2.81. Therefore, the first integral is

$$S(t)^2 \dot{\chi} = \alpha = \text{const.} \quad \Rightarrow \quad \dot{\chi} = \frac{\alpha}{S(t)^2}, \tag{3.1558}$$

which from the metric condition gives

$$\dot{t} = \pm \frac{1}{c}\sqrt{\beta + \frac{\alpha^2}{S(t)^2}}. \tag{3.1559}$$

b) For lightlike geodesics, $\beta = 0$. Thus, inserting $\beta = 0$ and $\alpha = S(t)^2\dot{\chi}$ into the result of a), we find that

$$\dot{t} = \pm \frac{1}{c}\frac{\alpha}{S(t)} = \pm \frac{1}{c} S(t)\dot{\chi} \quad \Rightarrow \quad \frac{\dot{\chi}}{\dot{t}} = \pm \frac{c}{S(t)} \quad \Rightarrow \quad \frac{d\chi}{dt} = \pm \frac{c}{S(t)}. \tag{3.1560}$$

Now, using the subsidiary condition on χ given in the problem and assuming propagation *forward* in time, we have the positive derivative, i.e.,

$$\frac{d\chi}{dt} = \frac{c}{S(t)} \quad \Rightarrow \quad d\chi = \frac{c}{S(t)} dt. \tag{3.1561}$$

Integrating leads to

$$\Delta\chi \equiv \chi_1 - \chi_0 = \int d\chi = \int_{t_0}^{t_1} \frac{c}{S(t)} dt = \int_0^T \frac{c}{S(t + t_0)} dt, \tag{3.1562}$$

where $t_1 \equiv t_0 + T$. The coordinate transformation between χ and r is given by $\chi = \arcsin r$, which means that for $r = r_0 = 0$, we have $\chi = \chi_0 = 0$, and thus, we find that

$$\Delta\chi(T) = \int_0^T \frac{c}{S(t+t_0)}\,dt \equiv \arcsin \Delta r(T)$$

$$\Leftrightarrow \quad \Delta r(T) = \sin \Delta\chi(T) = \sin\left(\int_0^T \frac{c}{S(t+t_0)}\,dt\right). \tag{3.1563}$$

Therefore, the distance in the r-coordinate of the Robertson–Walker metric that a light ray (emitted at universal time $t = t_0$ at $r = r_0 = 0$) travels in the universal time interval $[t_0, t_0 + T]$ is given by

$$\Delta r(T) = \sin\left(\int_0^T \frac{c}{S(t+t_0)}\,dt\right). \tag{3.1564}$$

2.91

The given standard Schwarzschild metric is

$$ds^2 = \left(1 - \frac{r_*}{r}\right)c^2dt^2 - \left(1 - \frac{r_*}{r}\right)^{-1}dr^2 - r^2d\Omega^2, \tag{3.1565}$$

where s is the path parameter, $r_* \equiv 2GM/c^2$, and $d\Omega^2 = d\theta^2 + \sin^2\theta d\phi^2$. In addition, the given initial conditions are

$$r = r_0, \quad \dot{t} = \beta, \quad \dot{r} = \alpha, \quad \dot{\theta} = 0, \quad \dot{\phi} = 0. \tag{3.1566}$$

We introduce the Lagrangian

$$\mathcal{L} = g_{\mu\nu}\dot{x}^\mu\dot{x}^\nu = \left(1 - \frac{r_*}{r}\right)c^2\dot{t}^2 - \left(1 - \frac{r_*}{r}\right)^{-1}\dot{r}^2 - r^2\dot{\theta}^2 - r^2\sin^2\theta\dot{\phi}^2. \tag{3.1567}$$

Since $\frac{\partial\mathcal{L}}{\partial t} = 0$ and $\frac{\partial\mathcal{L}}{\partial\phi} = 0$, we have two constants of motion stemming from the Euler–Lagrange equations for t and ϕ, namely

$$\frac{d}{ds}\left[2\left(1 - \frac{r_*}{r}\right)c^2\dot{t}\right] = 0 \quad \Rightarrow \quad 2\left(1 - \frac{r_*}{r}\right)c^2\dot{t} = E = \text{const.}, \tag{3.1568}$$

$$\frac{d}{ds}\left(-2r^2\sin^2\theta\dot{\phi}\right) = 0 \quad \Rightarrow \quad -2r^2\sin^2\theta\dot{\phi} = L = \text{const.} \tag{3.1569}$$

Using the initial conditions, we can determine E and L as

$$2\left(1 - \frac{r_*}{r_0}\right)c^2\beta = E, \quad -2r_0^2\sin^2\theta_0 \cdot 0 = L, \tag{3.1570}$$

which means that $L = 0$, independent of the value of θ_0 that is not known. Therefore, assuming $0 \le r \le r_0$ and $0 \le \theta \le \pi$, we have that $\dot{\phi} = 0$. Furthermore, the Euler–Lagrange equation for the θ-coordinate yields

$$\frac{\partial \mathcal{L}}{\partial \theta} = -r^2 \sin 2\theta \dot{\phi}^2, \quad \frac{d}{ds}\frac{\partial \mathcal{L}}{\partial \theta} = -2(2r\dot{r}\dot{\theta} + r^2\ddot{\theta})$$

$$\Rightarrow \quad \ddot{\theta} + 2r\dot{r}\dot{\theta} - \frac{1}{2}\sin 2\theta \dot{\phi}^2 = 0. \tag{3.1571}$$

Now, since $\dot{\phi} = 0$, we also have $\frac{\partial \mathcal{L}}{\partial \theta} = -r^2 \sin 2\theta \dot{\phi}^2 = -r^2 \sin 2\theta \cdot 0 = 0$, which means that we have another constant of motion stemming from the Euler–Lagrange equation for θ (due to the initial conditions), namely

$$\frac{d}{ds}\left(-2r^2\dot{\theta}\right) = 0 \quad \Rightarrow \quad -2r^2\dot{\theta} = A = \text{const.} \tag{3.1572}$$

Again, using the initial conditions, we can determine A as $-2r_0^2 \cdot 0 = A$, which also means that $A = 0$. Therefore, assuming $0 \le r \le r_0$ as before, we have $-2r^2\dot{\theta} = 0$, which leads to $\dot{\theta} = 0$. Thus, based on the initial conditions, the given standard Schwarzschild metric is reduced to

$$ds^2 = \left(1 - \frac{r_*}{r}\right) c^2 dt^2 - \left(1 - \frac{r_*}{r}\right)^{-1} dr^2. \tag{3.1573}$$

Now, using the constant of motion E, given by the initial conditions, we find that

$$\dot{t} = \frac{E}{2\left(1 - \frac{r_*}{r}\right)c^2} = \left(1 - \frac{r_*}{r}\right)^{-1} \frac{E}{2c^2} = \frac{dt}{ds} \quad \Rightarrow \quad dt = \left(1 - \frac{r_*}{r}\right)^{-1} \frac{E}{2c^2} ds. \tag{3.1574}$$

Inserting the expression for dt into the reduced standard Schwarzschild metric, we obtain

$$\begin{aligned} ds^2 &= \left(1 - \frac{r_*}{r}\right) c^2 \left(1 - \frac{r_*}{r}\right)^{-2} \frac{E^2}{4c^4} - \left(1 - \frac{r_*}{r}\right)^{-1} dr^2 \\ &= \left(1 - \frac{r_*}{r}\right)^{-1} \frac{E^2}{4c^2} ds^2 - \left(1 - \frac{r_*}{r}\right)^{-1} dr^2, \end{aligned} \tag{3.1575}$$

which can be written as

$$dr^2 = \left[\frac{E^2}{4c^2} - \left(1 - \frac{r_*}{r}\right)\right] ds^2 \quad \Rightarrow \quad \left(\frac{dr}{ds}\right) = \frac{E^2}{4c^2} - \left(1 - \frac{r_*}{r}\right). \tag{3.1576}$$

Since the spaceship is freely falling from r_0 *toward* the true singularity at $r = 0$ in a Schwarzschild black hole, we have $\dot{r} = \frac{dr}{ds} < 0$, which means that

$$\frac{dr}{ds} = -\sqrt{\frac{E^2}{4c^2} - \left(1 - \frac{r_*}{r}\right)} = \frac{dr}{c\,d\tau} \quad \Rightarrow \quad d\tau = -\frac{dr}{c\sqrt{\frac{E^2}{4c^2} - \left(1 - \frac{r_*}{r}\right)}}, \tag{3.1577}$$

where τ is the proper time. Integrating the expression for $d\tau$ from $r = r_0$ (at $\tau = 0$) to $\dot{r} = 0$, we find the proper time τ needed to reach the singularity $r = 0$ (when starting from $r = r_0 < r_*$) as

$$\tau = -\int_{r_0}^{0} \frac{dr}{c\sqrt{\frac{E^2}{4c^2} - \left(1 - \frac{r_*}{r}\right)}} = \int_{0}^{r_0} \frac{dr}{c\sqrt{\frac{E^2}{4c^2} - \left(1 - \frac{r_*}{r}\right)}}. \qquad (3.1578)$$

However, note that (see above) the constant of motion E can be written in terms of the initial conditions as $E = 2\left(1 - \frac{r_*}{r_0}\right)c^2\beta$, which means that the proper time τ is given by

$$\tau = \int_0^{r_0} \frac{dr}{c\sqrt{\left(1 - \frac{r_*}{r_0}\right)^2 c^2\beta^2 - \left(1 - \frac{r_*}{r}\right)}} \equiv \int_0^{r_0} f(r)\,dr. \qquad (3.1579)$$

Thus, we can finally identify the function $f(r)$ as

$$f(r) = \frac{1}{c\sqrt{\left(1 - \frac{r_*}{r_0}\right)^2 c^2\beta^2 - \left(1 - \frac{r_*}{r}\right)}}. \qquad (3.1580)$$

2.92
Write the geodesic equations in (x^0, r, θ, ϕ) coordinates in the plane $\theta = \frac{\pi}{2}$, namely

$$\ddot{x}^0 + \frac{\alpha}{r^2}\left(1 - \frac{\alpha}{r}\right)^{-1}\dot{x}^0\dot{r} = 0, \qquad (3.1581)$$

$$\ddot{r} + \frac{\alpha}{2r^2}\left(1 - \frac{\alpha}{r}\right)\left(\dot{x}^0\right)^2 - \frac{\alpha}{2r^2}\left(1 - \frac{\alpha}{r}\right)^{-1}\dot{r}^2 - r\left(1 - \frac{\alpha}{r}\right)\dot{\phi}^2 = 0, \qquad (3.1582)$$

$$\ddot{\phi} + \frac{2}{r}\dot{r}\dot{\phi} = 0, \qquad (3.1583)$$

where $\alpha = \frac{2GM}{c^2}$. For a derivation of the geodesic equations, see Problem 2.70. The geodesic equation for x^0 can be integrated at once, which yields

$$\dot{x}^0 = \frac{k}{1 - \frac{\alpha}{r}}, \quad \text{where } k \text{ is a constant.} \qquad (3.1584)$$

On a geodesic for a timelike object, it holds that

$$1 = g_{00}(\dot{x}^0)^2 + g_{rr}\dot{r}^2, \qquad (3.1585)$$

where $g_{00} = 1 - \frac{\alpha}{r}$ and $g_{rr} = -\left(1 - \frac{\alpha}{r}\right)^{-1}$. This implies that

$$\dot{r} = -\sqrt{\epsilon + \frac{\alpha}{r}} = \frac{dr}{ds} = \frac{1}{c}\frac{dr}{d\tau}, \qquad (3.1586)$$

where $\epsilon = k^2 - 1$. Note the minus sign (which is due to the fact that r is decreasing, i.e., the observer is freely falling *toward* the center of the black hole, and hence, $\dot{r} < 0$). So, we have

$$\frac{d\tau}{dr} = -\frac{1}{c\sqrt{\epsilon + \frac{\alpha}{r}}}. \tag{3.1587}$$

According to the hint, integration gives

$$\int \frac{dr}{\sqrt{\epsilon + \frac{\alpha}{r}}} = \frac{1}{\sqrt{\epsilon}}\sqrt{y^2 - \frac{\beta^2}{4}} - \frac{1}{\sqrt{\epsilon}}\frac{\beta}{2}\ln\left(y + \sqrt{y^2 - \frac{\beta^2}{4}}\right), \tag{3.1588}$$

where $\beta = \frac{\alpha}{\epsilon}$ and $y = r + \frac{\beta}{2}$. Thus, the (proper) time needed for the interval $r_0 \leq r \leq r_1$ is given by

$$\Delta\tau \equiv \tau_1 - \tau_0 = -\frac{1}{c}\int_{r_0}^{r_1} \frac{dr}{\sqrt{\epsilon + \frac{\alpha}{r}}} = -\frac{1}{c}\int_{r_0}^{r_1} \frac{dr}{\sqrt{k^2 - \left(1 - \frac{\alpha}{r}\right)}}, \tag{3.1589}$$

where $r_0 = r(\tau_0) = 10^{10}$ km and $r_1 = r(\tau_1) = \alpha$, i.e., the Schwarzschild horizon. Now, use the initial condition to calculate ϵ:

$$-v_0 = c\frac{dr}{dx^0} = c\frac{\dot{r}}{\dot{x}^0} \quad \text{at } \tau = \tau_0, \tag{3.1590}$$

which implies that

$$v_0 = \frac{c}{k}\sqrt{\epsilon + \frac{\alpha}{r_0}}\left(1 - \frac{\alpha}{r_0}\right) \Rightarrow k^2 = \left(1 - \frac{\alpha}{r_0}\right)\left[1 - \frac{v_0^2}{c^2}\left(1 - \frac{\alpha}{r_0}\right)^{-2}\right]^{-1}. \tag{3.1591}$$

Inserting k^2 into the (proper) time interval $\Delta\tau$, we obtain

$$\Delta\tau = -\frac{1}{c}\int_{r_0}^{r_1} \frac{dr}{\sqrt{\left(1 - \frac{\alpha}{r_0}\right)\left[1 - \frac{v_0^2}{c^2}\left(1 - \frac{\alpha}{r_0}\right)^{-2}\right]^{-1} - \left(1 - \frac{\alpha}{r}\right)}}, \tag{3.1592}$$

which can be computed analytically using the hint, but the result is too lengthy for it to be useful to display. Computing the integral numerically, the result is $\Delta\tau \simeq 5.85615 \cdot 10^8$ s, which corresponds to about 18.6 years. Thus, the proper time needed for the observer to reach the Schwarzschild horizon is about 18.6 years.

However, the quantities α, r_0, and v_0 are given, so the constant $\epsilon = k^2 - 1$ can be uniquely computed from them. Assuming $r_0 \gg \alpha$, qualitative estimates of the parameters are as follows

$$v_0 \simeq \frac{c}{k}\sqrt{\epsilon}, \quad \epsilon = k^2 - 1 \simeq \frac{v_0^2}{c^2}k^2, \quad k \simeq \frac{1}{\sqrt{1 - v_0^2/c^2}} \; (\sim 1),$$

$$\epsilon = k^2 - 1 \simeq \frac{v_0^2}{c^2 - v_0^2} \left(\sim \frac{v_0^2}{c^2}\right), \quad \text{and} \quad \beta \simeq \frac{2GM}{v_0^2}\left(1 - \frac{v_0^2}{c^2}\right)\left(\sim \frac{2GM}{v_0^2}\right), \tag{3.1593}$$

where limits within parentheses also hold if it is assumed that $v_0 \ll c$. Therefore, using the estimates of the parameters, we can approximate the integral for the (proper) time interval $\Delta\tau$ as

$$\Delta\tau \simeq -\frac{1}{v_0}\int_{r_0}^{r_1} \frac{dr}{\sqrt{1 + \frac{2GM}{v_0^2 r}}} \simeq \frac{r_0}{v_0}\left(\sqrt{1 + \frac{2GM}{v_0^2 r_0}} - \frac{2GM}{v_0^2 r_0}\operatorname{arsinh}\sqrt{\frac{v_0^2 r_0}{2GM}}\right)$$

$$\simeq 4.32267 \cdot 10^8 \text{ s} \sim 13.7 \text{ years}, \tag{3.1594}$$

which means that the approximation underestimates (by about 25 %) the true value of the proper time, but it gives the correct order of magnitude.

2.93

Due to the Schwarzschild metric components not depending on t, we can conclude that ∂_t is a Killing vector field. It follows that a freely falling particle with a world-line that is a geodesic has a constant of motion

$$k = g(\partial_t, \dot{\gamma}) = g_{00}\dot{x}^0. \tag{3.1595}$$

Since m and c are constants, it also follows that $p_0 = mck$ is a constant. Furthermore, the normalization of the 4-velocity implies that

$$\mathcal{L} = g_{\mu\nu}\dot{x}^\mu\dot{x}^\nu = g_{00}c^2\dot{t}^2 - g_{rr}\dot{r}^2 = 1, \tag{3.1596}$$

for a particle falling radially. With $E = p_0/c$, this implies that

$$1 = g_{00}(\dot{x}^0)^2 + g_{rr}\dot{r}^2 = \frac{1}{g_{00}}\left[\frac{E^2}{(mc^2)^2} - \dot{r}^2\right]$$

$$\implies \quad \dot{r} = -\sqrt{\left(\frac{E}{mc^2}\right)^2 - \left(1 - \frac{r_*}{r}\right)}, \tag{3.1597}$$

with the negative sign being due to r decreasing with proper time. Integrating from $r = 3r_*/2$ to $r = r_*$, we find that

$$\Delta s = -\int_{3r_*/r}^{r_*} \frac{dr}{\sqrt{\left(\frac{E}{mc^2}\right)^2 - \left(1 - \frac{r_*}{r}\right)}} = \int_{r_*}^{3r_*/r} \frac{dr}{\sqrt{\left(\frac{E}{mc^2}\right)^2 - \left(1 - \frac{r_*}{r}\right)}}. \tag{3.1598}$$

2.94

The worldline of the observer is given by

$$t = \beta\tau, \quad r = r_0, \quad \theta = \frac{\pi}{2}, \quad \varphi = \omega\tau, \tag{3.1599}$$

where τ is the proper time. To find the constant β, we note that normalization of the 4-velocity yields

$$g_{tt}\beta^2 + g_{\varphi\varphi}^2\omega^2 = 1, \tag{3.1600}$$

which implies

$$\beta = \sqrt{\frac{1 + r_0^2\omega^2}{1 - r_*/r_0}}, \tag{3.1601}$$

where $r_* = 2GM$ is the Schwarzschild radius. With the 4-velocity being $V = \beta\partial_t + \omega\partial_\varphi$, we generally find that

$$\begin{aligned} A = \nabla_V V &= \beta^2\nabla_t\partial_t + \beta\omega(\nabla_t\partial_\varphi + \nabla_\varphi\partial_t) + \omega^2\nabla_\varphi\partial_\varphi \\ &= (\beta^2\Gamma^a_{tt} + 2\beta\omega\Gamma^a_{t\varphi} + \omega^2\Gamma^a_{\varphi\varphi})\partial_a. \end{aligned} \tag{3.1602}$$

The relevant nonzero Christoffel symbols of the Schwarzschild metric are

$$\Gamma^r_{tt} = \frac{r_*(r - r_*)}{2r^3}, \quad \Gamma^r_{\varphi\varphi} = -(r - r_*)\sin^2(\theta), \quad \Gamma^\theta_{\varphi\varphi} = -\sin(\theta)\cos(\theta). \tag{3.1603}$$

However, $\theta = \pi/2$ leads to $\sin(\theta) = 1$ and $\cos(\theta) = 0$, resulting in

$$A = \left[\frac{r_*}{2}(r_0^{-2} + 3\omega^2) - r_0\omega^2\right]\partial_r. \tag{3.1604}$$

For the proper acceleration α, we use that $\alpha^2 = -g(A, A)$, and therefore,

$$\alpha^2 = \frac{r_0}{r_0 - r_*}\left[\frac{r_*}{2}(r_0^{-2} + 3\omega^2) - r_0\omega^2\right]^2, \tag{3.1605}$$

and consequently,

$$\alpha = \sqrt{\frac{r_0}{r_0 - r_*}}\left|\frac{r_*}{2}(r_0^{-2} + 3\omega^2) - r_0\omega^2\right|. \tag{3.1606}$$

Note that, for $r_* \ll r_0$, we find that $\alpha = 0$ when

$$\frac{GM}{r_0^2} = r_0\omega^2, \tag{3.1607}$$

i.e., when the classical gravitational acceleration is the same as the centripetal acceleration required to keep the orbit at $r = r_0$.

2.95

Using the Schwarzschild metric, we obtain the Lagrangian

$$\mathcal{L} = \left(1 - \frac{r_*}{r}\right)\dot{t}^2 - \left(1 - \frac{r_*}{r}\right)^{-1}\dot{r}^2, \tag{3.1608}$$

where $r_* = 2GM$ is the Schwarzschild radius and the dot means differentiation with respect to proper time τ. There exists a conservation law $\mathcal{L} = 1$ (remember that we set $c = 1$) and we have the Euler–Lagrange equation

$$\frac{d}{d\tau}\frac{\partial \mathcal{L}}{\partial \dot{t}} - \frac{\partial \mathcal{L}}{\partial t} = \frac{d}{d\tau}\left[2\left(1 - \frac{r_*}{r}\right)\dot{t}\right] = 0 \quad \Rightarrow \quad \dot{t} = C\left(1 - \frac{r_*}{r}\right)^{-1}, \tag{3.1609}$$

for some integration constant C. Combining the conservation law and the Euler–Lagrange equation yields

$$\begin{aligned}\mathcal{L} - 1 &= \left(1 - \frac{r_*}{r}\right)C^2\left(1 - \frac{r_*}{r}\right)^{-2} - \left(1 - \frac{r_*}{r}\right)^{-1}\dot{r}^2 - 1 \\ &= \left(1 - \frac{r_*}{r}\right)^{-1}(C^2 - \dot{r}^2) - 1 = 0,\end{aligned} \tag{3.1610}$$

i.e.,

$$\dot{r}^2 = C^2 - \left(1 - \frac{r_*}{r}\right). \tag{3.1611}$$

Separating this equation, we find that

$$d\tau = -\frac{dr}{\sqrt{C^2 - \left(1 - \frac{r_*}{r}\right)}}, \tag{3.1612}$$

where we fix the sign of the square root so that $\dot{r} < 0$, since the particle is falling inward. Thus, we obtain the proper time

$$\Delta\tau = -\int_{r_0}^{r_1} \frac{dr}{\sqrt{C^2 - \left(1 - \frac{r_*}{r}\right)}}, \tag{3.1613}$$

where r_0 and r_1 are the initial and final positions of the particle, respectively, and the integration constant C is determined by the initial velocity of the particle.

Furthermore, to compute the coordinate time, we have (again fixing $\dot{r} < 0$)

$$\frac{dr}{dt} = \frac{\dot{r}}{\dot{t}} = -\sqrt{C^2 - \left(1 - \frac{r_*}{r}\right)} \cdot \frac{1}{C}\left(1 - \frac{r_*}{r}\right) = -\left(1 - \frac{r_*}{r}\right)\sqrt{1 - \frac{1}{C^2}\left(1 - \frac{r_*}{r}\right)}. \tag{3.1614}$$

Thus, integrating, we obtain the coordinate time

$$\Delta t = -\int_{r_0}^{r_1} \frac{dr}{\left(1 - \frac{r_*}{r}\right)\sqrt{1 - \frac{1}{C^2}\left(1 - \frac{r_*}{r}\right)}}. \tag{3.1615}$$

Indeed, the integrand of the integral giving $\Delta\tau$ remains finite when $r \to r_*$, actually the limit is $-1/C$, and thus, $\Delta\tau$ is finite as $r_1 \to r_*$, which means that it only takes a finite proper time to reach the even horizon at r_*. However, the integrand of the integral giving Δt diverges as $-(\frac{r}{r_*} - 1)^{-1}$ when $r \to r_*$, and thus, Δt diverges as

$$-\int_{r_0}^{r_1} \frac{dr}{\frac{r}{r_*} - 1} = r_* \ln \frac{r_0/r_* - 1}{r_1/r_* - 1}, \tag{3.1616}$$

when $r_1 \to r_*$, which means that it takes an infinite amount of coordinate time to reach the event horizon at r_*.

In order to reach the same conclusions as above, we can choose $C = 1$ and find explicitly that

$$\Delta\tau = -\int_{r_0}^{r_1} \frac{dr}{\sqrt{1 - (1 - \frac{r_*}{r})}} = -\int_{r_0}^{r_1} \sqrt{\frac{r}{r_*}}\, dr = \frac{2}{3\sqrt{r_*}} \left(r_0^{3/2} - r_1^{3/2} \right)$$
$$\to \frac{2}{3} \left(r_0 \sqrt{\frac{r_0}{r_*}} - r_* \right) < \infty \quad \text{when } r_1 \to r_*, \tag{3.1617}$$

$$\Delta t = -\int_{r_0}^{r_1} \frac{dr}{(1 - \frac{r_*}{r}) \sqrt{1 - (1 - \frac{r_*}{r})}} = -\int_{r_0}^{r_1} \frac{1}{1 - \frac{r_*}{r}} \sqrt{\frac{r}{r_*}}\, dr$$
$$= \frac{2}{3} \left[(r_0 + 3r_*) \sqrt{\frac{r_0}{r_*}} - (r_1 + 3r_*) \sqrt{\frac{r_1}{r_*}} - 3r_* \left(\operatorname{artanh} \sqrt{\frac{r_0}{r_*}} - \operatorname{artanh} \sqrt{\frac{r_1}{r_*}} \right) \right]$$
$$\to \infty \quad \text{when } r_1 \to r_*, \tag{3.1618}$$

since $\lim_{x \to 1} \operatorname{artanh} x \to \infty$. The choice of $C = 1$ corresponds to $\dot{t} = (1 - \frac{r_*}{r})^{-1} \to 1$ as $r \to \infty$, i.e., the kinetic energy of the particle is chosen exactly such that it would be at rest at $r \to \infty$.

2.96

Initially, since the observer is moving tangentially, $\dot{r} = 0$. Let us consider the constants of motion for the observer due to ∂_t and ∂_φ being Killing vector fields and the observer following a timelike geodesic. Thus, we define

$$E = \frac{1}{2} g(\partial_t, \dot{\gamma})^2, \quad L = -g(\partial_\varphi, \dot{\gamma}), \tag{3.1619}$$

where $\dot{\gamma}$ is the observer's 4-velocity. We use a coordinate system such that $\theta = \pi/2$ and the initial tangent at $r = r_0$ is therefore

$$\dot{\gamma}_0 = \alpha \partial_t + \beta \partial_\varphi. \tag{3.1620}$$

The numbers α and β are determined from the normalization of the 4-velocity and that the velocity relative to the stationary frame is v_0 given as

$$g(\dot{\gamma}_0, \dot{\gamma}_0) = \alpha^2 \left(1 - \frac{r_*}{r_0}\right) - \beta^2 r_0^2 = 1, \quad \frac{1}{\sqrt{1 - v_0^2}} = g(V, \dot{\gamma}) = \alpha A \left(1 - \frac{r_*}{r_0}\right), \tag{3.1621}$$

where $V = A\partial_t$ is the 4-velocity of a local stationary observer at $r = r_0$. From the normalization of V follows that

$$g(V, V) = A^2 \left(1 - \frac{r_*}{r_0}\right) = 1 \implies A = \sqrt{\frac{r_0}{r_0 - r_*}}. \tag{3.1622}$$

Solving for α and β leads to

$$\alpha = \sqrt{\frac{r_0}{(1 - v_0^2)(r_0 - r_*)}}, \quad \beta = \frac{v_0}{r_0\sqrt{1 - v_0^2}}. \tag{3.1623}$$

The constant of motion L is now expressed as

$$L = -g(\partial_\varphi, \dot{\gamma}) = r_0^2 \beta = \frac{v_0 r_0}{\sqrt{1 - v_0^2}}. \tag{3.1624}$$

Examining the effective potential

$$V(r) = \frac{1}{2}\left(1 + \frac{L^2}{r^2}\right)\left(1 - \frac{r_*}{r}\right), \tag{3.1625}$$

there are two possible situations when the observer will not fall into the black hole

1. When $dV/dr \leq 0$, the observer will initially start moving outward. The position $r = r_0$ will then always be a turning point meaning that $r \geq r_0$ for the entire solution. In the case of $dV/dr = 0$, the observer will move in a circular orbit.
2. When $dV/dr > 0$, the observer will start moving toward the black hole. The requirement not to fall into the black hole is then that there is a turning point at a smaller value of r, i.e., that $V(r) = V(r_0)$ for some $r < r_0$. There will then necessarily be another turning point, which is not accessible, at an even smaller r. The border of this possibility occurs when both of the turning points coincide, i.e., when $V(x) - V(x_0)$ has a double root different from x_0.

To see when the first case to applies, we compute dV/dx with $x = 1/r$. As $dx/dr < 0$, $dV/dr \leq 0$ is equivalent with $dV/dx = (dV/dr)(dr/dx) \geq 0$. We find that

$$\left.\frac{dV}{dx}\right|_{x=x_0} = \frac{v_0^2 r_0}{1 - v_0^2}\left(1 - \frac{r_*}{r_0}\right) - \frac{r_*}{2(1 - v_0^2)} \geq 0 \implies v_0^2 \geq \frac{r_*}{2(r_0 - r_*)}. \tag{3.1626}$$

This puts a lower limit on the velocity v_0 for which case 1 applies.

Case 2 is a bit more complicated. In order for three turning points to exist, we must have three real solutions to $V(x) - V(x_0) = 0$, where the left-hand side is a third degree polynomial. Luckily, we know that $x = x_0$ is a root of the polynomial and can therefore be factored out and we are left with having to have two real roots for a second degree polynomial. After some algebra, we find that the requirement on v_0 for all roots to be real is

$$v_0 \geq \frac{2r_*^2}{r_0 + R}. \tag{3.1627}$$

However, we also need to ensure that $r = r_0$ is the turning point at the largest r. This is the case whenever $dV/dx < 0$ and the midpoint of the other solutions is smaller than r_0. The midpoint $1/\bar{r}$ is found to be located at

$$\frac{1}{\bar{r}} = \frac{1}{2}\left(\frac{1}{r_*} - \frac{1}{r_0}\right). \tag{3.1628}$$

Requiring that $r_0 > \bar{r}$ then leads to

$$r_0 > 3r_*. \tag{3.1629}$$

Thus, in the region $r_0 \leq 3r_*$, case 1 always applies and for $r_0 > 3r_*$, we find that the requirement not to fall into the black hole is given by

$$v_0^2 \geq \min\left(\frac{4r_*^2}{(r_0 + r_*)^2}, \frac{r_*}{2(r_0 - r_*)}\right) = \frac{4r_*^2}{(r_0 + r_*)^2}. \tag{3.1630}$$

(The expressions are equal when $r_0 = 3r_*$ and otherwise the first expression is always smaller.) The solution to the given problem is therefore that the minimal velocity v_0 is given by

$$v_{0,min} = \begin{cases} \sqrt{\frac{r_*}{2(r_0 - r_*)}}, & r_0 \leq 3r_*, \\ \frac{2r_*}{r_0 + r_*}, & r_0 \geq 3r_*. \end{cases} \tag{3.1631}$$

b) For $v_0 = 0$, we have the effective potential

$$V(r) = \frac{1}{2}\left(1 - \frac{r_*}{r}\right), \tag{3.1632}$$

since $L = 0$. In addition, the initial 4-velocity is equal to that of the stationary observer, which was found in a), i.e.,

$$\dot{\gamma}_0 = \sqrt{\frac{r_0}{r_0 - r_*}}\partial_t. \tag{3.1633}$$

Evaluated at $r = r_0$, the constant of motion is therefore

$$E = \frac{1}{2}\left(1 - \frac{r_*}{r_0}\right). \tag{3.1634}$$

The equation of motion for the r-coordinate is then

$$\dot{r}^2 = 2E - 2V(r) = \frac{r_*}{r} - \frac{r_*}{r_0} \quad \Longrightarrow \quad \frac{-dr}{\sqrt{\frac{r_*}{r} - \frac{r_*}{r_0}}} = d\tau, \tag{3.1635}$$

where we have used that dr is negative to omit the positive root. The proper time to reach the singularity is given by integrating this from r_0 to 0 and we find

$$\tau = -\int_{r_0}^{0} \frac{dr}{\sqrt{\frac{r_*}{r} - \frac{r_*}{r_0}}} = \frac{1}{\sqrt{r_*}} \int_{x_0}^{\infty} \frac{dx}{x^2\sqrt{x - x_0}} = \frac{\pi r_0}{2}\sqrt{\frac{r_0}{r_*}}. \tag{3.1636}$$

Note that the integral is applicable even across the Schwarzschild event horizon as its form removes the coordinate singularity that appears there. The integral itself therefore does not suffer from the coordinate singularity and behaves nicely all the way to the singularity at $r = 0$.

2.97

When the engines are turned off, the satellite will follow a geodesic. Assuming $\theta = \pi/2 = \text{const.}$, the Lagrangian is given by

$$\mathcal{L} = g_{\mu\nu}\dot{x}^\mu \dot{x}^\nu = \left(1 - \frac{r_*}{r}\right)\dot{t}^2 - \left(1 - \frac{r_*}{r}\right)^{-1}\dot{r}^2 - r^2\dot{\phi}^2. \tag{3.1637}$$

No t and ϕ dependencies give two constants of motions, coming from the Euler–Lagrange equations, i.e.,

$$\kappa = \left(1 - \frac{r_*}{r}\right)\dot{t}, \quad \ell = r^2\dot{\phi} = R^2 B. \tag{3.1638}$$

We also get a third relation from setting $\mathcal{L} = 1$, namely

$$1 = \left(1 - \frac{r_*}{r}\right)\dot{t}^2 - \left(1 - \frac{r_*}{r}\right)^{-1}\dot{r}^2 - r^2\dot{\phi}^2. \tag{3.1639}$$

Using the first and second equations, we can rewrite the third equation in terms of just radial dependence (with $\dot{r} = 0$):

$$1 = \left(1 - \frac{r_*}{r}\right)^{-1}\kappa^2 - r^{-2}\ell^2 = \left(1 - \frac{r_*}{r}\right)^{-1}\kappa^2 - \frac{R^4 B^2}{r^2}$$

$$\Rightarrow \quad \kappa^2 = \left(1 - \frac{r_*}{r}\right)\left(1 + \frac{R^4 B^2}{r^2}\right). \tag{3.1640}$$

Normally, these three equations are sufficient, and the third equation is equivalent to the equation of motion that one obtains from varying the Lagrangian with respect to r, but it happens that there is an exception to this case and that is precisely when

$\dot{r} = 0$. Varying the Lagrangian with respect to r and neglecting all time derivatives of r gives the equation

$$\frac{r_*}{r^2}\left(1-\frac{r_*}{r}\right)^{-2}\kappa^2 - 2r\left(\frac{R^2B}{r^2}\right)^2 = 0 \quad \Rightarrow \quad \kappa^2 = \frac{2r^3}{r_*}\left(1-\frac{r_*}{r}\right)^2\left(\frac{R^2B}{r^2}\right)^2. \tag{3.1641}$$

Now, at $r = R$, we find B according to

$$\left(1-\frac{r_*}{R}\right)(1+R^2B^2) = \frac{2R^3B^2}{r_*}\left(1-\frac{r_*}{R}\right)^2 \quad \Rightarrow \quad B^2 = R^{-2}\left(\frac{2R}{r_*}-3\right)^{-1}. \tag{3.1642}$$

We make an interesting observation. We have a minimal radius for which we can have a geodesic circular orbit around a star or a black hole, i.e., $R = 3r_*/2$. This orbit will be unstable. Stable circular orbits only exist when the radius R is larger than $3r_*/2$, i.e., $R > 3r_*/2$. For the case of the Earth, $R > R_0 \gg 3r_*/2$.

2.98

The satellite's orbit is at constant radial distance r from the Earth, which means that $dr = 0$.

a) Now, the proper time for the satellite is given by

$$d\tau^2 = ds^2. \tag{3.1643}$$

Using the Schwarzschild metric with $dr = 0$ and $d\theta = 0$ (since $\theta = \pi/2$) for the satellite's orbit around the Earth and that $v = r d\phi/dt$, we find that

$$\begin{aligned} d\tau^2 &= \left(1-\frac{r_*}{r}\right)dt^2 - r^2 d\phi^2 = \left(1-\frac{r_*}{r}\right)dt^2 - \left(r\frac{d\phi}{dt}\right)^2 dt^2 \\ &= \left[\left(1-\frac{r_*}{r}\right) - v^2\right]dt^2, \end{aligned} \tag{3.1644}$$

which implies that

$$d\tau = \pm\sqrt{\left(1-\frac{r_*}{r}\right) - v^2}\, dt, \tag{3.1645}$$

where the square root is independent of time t. Thus, using $dt/d\tau > 0$ and integrating from $t = 0$ to $t = T = 2\pi r/v$, we obtain the proper time τ for the satellite to complete one orbit around the Earth as

$$\tau = \int_0^T \sqrt{\left(1-\frac{r_*}{r}\right) - v^2}\, dt = T\sqrt{\left(1-\frac{r_*}{r}\right) - v^2} = \frac{2\pi r}{v}\sqrt{\left(1-\frac{2GM}{r}\right) - v^2}. \tag{3.1646}$$

b) The gravitational potential at the satellite is given by

$$\Phi_s = -\frac{1}{2}\frac{r_*}{r} = -\frac{GM}{r}. \tag{3.1647}$$

Now, we calculate the ratio between the coordinate time $t = T$ and the proper time for the satellite τ. Then, we series expand this ratio for small v and r_*/r. Therefore, using the result in a) for T/τ, we find that

$$\frac{T}{\tau} = \frac{1}{\sqrt{\left(1 - \frac{r_*}{r}\right) - v^2}} \simeq \frac{1}{1 - \frac{r_*}{2r} - \frac{v^2}{2}} \simeq 1 - \Phi_s + \frac{v^2}{2}. \tag{3.1648}$$

Thus, we obtain

$$\frac{T}{\tau} - 1 \simeq \frac{v^2}{2} - \Phi_s, \tag{3.1649}$$

which is what we wanted to show.

2.99

a) The worldline of the satellite will be given by

$$t = \alpha s, \quad r = r_0, \quad \varphi = \beta s, \tag{3.1650}$$

as the 4-velocity $U = \alpha \partial_t + \beta \partial_\varphi$, with α and β constant. It follows that the 4-acceleration is

$$A = \nabla_U U = \left(\alpha^2 \Gamma^a_{tt} + 2\alpha\beta\Gamma^a_{t\varphi} + \beta^2 \Gamma^a_{\varphi\varphi}\right)\partial_a = 0, \tag{3.1651}$$

since the satellite is in free fall. Given the Christoffel symbols of the Schwarzschild metric, the only nontrivial component of this equation is the $a = r$ component

$$\alpha^2 \Gamma^r_{tt} + 2\alpha\beta\Gamma^r_{t\varphi} + \beta^2 \Gamma^r_{\varphi\varphi} = \alpha^2 \frac{r_*(r_0 - r_*)}{2r_0^3} - \beta^2 (r_0 - r_*) = 0. \tag{3.1652}$$

This therefore leads to

$$\beta = \alpha \sqrt{\frac{r_*}{2r_0^3}}, \tag{3.1653}$$

where we have chosen coordinates such that φ increases with proper time. Also requiring that the 4-velocity is normalized, i.e.,

$$g(U,U) = \left(1 - \frac{r_*}{r_0}\right)\alpha^2 - r_0^2 \beta^2 = 1, \tag{3.1654}$$

now implies that

$$\alpha = \frac{1}{\sqrt{1 - \frac{3r_*}{2r_0}}}, \quad \beta = \frac{1}{r_0}\sqrt{\frac{r_*}{2r_0 - 3r_*}}. \tag{3.1655}$$

b) For the satellite to complete a full lap, φ needs to change by 2π. The proper time Δs it takes for the satellite to travel an angle $\Delta\varphi$ is given by

$$\frac{\Delta\varphi}{\Delta s} = \dot{\varphi} \quad \Longrightarrow \quad \Delta s = \frac{\Delta\varphi}{\dot{\varphi}} = \frac{\Delta\varphi}{\beta}. \tag{3.1656}$$

For $\Delta\varphi = 2\pi$, we therefore obtain

$$\Delta s = 2\pi r_0 \sqrt{\frac{2r_0 - 3r_*}{r_*}}. \tag{3.1657}$$

c) The stationary observer has 4-velocity $V = \alpha_0 \partial_t$. From $g(V,V) = 1$ follows that

$$\alpha_0 = \sqrt{\frac{r_0}{r_0 - r_*}}. \tag{3.1658}$$

This implies that the relative gamma factor between the satellite and the observer is given by

$$\gamma = g(U,V) = \left(1 - \frac{r_*}{r_0}\right)\alpha\alpha_0 = \sqrt{\frac{2r_0 - 2r_*}{2r_0 - 3r_*}} = \frac{1}{\sqrt{1 - v^2}}, \tag{3.1659}$$

where v is the relative speed. Solving for v results in

$$v = \sqrt{\frac{r_*}{2(r_0 - r_*)}}. \tag{3.1660}$$

2.100

a) The static observer (1) has constant spatial coordinates. It is therefore convenient to use a parametrization of its worldline in terms of its proper time s_1, which will be proportional to the coordinate time t

$$t_1 = \alpha s_1, \quad r_1 = r_0, \quad \varphi_1 = \varphi_0. \tag{3.1661}$$

The value of α can be determined through the requirement that the 4-velocity has norm one, i.e.,

$$1 = \left(1 - \frac{r_*}{r_1}\right)\dot{t}^2 = \left(1 - \frac{r_*}{r_0}\right)\alpha^2 \quad \Longrightarrow \quad \alpha = \frac{1}{\sqrt{1 - \frac{r_*}{r_1}}}. \tag{3.1662}$$

The 4-acceleration is defined as $A = \nabla_V V$. For the worldline of (1), we have $V = \alpha\partial_t$, and therefore,

$$A_1 = \alpha^2 \nabla_t \partial_t = \alpha^2 \Gamma^\mu_{tt} \partial_\mu = \alpha^2 \frac{r_*(r_0 - r_*)}{2r_0^3} \partial_r = \frac{r_*}{2r_0^2} \partial_r, \tag{3.1663}$$

where we have used the expression for Γ^r_{tt}, which is the only nonzero Christoffel symbol on the form Γ^μ_{tt}. The proper acceleration a is given by $-a^2 = g(A, A) = A^2$, and therefore,

$$a_1^2 = -A_1^2 = -\left(\frac{r_*}{2r_0^2}\right)^2 g_{rr} = \frac{r_*^2}{4r_0^4(1 - r_*/r_0)} \implies a_1 = \frac{r_*}{2r_0^2\sqrt{1 - r_*/r_0}}. \tag{3.1664}$$

We can note that, for $r_0 \gg r_*$ and with $r_* = 2MG$, this becomes

$$a_1 \simeq \frac{2MG}{2r_0^2} = \frac{MG}{r_0^2}, \tag{3.1665}$$

which is the expected result from Newtonian gravity.

For the circularly orbiting observer (2), the worldline can be parametrized as

$$t_2 = \beta s_2, \quad r_2 = r_0, \quad \varphi_2 = \gamma s_2, \tag{3.1666}$$

where β and γ are constants and s_2 is the proper time of the observer. By definition, this observer has zero 4-acceleration $a = 0$, which can be used to fix the constants. The 4-velocity takes the form $V_2 = \beta \partial_t + \gamma \partial_\varphi$ and the 4-acceleration is therefore given by

$$\begin{aligned} A_2 = \nabla_{V_2} V_2 &= \beta^2 \nabla_t \partial_t + \beta\gamma(\nabla_t \partial_\varphi + \nabla_\varphi \partial_t) + \gamma^2 \nabla_\varphi \partial_\varphi \\ &= \left[\beta^2 \Gamma^\mu_{tt} + 2\beta\gamma \Gamma^\mu_{t\varphi} + \gamma^2 \Gamma^\mu_{\varphi\varphi}\right] \partial_\mu = 0. \end{aligned} \tag{3.1667}$$

The entire 4-acceleration being zero requires that each component is zero separately. However, all of the components except the r-component are trivially vanishing and we are left with

$$\beta^2 \Gamma^r_{tt} + 2\beta\gamma \Gamma^r_{t\varphi} + \gamma^2 \Gamma^r_{\varphi\varphi} = \beta^2 \frac{r_*(r - r_*)}{2r_0^3} - \gamma^2 (r - r_*) = 0, \tag{3.1668}$$

leading to

$$\gamma = \pm\beta \sqrt{\frac{r_*}{2r_0^3}}, \tag{3.1669}$$

where the $\pm$ just defines the direction of the orbit and we will pick the positive sign in the future. The normalization is again set by the requirement that $V^2 = 1$, leading to

$$1 = \left(1 - \frac{r_*}{r_0}\right)\beta^2 - r_0^2\gamma^2 = \beta^2\left(1 - \frac{r_*}{r_0} - r_0^2 \frac{r_*}{2r_0^3}\right) = \beta^2\left(1 - \frac{3r_*}{2r_0}\right), \tag{3.1670}$$

and therefore, we find that

$$\beta = \frac{1}{\sqrt{1 - 3r_*/2r_0}}. \tag{3.1671}$$

The parametrizations of the worldlines are therefore

$$t_1 = \frac{s_1}{\sqrt{1 - r_*/r_0}}, \quad r_1 = r_0, \quad \varphi_1 = \varphi_0, \tag{3.1672}$$

and

$$t_2 = \frac{s_2}{\sqrt{1 - 3r_*/2r_0}}, \quad r_2 = r_0, \quad \varphi_2 = \frac{s_2/r_0}{\sqrt{2r_0/r_* - 3}}, \tag{3.1673}$$

respectively, while the proper accelerations are given by

$$a_1 = \frac{r_*}{2r_0^2\sqrt{1 - r_*/r_0}} \quad \text{and} \quad a_2 = 0. \tag{3.1674}$$

b) The differences in the proper times will both be proportional to the difference in the coordinate time t according to

$$\Delta s_1 = \frac{\Delta t}{\alpha} \quad \text{and} \quad \Delta s_2 = \frac{\Delta t}{\beta}. \tag{3.1675}$$

Therefore, we find that

$$\frac{\Delta s_1}{\Delta s_2} = \frac{\beta}{\alpha} = \sqrt{\frac{r_0 - r_*}{r_0 - 3r_*/2}} \equiv \rho. \tag{3.1676}$$

In particular, we can see that this expression has the correct limits as $\rho \to 1$ when $r_0 \to \infty$ and $\rho \to \infty$ as $r_0 \to 3r_*/2$. The latter result is expected as the circular geodesic at $r_0 = 3r_*/2$ is lightlike.

c) The relative gamma factor $\tilde{\gamma} = 1/\sqrt{1 - v^2}$, where v is the relative speed, is given by the inner product between the two 4-velocities, i.e.,

$$\tilde{\gamma} = g(V_1, V_2) = \left(1 - \frac{r_*}{r_0}\right)\alpha\beta = \sqrt{\frac{2(r_0 - r_*)}{2r_0 - 3r_*}} \equiv \sqrt{\frac{2(1 - x)}{2 - 3x}}, \tag{3.1677}$$

where $x = r_*/r_0$. Squaring this relation leads to

$$1 - v^2 = \frac{2 - 3x}{2(1 - x)} \quad \Longrightarrow \quad v = \sqrt{\frac{x}{2(1 - x)}} = \sqrt{\frac{r_*}{2(r_0 - r_*)}}. \tag{3.1678}$$

Note that we again recover the classical orbital velocity

$$v = \sqrt{\frac{MG}{r_0}}, \tag{3.1679}$$

in the limit where $r_0 \gg r_* = 2MG$.

2.101

a) The proper time τ can be found from

$$d\tau^2 = ds^2. \tag{3.1680}$$

Using the Kerr metric with $dr = 0$ and $d\theta = 0$ (since $r = R$ and $\theta = \pi/2$) for the given orbit, we get

$$\begin{aligned} d\tau^2 &= \left(1 - \frac{r_*}{R}\right) dt^2 + \frac{2ar_*}{R} dt d\phi - \left(R^2 + a^2 + \frac{a^2 r_*}{R}\right) d\phi^2 \\ &= \left[\left(1 - \frac{r_*}{R}\right) + \frac{2ar_*}{R}\frac{d\phi}{dt} - \left(R^2 + a^2 + \frac{a^2 r_*}{R}\right)\left(\frac{d\phi}{dt}\right)^2\right] dt^2 \\ &= \left[\left(1 - \frac{r_*}{R}\right) + \frac{2ar_*}{R^2} v - \left(1 + \frac{a^2}{R^2} + \frac{a^2 r_*}{R^3}\right) v^2\right] dt^2, \end{aligned} \tag{3.1681}$$

since $\rho^2 = R^2$ and $v = R d\phi/dt$. This implies that

$$d\tau = \pm\sqrt{\left(1 - \frac{r_*}{R}\right) + \frac{2ar_*}{R^2} v - \left(1 + \frac{a^2}{R^2} + \frac{a^2 r_*}{R^3}\right) v^2}\, dt, \tag{3.1682}$$

where the square root is independent of time t. Thus, using $dt/d\tau > 0$ and integrating from $t = t_0 = 0$ to $t = T = 2\pi R/v$, we obtain the proper time

$$\tau = T\sqrt{\left(1 - \frac{r_*}{R}\right) + \frac{2ar_*}{R^2} v - \left(1 + \frac{a^2}{R^2} + \frac{a^2 r_*}{R^3}\right) v^2}. \tag{3.1683}$$

b) Solving the result in a) for T/τ and series expanding in v and r_*/R up to orders v^2 and r_*/R (assuming $v^2 \sim r_*/R$), we obtain

$$\begin{aligned} \frac{T}{\tau} &= \frac{1}{\sqrt{\left(1 - \frac{r_*}{R}\right) + \frac{2ar_*}{R^2} v - \left(1 + \frac{a^2}{R^2} + \frac{a^2 r_*}{R^3}\right) v^2}} \\ &\simeq 1 + \frac{1}{2}\left(1 + \frac{a^2}{R^2}\right) v^2 + \frac{1}{2R} r_*. \end{aligned} \tag{3.1684}$$

2.102

a) If we assume a fixed time $t = t_0$, we observe that the metric is just a flat metric, i.e.,

$$ds^2 = -a(t_0)^2 e^{2Ct_0} (dx^2 + dy^2 + dz^2), \tag{3.1685}$$

where $x = \rho \sin\theta \cos\phi$, $y = \rho \sin\theta \sin\phi$, and $z = \rho \cos\theta$ are expressed in the spherical coordinates ρ, θ, and ϕ. Note that $a(t_0)e^{Ct_0}$ is a constant, since t_0 is fixed and C is a constant.

b) We choose to use the Cartesian coordinates x, y, and z. In order to find a geodesic in the spacetime, we use the Lagrangian

$$\mathcal{L} = \dot{t}^2 - a(t)^2\left(\dot{x}^2 + \dot{y}^2 + \dot{z}^2\right) = 1, \tag{3.1686}$$

where dot means differentiation with respect to proper time τ. Now, the metric does not depend on x, y, and z, so using the Euler–Lagrange equations $\frac{d}{d\tau}\left(\frac{\partial \mathcal{L}}{\partial \dot{x}^i}\right) - \frac{\partial \mathcal{L}}{\partial x^i} = 0$, we have three constants of motion, which are given as follows

$$\frac{d}{d\tau}(a(t)^2\dot{x}^i) = 0 \quad \Rightarrow \quad a(t)^2\dot{x}^i = k_i, \tag{3.1687}$$

where k_i, $i = 1,2,3$, are the constants of motion. Inserting the three constants of motion into the Lagrangian, we obtain

$$\dot{t}^2 - \frac{1}{a(t)^2}(k_1^2 + k_2^2 + k_3^2) = 1, \tag{3.1688}$$

where $a(t) = a_0 e^{Ct}$. Then, rewrite the initial conditions, we find

$$\begin{aligned}\left.\frac{dx}{d\tau}\right|_{\tau=0} &= \left.\left(\sin\theta\cos\phi\frac{d\rho}{d\tau} + \rho\cos\theta\cos\phi\frac{d\theta}{d\tau} - \rho\sin\theta\sin\phi\frac{d\phi}{d\tau}\right)\right|_{\tau=0}\\ &= \sin 0\cdot\cos 0\cdot\frac{1}{4} + 1\cdot\cos 0\cdot\cos 0\cdot\frac{1}{4} - 1\cdot\sin 0\cdot\sin 0\cdot\frac{1}{4} = \frac{1}{4},\end{aligned} \tag{3.1689}$$

$$\begin{aligned}\left.\frac{dy}{d\tau}\right|_{\tau=0} &= \left.\left(\sin\theta\sin\phi\frac{d\rho}{d\tau} + \rho\cos\theta\sin\phi\frac{d\theta}{d\tau} + \rho\sin\theta\cos\phi\frac{d\phi}{d\tau}\right)\right|_{\tau=0}\\ &= \sin 0\cdot\sin 0\cdot\frac{1}{4} + 1\cdot\cos 0\cdot\sin 0\cdot\frac{1}{4} + 1\cdot\sin 0\cdot\cos 0\cdot\frac{1}{4} = 0,\end{aligned} \tag{3.1690}$$

$$\left.\frac{dz}{d\tau}\right|_{\tau=0} = \left.\left(\cos\theta\frac{d\rho}{d\tau} - \rho\sin\theta\frac{d\theta}{d\tau}\right)\right|_{\tau=0} = \cos 0\cdot\frac{1}{4} - 1\cdot\sin 0\cdot\frac{1}{4} = \frac{1}{4}, \tag{3.1691}$$

which lead to

$$a(t)^2\dot{x}\big|_{\tau=0} = \frac{a_0^2}{4} = k_1, \quad a(t)^2\dot{y}\big|_{\tau=0} = 0 = k_2, \quad a(t)^2\dot{z}\big|_{\tau=0} = \frac{a_0^2}{4} = k_3, \tag{3.1692}$$

since $t|_{\tau=0} = 0$ and $a(0) = 0$, which yield

$$k_1 = k_3 = \frac{a_0^2}{4}, \quad k_2 = 0, \tag{3.1693}$$

and imply that $k_1^2 + k_2^2 + k_3^2 = a_0^4/8$. Thus, we obtain

$$\dot{t}^2 - \frac{1}{a_0^2 e^{2Ct}}\frac{a_0^4}{8} = 1 \quad \Rightarrow \quad \dot{t}^2 = 1 + \frac{a_0^2}{8}e^{-2Ct} \quad \Rightarrow \quad \frac{dt}{d\tau} = \pm\sqrt{1 + \frac{a_0^2}{8}e^{-2Ct}}, \tag{3.1694}$$

which can be separated to

$$d\tau = \pm \frac{dt}{\sqrt{1 + \frac{a_0^2}{8} e^{-2Ct}}}. \tag{3.1695}$$

Finally, using $\dot{t} = dt/d\tau > 0$, we calculate the proper time $\Delta\tau$ for the free-falling particle between the coordinate times $t = 0$ and $t = t_1$

$$\Delta\tau = \int_0^{t_1} \frac{dt}{\sqrt{1 + \frac{a_0^2}{8} e^{-2Ct}}} = \frac{1}{C} \left(\operatorname{artanh} \frac{1}{\sqrt{1 + \frac{a_0^2}{8} e^{-2Ct_1}}} - \operatorname{artanh} \frac{1}{\sqrt{1 + \frac{a_0^2}{8}}} \right). \tag{3.1696}$$

2.103

a) Let the initial and final times be $t = 0$ and $t = T_F$, respectively. Assuming that $v = a(t) R_0 \frac{d\phi}{dt} = \text{const.}$ and using

$$d\tau = \sqrt{1 - a(t)^2 R_0^2 \left(\frac{d\phi}{dt} \right)^2} \, dt = \sqrt{1 - v^2} \, dt, \tag{3.1697}$$

we obtain the proper time from $t = 0$ to $t = T_F$ as

$$\tau = \int_0^{T_F} \sqrt{1 - v^2} \, dt = \sqrt{1 - v^2} T_F. \tag{3.1698}$$

Now, we need to find T_F. From the fact that v is a constant and using $v = e^t R_0 \frac{d\phi}{dt}$, we can calculate how ϕ depends on t:

$$e^{-t} dt = \frac{R_0}{v} d\phi \quad \Rightarrow \quad -\left(e^{-T_F} - 1\right) = \frac{R_0}{v} 2\pi, \tag{3.1699}$$

which implies that

$$T_F = \ln \left(1 - \frac{2\pi R_0}{v} \right)^{-1}. \tag{3.1700}$$

b) No, T_F becomes infinite if $1 - 2\pi R_0/v = 0$.

2.104

The Robertson–Walker metric with zero curvature is

$$ds^2 = dt^2 - a(t)^2 (dx^2 + dy^2 + dz^2). \tag{3.1701}$$

The Lagrangian is given by

$$\mathcal{L} = \dot{t}^2 - a(t)^2 (\dot{x}^2 + \dot{y}^2 + \dot{z}^2) = 1. \tag{3.1702}$$

The Euler–Lagrange equation with respect to x leads to

$$\frac{d}{d\tau}\left(a(t)^2\dot{x}\right) = 0 \quad \Rightarrow \quad a(t)^2\dot{x} = c_1, \tag{3.1703}$$

and together with the initial conditions $\dot{x}(\tau = 0) = A$ and $t(\tau = 0) = 0$, we obtain $c_1 = a(0)^2 A$, and thus, in total, we find that

$$\dot{x} = \frac{a(0)^2 A}{a(t)^2}. \tag{3.1704}$$

Similarly, using the same procedure, we can conclude that $\dot{y} = \dot{z} = 0$. In principle, we can determine t as a function of τ from the Lagrangian

$$\mathcal{L} = \dot{t}^2 - \frac{a(0)^4 A^2}{a(t)^2} = 1 \quad \Rightarrow \quad \dot{t}^2 = 1 + \frac{a(0)^4 A^2}{a(t)^2} \quad \Rightarrow \quad \dot{t} = \sqrt{1 + \frac{a(0)^4 A^2}{a(t)^2}}. \tag{3.1705}$$

The equation $\dot{x} = a(0)^2 A / a(t)^2$ can now be rearranged as

$$\frac{a(t)^2}{a(0)^2 A} dx = d\tau. \tag{3.1706}$$

Integrating both sides, we obtain

$$\tau = \int_{X_0}^{X_D} \frac{a(t)^2}{a(0)^2 A} dx. \tag{3.1707}$$

2.105

a) The Schwarzschild metric is given by

$$ds^2 = \left(1 - \frac{r_*}{r}\right) dt^2 - \left(1 - \frac{r_*}{r}\right)^{-1} dr^2 - r^2 d\Omega^2. \tag{3.1708}$$

See Figure 3.15 for the setup of observers A and B. We will assume motion in the plane $\theta = \frac{\pi}{2}$ and so $d\theta = 0$ and $d\Omega^2 = d\varphi^2$. From the symmetries of the spacetime (Killing vector fields ∂_t and ∂_φ), we have the conserved quantities

$$E = g(\partial_t, \dot{\gamma}) = \left(1 - \frac{r_*}{r}\right)\dot{t}, \quad L = -g(\partial_\varphi, \dot{\gamma}) = r^2 \dot{\varphi}. \tag{3.1709}$$

Figure 3.15 Setup of the black hole ("BH") and observers A and B.

Observer A is not in geodesic motion but is stationary. The worldline of A can therefore be parametrized by

$$t = \alpha_0 s, \quad r = r_0, \tag{3.1710}$$

where s is the proper time. Normalizing this such that $V_A = \dot{t}\partial_t + \dot{r}\partial_r = \alpha_0 \partial_t$ satisfies $V_A^2 = 1$, we find that

$$\alpha^2 g_{tt} = \alpha_0^2 \left(1 - \frac{r_*}{r_0}\right) = 1 \quad \Rightarrow \quad \alpha_0 = \frac{1}{\sqrt{1 - r_*/r_0}}. \tag{3.1711}$$

For observer B, the 4-velocity is purely radial and at $r = r_1$ is given by $V_{B0} = \alpha_1 \partial_t$. Normalization of V_{B0} leads to

$$\alpha_1^2 g(\partial_t, \partial_t) = \alpha_1^2 \left(1 - \frac{r_*}{r_1}\right) = 1 \quad \Rightarrow \quad \alpha_1 = \frac{1}{\sqrt{1 - r_*/r_1}}. \tag{3.1712}$$

At radius r, the 4-velocity of B is given by

$$V_B = \alpha_1(r)\partial_t - \beta(r)\partial_r = \dot{t}\partial_t + \dot{r}\partial_r. \tag{3.1713}$$

Since B is freely falling and ∂_t is a Killing vector field, it follows that

$$g(V_B, \partial_t) = g_{tt}\alpha_1(r) = \left(1 - \frac{r_*}{r}\right)\alpha_1(r) = E, \tag{3.1714}$$

is a constant of motion. From $r = r_1$, we find

$$E = \frac{1 - \frac{r_*}{r_1}}{\sqrt{1 - r_*/r_1}} = \sqrt{1 - r_*/r_1} \quad \Rightarrow \quad \alpha_1(r) = \frac{\sqrt{1 - r_*/r_1}}{1 - \frac{r_*}{r}}. \tag{3.1715}$$

The relative γ factor between A and B as they pass each other is given by

$$\begin{aligned} \gamma = g(V_A, V_B) = \alpha_0 \alpha_1(r_0) g_{tt} &= \frac{1}{\sqrt{1 - r_*/r_0}} \frac{\sqrt{1 - r_*/r_1}}{1 - \frac{r_*}{r_0}} \left(1 - \frac{r_*}{r_0}\right) \\ &= \sqrt{\frac{1 - r_*/r_1}{1 - r_*/r_0}} \equiv \frac{1}{\sqrt{1 - v^2}}, \end{aligned} \tag{3.1716}$$

where v is the relative velocity between A and B by definition. Solving for v, we obtain

$$\begin{aligned} v = \sqrt{1 - \frac{1}{\gamma^2}} &= \sqrt{1 - \frac{1 - r_*/r_0}{1 - r_*/r_1}} = \frac{1}{\sqrt{1 - r_*/r_1}} \sqrt{1 - \frac{r_*}{r_1} - 1 + \frac{r_*}{r_0}} \\ &= \sqrt{\frac{r_*}{1 - r_*/r_1}} \sqrt{\frac{1}{r_0} - \frac{1}{r_1}}. \end{aligned} \tag{3.1717}$$

Note that, while $\beta(r)$ can be computed from the normalization of V_B, it is not necessary to find the relative velocity.

b) From the normalization of B's 4-velocity, we find that

$$\left(1-\frac{r_*}{r}\right)\dot{t}^2-\left(1-\frac{r_*}{r}\right)^{-1}\dot{r}^2=\left(1-\frac{r_*}{r}\right)^{-1}E^2-\left(1-\frac{r_*}{r}\right)^{-1}\dot{r}^2=1$$

$$\Rightarrow \quad E^2-\dot{r}^2=1-\frac{r_*}{r}, \tag{3.1718}$$

where $\dot{r}=\frac{dr}{d\tau}$. Solving for $\dot{r}$ leads to

$$\dot{r}=-\sqrt{E^2-1+\frac{r_*}{r}}=-\sqrt{\frac{r_*}{r}-\frac{r_*}{r_1}}. \tag{3.1719}$$

This is a separable differential equation, which integrates to

$$\tau=-\int_{r_1}^{r_0}\frac{dr}{\sqrt{\frac{r_*}{r}-\frac{r_*}{r_1}}}=\int_{r_0}^{r_1}\sqrt{\frac{rr_1}{r_*(r_1-r)}}\,dr, \tag{3.1720}$$

which is therefore the proper time for B between r_1 and r_0.

2.106

a) For an observer with $x=x_0$, we have $dx=0$, and therefore, we find that

$$ds^2=x_0^2dt^2 \quad\Rightarrow\quad \frac{dt}{ds}=+\frac{1}{x_0} \quad\Rightarrow\quad t=\frac{s}{x_0}+t_0. \tag{3.1721}$$

The worldline can therefore be described by (choosing $t_0=0$)

$$x=x_0, \quad t=\frac{s}{x_0}, \tag{3.1722}$$

where s is the proper time. The 4-acceleration A is defined by $A=\nabla_{\dot{\gamma}}\dot{\gamma}=(\ddot{\chi}^\mu+\Gamma^\mu_{\nu\sigma}\dot{\chi}^\nu\dot{\chi}^\sigma)\partial_\mu$. For our worldline, we have $\dot{x}=0$, $\dot{t}=\frac{1}{x_0}$, and $\ddot{x}=\ddot{t}=0$, and therefore, we obtain

$$A=\Gamma^a_{tt}x_0^{-2}\partial_a. \tag{3.1723}$$

We know that

$$\begin{aligned}\Gamma^a_{tt}&=\frac{1}{2}g^{ab}(\partial_t g_{tb}+\partial_t g_{tb}-\partial_b g_{tt})\\&=\frac{1}{2}g^{ab}(0+0-\partial_b g_{tt})=-\frac{1}{2}g^{ab}\partial_b g_{tt}=-\frac{1}{2}g^{ab}\partial_b(x^2).\end{aligned} \tag{3.1724}$$

Thus, we identify $\Gamma^t_{tt}=0$ and $\Gamma^x_{tt}=-\frac{1}{2}g^{xx}\partial_x x^2=-\frac{1}{2}(-1)2x=x$. Therefore, for our worldline, we have

$$A=x_0^{-1}\partial_x. \tag{3.1725}$$

The proper acceleration α is given by $\alpha^2 = -A^2 = -g(A, A) = -x_0^{-2} g(\partial_x, \partial_x) = x_0^{-2}$, which implies that

$$\alpha = \frac{1}{x_0}. \tag{3.1726}$$

b) We have the Killing vector field $K = \partial_t$. Since a free-falling observer moves along a geodesic, we have the constant of motion

$$Q = g(\partial_t, \dot{\gamma}) = x^2 \dot{t} \quad \Rightarrow \quad \dot{t} = \frac{Q}{x^2}. \tag{3.1727}$$

We also know that $g(\dot{\gamma}, \dot{\gamma}) = x^2 \dot{t}^2 - \dot{x}^2 = \frac{Q^2}{x^2} - \dot{x}^2 = 1$. This leads to

$$\dot{x} = \sqrt{\frac{Q^2}{x^2} - 1} = \frac{dx}{ds}. \tag{3.1728}$$

Since we start with $\frac{dx}{dt} = 0$, we know that $Q^2 = x_0^2 \to Q = +x_0$. To compute the proper time s to reach $x = 0$, we need to compute the integral

$$\begin{aligned}
s &= \int ds = \int_0^{x_0} \frac{ds}{dx} dx = \int_0^{x_0} \frac{dx}{\sqrt{\frac{x_0^2}{x^2} - 1}} \\
&= \int_0^{x_0} \frac{x\,dx}{\sqrt{x_0^2 - x^2}} = \{x = x_0 \sin\theta, dx = x_0 \cos\theta d\theta\} \\
&= x_0 \int_0^{\arcsin 1} \frac{\sin\theta}{\sqrt{1 - \sin^2\theta}} \cos\theta\, d\theta \\
&= x_0 \int_0^{\arcsin 1} \sin\theta\, d\theta = -x_0 \left[\cos\theta\right]_0^{\arcsin 1} = x_0.
\end{aligned} \tag{3.1729}$$

Note that the given coordinates are Rindler coordinates on Minkowski space. We can also easily deduce $s = x_0$ by transforming to Minkowski coordinates.

2.107

Using the first relation $uv = (2\mu - r)e^{(r-2\mu)/2\mu}$, where $u < 0$ and $v > 0$, we obtain by differentiation and Leibniz' rule

$$u\,dv + v\,du = -r\,dr e^{(r-2\mu)/2\mu} + (2\mu - r)\frac{1}{2\mu} e^{(r-2\mu)/2\mu} dr = -\frac{1}{2\mu} e^{(r-2\mu)/2\mu} r\,dr. \tag{3.1730}$$

Similarly, using the second relation $t = 2\mu \ln(-v/u)$, we obtain

$$\frac{1}{2\mu} dt = \frac{1}{v} dv - \frac{1}{u} du, \tag{3.1731}$$

which implies that

$$u\,dv - v\,du = \frac{1}{2\mu}uv\,dt = \frac{1}{2\mu}(2\mu - r)e^{(r-2\mu)/2\mu}dt. \tag{3.1732}$$

Solving for du and dv, we find that

$$2u\,dv = \left[\frac{1}{2\mu}(2\mu - r)dt - \frac{1}{2\mu}rdr\right]e^{(r-2\mu)/2\mu}, \tag{3.1733}$$

$$2v\,du = \left[-\frac{1}{2\mu}(2\mu - r)dt - \frac{1}{2\mu}rdr\right]e^{(r-2\mu)/2\mu}. \tag{3.1734}$$

Thus, we have

$$\begin{aligned} 4\,du\,dv &= \frac{1}{uv}2u\,dv \cdot 2v\,du \\ &= \frac{1}{2\mu - r}\left[-\frac{1}{4\mu^2}(2\mu - r)^2dt^2 + \frac{1}{4\mu^2}r^2dr^2\right]e^{(r-2\mu)/2\mu}, \end{aligned} \tag{3.1735}$$

and finally, assuming $\mu \equiv GM$, we obtain the equivalence between the Kruskal–Szekeres metric and the standard Schwarzschild metric as

$$\begin{aligned} \frac{16\mu^2}{r}e^{-(r-2\mu)/2\mu}dudv &= \frac{4\mu^2}{r(2\mu - r)}\left[-\frac{1}{4\mu^2}(2\mu - r)^2dt^2 + \frac{1}{4\mu^2}r^2dr^2\right] \\ &= -\frac{2\mu - r}{r}dt^2 + \frac{r}{2\mu - r}dr^2 \\ &= \left(1 - \frac{2\mu}{r}\right)dt^2 - \left(1 - \frac{2\mu}{r}\right)^{-1}dr^2 \\ &= \left(1 - \frac{2GM}{r}\right)dx^0 - \left(1 - \frac{2GM}{r}\right)^{-1}dr^2. \end{aligned} \tag{3.1736}$$

2.108
The metric is then given by

$$ds^2 = \frac{16\mu^2}{r}e^{(2\mu - r)/2\mu}dudv - r^2d\Omega^2, \tag{3.1737}$$

where $\mu = GM/c^2$ and r (as well as the time $t = t(u, v)$, see below) is a function of u and v. The coordinate r is defined by the equation

$$uv = (2\mu - r)e^{(r-2\mu)/2\mu}. \tag{3.1738}$$

Note that $f(x) = xe^{x/a}$ is monotonically increasing when $x > -a$ (and $f(x) > -a/e$), and therefore, $y = f(x)$ has a unique solution x for any $y > -a/e$. We treat u as a kind of universal time and a timelike vector is future directed if its

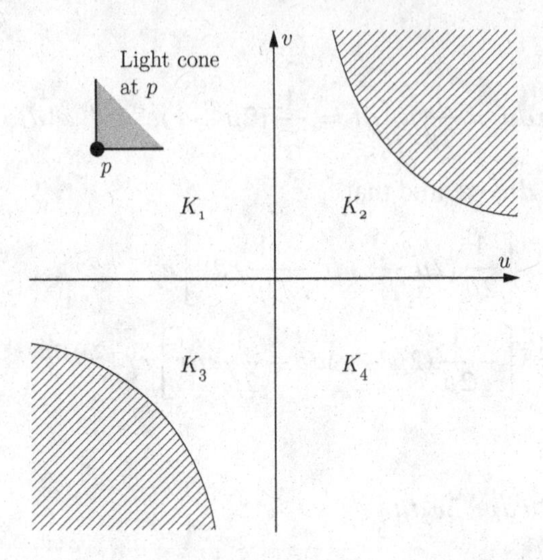

Figure 3.16 The four regions K_1, K_2, K_3, and K_4 of Kruskal–Szekeres coordinates u and v.

projection to ∂_u is positive. The orientation (needed in integration) is defined by the ordering (u, v, θ, ϕ) of coordinates. Note that the radial null lines (radial light rays) are given by $du = 0$ or $dv = 0$.

The Kruskal–Szekeres spacetime can be divided into four regions (see Figure 3.16): region K_1 consists of points $u < 0, v > 0$, region K_2 of points $u, v > 0$, region K_3 of points $u, v < 0$, and finally, region K_4 of points $u > 0$, $v < 0$. The boundaries between these regions are nonsingular points for the metric. The only singularities are at the boundary $uv = 2\mu/e$. The region K_1 is equivalent to the outer region of a Schwarzschild spacetime. This is seen by performing the coordinate transformation $(u, v, \theta, \phi) \mapsto (t, r, \theta, \phi)$, where $r = r(u, v)$ as above and the Schwarzschild time is $t = 2\mu \ln(-v/u)$. With a similar coordinate transformation, the region K_4 is also seen to be equivalent to the outer Schwarzschild solution. The region K_2 is equivalent with the Schwarzschild black hole. This equivalence is obtained through the coordinate transformation $(u, v, \theta, \phi) \mapsto (t, r, \theta, \phi)$, where $r = r(u, v)$ is the same as before but now $t = 2\mu \ln(v/u)$. The region K_3 is called a "*white hole*."

It is easy to construct smooth timelike curves, which go from either K_1 or K_4 to the black hole K_2. However, we will prove that once an observer falls into the black hole K_2, there is no way to go back to the "normal" regions K_1 and K_4. Analogously, everything escapes the "white hole" K_3.

Let $x(t)$ be the timelike path of an observer. Then, along the path

$$\frac{dr}{dt} = \frac{\partial r}{\partial u}\frac{du}{dt} + \frac{\partial r}{\partial v}\frac{dv}{dt} = \frac{r}{8\mu^2}e^{(r-2\mu)/2\mu}\left[\frac{\partial r}{\partial u}g(\partial_v, x'(t)) + \frac{\partial r}{\partial v}g(\partial_u, x'(t))\right] < 0, \tag{3.1739}$$

since $x(t)$ is timelike and in K_2 it holds that $r\frac{\partial r}{\partial u} = -2\mu v e^{(2\mu-r)/2\mu} < 0$ and similarly for the v-coordinate.

The boundary between K_2 and the normal regions is $r = 2\mu$ (i.e., $u = 0$ or $v = 0$). The function $r(x(t))$ was seen to be decreasing, and therefore, the path $x(t)$ can never hit the boundary $r = 2\mu$. However, the observer entering K_2 has a deplorable future, since it will eventually hit the true singularity $r = 0$, again using the monotonicity of the function $r(x(t))$.

Note that there is also another singularity, the outer boundary of K_3. Nevertheless, this is of no great concern, since it is in the past; no future directed timelike curve can enter that singularity.

2.109

For $s = 0$, we have $r = 2\mu$, $v = v_0$, $u = 0$ and the constant of motion is then given by

$$\frac{\dot{u}(0)v_0}{2\mu} = \frac{E}{2\mu}. \tag{3.1740}$$

It follows that, at any point,

$$E = \frac{2\mu}{r}e^{\frac{2\mu-r}{2\mu}}(\dot{u}v - \dot{v}u). \tag{3.1741}$$

Taking r as a new parameter of the geodesic and differentiating uv with respect to r, we obtain

$$\frac{d(uv)}{dr} = \frac{ds}{dr}(\dot{u}v + \dot{v}u) = \frac{d}{dr}(2\mu - r)e^{\frac{r-2\mu}{2\mu}} = -\frac{r}{2\mu}e^{\frac{r-2\mu}{2\mu}}. \tag{3.1742}$$

We now have the linear system

$$\dot{u}v - \dot{v}u = E\frac{r}{2\mu}e^{\frac{r-2\mu}{2\mu}}, \tag{3.1743}$$

$$\dot{u}v + \dot{v}u = -\frac{1}{\left(\frac{ds}{dr}\right)}\frac{r}{2\mu}e^{\frac{r-2\mu}{2\mu}}, \tag{3.1744}$$

for $\dot{u}v$ and $\dot{v}u$. Solving this system, we obtain

$$\dot{u}v = \frac{r}{4\mu}e^{\frac{r-2\mu}{2\mu}}\left[E - \frac{1}{\left(\frac{ds}{dr}\right)}\right], \quad \dot{v}u = -\frac{r}{4\mu}e^{\frac{r-2\mu}{2\mu}}\left[E + \frac{1}{\left(\frac{ds}{dr}\right)}\right]. \tag{3.1745}$$

In addition, we also have the requirement that

$$g_{ij}\dot{x}^i\dot{x}^j = 1 \quad \Rightarrow \quad \frac{16\mu^2}{r}e^{\frac{2\mu-r}{2\mu}}\frac{\dot{u}v\dot{v}u}{uv} = 1. \tag{3.1746}$$

Inserting our expressions for $\dot{u}v$, $\dot{v}u$, and the relation between uv and r, this is equivalent to

$$\frac{r}{2\mu - r}\left[\frac{1}{\left(\frac{ds}{dr}\right)^2} - E^2\right] = 1. \tag{3.1747}$$

Solving for ds/dr, we obtain

$$\frac{ds}{dr} = -\frac{1}{\sqrt{E^2 - 1 + \frac{2\mu}{r}}}, \tag{3.1748}$$

where we have used the negative root since the r-coordinate decreases when the proper time increases.

It follows that the proper time Δs to reach the singularity is given by

$$\Delta s = \int_0^{\Delta s} ds = \int_{2\mu}^{0} \frac{ds}{dr}\, dr = \int_0^{2\mu} \frac{dr}{\sqrt{E^2 - 1 + \frac{2\mu}{r}}}, \tag{3.1749}$$

and thus, we find that

$$f(r) = \frac{1}{\sqrt{E^2 - 1 + \frac{2\mu}{r}}}. \tag{3.1750}$$

2.110

a) The scalar quantity $R_{\mu\nu\alpha\beta}R^{\mu\nu\alpha\beta}$ has a singularity at $r = 0$. From this follows that the singularity at $r = 0$ is a physical singularity and not simply due to a bad choice of coordinates (which is the case for the coordinate singularity at $r = r_*$).

b) For a radial light signal, we have $d\theta = d\phi = 0$ and $ds^2 = 0$. It follows that

$$dr = \pm\left(1 - \frac{r_*}{r}\right) dt. \tag{3.1751}$$

For $r > r_*$, we can see from the metric that t is the time-coordinate since $g_{tt} > 0$. The forward light cone is therefore given by

$$\frac{dr}{dt} = \pm\left(1 - \frac{r_*}{r}\right), \tag{3.1752}$$

i.e., one side of the cone going radially outward and another radially inward. For $r < r_*$, we instead have $g_{rr} > 0$, and thus, r represents the time-coordinate with time increasing in the negative r-direction. Thus, for this case, we have

$$\frac{dt}{dr} = \pm\left(1 - \frac{r_*}{r}\right)^{-1}, \tag{3.1753}$$

i.e., the t-coordinate (which is now a spatial coordinate) can both decrease or increase with time (r).

c) In Kruskal–Szekeres coordinates, we still have $d\Omega = 0$ and $ds^2 = 0$ for the radial light cone. It follows that

$$dU = \pm dV. \tag{3.1754}$$

Hence, the light cones are straight lines with slope 1 in these coordinates.

2.111

a) The equation of motion for a massive test particle m (at position $\mathbf{r}$) in a gravitational potential $\Phi(r) = -GM/r$ according to Newton's mechanics (i.e., Newton's second law) is given by

$$\mathbf{F}(\mathbf{r}) = m\ddot{\mathbf{r}} = -m\nabla\Phi(r) = mGM\nabla\frac{1}{r} = -\frac{mGM}{r^2}\mathbf{e}_r, \tag{3.1755}$$

where M is the total point mass of a spherically symmetric source (located at the origin) giving rise to the gravitational potential $\Phi(r)$ (solving Newton's field equation in differential form, i.e., $\nabla^2\Phi = 4\pi G\rho$ with ρ being the mass density function), or without the massive test particle, the equation of motion in differential form is

$$\frac{d^2\mathbf{r}}{dt^2} = -\nabla\Phi(r), \tag{3.1756}$$

whereas the equations of motion according to general relativity are given by the geodesic equations $\ddot{x}^\mu + \Gamma^\mu_{\nu\lambda}\dot{x}^\nu\dot{x}^\lambda = 0$, or more explicitly,

$$\frac{d^2x^\mu}{d\sigma^2} + \Gamma^\mu_{\nu\lambda}\frac{dx^\nu}{d\sigma}\frac{dx^\lambda}{d\sigma} = 0, \tag{3.1757}$$

where σ is the curve parameter and $\Gamma^\mu_{\nu\lambda}$ are the Christoffel symbols defined through

$$g_{\alpha\mu}\Gamma^\mu_{\nu\lambda} = \frac{1}{2}\left(\frac{\partial g_{\nu\alpha}}{\partial x^\lambda} + \frac{\partial g_{\lambda\alpha}}{\partial x^\nu} - \frac{\partial g_{\nu\lambda}}{\partial x^\alpha}\right), \tag{3.1758}$$

with $g_{\mu\nu}$ being the metric. Thus, the Christoffel symbols are combinations of first-order derivatives of the metric.

In the Newtonian limit, we wish to derive the equation of motion according to Newton's mechanics from the equations of motion according to general relativity. Thus, we compute the geodesic equations using the metric $g_{\mu\nu} = \eta_{\mu\nu} + h_{\mu\nu}$ in the linear approximation, i.e., we neglect higher-order terms in $h_{\mu\nu}$. Note that $(\eta_{\mu\nu}) = \operatorname{diag}(1, -1, -1, -1)$. For small velocities, the time component $\dot{x}^0(\sigma)$ of the 4-velocity is much larger than the spatial components. For this reason, we can approximate the geodesic equations as

$$\frac{d^2x^\mu}{d\sigma^2} + \Gamma^\mu_{00}\left(\frac{dx^0}{d\sigma}\right)^2 = 0. \tag{3.1759}$$

In the linear approximation, we have

$$\Gamma^0_{00} = \frac{1}{c^2}\partial^0\Phi, \quad \Gamma^i_{00} = -\frac{1}{c^2}\partial^i\Phi. \tag{3.1760}$$

Thus, the geodesic equations become

$$\ddot{x}^0 + \frac{1}{c^2}\partial^0\Phi(\dot{x}^0)^2 = 0, \tag{3.1761}$$

$$\ddot{x}^i - \frac{1}{c^2}\partial^i\Phi(\dot{x}^0)^2 = 0. \tag{3.1762}$$

In the frame, where the source is at rest, the first equation says that we can choose the time t as the curve parameter, i.e., $x^0(\sigma) = \sigma = ct$, and then the second equation becomes

$$\frac{d^2x^i}{dt^2} = \partial^i\Phi = -\partial_i\Phi \quad \Leftrightarrow \quad \frac{d^2\mathbf{x}}{dt^2} = -\nabla\Phi. \tag{3.1763}$$

The right-hand side (after multiplication by the mass m of the test particle) is the gravitational force of the source on m, so this equation is just Newton's second law, i.e., $m\ddot{\mathbf{r}} = -m\nabla\Phi(r)$.

b) Tidal forces in Newtonian physics are due to differences in gravitational accelerations in neighboring points. Since gravitational accelerations are proportional to the derivatives of the gravitational potential, the differences in gravitational accelerations are second-order derivatives of the gravitational potential, and therefore proportional to r^{-3}, since the gravitational potential itself is proportional to r^{-1}. In general relativity, tidal forces are related to the so-called geodesic deviation, which describes how nearby geodesics separate or converge. They are proportional to second-order derivatives of the metric, since the relativistic gravitational potential in the weak field limit appears as the perturbations to the metric. Thus, tidal forces in general relativity are proportional to the curvature (which is proportional to second-order derivatives of the metric).

2.112

a) In the Newtonian limit (where $|h_{\mu\nu}| \ll 1$ everywhere in spacetime), we have the metric tensor $g_{\mu\nu}$, the Christoffel symbols $\Gamma^\lambda_{\mu\nu}$, the Riemann curvature tensor $R^\mu{}_{\nu\lambda\rho}$, the Ricci tensor $R_{\mu\nu}$, the Ricci scalar R, and the Einstein tensor $G_{\mu\nu}$ as

$$g_{\mu\nu} = \eta_{\mu\nu} + h_{\mu\nu}, \tag{3.1764}$$

$$\begin{aligned}\Gamma^\lambda_{\mu\nu} &= \frac{1}{2}g^{\lambda\rho}\left(\partial_\mu g_{\nu\rho} + \partial_\nu g_{\mu\rho} - \partial_\rho g_{\mu\nu}\right)\\ &\simeq \frac{1}{2}\eta^{\lambda\rho}\left(\partial_\mu h_{\nu\rho} + \partial_\nu h_{\mu\rho} - \partial_\rho h_{\mu\nu}\right) \equiv \Gamma^{(1)\lambda}{}_{\mu\nu},\end{aligned} \tag{3.1765}$$

$$\begin{aligned}
R^{\mu}{}_{\nu\lambda\rho} &= \partial_\lambda \Gamma^{\mu}_{\rho\nu} - \partial_\rho \Gamma^{\mu}_{\lambda\nu} + \Gamma^{\mu}_{\lambda\sigma}\Gamma^{\sigma}_{\rho\nu} - \Gamma^{\mu}_{\rho\sigma}\Gamma^{\sigma}_{\lambda\nu} \simeq \partial_\lambda \Gamma^{(1)\mu}{}_{\rho\nu} - \partial_\rho \Gamma^{(1)\mu}{}_{\lambda\nu} \\
&= \frac{1}{2}\eta^{\mu\sigma}\left(\partial_\lambda\partial_\rho h_{\nu\sigma} + \partial_\nu\partial_\lambda h_{\rho\sigma} - \partial_\lambda\partial_\sigma h_{\rho\nu} - \partial_\lambda\partial_\rho h_{\nu\sigma} - \partial_\nu\partial_\rho h_{\lambda\sigma} + \partial_\rho\partial_\sigma h_{\lambda\nu}\right) \\
&= \frac{1}{2}\left(\partial_\nu\partial_\lambda h^{\mu}_{\rho} + \partial^{\mu}\partial_\rho h_{\lambda\nu} - \partial_\nu\partial_\rho h^{\mu}_{\lambda} - \partial^{\mu}\partial_\lambda h_{\rho\nu}\right) \\
&= \frac{1}{2}\eta^{\mu\sigma}\left(\partial_\nu\partial_\lambda h_{\rho\sigma} + \partial_\sigma\partial_\rho h_{\nu\lambda} - \partial_\nu\partial_\rho h_{\lambda\sigma} - \partial_\sigma\partial_\lambda h_{\nu\rho}\right) \equiv R^{(1)\mu}{}_{\nu\lambda\rho}, \qquad (3.1766)
\end{aligned}$$

$$\begin{aligned}
R_{\mu\nu\lambda\rho} &= g_{\mu\sigma} R^{\sigma}{}_{\nu\lambda\rho} \simeq \eta_{\mu\sigma} R^{(1)\sigma}{}_{\nu\lambda\rho} \\
&= \frac{1}{2}\left(\partial_\nu\partial_\lambda h_{\mu\rho} + \partial_\mu\partial_\rho h_{\nu\lambda} - \partial_\nu\partial_\rho h_{\mu\lambda} - \partial_\mu\partial_\lambda h_{\nu\rho}\right) \equiv R^{(1)}_{\mu\nu\lambda\rho}, \qquad (3.1767)
\end{aligned}$$

$$\begin{aligned}
R_{\mu\nu} &= g^{\lambda\rho} R_{\rho\mu\lambda\nu} = R^{\lambda}{}_{\mu\lambda\nu} \simeq \eta^{\lambda\rho} R^{(1)}_{\rho\mu\lambda\nu} \\
&= \frac{1}{2}\left(\partial_\lambda\partial_\mu h^{\lambda}_{\nu} + \partial_\lambda\partial_\nu h^{\lambda}_{\mu} - \partial_\mu\partial_\nu h - \Box h_{\mu\nu}\right) \equiv R^{(1)}_{\mu\nu}, \qquad (3.1768)
\end{aligned}$$

$$R = g^{\mu\nu} R_{\mu\nu} = R^{\mu}_{\mu} \simeq \eta^{\mu\nu} R^{(1)}_{\mu\nu} = \partial^{\mu}\partial^{\nu} h_{\mu\nu} - \Box h \equiv R^{(1)}, \qquad (3.1769)$$

$$\begin{aligned}
G_{\mu\nu} &= R_{\mu\nu} - \frac{1}{2} g_{\mu\nu} R \simeq R^{(1)}_{\mu\nu} - \frac{1}{2}\eta_{\mu\nu} R^{(1)} \\
&= \frac{1}{2}\left(\partial_\lambda\partial_\mu h^{\lambda}_{\nu} + \partial_\lambda\partial_\nu h^{\lambda}_{\mu} - \partial_\mu\partial_\nu h - \Box h_{\mu\nu} - \eta_{\mu\nu}\partial^{\alpha}\partial^{\beta} h_{\alpha\beta} + \eta_{\mu\nu}\Box h\right) \equiv G^{(1)}_{\mu\nu}, \\
&\qquad (3.1770)
\end{aligned}$$

where $h_{\mu\nu} = h_{\nu\mu}$ and $h \equiv h^{\mu}_{\mu}$. Note that in the weak field limit the Christoffel symbols are first-order derivatives of $h_{\mu\nu}$, whereas the Riemann curvature tensor, the Ricci tensor, the Ricci scalar, and the Einstein tensor are all second-order derivatives of $h_{\mu\nu}$.

Thus, the solution to this problem, i.e., the Ricci tensor in the linear approximation for a metric $g_{\mu\nu} = \eta_{\mu\nu} + h_{\mu\nu}$, is given by

$$R^{(1)}_{\mu\nu} = \frac{1}{2}\left(\partial_\lambda\partial_\mu h^{\lambda}_{\nu} + \partial_\lambda\partial_\nu h^{\lambda}_{\mu} - \partial_\mu\partial_\nu h - \Box h_{\mu\nu}\right). \qquad (3.1771)$$

b) Consider the coordinate transformation $x^{\mu} \mapsto x'^{\mu} = x^{\mu} + \chi^{\mu}$, where $|\partial_\nu \chi^{\mu}| \ll 1$. Differentiating this coordinate transformation yields

$$\partial_\nu x'^{\mu} = \partial_\nu x^{\mu} + \partial_\nu \chi^{\mu} = \delta^{\mu}_{\nu} + \partial_\nu \chi^{\mu}. \qquad (3.1772)$$

Since $|\partial_\nu \chi^{\mu}| \ll 1$, we have the inverse coordinate transformation, namely

$$\partial'_\nu x^{\mu} \simeq \delta^{\mu}_{\nu} - \partial_\nu \chi^{\mu}. \qquad (3.1773)$$

Using this inverse coordinate transformation for the metric tensor and the definition $g^{(1)}_{\mu\nu} = \eta_{\mu\nu} + h_{\mu\nu}$, we find that

$$g'_{\mu\nu} = \partial'_\mu x^{\alpha}\partial'_\nu x^{\beta} g_{\alpha\beta} \simeq \delta^{\alpha}_{\mu}\delta^{\beta}_{\nu} g_{\alpha\beta} - \partial_\mu \chi^{\alpha}\eta_{\alpha\nu} - \partial_\nu \chi^{\beta}\eta_{\mu\beta} = g_{\mu\nu} - \partial_\mu \chi_\nu - \partial_\nu \chi_\mu. \qquad (3.1774)$$

Finally, using again the definition $g_{\mu\nu} = \eta_{\mu\nu} + h_{\mu\nu}$, we obtain

$$h_{\mu\nu} \mapsto h'_{\mu\nu} = h_{\mu\nu} - \partial_\mu \chi_\nu - \partial_\nu \chi_\mu, \tag{3.1775}$$

which is the gauge transformation of $h_{\mu\nu}$.

c) Consider a coordinate system in which the harmonic (or Lorenz) gauge condition holds, i.e., $\partial^\mu \bar{h}_{\mu\nu} = 0$, where

$$\bar{h}_{\mu\nu} \equiv h_{\mu\nu} - \frac{1}{2} h \eta_{\mu\nu}. \tag{3.1776}$$

Note that $\bar{h}^\mu_\mu = -h^\mu_\mu$ (or $\bar{h} = -h$). The tensor $\bar{h}_{\mu\nu}$ has the following gauge transformation

$$\bar{h}_{\mu\nu} \mapsto \bar{h}'_{\mu\nu} = \bar{h}_{\mu\nu} - \partial_\mu \chi_\nu - \partial_\nu \chi_\mu + (\partial_\lambda \chi^\lambda)\eta_{\mu\nu}. \tag{3.1777}$$

Inserting the definition of $\bar{h}_{\mu\nu}$ into the harmonic gauge condition, we have

$$\partial^\mu h_{\mu\nu} = \frac{1}{2}\partial_\nu h^\mu_\mu = \frac{1}{2}\partial_\nu h, \tag{3.1778}$$

which implies that

$$R^{(1)}_{\mu\nu} = -\frac{1}{2}\Box h_{\mu\nu}, \quad R^{(1)} = -\frac{1}{2}\Box h, \tag{3.1779}$$

and thus, we obtain

$$G^{(1)}_{\mu\nu} = R^{(1)}_{\mu\nu} - \frac{1}{2}R^{(1)}\eta_{\mu\nu} = -\frac{1}{2}\Box\left(h_{\mu\nu} - \frac{1}{2}h\eta_{\mu\nu}\right) = -\frac{1}{2}\Box\bar{h}_{\mu\nu}. \tag{3.1780}$$

This means that the linearized Einstein equations turn into standard wave equations of the form

$$\Box\bar{h}_{\mu\nu} = -\frac{16\pi G}{c^4}T_{\mu\nu}. \tag{3.1781}$$

2.113

a) Considering the metric corresponding to a weak gravitational potential $\Phi(\mathbf{x})$ and setting $c = 1$, we have the Lagrangian

$$\mathcal{L} = [1 - 2\Phi(\mathbf{x})]\,\dot{t}^2 - [1 + 2\Phi(\mathbf{x})]\,(\dot{x}^2 + \dot{y}^2 + \dot{z}^2), \tag{3.1782}$$

where $\Phi(\mathbf{x})$ is independent of time t. Using the Euler–Lagrange equations, we find that

$$\frac{d}{d\tau}\frac{\partial \mathcal{L}}{\partial \dot{t}} - \frac{\partial \mathcal{L}}{\partial t} = 0 \quad \Rightarrow \quad \frac{d}{d\tau}\left[(1 - 2\Phi(\mathbf{x}))\dot{t}\right] = 0 \quad \Rightarrow \quad (1 - 2\Phi(\mathbf{x}))\dot{t} = \text{const.}, \tag{3.1783}$$

$$\frac{d}{d\tau}\frac{\partial \mathcal{L}}{\partial \dot{x}^i} - \frac{\partial \mathcal{L}}{\partial x^i} = 0$$

$$\Rightarrow \quad \ddot{x}^i + \frac{2}{1+2\Phi(\mathbf{x})}\left(\Phi'_x \dot{x} + \Phi'_y \dot{y} + \Phi'_z \dot{z}\right)\dot{x}^i - \frac{\Phi'_{x^i}}{1+2\Phi(\mathbf{x})}\left(\dot{t}^2 + \dot{x}^2 + \dot{y}^2 + \dot{z}^2\right) = 0. \tag{3.1784}$$

However, since the metric only holds in the weak-field limit (where only lowest-order terms of Φ are kept), we should write the Euler–Lagrange equations in this limit and have

$$(1 - 2\Phi(\mathbf{x}))\dot{t} = \text{const.}, \tag{3.1785}$$

$$\ddot{x}^i + 2\left(\Phi'_x \dot{x} + \Phi'_y \dot{y} + \Phi'_z \dot{z}\right)\dot{x}^i - \Phi'_{x^i}\left(\dot{t}^2 + \dot{x}^2 + \dot{y}^2 + \dot{z}^2\right) = 0. \tag{3.1786}$$

Furthermore, we apply the nonrelativistic limit, which says that $\dot{x}^i \ll \dot{t}$, since it holds for a slowly moving massive particle that $\tau \approx t$ and $dx^i/dt \ll 1$, which implies that $\dot{x}^i \ll \dot{t} \approx 1$. Thus, we obtain the simple equations

$$\ddot{t} \simeq 0, \quad \ddot{x}^i - \Phi'_{x^i} \simeq 0, \tag{3.1787}$$

which are the geodesic equations in the nonrelativistic and weak-field limits. The first equation means that energy (or $\dot{t}$) is conserved (the integration constant is chosen such that $\dot{t} = 1$, which then coincides with $\mathcal{L} = 1$ in the nonrelativistic limit), whereas multiplying the second equation with the mass m of the massive test particle leads to

$$m\ddot{x}^i = m\Phi'_{x^i} = m\frac{\partial \Phi}{\partial x^i} = m\partial_i \Phi = -m\partial^i \Phi, \tag{3.1788}$$

which is just Newton's second law, i.e., $m\mathbf{a} = \mathbf{F} = -m\nabla\Phi(\mathbf{x})$, where the force $\mathbf{F} = -m\nabla\Phi(\mathbf{x})$ is that of a weak gravitational potential $\Phi(\mathbf{x})$. Compare the solution to Problem 2.111 a).

b) Using the metric for the weak gravitational potential $\Phi(\mathbf{x})$, we have

$$g_{00}(\mathbf{x}) = \left[c^2 - 2\Phi(\mathbf{x})\right]\frac{1}{c^2} = 1 - \frac{2}{c^2}\Phi(\mathbf{x}). \tag{3.1789}$$

Now, assuming $\Phi(\mathbf{x}) = -gz$, we find that

$$g_{00}(z) = 1 + \frac{2}{c^2}gz. \tag{3.1790}$$

Then, using the formula for the gravitational redshift to compute the photon's gravitational redshift, we obtain

$$\frac{\omega'}{\omega} = \sqrt{\frac{1 + \frac{2}{c^2} g \cdot 0}{1 + \frac{2}{c^2} g \cdot h}} \simeq 1 - \frac{gh}{c^2}$$

$$\Rightarrow \quad z = \frac{\lambda'}{\lambda} - 1 = \frac{\omega}{\omega'} - 1 = \sqrt{1 + \frac{2gh}{c^2}} - 1 \simeq \frac{gh}{c^2}, \tag{3.1791}$$

where $h > 0$. Thus, a photon emitted at a lower gravitational potential with an angular frequency ω is received at a higher gravitational potential with a smaller angular frequency ω', i.e., it is redshifted, although the emitter and the receiver are not in relative motion.

Finally, consider a particular frame to be at rest when the photon starts its upward climb in the gravitational potential and falls freely after that. Since the photon climbs from $z = 0$ to $z = h > 0$, i.e., a distance h, it takes the time $\Delta t = h/c$ to arrive at the higher gravitational potential. During this time, the frame has obtained the velocity $v = g\Delta t = gh/c$ downward relative to the higher gravitational potential. Therefore, using formula for the Doppler shift, the photon's angular frequency ω_{ff} relative to the freely falling frame is

$$\frac{\omega_{\text{ff}}}{\omega'} = \sqrt{\frac{1 - (-gh/c^2)}{1 + (-gh/c^2)}} \simeq 1 + \frac{gh}{c^2}, \tag{3.1792}$$

which means that

$$\omega_{\text{ff}} \simeq \omega' \left(1 + \frac{gh}{c^2}\right) \simeq \omega \left(1 - \frac{gh}{c^2}\right)\left(1 + \frac{gh}{c^2}\right) \simeq \omega. \tag{3.1793}$$

Thus, there is no redshift in a freely falling frame, confirming that it is a local inertial frame. This is the essence of Einstein's equivalence principle, which means that the gravitational field can be transformed away by transforming to an appropriate accelerating ("freely falling") frame of reference.

2.114

a) Consider two massive particles moving freely on two close paths, where one particle has the spacetime path $x^\mu(\tau)$ and the other one has the spacetime path $x^\mu(\tau) + s^\mu(\tau)$. Therefore, the two particles are separated by the displacement vector $s^\mu(\tau)$, which is small. The geodesic equations (or the equations of motion) for the two particles are given by

$$\ddot{x}^\mu + \Gamma^\mu_{\alpha\beta}(x)\dot{x}^\alpha \dot{x}^\beta = 0, \quad (\ddot{x}^\mu + \ddot{s}^\mu) + \Gamma^\mu_{\alpha\beta}(x+s)(\dot{x}^\alpha + \dot{s}^\alpha)(\dot{x}^\beta + \dot{s}^\beta) = 0. \tag{3.1794}$$

Since s^μ is small, we can series expand the Christoffel symbols $\Gamma^\mu_{\alpha\beta}(x+s)$ in the second equation as

$$\Gamma^\mu_{\alpha\beta}(x+s) \simeq \Gamma^\mu_{\alpha\beta}(x) + \partial_\lambda \Gamma^\mu_{\alpha\beta}(x) s^\lambda. \tag{3.1795}$$

Thus, from the difference between the two geodesics equations, we have to first order in s

$$\ddot{s}^\mu \simeq -2\Gamma^\mu_{\alpha\beta}(x)\dot{s}^\alpha \dot{x}^\beta - \partial_\lambda \Gamma^\mu_{\alpha\beta}(x) s^\lambda \dot{x}^\alpha \dot{x}^\beta, \tag{3.1796}$$

where we used $\Gamma^\mu_{\alpha\beta}(x) = \Gamma^\mu_{\beta\alpha}(x)$. Using the definition of the parallel transport of a vector V^μ along a path $x^\mu(\tau)$, i.e.,

$$\frac{DV^\mu}{D\tau} = \dot{V}^\mu + \Gamma^\mu_{\alpha\beta} V^\alpha \dot{x}^\beta = 0, \tag{3.1797}$$

we have directly the first derivative of s^μ

$$\frac{Ds^\mu}{D\tau} = \dot{V}^\mu + \Gamma^\mu_{\alpha\beta} V^\alpha \dot{x}^\beta, \tag{3.1798}$$

and we can also find the second derivative of s^μ as

$$\begin{aligned}\frac{D^2 s^\mu}{D\tau^2} &= \frac{D}{D\tau}\frac{Ds^\mu}{D\tau} = \frac{d}{d\tau}\frac{Ds^\mu}{D\tau} + \Gamma^\mu_{\alpha\beta}\frac{Ds^\alpha}{D\tau}\dot{x}^\beta \\ &= \frac{d}{d\tau}\left(\dot{s}^\mu + \Gamma^\mu_{\alpha\beta} s^\alpha \dot{x}^\beta\right) + \Gamma^\mu_{\alpha\beta}\left(\dot{s}^\alpha + \Gamma^\alpha_{\lambda\rho} s^\lambda \dot{x}^\rho\right)\dot{x}^\beta \\ &= \ddot{s}^\mu + \partial_\lambda \Gamma^\mu_{\alpha\beta}\dot{x}^\lambda s^\alpha \dot{x}^\beta + \Gamma^\mu_{\alpha\beta}\dot{s}^\alpha \dot{x}^\beta + \Gamma^\mu_{\alpha\beta} s^\alpha \ddot{x}^\beta + \Gamma^\mu_{\alpha\beta}\dot{s}^\alpha \dot{x}^\beta + \Gamma^\mu_{\alpha\beta}\Gamma^\alpha_{\lambda\rho} s^\lambda \dot{x}^\rho \dot{x}^\beta.\end{aligned} \tag{3.1799}$$

In order to simplify the second derivative, we use the above-derived expressions for $\ddot{s}^\mu$ and $\ddot{x}^\beta$ and obtain

$$\begin{aligned}\frac{D^2 s^\mu}{D\tau^2} &= -2\Gamma^\mu_{\alpha\beta}\dot{s}^\alpha \dot{x}^\beta - \partial_\lambda \Gamma^\mu_{\alpha\beta} s^\lambda \dot{x}^\alpha \dot{x}^\beta + \partial_\lambda \Gamma^\mu_{\alpha\beta}\dot{x}^\lambda s^\alpha \dot{x}^\beta + \Gamma^\mu_{\alpha\beta}\dot{s}^\alpha \dot{x}^\beta \\ &\quad - \Gamma^\mu_{\alpha\beta}\Gamma^\beta_{\lambda\rho} s^\alpha \dot{x}^\lambda \dot{x}^\rho + \Gamma^\mu_{\alpha\beta}\dot{s}^\alpha \dot{x}^\beta + \Gamma^\mu_{\alpha\beta}\Gamma^\alpha_{\lambda\rho} s^\lambda \dot{x}^\rho \dot{x}^\beta \\ &= -\partial_\lambda \Gamma^\mu_{\alpha\beta} s^\lambda \dot{x}^\alpha \dot{x}^\beta + \partial_\lambda \Gamma^\mu_{\alpha\beta}\dot{x}^\lambda s^\alpha \dot{x}^\beta - \Gamma^\mu_{\alpha\beta}\Gamma^\beta_{\lambda\rho} s^\alpha \dot{x}^\lambda \dot{x}^\rho + \Gamma^\mu_{\alpha\beta}\Gamma^\alpha_{\lambda\rho} s^\lambda \dot{x}^\rho \dot{x}^\beta. \\ &= -(\partial_\lambda \Gamma^\mu_{\alpha\beta} - \partial_\alpha \Gamma^\mu_{\lambda\beta} + \Gamma^\mu_{\lambda\rho}\Gamma^\rho_{\alpha\beta} - \Gamma^\mu_{\rho\beta}\Gamma^\rho_{\lambda\alpha}) s^\lambda \dot{x}^\alpha \dot{x}^\beta,\end{aligned} \tag{3.1800}$$

where in the last step we relabeled some of the summation indices. It holds that $\partial_\alpha \Gamma^\mu_{\lambda\beta} s^\lambda \dot{x}^\alpha \dot{x}^\beta = \partial_\beta \Gamma^\mu_{\lambda\alpha} s^\lambda \dot{x}^\alpha \dot{x}^\beta$ and that $\Gamma^\mu_{\alpha\beta} = \Gamma^\mu_{\beta\alpha}$. Thus, we can identify the Riemann curvature tensor as

$$R^\mu{}_{\alpha\lambda\beta} = \partial_\lambda \Gamma^\mu_{\beta\alpha} - \partial_\beta \Gamma^\mu_{\lambda\alpha} + \Gamma^\mu_{\lambda\rho}\Gamma^\rho_{\beta\alpha} - \Gamma^\mu_{\beta\rho}\Gamma^\rho_{\lambda\alpha}, \tag{3.1801}$$

and therefore, we have derived the equation of geodesic deviation, namely

$$\frac{D^2 s^\mu}{D\tau^2} = -R^\mu{}_{\alpha\lambda\beta} s^\lambda \dot{x}^\alpha \dot{x}^\beta = -R^\mu{}_{\alpha\nu\beta} s^\nu \frac{dx^\alpha}{d\tau}\frac{dx^\beta}{d\tau}. \tag{3.1802}$$

b) In the Newtonian limit, we can approximate the metric by $g_{\mu\nu} = \eta_{\mu\nu} + h_{\mu\nu}$, where $|h_{\mu\nu}| \ll 1$ is small, and therefore, we have the metric tensor $g_{\mu\nu}$, the Christoffel symbols $\Gamma^{\lambda}_{\mu\nu}$, and the Riemann curvature tensor $R^{\mu}{}_{\nu\lambda\rho}$ as

$$g_{\mu\nu} \simeq \eta_{\mu\nu} + h_{\mu\nu}, \tag{3.1803}$$

$$\begin{aligned}\Gamma^{\lambda}_{\mu\nu} &= \frac{1}{2} g^{\lambda\rho} \left(\partial_\mu g_{\nu\rho} + \partial_\nu g_{\mu\rho} - \partial_\rho g_{\mu\nu}\right) \\ &\simeq \frac{1}{2} \eta^{\lambda\rho} \left(\partial_\mu h_{\nu\rho} + \partial_\nu h_{\mu\rho} - \partial_\rho h_{\mu\nu}\right) \equiv \Gamma^{(1)\lambda}{}_{\mu\nu},\end{aligned} \tag{3.1804}$$

$$\begin{aligned}R^{\mu}{}_{\nu\lambda\rho} &= \partial_\lambda \Gamma^{\mu}_{\rho\nu} - \partial_\rho \Gamma^{\mu}_{\lambda\nu} + \Gamma^{\mu}_{\lambda\sigma}\Gamma^{\sigma}_{\rho\nu} - \Gamma^{\mu}_{\rho\sigma}\Gamma^{\sigma}_{\lambda\nu} \simeq \partial_\lambda \Gamma^{(1)\mu}{}_{\rho\nu} - \partial_\rho \Gamma^{(1)\mu}{}_{\lambda\nu} \\ &= \frac{1}{2} \eta^{\mu\sigma} \left(\partial_\lambda \partial_\rho h_{\nu\sigma} + \partial_\nu \partial_\lambda h_{\rho\sigma} - \partial_\lambda \partial_\sigma h_{\rho\nu} - \partial_\lambda \partial_\rho h_{\nu\sigma} - \partial_\nu \partial_\rho h_{\lambda\sigma} + \partial_\rho \partial_\sigma h_{\lambda\nu}\right) \\ &= \frac{1}{2} \left(\partial_\nu \partial_\lambda h^{\mu}_{\rho} + \partial^{\mu} \partial_\rho h_{\lambda\nu} - \partial_\nu \partial_\rho h^{\mu}_{\lambda} - \partial^{\mu} \partial_\lambda h_{\rho\nu}\right) \\ &= \frac{1}{2} \eta^{\mu\sigma} \left(\partial_\nu \partial_\lambda h_{\rho\sigma} + \partial_\sigma \partial_\rho h_{\nu\lambda} - \partial_\nu \partial_\rho h_{\lambda\sigma} - \partial_\sigma \partial_\lambda h_{\nu\rho}\right) \equiv R^{(1)\mu}{}_{\nu\lambda\rho}.\end{aligned} \tag{3.1805}$$

Choosing $\mu = i$, $\nu = 0$, $\lambda = j$, and $\rho = 0$, we have

$$\begin{aligned}R^{i}{}_{0j0} \simeq R^{(1)i}{}_{0j0} &= \frac{1}{2} \eta^{i\sigma} \left(\partial_0 \partial_j h_{0\sigma} + \partial_\sigma \partial_0 h_{0j} - \partial_0 \partial_0 h_{j\sigma} - \partial_\sigma \partial_j h_{00}\right) \\ &= \frac{1}{2} \left(-\partial_j \partial_0 h_{0i} - \partial_i \partial_0 h_{0j} + \partial_0 \partial_0 h_{ji} + \partial_i \partial_j h_{00}\right).\end{aligned} \tag{3.1806}$$

Furthermore, in the Newtonian limit, we have the static field condition, i.e., $\partial_0 g_{\alpha\beta} = 0$, which means that all time derivatives of the metric can be neglected, and thus, we find that

$$R^{i}{}_{0j0} \simeq \frac{1}{2} \partial_i \partial_j h_{00}. \tag{3.1807}$$

Now, $h_{00} = 2\Phi/c^2$ (see Problem 2.112), so we finally obtain

$$R^{i}{}_{0j0} \simeq \frac{1}{c^2} \partial_i \partial_j \Phi = \frac{1}{c^2} \frac{\partial^2 \Phi}{\partial x^j \partial x^i}, \tag{3.1808}$$

which, using the result found in a) with $\tau = t$, setting $c = 1$, and again choosing $\mu = i$, leads to the equation for the tidal acceleration in Newton's theory of gravitation, namely

$$\frac{d^2 s^i}{dt^2} = -\frac{\partial^2 \Phi}{\partial x^i \partial x^j} s^j, \tag{3.1809}$$

which is what we wanted to show. In words, the acceleration of the separation between the trajectories of the two particles is given by a tensor of the second derivatives of the gravitational potential.

2.115

a) The metric in Minkowski space in Cartesian spatial coordinates is given by

$$ds^2 = c^2 dt^2 - dx_0^2 - dy_0^2 - dz_0^2. \tag{3.1810}$$

Consider the motion of a free particle in the lab frame described by a cylindrical coordinate system $x = r\cos\varphi$, $y = r\sin\varphi$, and $z = z$, but also in the rest frame of the particle that is rotating with constant angular velocity ω around the z-axis relative to the lab frame. The spatial cylindrical coordinates in the rest frame of the particle are given by $x = r_0\cos\varphi_0$, $y = r_0\sin\varphi_0$, and $z = z_0$ and the two frames are related to each other by the following coordinate transformations

$$r_0 = r, \quad \varphi_0 = \varphi - \omega t, \quad z_0 = z. \tag{3.1811}$$

Without loss of generality, we can set $c = 1$. In the rest frame of the particle, the metric becomes

$$\begin{aligned} ds^2 &= dt^2 - dr_0^2 - r_0^2 d\varphi_0^2 - dz_0^2 = dt^2 - dr^2 - r^2(d\varphi - \omega\, dt)^2 - dz^2 \\ &= (1-\omega^2)dt^2 - dr^2 - r^2 d\varphi^2 - dz^2 + 2r^2\omega\, dt\, d\varphi, \end{aligned} \tag{3.1812}$$

which in turn implies the Lagrangian

$$\mathcal{L} = g_{\mu\nu}dx^\mu dx^\nu = (1-\omega^2 r^2)\dot{t}^2 + 2\omega r^2 \dot{t}\dot{\varphi} - \dot{r}^2 - r^2\dot{\varphi}^2 - \dot{z}^2. \tag{3.1813}$$

Thus, we can immediately read off the nonzero components of the metric, namely

$$g_{tt} = 1 - \omega^2 r^2, \quad g_{t\varphi} = g_{\varphi t} = \omega r^2, \quad g_{rr} = -1, \quad g_{\varphi\varphi} = -r^2, \quad g_{zz} = -1. \tag{3.1814}$$

Now, using the Euler–Lagrange equations, i.e., $\frac{d}{d\tau}\frac{\partial\mathcal{L}}{\partial\dot{x}^\mu} - \frac{\partial\mathcal{L}}{\partial x^\mu} = 0$, we find that

$$\begin{aligned} &\frac{\partial\mathcal{L}}{\partial t} = 0, \quad \frac{\partial\mathcal{L}}{\partial\dot{t}} = 2(1-\omega^2 r^2)\dot{t} + 2\omega r^2\dot{\varphi} \\ &\Rightarrow \quad \ddot{t} - \frac{\omega r}{1-\omega^2 r^2}\left(2\omega\dot{t}\dot{r} - 2\dot{r}\dot{\varphi} - r\ddot{\varphi}\right) = 0, \end{aligned} \tag{3.1815}$$

$$\begin{aligned} &\frac{\partial\mathcal{L}}{\partial r} = -2\omega^2 r\dot{t}^2 + 4\omega r\dot{t}\dot{\varphi} - 2r\dot{\varphi}^2, \quad \frac{\partial\mathcal{L}}{\partial\dot{r}} = -2\dot{r} \\ &\Rightarrow \quad \ddot{r} - r\dot{\varphi}^2 - \omega r\left(\omega\dot{t}^2 - 2\dot{t}\dot{\varphi}\right) = 0, \end{aligned} \tag{3.1816}$$

$$\frac{\partial\mathcal{L}}{\partial\varphi} = 0, \quad \frac{\partial\mathcal{L}}{\partial\dot{\varphi}} = 2\omega r^2\dot{t} - 2r^2\dot{\varphi} \quad \Rightarrow \quad \ddot{\varphi} + \frac{2}{r}\dot{r}\dot{\varphi} - \omega\left(\frac{2}{r}\dot{t}\dot{r} + \ddot{t}\right) = 0, \tag{3.1817}$$

$$\frac{\partial\mathcal{L}}{\partial z} = 0, \quad \frac{\partial\mathcal{L}}{\partial\dot{z}} = -2\dot{z} \quad \Rightarrow \quad \frac{d}{d\tau}\dot{z} = 0. \tag{3.1818}$$

Since $\frac{\partial \mathcal{L}}{\partial t} = \frac{\partial \mathcal{L}}{\partial \varphi} = \frac{\partial \mathcal{L}}{\partial z} = 0$, we basically have three constants of motion, i.e., $\frac{\partial \mathcal{L}}{\partial \dot{t}} =$ const., $\frac{\partial \mathcal{L}}{\partial \dot{\varphi}} =$ const., and $\frac{\partial \mathcal{L}}{\partial \dot{z}} =$ const. However, we observe that two of the Euler–Lagrange equations contain both $\ddot{t}$ and $\ddot{\varphi}$, and therefore, solving for $\ddot{t}$ and $\ddot{\varphi}$, we obtain the geodesic equations

$$\ddot{t} = 0, \quad \ddot{r} - \omega^2 r\dot{t}^2 + 2\omega r\dot{t}\dot{\varphi} - r\dot{\varphi}^2 = 0, \quad \ddot{\varphi} - \frac{2\omega}{r}\dot{t}\dot{r} + \frac{2}{r}\dot{r}\dot{\varphi} = 0, \quad \ddot{z} = 0. \tag{3.1819}$$

Thus, comparing the geodesic equations with the general formula for the geodesic equations, i.e., $\ddot{x}^\mu + \Gamma^\mu_{\nu\lambda}\dot{x}^\nu\dot{x}^\lambda = 0$, and using $\Gamma^\mu_{\nu\lambda} = \Gamma^\mu_{\lambda\nu}$, we can read off the eight nonzero Christoffel symbols as

$$\Gamma^r_{tt} = -\omega^2 r, \quad \Gamma^r_{t\varphi} = \Gamma^r_{\varphi t} = \omega r, \quad \Gamma^r_{\varphi\varphi} = -r,$$
$$\Gamma^\varphi_{tr} = \Gamma^\varphi_{rt} = -\frac{\omega}{r}, \quad \Gamma^\varphi_{r\varphi} = \Gamma^\varphi_{\varphi r} = \frac{1}{r}. \tag{3.1820}$$

Note that in the case we set $\omega = 0$, the Christoffel symbols reduce to three, i.e., $\Gamma^r_{\varphi\varphi} = -r$ and $\Gamma^\varphi_{r\varphi} = \Gamma^\varphi_{\varphi r} = \frac{1}{r}$, which are the Christoffel symbols in nonrotating cylinder coordinates.

To summarize, in the rest frame of a free particle rotating with constant angular velocity ω around the z-coordinate axis, the Christoffel symbols and the geodesic equations are

$$\Gamma^r_{tt} = -\omega^2 r, \quad \Gamma^r_{t\varphi} = \Gamma^r_{\varphi t} = \omega r, \quad \Gamma^r_{\varphi\varphi} = -r,$$
$$\Gamma^\varphi_{tr} = \Gamma^\varphi_{rt} = -\frac{\omega}{r}, \quad \Gamma^\varphi_{r\varphi} = \Gamma^\varphi_{\varphi r} = \frac{1}{r},$$
$$\ddot{t} = 0, \quad \ddot{r} - \omega^2 r\dot{t}^2 + 2\omega r\dot{t}\dot{\varphi} - r\dot{\varphi}^2 = 0, \quad \ddot{\varphi} - \frac{2\omega}{r}\dot{t}\dot{r} + \frac{2}{r}\dot{r}\dot{\varphi} = 0, \quad \ddot{z} = 0. \tag{3.1821}$$

b) Now, in the Newtonian limit, the approximate geodesic equations (see also Problem 2.111) are given by

$$\frac{d^2 x^\mu}{d\sigma^2} + \Gamma^\mu_{00}\left(\frac{dx^0}{d\sigma}\right)^2 = 0, \tag{3.1822}$$

since for small velocities, the component $\dot{x}^0(\sigma)$ of the 4-velocity is much larger than the corresponding spatial components, and where σ is the curve parameter. Using the results found in a), we only have one nonzero Christoffel symbol on the form Γ^μ_{00}, which is $\Gamma^r_{00} = \Gamma^r_{tt} = -\omega^2 r$, and therefore, the approximate geodesic equations are reduced to

$$\frac{d^2 t}{d\sigma^2} = 0, \quad \frac{d^2 r}{d\sigma^2} - \omega^2 r\left(\frac{dt}{d\sigma}\right)^2 = 0. \tag{3.1823}$$

In the rest frame of the particle, the first equation means that we can choose the time t as the curve parameter σ, i.e., $t = \sigma$, and then the second equation becomes

$$\ddot{r} - \omega^2 r = 0. \tag{3.1824}$$

Multiplying this equation with the mass m of the particle, we obtain

$$m\ddot{r} = m\omega^2 r \equiv F_{\text{centrifugal}}, \tag{3.1825}$$

where $F_{\text{centrifugal}} = m\omega^2 r$ is the *centrifugal force* that appears to act on all objects when viewed in a rotating frame.

Thus, in the nonrelativistic limit, Newton's equation of motion in the rotating frame of the free particle is given by

$$m\ddot{r} = m\omega^2 r. \tag{3.1826}$$

Note that this equation is expressed in the radial coordinate of either the lab frame or the rest frame of the particle, since $r = r_0$ [see a)].

2.116

a) In spherical coordinates the Lagrangian is

$$\mathcal{L} = \frac{1}{2}m\dot{\mathbf{r}}^2 + \frac{GMm}{r} = \frac{1}{2}m(\dot{r}^2 + r^2\dot{\varphi}^2) + \frac{GMm}{r}, \tag{3.1827}$$

where it is assumed that $\theta = \pi/2$ and $\dot{\theta} = 0$ with dot meaning differentiation with respect to time t. This yields the Euler–Lagrange equations

$$\frac{d}{dt}\frac{\partial \mathcal{L}}{\partial \dot{r}} - \frac{\partial \mathcal{L}}{\partial r} = -mr\dot{\varphi}^2 + \frac{GMm}{r^2} = 0, \tag{3.1828}$$

$$\frac{d}{dt}\frac{\partial \mathcal{L}}{\partial \dot{\varphi}} - \frac{\partial \mathcal{L}}{\partial \varphi} = \frac{d}{dt}(mr^2\dot{\varphi}) = 0, \tag{3.1829}$$

where it holds that $\dot{r} = \ddot{r} = 0$, since it is assumed that $r = r_0 = \text{const}$. The second equation is solved by $L = mr^2\dot{\varphi} = \text{constant}$, i.e., $\dot{\varphi} = L/(mr^2)$. Inserting this into the first equation, we obtain

$$-mr\left(\frac{L}{mr^2}\right)^2 + \frac{GMm}{r^2} = 0 \quad \Rightarrow r = \frac{L^2}{GMm^2} \equiv r_0, \tag{3.1830}$$

which is the trajectory of the planet.

Note that r_0 is exactly the radius where the centrifugal force $mr^2\dot{\varphi}$ is balanced by the attractive gravitational force GMm/r^2, i.e., $mr^2\dot{\varphi} = GMm/r^2$, which together with $L = mr^2\dot{\varphi}$ gives an elementary derivation of the trajectory of the planet.

b) The result in a) can be summarized as follows: It has been shown that $m\ddot{r} = -m\Phi'_{\text{eff}}(r)$ with the effective gravitational potential

$$m\Phi_{\text{eff}}(r) = -\frac{GMm}{r} + \frac{1}{2}mr^2\dot{\varphi}^2 = m\left(-\frac{GM}{r} + \frac{L^2}{2m^2r^2}\right). \tag{3.1831}$$

The stationary orbit $r = r_0$ can thus be obtained as the solution to $\Phi'_{\text{eff}}(r) = 0$. This solution $r = r_0$ is a minimum of $\Phi_{\text{eff}}(r)$, and therefore, it corresponds to a stable orbit.

There exists a natural generalization of the effective potential $\Phi_{\text{eff}}(r)$ to general relativity, namely

$$\Phi(r) = -\frac{r_*}{2r} + \frac{L^2}{2m^2r^2} - \frac{r_*L^2}{2m^2r^3}, \tag{3.1832}$$

where $r_* = 2GM$ is the Schwarzschild radius, which can be written as

$$\Phi(r) = -\frac{r_*}{2r} + \frac{r_0r_*}{4r^2} - \frac{r_0r_*^2}{4r^3}, \tag{3.1833}$$

where $r_0 \equiv L^2/(GMm^2)$, which is the solution to the problem in a). The general relativity generalization of the orbit r_0 can be computed by solving $\Phi'(r) = 0$, which yields

$$r^2 - r_0r + \frac{3}{2}r_0r_* = 0. \tag{3.1834}$$

Thus, we have two possible solutions

$$r_\pm = \frac{r_0}{2}\left(1 \pm \sqrt{1 - \frac{6r_*}{r_0}}\right). \tag{3.1835}$$

The solution r_+ corresponds to a minimum of $\Phi_{\text{eff}}(r)$ and is thus a stable orbit generalizing the Newtonian solution found in a), whereas the solution r_- corresponds to a maximum and is thus unstable, but $r_- \to 0$ in the Newtonian limit $r_* \to 0$. Therefore, the generalization of the orbit r_0 in a) to general relativity is

$$r_+ = \frac{r_0}{2}\left(1 + \sqrt{1 - \frac{6r_*}{r_0}}\right). \tag{3.1836}$$

2.117

a) We need to assume the weak-field limit where the metric can be written

$$g_{\mu\nu} = \eta_{\mu\nu} + h_{\mu\nu}, \tag{3.1837}$$

where $\eta_{\mu\nu}$ is the Minkowski metric and $h_{\mu\nu}$ is a small perturbation. We will only need to determine the Riemann tensor to first order in $h_{\mu\nu}$. Furthermore, we assume that the source of the gravitational field is not large and slowly moving, hence only the component $T^{00} = c^2\rho$ gives a relevant contribution to the field. Furthermore, we assume that in the Newtonian limit $\rho \gg \frac{p}{c^2}$ hence we can assume $T^{\mu\nu} = \rho U^\mu U^\nu$.

b) Einstein's equations are

$$G^{\mu\nu} = 8\pi \frac{G}{c^4} T^{\mu\nu}, \tag{3.1838}$$

where $G^{\mu\nu} = R^{\mu\nu} - \frac{1}{2} R g^{\mu\nu}$. For the metric in $g_{\mu\nu} = \eta_{\mu\nu} + h_{\mu\nu}$ the Christoffel symbols are given by

$$\Gamma^{\omega}_{\mu\nu} = \frac{1}{2}\eta^{\omega\rho}(\partial_\mu h_{\nu\rho} + \partial_\nu h_{\mu\rho} - \partial_\rho h_{\mu\nu}). \tag{3.1839}$$

Since the lowest order of $h_{\mu\nu}$ in the Christoffel symbols is the first, the Riemann tensor of first order will therefore only contain the derivative terms

$$R^{\lambda}{}_{\omega\mu\nu} = \frac{1}{2}\left(\partial_\mu \Gamma^{\lambda}_{\nu\omega} - \partial_\nu \Gamma^{\lambda}_{\mu\omega}\right) = \frac{1}{2}\left(\partial_\omega \partial_\mu h^{\lambda}_{\nu} + \partial^{\lambda}\partial_\nu h_{\omega\mu} - \partial_\omega \partial_\nu h^{\lambda}_{\mu} - \partial^{\lambda}\partial_\mu h_{\omega\nu}\right). \tag{3.1840}$$

Using the gauge condition $\partial^{\lambda} h_{\mu\lambda} = \partial_\mu h/2$, we can write the Ricci tensor and Ricci scalar as

$$R_{\mu\nu} = -\frac{1}{2}\Box h_{\mu\nu}, \quad R = \frac{1}{2}\Box h, \quad \Box = \partial_\mu \partial^\mu. \tag{3.1841}$$

Einstein's equations can now be written as

$$-\frac{1}{2}\Box\left(h^{\mu\nu} - \frac{1}{2}h\eta^{\mu\nu}\right) = 8\pi\frac{G}{c^4}T^{\mu\nu}. \tag{3.1842}$$

Since only the 00-component gives a sizable contribution, we obtain

$$\Box\left(h^{00} - \frac{1}{2}h\right) = -16\pi\frac{G}{c^2}\rho. \tag{3.1843}$$

Since the source was moving slowly we can drop the 0-component of the derivative $\partial_0 = \frac{1}{c}\partial_t$. Thus,

$$\nabla^2\left(h^{00} - \frac{1}{2}h\right) = 16\pi\frac{G}{c^2}\rho. \tag{3.1844}$$

Hence, with

$$h^{00} - \frac{1}{2}h = \frac{4}{c^2}\phi, \tag{3.1845}$$

we find $\nabla^2\phi = 4\pi G\rho$.

2.118

The metric outside of the Earth is given by

$$ds^2 = (1 + 2\Phi)dt^2 + (1 + 2\Phi)^{-1}dr^2 - r^2 d\Omega^2, \tag{3.1846}$$

where $\Phi = -\frac{GM}{r}$ is the gravitational potential and $d\Omega^2 = d\theta^2 + \sin^2\theta d\phi^2$. A satellite is orbiting at a distance R_1 from the surface. We are interested the eigentime for the satellite to complete a full orbit around the Earth. Therefore, we have $R = R_E + R_1$ and due to the spherical symmetry we can assume $\theta = \frac{\pi}{2}$. Thus, the metric can be written as

$$ds^2 = (1+2\Phi)dt^2 - (1+2\Phi)^{-1}dr^2 - R^2 d\phi^2. \tag{3.1847}$$

The Lagrangian is given by

$$\mathcal{L} = 1 = (1+2\Phi)\dot{t}^2 - (1+2\Phi)^{-1}\dot{r}^2 - R^2\dot{\phi}^2. \tag{3.1848}$$

The Euler–Lagrange equations are then given by

$$\ddot{r} - \frac{2GM}{r^2}(1+2\Phi)^{-1}\dot{r}^2 - \frac{GM}{r^2}(1+2\Phi)\dot{t}^2 + r(1+2\Phi)\dot{\phi}^2 = 0, \tag{3.1849}$$

$$\frac{d}{d\tau}(\dot{t}) = 0, \tag{3.1850}$$

$$\frac{d}{d\tau}(\dot{\phi}) = 0. \tag{3.1851}$$

The last two equations can be integrated to give

$$\dot{t} = \alpha, \tag{3.1852}$$

$$\dot{\phi} = \beta, \tag{3.1853}$$

where α and β are constants. From the first equation, we can determine a relation between $\dot{t}$ and $\dot{\phi}$, using $\dot{r} = 0$ and $\ddot{r} = 0$. Inserting the expressions for $\dot{t}$ and $\dot{\phi}$, we get

$$R\beta^2 - \frac{GM}{R^2}\alpha^2 = 0. \tag{3.1854}$$

Furthermore, the Lagrangian can be reduced, using $\dot{r} = 0$, to

$$\mathcal{L} = 1 = (1+2\Phi)\alpha^2 - R^2\beta^2, \tag{3.1855}$$

which can be solved for $\alpha = \frac{1}{\sqrt{1+\frac{3}{2}\Phi}}$ and $\beta = \sqrt{\frac{\Phi}{R(2+3\Phi)}}$, and thus, from $\frac{d\phi}{d\tau} = \beta$, we get

$$2\pi = \beta\Delta\tau \quad \Rightarrow \quad \Delta\tau = 2\pi\sqrt{\frac{R(2+3\Phi)}{\Phi}}. \tag{3.1856}$$

We can also determine t in terms of τ as

$$\Delta t = \alpha\Delta\tau = \frac{\Delta\tau}{\sqrt{1+\frac{3}{2}\Phi}}. \tag{3.1857}$$

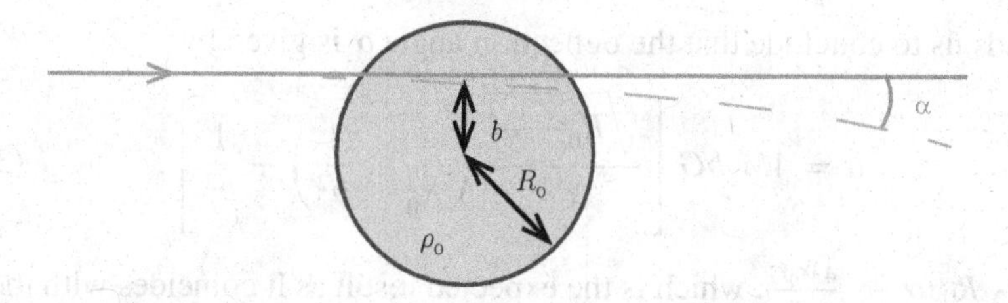

Figure 3.17 Setup of the spherical body of radius R_0 and impact parameter b and deflection angle α of the neutrino, respectively.

2.119

See Figure 3.17 for the setup of the problem. The deflection angle due to gravitational lensing in the weak-field limit is given by

$$\alpha = \left| 2 \int (\nabla \phi)_\perp \, dt \right|, \tag{3.1858}$$

where ϕ is the Newtonian gravitational potential and $(\nabla\phi)_\perp$ is the component of its gradient which is perpendicular to the zeroth order direction of motion. We pick a coordinate system such that the zeroth order worldline of the neutrinos is given by $\boldsymbol{x}(t) = t\boldsymbol{e}_1 + b\boldsymbol{e}_2$. It follows that

$$\alpha = 2 \left| \int_{-\infty}^{\infty} \boldsymbol{e}_2 \cdot \nabla\phi \, dt \right| = 4 \left| \int_0^{\infty} \boldsymbol{e}_2 \cdot \nabla\phi \, dt \right|. \tag{3.1859}$$

The gradient $\nabla\phi$ is given by

$$\nabla\phi = \boldsymbol{e}_r \partial_r \phi = \frac{GM(r)}{r^2} \boldsymbol{e}_r, \tag{3.1860}$$

where $M(r)$ is the mass inside the radius r. This mass is given by

$$M(r) = \begin{cases} M_0 \frac{r^3}{R_0^3}, & r < r_0 \\ M_0, & r \geq r_0 \end{cases}. \tag{3.1861}$$

Noting that $r^2 = t^2 + b^2$, it follows that

$$\begin{aligned} \int_0^{\infty} \boldsymbol{e}_2 \cdot \nabla\phi \, dt &= \int_0^{\infty} \frac{GbM(r)}{r^3} dt = GM_0 \int_0^{\sqrt{R_0^2 - b^2}} \frac{b}{R_0^3} dt + GM_0 \int_{\sqrt{R_0^2 - b^2}}^{\infty} \frac{b}{r^3} dt \\ &= M_0 bG \left[\frac{\sqrt{R_0^2 - b^2}}{R_0^3} + \frac{1}{b^2} - \frac{1}{b^2} \frac{\sqrt{R_0^2 - b^2}}{R_0} \right] \\ &= M_0 bG \left[\frac{\sqrt{R_0^2 - b^2}}{R_0} \left(\frac{1}{R_0^2} - \frac{1}{b^2} \right) + \frac{1}{b^2} \right]. \end{aligned} \tag{3.1862}$$

This leads us to conclude that the deflection angle α is given by

$$\alpha = 4M_0bG\left[\frac{\sqrt{R_0^2 - b^2}}{R_0}\left(\frac{1}{R_0^2} - \frac{1}{b^2}\right) + \frac{1}{b^2}\right]. \tag{3.1863}$$

For $b \to R_0$, $\alpha \to \frac{4M_0G}{b}$, which is the expected result as it coincides with the result for the deflection outside of a spherical mass distribution with total mass M_0.

2.120

The Newtonian gravitational potential Φ satisfies Poisson's equation

$$\nabla^2\Phi = 4\pi G\rho \tag{3.1864}$$

in the region $r > r_s$. Due to the symmetry of the problem, we will assume that Φ is spherically symmetric and introduce the function $f(r) = r\Phi(r)$, with which the differential equation for the gravitational potential takes the form

$$f''(r) = 4\pi rG\rho(r). \tag{3.1865}$$

With the assumption of the NFW halo profile for $\rho(r)$, we find that

$$f''(r) = \frac{4\pi Gk}{r^2} \quad\Longrightarrow\quad f(r) = -4\pi Gk\ln\left(\frac{r}{R}\right) + Cr, \tag{3.1866}$$

where C and R are integration constants. The resulting expression for the gravitational potential is therefore

$$\Phi(r) = -\frac{4\pi Gk}{r}\ln\left(\frac{r}{R}\right) + C. \tag{3.1867}$$

The gradient of the potential is now

$$\nabla\Phi = \mathbf{e}_r\Phi'(r) = \frac{4\pi Gk}{r^2}\ln\left(\frac{r}{eR}\right) = \mathbf{e}_r\left[\frac{4\pi Gk}{r^2}\ln\left(\frac{r}{r_s}\right) + \frac{K}{r^2}\right], \tag{3.1868}$$

where we have introduced the new constant $K = 4\pi Gk\ln(r_s/eR)$. The constant K can be related to the mass M_0 enclosed within $r < r_s$ according to the relation

$$\Phi'(r_s) = \frac{GM_0}{r_s^2} = \frac{K}{r_s^2} \quad\Longrightarrow\quad K = GM_0. \tag{3.1869}$$

Letting the zeroth order worldline of the light signal be given by $\mathbf{x}(t) = t\mathbf{e}_1 + r_0\mathbf{e}_2$, the first-order lensing is given by

$$\theta \simeq 2\int \mathbf{e}_2\cdot\nabla\Phi\,dt = 2\int_{-\infty}^{\infty}\frac{GM_0r_0}{\sqrt{r_0^2+t^2}^{\,3}}dt + 4\pi Gk\int_{-\infty}^{\infty}\frac{\ln[(t^2+r_0^2)/r_s^2]}{\sqrt{r_0^2+t^2}^{\,3}}dt. \tag{3.1870}$$

The first of these integrals is the same integral that we have encountered for the gravitational lensing of a point source. The second integral can be solved through a lengthy process involving substitutions and partial integrations to give the final result

$$\theta \simeq \frac{4GM_0}{r_0} + \frac{16\pi Gk}{r_0} \ln\left(\frac{er_0}{2r_s}\right) \simeq \frac{4GM_0}{r_0} + \frac{16\pi Gk}{r_0}\left[\ln\left(\frac{r_0}{r_s}\right) + 0.31\right], \tag{3.1871}$$

where $e = 2.71828\ldots$ is the base of the natural logarithm.

2.121

a) Consider two observers A and B located at spacetime points x_A^μ and x_B^μ, respectively. Suppose that A is sending out light waves at the rate n per coordinate time interval Δ. In the (local) rest frame of an observer, the time interval Δ is related to the proper time interval by the factor $\sqrt{g_{00}}$. It follows that the relation between the number of light waves received by B per unit proper time and the number of light waves emitted by A per unit proper time is

$$\frac{\nu_B}{\nu_A} = \sqrt{\frac{g_{00}(x_A)}{g_{00}(x_B)}}, \quad \frac{\lambda_B}{\lambda_A} = \sqrt{\frac{g_{00}(x_B)}{g_{00}(x_A)}}, \tag{3.1872}$$

where ν_A (λ_A) is the frequency (wavelength) of the light emitted by A and ν_B (λ_B) is the frequency (wavelength) of the light received by B. Note that a frequency ν and its corresponding wavelength λ are related as $\nu\lambda = c$. The light is traveling along a *null geodesic* (i.e., a geodesic such that the tangent vector at each point is lightlike) from A to B, but we do not need the explicit solution of the geodesic equations.

b) In general, if the radial coordinate of B is $r_B \gg r_*$ in the Schwarzschild spacetime with $g_{00} = 1 - r_*/r$, then

$$\frac{\nu_B}{\nu_A} \simeq \sqrt{1 - \frac{r_*}{r_A}}, \quad \frac{\lambda_B}{\lambda_A} \simeq \frac{1}{\sqrt{1 - \frac{r_*}{r_A}}}. \tag{3.1873}$$

Furthermore, if it holds for the radial coordinate of A that $r_A \gg r_*$, then

$$\frac{\nu_B}{\nu_A} \simeq 1 - \frac{1}{2}\frac{r_*}{r_A}, \quad \frac{\lambda_B}{\lambda_A} \simeq 1 + \frac{1}{2}\frac{r_*}{r_A}. \tag{3.1874}$$

This means that the frequency observed by B is actually smaller than the frequency emitted by A and the wavelength observed by B is longer than the wavelength emitted by A, i.e., the light is *redshifted*. The redshift z is defined as $z \equiv \lambda_B/\lambda_A - 1$.

Finally, assume that light is emitted by A at r_A (e.g., from the surface of a star) and the light is observed by B far away (on Earth), i.e., $r_B \equiv r_\infty \gg r_*$, then the formula for the *gravitational redshift* is given by

$$z_\infty \equiv \frac{\lambda_\infty}{\lambda_A} - 1 \simeq \frac{1}{\sqrt{1 - \frac{r_*}{r_A}}} - 1 = \frac{1}{\sqrt{1 - \frac{2GM}{c^2 r_A}}} - 1. \tag{3.1875}$$

Note that for light emitted at $r \to r_*$, the redshift grows to infinity, whereas for light emitted from $r_A \gg r_*$, there is no redshift. In the Newtonian limit, when r_A is sufficiently large compared to r_*, i.e., $r_A \gg r_*$, the gravitational redshift can be approximated as

$$z_\infty \simeq \frac{1}{2}\frac{r_*}{r_A} = \frac{GM}{c^2 r_A}. \tag{3.1876}$$

White dwarfs like Sirius B and 40 Eridani B do show gravitational redshifts in the range between 10^{-4} and 10^{-5}, which are of the right order of magnitude. More reliable and quantitatively accurate measurements are possible only in terrestrial experiments. For example, in 1960, Pound and Rebka measured the change of frequency of a γ-ray photon emitted by an excited iron nucleus as it fell from a height of 18–21 m. When the photon falls from a height h, the change in the Newtonian gravitational potential is gh, where g is the acceleration due to gravity on the Earth's surface. Since $g_{00} - 1$ is approximately given by the Newtonian gravitational potential, the frequency increases by a factor $1 + gh/c^2$. The fraction gh/c^2 is small, about 10^{-15}, but it can still be measured, confirming the gravitational redshift effect.

2.122

The surface $(ct)^2 - x^2 - y^2 = -K^2$, where $K > 0$, can be written as $(ct)^2 - r^2 = -K^2$ if we introduce polar coordinates such that $(t, x, y) = (t, r\cos\phi, r\sin\phi)$. Thus, we have

$$r = \sqrt{c^2t^2 + K^2}. \tag{3.1877}$$

Therefore, the position vector is $(t, x, y) = \left(t, \sqrt{c^2t^2 + K^2}\cos\phi, \sqrt{c^2t^2 + K^2}\sin\phi\right)$. Differentiating the position vector, we obtain

$$\mathbf{u} \equiv \frac{\partial}{\partial t}(t, x, y) = \left(1, \frac{c^2 t}{r}\cos\phi, \frac{c^2 t}{r}\sin\phi\right), \tag{3.1878}$$

$$\mathbf{v} \equiv \frac{\partial}{\partial \phi}(t, x, y) = (0, -r\sin\phi, r\cos\phi). \tag{3.1879}$$

Using $\mathbf{u}$ and $\mathbf{v}$, we find the components of the metric as

$$g_{tt} = \mathbf{u}\cdot\mathbf{u} = c^2 - \frac{c^4t^2}{r^2} = \frac{c^2K^2}{c^2t^2 + K^2} > 0, \tag{3.1880}$$

$$g_{\phi\phi} = \mathbf{v}\cdot\mathbf{v} = -r^2 = -(c^2t^2 + K^2) < 0, \tag{3.1881}$$

$$g_{t\phi} = g_{\phi t} = \mathbf{u}\cdot\mathbf{v} = 0. \tag{3.1882}$$

Since the metric components are independent of ϕ, the vector field ∂_ϕ is a Killing vector field and with $N = \alpha\partial_t + \beta\partial_\phi$ being the 4-frequency of a light pulse sent from A at time coordinate t to B at time coordinate t', this implies that

$$-g(\partial_\phi, N) = -\beta g_{\phi\phi} = k, \tag{3.1883}$$

is a constant. Furthermore, as the 4-frequency is a null vector, we find

$$g(N,N) = g_{tt}\alpha^2 + g_{\phi\phi}\beta^2 = 0 \quad \Longrightarrow \quad \alpha = \sqrt{\frac{-\beta^2 g_{\phi\phi}}{g_{tt}}} = \frac{k}{\sqrt{-g_{tt}g_{\phi\phi}}}. \tag{3.1884}$$

For any comoving observer with constant ϕ, the 4-velocity of the observer is given by $V = \gamma\partial_t$. Normalizing this to one we obtain

$$g(V,V) = \gamma^2 g_{tt} = 1 \quad \Longrightarrow \quad \gamma = \frac{1}{\sqrt{g_{tt}}}. \tag{3.1885}$$

The frequency ω measured by such an observer is given by the inner product of the 4-frequency with the 4-velocity, i.e.,

$$\omega = g(N,V) = \alpha\gamma g_{tt} = \frac{k}{\sqrt{-g_{tt}g_{\phi\phi}}}\frac{1}{\sqrt{g_{tt}}}g_{tt} = \frac{k}{\sqrt{-g_{\phi\phi}}}. \tag{3.1886}$$

It follows that the frequencies of the light sent from A at time t and received by B at time t' are related as

$$\frac{\omega_B}{\omega_A} = \frac{\frac{k}{\sqrt{-g_{\phi\phi}(t')}}}{\frac{k}{\sqrt{-g_{\phi\phi}(t)}}} = \sqrt{\frac{g_{\phi\phi}(t')}{g_{\phi\phi}(t)}} = \sqrt{\frac{c^2t'^2 + K^2}{c^2t^2 + K^2}}. \tag{3.1887}$$

2.123

Assume that the light signal has 4-frequency N. Since ∂_t is a Killing vector field of the Schwarzschild metric, we know that

$$g(N, \partial_t) = g_{00}N^t = E, \tag{3.1888}$$

is a constant. Furthermore, since the 4-frequency is a null vector, it holds that

$$g(N,N) = g_{00}(N^t)^2 + g_{rr}(N^r)^2 = \frac{E^2}{g_{00}} + g_{rr}(N^r)^2 = 0 \quad \Longrightarrow \quad N^r = -E, \tag{3.1889}$$

where we have used that $g_{rr} = -1/g_{00}$ and that the light is moving radially toward smaller r. When the light signal is emitted from $r \to \infty$, we have $g_{00} \to 1$ and thus $E = \omega_0$, where ω_0 is the frequency observed by a stationary observer at infinity. The 4-frequency is therefore generally given by

$$N = \frac{E}{g_{00}}\partial_t - E\partial_r. \tag{3.1890}$$

For the 4-velocity V of the space ship, we know that

$$V = \frac{dt}{d\tau}\partial_t + \frac{dr}{d\tau}\partial_r = \frac{dt}{d\tau}(\partial_t + v_c\partial_r), \tag{3.1891}$$

where $v_c = dr/dt = -0.1$ is the coordinate velocity. Normalizing the 4-velocity to one, we find that

$$1 = g(V,V) = \left(\frac{dt}{d\tau}\right)^2 \left[g_{00} + g_{rr}v_c^2\right] \implies \frac{dt}{d\tau} = \frac{1}{\sqrt{g_{00} + g_{rr}v_c^2}}. \tag{3.1892}$$

The frequency observed by the space ship is now given by

$$\omega = g(N,V) = g_{00}\frac{E}{g_{00}}\frac{dt}{d\tau} - g_{rr}Ev_c\frac{dt}{d\tau} = \omega_0\frac{1 - g_{rr}v_c}{\sqrt{g_{00} + g_{rr}v_c^2}}. \tag{3.1893}$$

Using that $g_{rr} = -1/g_{00}$ now leads to

$$\omega = \omega_0\frac{1 + v_c/g_{00}}{\sqrt{g_{00} - v_c^2/g_{00}}} = \frac{\omega_0}{\sqrt{g_{00}}}\frac{g_{00} + v_c}{\sqrt{g_{00}^2 - v_c^2}} = \frac{\omega_0}{\sqrt{g_{00}}}\sqrt{\frac{g_{00} + v_c}{g_{00} - v_c}}. \tag{3.1894}$$

Using that $\lambda \propto 1/\omega$ and inserting $g_{00} = 1 - 2GM/r$ the yields

$$\lambda = \lambda_0\sqrt{1 - \frac{2GM}{r}}\sqrt{\frac{1 - \frac{2GM}{r} - v_c}{1 - \frac{2GM}{r} + v_c}}. \tag{3.1895}$$

Inserting the values provided in the problem, we find that

$$\lambda \simeq 4\,420 \text{ Å}. \tag{3.1896}$$

2.124

The gravitational redshift observed far away for light emitted from the surface of a star (when the mass of the star is $M = 2 \cdot 10^{30}$ kg and $r_{\text{star}} = 7 \cdot 10^8$ m) is given by

$$z = \frac{\lambda_\infty - \lambda_{\text{star}}}{\lambda_{\text{star}}} \simeq \frac{GM}{c^2 r_{\text{star}}} \approx 2 \cdot 10^{-6}. \tag{3.1897}$$

See Problem 2.121 for a derivation of the formula for the gravitational redshift.

2.125

The Schwarzschild metric is given by

$$ds^2 = \left(1 - \frac{2GM}{c^2 r}\right)(dx^0)^2 - \left(1 - \frac{2GM}{c^2 r}\right)^{-1} dr^2 - r^2 d\Omega^2. \tag{3.1898}$$

From this metric, one obtains

$$g_{00}(r) = 1 - \frac{2GM}{c^2 r}. \tag{3.1899}$$

Now, one finds the ratio between the observed frequency ν' and the emitted frequency ν as (see the discussion in the solution to Problem 2.121)

$$\frac{\nu'}{\nu} = \sqrt{\frac{g_{00}(r)}{g_{00}(r')}} = \sqrt{\frac{1 - \frac{2GM}{c^2 r}}{1 - \frac{2GM}{c^2 r'}}}, \tag{3.1900}$$

where r is the solar radius and r' is the average Sun-Earth distance.

Then, since $r' \gg r$ and $r \gg 2GM/c^2$, one has

$$\frac{\nu'}{\nu} \simeq \sqrt{1 - \frac{2GM}{c^2 r}} \simeq 1 - \frac{GM}{c^2 r}. \tag{3.1901}$$

Finally, one obtains the magnitude and sign of the relative frequency shift $\Delta\nu/\nu$ of the spectral line as

$$\frac{\Delta\nu}{\nu} \equiv \frac{\nu' - \nu}{\nu} = \frac{\nu'}{\nu} - 1 \simeq -\frac{GM}{c^2 r} \approx -1.9 \cdot 10^{-6}. \tag{3.1902}$$

In addition, one can note that the *redshift* z is defined as and given by

$$z \equiv \frac{\Delta\lambda}{\lambda} \equiv \frac{\lambda' - \lambda}{\lambda} = \frac{\lambda'}{\lambda} - 1 = \{\lambda\nu = c\} = \frac{\nu}{\nu'} - 1 \simeq \frac{GM}{c^2 r} \approx 1.9 \cdot 10^{-6}. \tag{3.1903}$$

2.126

When the spaceship is sending the light signal to the Earth, there are two effects: One is the redshift due to the Doppler effect, the other is a blueshift due to the gravitational pull from the Earth.

Let the frequency of the emitted light be ν. The redshift is then given by the z-factor, which is

$$z_r = \frac{\nu}{\nu_{\text{obs}}} - 1 \simeq \frac{v}{c}, \tag{3.1904}$$

where ν_{obs} is the observed frequency and v is the velocity of the spaceship. Using $v = 100$ m/s and $c = 3 \cdot 10^8$ m/s, we obtain for the redshift $z_r \approx 3.3 \cdot 10^{-7}$.

For the blueshift due the the mass of the Earth, we use the Schwarzschild metric. Thus, we have

$$z_b = \frac{\nu}{\nu_{\text{obs}}} - 1 = \sqrt{\frac{g_{00}(R)}{g_{00}(R+h)}} - 1, \quad g_{00}(r) = 1 - \frac{2GM}{c^2 r}, \tag{3.1905}$$

where R is the radius of the Earth and h is the altitude of the spaceship. Since h/R ($\approx 0.16 \ll 1$) is small, we obtain

$$z_b \simeq -\frac{GM}{c^2}\left(\frac{1}{R} - \frac{1}{R+h}\right) \simeq -\frac{GMh}{c^2 R^2} = -\frac{gh}{c^2}, \tag{3.1906}$$

where g is the acceleration at the Earth, i.e., $g \simeq 9.8 \text{ m/s}^2$. Since $h = 10^6$ m, we obtain for the blueshift $z_b \approx -1.1 \cdot 10^{-10}$.

In conclusion, the redshift is much larger (i.e., 3 000 times) than the blueshift. Thus, in this case, the Doppler effect is the most important physical effect.

2.127

The gravitational potential of the Sun at the Earth is given by

$$\phi_\odot = -\frac{GM_\odot}{r}, \tag{3.1907}$$

where r is the distance between the Sun and the Earth. Thus, the gravitational potential is

$$\phi_\odot = -\frac{1.3 \cdot 10^{20} \text{ m}^3/s^2}{1.5 \cdot 10^{11} \text{ m}} \simeq -9 \cdot 10^8 \text{ m}^2/s^2, \tag{3.1908}$$

which is an order of magnitude larger than the gravitational potential of the Earth at its surface. Thus, we can neglect the influence of the gravitational field of the Earth. The relation between the frequencies at ($\nu_\oplus$) and far away from (ν_∞) the Earth will be

$$\frac{\nu_\infty}{\nu_\oplus} \simeq \sqrt{1 + \frac{2\phi_\odot}{c^2}} \simeq 1 + \frac{\phi_\odot}{c^2}, \tag{3.1909}$$

for small $\phi_\odot/c^2$. See the general discussion of this relation in the solution to Problem 2.125. The redshift parameter z is given by

$$z = \frac{\lambda_\oplus - \lambda_\infty}{\lambda_\infty} = \frac{\nu_\infty}{\nu_\oplus} - 1 \simeq \frac{\phi_\odot}{c^2} \simeq -\frac{9 \cdot 10^8 \text{ m}^2/s^2}{(3 \cdot 10^8 \text{ m/s})^2} = -10^{-8}, \tag{3.1910}$$

(i.e., a blueshift, since $z < 0$).

2.128

We call the free-falling observer A and the observer at infinity B. The motion of the free-falling observer is governed by the radial differential equation

$$E - \frac{\dot{r}^2}{2} = \frac{1}{2}\left(1 - \frac{r_*}{r}\right), \tag{3.1911}$$

where E is a constant of motion given by $\sqrt{2E} = g(\partial_t, U)$, where U is the 4-velocity of the observer A. Note that the constant of motion $g(\partial_\varphi, U) = 0$, since the motion of A is assumed to be purely radial. In order to have a velocity just large enough to escape to infinity, we require that $\dot{r} \to 0$ as $r \to \infty$, leading to $E = 1/2$ and therefore

$$\dot{r} = \sqrt{\frac{r_*}{r}}. \tag{3.1912}$$

From the expression for E in terms of $g(\partial_t, U)$, we also find

$$1 = g(\partial_t, U) = g_{tt}\dot{t} \implies \dot{t} = \frac{r}{r - r_*}. \tag{3.1913}$$

The 4-velocity of A is therefore

$$U = \frac{r}{r - r_*}\partial_t + \sqrt{\frac{r_*}{r}}\partial_r. \tag{3.1914}$$

We now let the 4-frequency of the light be given by

$$N = \alpha\partial_t + \beta\partial_r. \tag{3.1915}$$

Since the 4-frequency is lightlike, we must have

$$g(N, N) = \alpha^2\frac{r - r_*}{r} - \beta^2\frac{r}{r - r_*} = 0 \implies \beta = \alpha\frac{r - r_*}{r}. \tag{3.1916}$$

The frequency emitted by A at $r = r_A$ is given by

$$f_0 = g(N, U) = g_{tt}\alpha_A\dot{t} + g_{rr}\dot{r}\beta_A = \alpha_A\left(1 - \sqrt{\frac{r_*}{r_A}}\right), \tag{3.1917}$$

leading to

$$\alpha_A = \frac{f_0}{1 - \sqrt{R/r_A}}. \tag{3.1918}$$

Furthermore, we know that N is tangent to and parallel along the lightlike geodesic describing the worldline of the light signal. Since ∂_t is a Killing vector field of the Schwarzschild spacetime, we find that

$$g(\partial_t, N) = \frac{r - r_*}{r}\alpha = \frac{r_A r_*}{r_A}\alpha_A = \frac{r_A - r_*}{r_A}\frac{f_0}{1 - \sqrt{r_*/r_A}} \tag{3.1919}$$

is a constant. For observer B at rest at infinity, the 4-velocity is $V = \partial_t$ and the frequency observed by B is therefore

$$f = g(V, N) = g(\partial_t, N) = \frac{f_0}{r_A}\frac{r_A - r_*}{1 - \sqrt{r_*/r_A}}. \tag{3.1920}$$

2.129

Light signals follow geodesics and their 4-frequency N is proportional to the tangent of the affinely parametrized geodesic. Since $K = \partial_t$ is a Killing vector field, we know that

$$F = g(\partial_t, N) = g_{tt}N^t = x^2 N^t, \tag{3.1921}$$

is constant, which means that $N^t = F/x^2$. Since the light signals follow a lightlike worldline:

$$g(N,N) = x^2(N^t)^2 - (N^x)^2 = \frac{F^2}{x^2} - (N^x)^2 = 0 \quad \Rightarrow \quad N^x = +\frac{F}{x}. \tag{3.1922}$$

The falling observer is also following a geodesic, so we also know that

$$Q = g(\partial_t, \dot{\gamma}_0) = g_{tt}\dot{t} = x^2\dot{t}, \tag{3.1923}$$

is constant along its worldline γ_0. Furthermore, $g(\dot{\gamma}_0, \dot{\gamma}_0) = \frac{Q^2}{x^2} - \dot{x}^2 = 1$. Since the observer falls from $x = x_0$, we find that $Q = +x_0$. We can conclude that $\dot{t} = \frac{x_0}{x^2}$, $\dot{x} = -\sqrt{\frac{x_0 r}{x^2} - 1}$ for the falling observer. The emitted frequency f_e is given by

$$\begin{aligned} f_e = g(\dot{\gamma}_0, N)|_{x=x_e} = x_e^2\dot{t}N^t - \dot{x}N^x &= x_e^2\frac{x_0}{x_e^2}\frac{F}{x_e^2} + \sqrt{\frac{x_0^2}{x_e^2} - 1}\frac{F}{x_e} \\ &= \frac{F}{x_e^2}\left[x_0 + \sqrt{x_0^2 - x_e^2}\right]. \end{aligned} \tag{3.1924}$$

The observer at $x = x_1$ with worldline γ_1 has $\dot{x} = 0$ and therefore $g(\dot{\gamma}_1, \dot{\gamma}_1) = x_1^2\dot{t}^2 = 1$, which means that $\dot{t} = x_1^{-1}$. The observed frequency is therefore

$$f_{\text{obs}} = g(\dot{\gamma}_1, N) = x_1^2\frac{1}{x_1}N^t = \frac{F}{x_1}, \tag{3.1925}$$

and the ratio f_e/f_{obs} is, thus, given by

$$\frac{f_e}{f_{\text{obs}}} = \frac{x_1}{x_e^2}\left[x_0 + \sqrt{x_0^2 - x_e^2}\right]. \tag{3.1926}$$

The redshift z is therefore

$$z = \frac{f_e}{f_{\text{obs}}} - 1 = \frac{x_1}{x_e^2}\left[x_0 + \sqrt{x_0^2 - x_e^2}\right] - 1. \tag{3.1927}$$

2.130

Consider the Robertson–Walker metric for $d\Omega = 0$, i.e.,

$$ds^2 = c^2dt^2 - \frac{S(t)^2}{1 - kr^2}dr^2. \tag{3.1928}$$

Using $ds^2 = 0$, which holds for the path of a light signal, we find that

$$0 = c^2dt^2 - \frac{S(t)^2}{1 - kr^2}dr^2 \quad \Rightarrow \quad \frac{c}{S(t)}dt = \frac{1}{\sqrt{1 - kr^2}}dr, \tag{3.1929}$$

where we assumed that $dr/dt > 0$, i.e., propagation *forward* in time. Since the observers are at rest with respect to r, we must have

$$\int_{t_0}^{t_1} \frac{dt}{S(t)} = \int_{t_0+\epsilon}^{t_1+\epsilon'} \frac{dt}{S(t)}, \tag{3.1930}$$

from which, in the limit $\epsilon \to 0$, we obtain

$$\frac{\epsilon}{S(t_0)} \simeq \frac{\epsilon'}{S(t_1)}. \tag{3.1931}$$

Thus, the cosmological redshift z is given by

$$1 + z \equiv \frac{\epsilon}{\epsilon'} = \frac{S(t_0)}{S(t_1)} \quad \Rightarrow \quad z = \frac{S(t_0)}{S(t_1)} - 1. \tag{3.1932}$$

2.131

The Robertson–Walker metric for $k = 1$ is given by

$$ds^2 = c^2 dt^2 - S(t)^2 \left[d\chi^2 + \sin^2 \chi \, (d\theta^2 + \sin^2 \theta d\phi^2)\right]. \tag{3.1933}$$

Construct the Lagrangian

$$\mathscr{L} = c^2 \dot{t}^2 - S(t)^2 \left[\dot{\chi}^2 + \sin^2 \chi \, (\dot{\theta}^2 + \sin^2 \theta \, \dot{\phi}^2)\right], \tag{3.1934}$$

where the dot indicates differentiation with respect to the path parameter s. Using Euler–Lagrange equations yields the differential equations for the geodesics, i.e.,

$$c^2 \ddot{t} + S(t) S'(t) \left[\dot{\chi}^2 + \sin^2 \chi \, (\dot{\theta}^2 + \sin^2 \theta \, \dot{\phi}^2)\right] = 0, \tag{3.1935}$$

$$\frac{d}{ds}\left[S(t)^2 \, \dot{\chi}\right] - \frac{1}{2} S(t)^2 \sin 2\chi \, (\dot{\theta}^2 + \sin^2 \theta \, \dot{\phi}^2) = 0, \tag{3.1936}$$

$$\frac{d}{ds}\left[S(t)^2 \sin^2 \chi \, \dot{\theta}\right] - \frac{1}{2} S(t)^2 \sin 2\theta \sin^2 \chi \, \dot{\phi}^2 = 0, \tag{3.1937}$$

$$\frac{d}{ds}\left[S(t)^2 \sin^2 \chi \sin^2 \theta \, \dot{\phi}\right] = 0. \tag{3.1938}$$

2.132

a) The metric in the χ' coordinates becomes

$$\begin{aligned} g'_{ab} &= \frac{\partial \chi^c}{\partial \chi'^a} \frac{\partial \chi^d}{\partial \chi'^b} g_{cd} \simeq \left(\delta^c_a - \varepsilon \partial_a \xi^c\right)\left(\delta^d_b - \varepsilon \partial_b \xi^d\right)\left(\eta_{cd} + \varepsilon h_{cd}\right) \\ &\simeq \eta_{ab} + \varepsilon (h_{ab} - \partial_a \xi_b - \partial_b \xi_a), \end{aligned} \tag{3.1939}$$

where $\simeq$ denotes equality up to linear order in ε. This is equivalent to

$$g'_{ab} = \eta_{ab} + \varepsilon h'_{ab}, \tag{3.1940}$$

with

$$h'_{ab} = h_{ab} - \partial_a \xi_b - \partial_b \xi_a. \tag{3.1941}$$

Note that, since the term including the perturbations is linear in ε already, the indices of ξ^a are can be raised and lowered with the Minkowski metric without affecting this statement and the lowering of the indices inside the partial derivatives is consistent to linear order in ε. Thus, we are still in the weak-field regime after the coordinate change with the new metric perturbation given above. Furthermore, we find that

$$g^{ab}\nabla_a\nabla_b\chi'^c = g^{ab}\nabla_a\nabla_b(\chi^c + \varepsilon\xi^c) = 0, \tag{3.1942}$$

as long as both χ^c and ξ^c are harmonic functions. The coordinate change therefore also preserves the harmonic gauge condition.

b) In the harmonic gauge, we find that

$$g^{ab}\nabla_a\nabla_b\chi^c = g^{ab}\nabla_a\delta^c_b = g^{ab}\Gamma^c_{ab} = 0, \tag{3.1943}$$

where we note that $\nabla_b\chi^c = \partial_b\chi^c = \delta^c_b$ and c is just a counter for the coordinate functions, not a tensor index. In the weak-field limit, we also have

$$\Gamma^c_{ab} \simeq \frac{\varepsilon}{2}\eta^{cd}(\partial_a h_{db} + \partial_b h_{ad} - \partial_d h_{ab}) = \frac{\varepsilon}{2}\left(\partial_a h^c_b + \partial_b h^c_a - \partial^c h_{ab}\right). \tag{3.1944}$$

Contracting this with g^{ab} now leads to

$$\begin{aligned} g^{ab}\Gamma^c_{ab} &\simeq \frac{\varepsilon}{2}\eta^{ab}\left(\partial_a h^c_b + \partial_b h^c_a - \partial^c h_{ab}\right) \\ &= \varepsilon\left(\partial_a h^{ac} - \frac{1}{2}\partial^c h\right) = \varepsilon\partial_a\left(h^{ac} - \frac{1}{2}\eta^{ac}h\right) \\ &= \varepsilon\partial_a\bar{h}^{ac} = 0. \end{aligned} \tag{3.1945}$$

By lowering the index c follows that $\partial^a\bar{h}_{ac} = 0$.

c) From a) follows that

$$\bar{h}'_{ab} = \bar{h}_{ab} - \partial_a\xi_b - \partial_b\xi_a + \eta_{ab}\partial_c\xi^c. \tag{3.1946}$$

With $\xi^a = iC^a\exp(ik\cdot\chi)$, where $k\cdot\chi = k_c\chi^c$, the partial derivatives of ξ_a are given by

$$\partial_b\xi_a = \eta_{ac}\partial_b\xi^c = \eta_{ac}iC^c ik_b\exp(ik\cdot\chi) = -C_a k_b\exp(ik\cdot\chi), \tag{3.1947}$$

and thus, we obtain

$$A'_{ab} = A_{ab} + C_a k_b + k_a C_b - \eta_{ab}k_c C^c. \tag{3.1948}$$

d) From the harmonic gauge condition, we find that

$$\partial^a \bar{h}_{ab} = A_{ab} i k^a \exp(ik \cdot \chi) = 0, \tag{3.1949}$$

i.e., with the given $k_0 = k_3 = 1$ (or, equivalently, $k^0 = -k^3 = 1$),

$$A_{0b}k^0 + A_{3b}k^3 = A_{0b} - A_{3b} = 0 \implies A_{0b} = A_{3b}, \tag{3.1950}$$

for all b. Due to the symmetry of A, this also means that $A_{b0} = A_{b3}$. This leads to $A'_{00} = A'_{03} = A'_{30} = A'_{33}$ and it is therefore sufficient to require that $A'_{00} = 0$. Furthermore, we also find that

$$A'_{00} = A_{00} + 2C_0 - C_0 + C_3 = A_{00} + C_0 + C_3 = 0, \tag{3.1951}$$

$$(i \neq 3) \quad A'_{0i} = A_{0i} + C_0 k_i + k_0 C_i = A_{0i} + C_i = 0, \tag{3.1952}$$

$$A'^a_{\ a} = A^a_{\ a} - 2k \cdot C = A^a_{\ a} - 2(C_0 - C_3) = 0. \tag{3.1953}$$

This is a linear set of equations for the four constants C^a with the solution

$$C_0 = \frac{1}{4}A^a_{\ a} - \frac{1}{2}A_{00}, \quad C_3 = -\frac{1}{4}A^a_{\ a} - \frac{1}{2}A_{00}, \quad C_1 = -A_{01}, \quad C_2 = -A_{02}. \tag{3.1954}$$

2.133

a) Consider the quantities $h_{\mu\nu}$ to be perturbations or deviations of the components of the metric tensor away from flat spacetime. In general, for $h_{\mu\nu}$, the Lorenz gauge condition is given by

$$\partial_\nu h^\nu{}_\mu - \frac{1}{2}\partial_\mu h^\nu{}_\nu = 0. \tag{3.1955}$$

Cf. the analogy to this gauge condition in electromagnetism. If the Lorenz gauge condition is fulfilled, then Einstein's equations in vacuum reduce to $\Box \bar{h}_{\mu\nu} = 0$, where $\bar{h}_{\mu\nu} = h_{\mu\nu} - \eta_{\mu\nu}h/2$ with $h = \eta^{\mu\nu}h_{\mu\nu}$. This gauge condition does not completely fix the coordinates, so in order to do so, further conditions can be imposed, e.g., $h_{i0} = 0$ and $\eta^{\mu\nu}h_{\mu\nu} = 0$, which are collectively referred to as the transverse traceless gauge (or TT gauge), where $h_{\mu\nu} = \bar{h}_{\mu\nu}$. In particular, in the TT gauge, the Lorenz gauge condition reduces to $\partial_\nu h^\nu{}_\mu = 0$. The symmetric tensor $h_{\mu\nu}$ has ten independent components and the TT gauge contains eight conditions, which in fact means that only two components of $h_{\mu\nu}$ are independent. The two independent components of $h_{\mu\nu}$ are normally denoted h_+ and $h_\times$, which correspond to the two independent polarization states of a gravitational wave.

b) First, let us consider two particles, which are influenced by a plus-polarized gravitational wave along the z-direction such that

$$(h_{\mu\nu}) = \mathrm{diag}(0, h_+, -h_+, 0), \quad h_+ = h_0 \sin\left[2\pi f(t - z)\right], \quad h_\times = 0, \tag{3.1956}$$

where $h_0 \ll 1$ is the amplitude and f is the frequency.

Then, in fact, the geodesic equation for the displacement vector S_μ is given by

$$\frac{d^2 S_\mu}{dt^2} = \frac{1}{2}\frac{d^2 h_{\mu\nu}}{dt^2} S^\nu, \tag{3.1957}$$

where $h_{\mu\nu}$ is the given gravitational wave. We observe that this gravitational wave affects neither S_0 nor S_3. Thus, the only effect on the geodesics is taking place in the x- and y-directions. Without loss of generality, we can therefore assume that $z = 0$. In this case, the gravitational wave is plus-polarized only, i.e., $h_\times = 0$, so that the geodesic equation simplifies to two equations for $S_1 = -S^1$ and $S_2 = -S^2$, namely

$$\frac{d^2 S_1}{dt^2} = \frac{(2\pi f)^2}{2} h_0 \sin(2\pi f t)\, S_1, \quad \frac{d^2 S_2}{dt^2} = -\frac{(2\pi f)^2}{2} h_0 \sin(2\pi f t)\, S_2, \tag{3.1958}$$

which can be solved perturbatively in h_0. Up to first order in h_0, we obtain

$$S_1(t) = S_1(0)\left[1 - \frac{1}{2} h_0 \sin(2\pi f t) + \cdots\right], \tag{3.1959}$$

$$S_2(t) = S_2(0)\left[1 + \frac{1}{2} h_0 \sin(2\pi f t) + \cdots\right]. \tag{3.1960}$$

Next, the measured distance $\Delta x \equiv S_1(t)$ between the two particles, which was initially the distance $\Delta x_0 \equiv S_1(0)$ along the x-direction, will be

$$\frac{\Delta x}{\Delta x_0} = \frac{S_1(t)}{S_1(0)} \simeq 1 - \frac{1}{2} h_0 \sin(2\pi f t), \tag{3.1961}$$

which means that the relative distance $\delta x \equiv \Delta x - \Delta x_0$ between the two particles oscillate with f. This does not mean that the positions of the particle coordinates change, but the coordinates themselves oscillate.

Finally, assuming d to be the measured distance Δx and L the initial distance Δx_0 between the two particles, we obtain

$$\frac{d}{L} = 1 - \frac{1}{2} h_0 \sin(2\pi f t) \quad \Rightarrow \quad d = \left[1 - \frac{1}{2} h_0 \sin(2\pi f t)\right] L, \tag{3.1962}$$

which is what we wanted to show.

2.134

First, consider the energy–momentum tensor $T^{\mu\nu}$. Using $T^{\mu\nu}$ and its conservation law, i.e., $\nabla_\nu T^{\mu\nu} = 0$, we can write

$$T^{\mu\nu} = \partial_\lambda (T^{\mu\lambda} x^\nu) - (\partial_\lambda T^{\mu\lambda}) x^\nu. \tag{3.1963}$$

Note that in the linearized approximation, $\nabla_\nu T^{\mu\nu} = 0$ reduces to $\partial_\nu T^{\mu\nu} = 0$. For the spatial components of $T^{\mu\nu}$, using the fact that $\partial_0(T^{i0}x^j) - (\partial_0 T^{i0})x^j = 0$, we have

$$T^{ij} = \partial_k(T^{ik}x^j) - (\partial_k T^{ik})x^j = \partial_k(T^{ik}x^j) - [(\partial_\nu T^{i\nu})x^j - (\partial_0 T^{i0})x^j], \quad (3.1964)$$

since $\partial_\nu T^{i\nu} = \partial_0 T^{i0} + \partial_k T^{ik} = 0$, and therefore, we find that

$$T^{ij} = \partial_k(T^{ik}x^j) + (\partial_0 T^{i0})x^j = \partial_k(T^{ik}x^j) + \partial_0(T^{i0}x^j), \quad (3.1965)$$

where we used $\partial_0(T^{i0}x^j) = (\partial_0 T^{i0})x^j + T^{i0}\partial_0 x^j = (\partial_0 T^{i0})x^j$. Similarly, for the $T^{i0}x^j$ term, we use the same rewriting technique, i.e.,

$$\begin{aligned} T^{i0}x^j &= \partial_k(T^{k0}x^j x^i) - \partial_k(T^{k0}x^j)x^i = \partial_k(T^{k0}x^j x^i) - (\partial_k T^{k0})x^j x^i - T^{k0}(\partial_k x^j)x^i \\ &= \partial_k(T^{k0}x^j x^i) - (\partial_k T^{k0})x^j x^i - T^{j0}x^i. \end{aligned} \quad (3.1966)$$

Now, symmetrizing the tensor structure $T^{i0}x^j$, we have

$$\begin{aligned} \widehat{T^{i0}x^j} &\equiv \frac{1}{2}\left(T^{i0}x^j + T^{j0}x^i\right) \\ &= \frac{1}{2}\big[\partial_k(T^{k0}x^j x^i) + \partial_k(T^{k0}x^i x^j) - (\partial_k T^{k0})x^j x^i \\ &\quad - (\partial_k T^{k0})x^i x^j - T^{j0}x^i - T^{i0}x^j\big] \\ &= \partial_k(T^{k0}x^i x^j) - (\partial_k T^{k0})x^i x^j - \frac{1}{2}(T^{i0}x^j + T^{j0}x^i) \\ &= \partial_k(T^{k0}x^i x^j) - (\partial_k T^{k0})x^i x^j - \widehat{T^{i0}x^j}, \end{aligned} \quad (3.1967)$$

which leads to

$$\widehat{T^{i0}x^j} = \frac{1}{2}\partial_k(T^{k0}x^i x^j) - \frac{1}{2}(\partial_k T^{k0})x^i x^j. \quad (3.1968)$$

Again, using $\partial_\nu T^{\nu 0} = \partial_0 T^{00} + \partial_k T^{k0} = 0$, we find that

$$\widehat{T^{i0}x^j} = \frac{1}{2}\partial_k(T^{k0}x^i x^j) + \frac{1}{2}(\partial_0 T^{00})x^i x^j. \quad (3.1969)$$

Since the energy–momentum tensor is symmetric, i.e., $T^{\mu\nu} = T^{\nu\mu}$, we can then combine the expressions for T^{ij} and $\widehat{T^{i0}x^j}$, and thus, we obtain

$$\begin{aligned} T^{ij} = \widehat{T^{ij}} &\equiv \frac{1}{2}(T^{ij} + T^{ji}) \\ &= \frac{1}{2}\left[\partial_k(T^{ik}x^j) + \partial_k(T^{jk}x^i)\right] + \frac{1}{2}\left[\partial_0(T^{i0}x^j) + \partial_0(T^{j0}x^i)\right] \end{aligned}$$

$$= \frac{1}{2}\partial_k(T^{ik}x^j + T^{jk}x^i) + \partial_0\widehat{T^{i0}x^j}$$

$$= \frac{1}{2}\partial_k(T^{ik}x^j + T^{jk}x^i) + \frac{1}{2}\partial_0\partial_k(T^{k0}x^ix^j) + \frac{1}{2}\partial_0[(\partial_0 T^{00})x^ix^j]. \quad (3.1970)$$

In addition, using that $\partial_0\partial_k = \partial_k\partial_0$ and $\partial_0[(\partial_0 T^{00})x^ix^j] = (\partial_0{}^2T^{00})x^ix^j$, we have

$$T^{ij} = \frac{1}{2}\partial_k(T^{ik}x^j + T^{jk}x^i) + \frac{1}{2}\partial_k\partial_0(T^{k0}x^ix^j) + \frac{1}{2}(\partial_0{}^2T^{00})x^ix^j$$

$$= \partial_k\left[\frac{1}{2}(T^{ik}x^j + T^{jk}x^i) + \frac{1}{2}\partial_0(T^{k0}x^ix^j)\right] + \frac{1}{2}(\partial_0{}^2T^{00})x^ix^j. \quad (3.1971)$$

Next, using the definition

$$\bar{h}_{\mu\nu}(t,\mathbf{x}) \sim \frac{4}{r}\int T_{\mu\nu}(t-r,\mathbf{x}')\,d^3x', \quad (3.1972)$$

and choosing the spatial components, we have

$$\bar{h}_{ij} \sim \frac{4}{r}\int T'_{ij}\,d^3x'. \quad (3.1973)$$

Inserting the symmetrized version of the energy–momentum tensor, we obtain

$$\bar{h}_{ij} \sim \frac{4}{r}\int \frac{1}{2}(\partial_0{}^2T'_{00})x'_ix'_j\,d^3x', \quad (3.1974)$$

since we can drop the terms with total derivatives with respect to the spatial components. Therefore, we find that

$$\bar{h}_{ij} \sim \frac{2}{r}\int (\partial_0{}^2T'_{00})x'_ix'_j\,d^3x', \quad (3.1975)$$

and using the fact that $T_{00} = T^{00} = \rho$, we finally obtain

$$\bar{h}_{ij} \sim \frac{2}{r}\int \frac{\partial^2\rho'}{\partial t^2}x'_ix'_j\,d^3x' = \frac{2}{r}\frac{d^2}{dt^2}\int \rho' x'_ix'_j\,d^3x' = \frac{2}{r}\frac{d^2}{dt^2}\int \rho(t-r,\mathbf{x}')x'_ix'_j\,d^3x', \quad (3.1976)$$

where we used Leibniz's integral rule twice.

2.135

For the estimates in this problem, we use that the amplitude of the gravitational waves is approximated by the relation

$$h \propto \frac{M^2}{rR} \sim \frac{M^{5/3}}{r}\left(\frac{4\pi}{P}\right)^{2/3}, \quad (3.1977)$$

where M is the mass of the binary, r is the distance to the binary, and P is the period of one orbit of the binary, i.e., the orbital period. In conventional units, the amplitude of the gravitational waves can be written as [see, e.g., K.D. Kokkotas, *Gravitational Waves*, Acta Phys. Polon. B **38**, 3 891 (2007)]

$$h \sim 5 \cdot 10^{-22} \left(\frac{M}{2.8M_\odot}\right)^{5/3} \left(\frac{f}{100\ \text{Hz}}\right)^{2/3} \left(\frac{15\ \text{Mpc}}{r}\right), \tag{3.1978}$$

where $f \equiv 1/P$ is the frequency.

a) Using $M \simeq 2.8M_\odot$ (where $M_\odot \simeq 1.988435 \cdot 10^{30}$ kg is the solar mass), $r = 5$ kpc, and $P = 1$ h (which means that $f = 1/3\,600$ Hz), we find that

$$h \sim 5 \cdot 10^{-22} \cdot 1^{5/3} \cdot \left(\frac{1}{3.6 \cdot 10^5}\right)^{2/3} \cdot 3 \cdot 10^3 \simeq 3 \cdot 10^{-22} \sim 10^{-22}. \tag{3.1979}$$

b) Again, using $M \simeq 2.8M_\odot$, but $r = 15$ Mpc and $P = 0.02$ s (which means that $f = 50$ Hz), we find that

$$h \sim 5 \cdot 10^{-22} \cdot \left(\frac{1}{2}\right)^{2/3} \cdot 1 \simeq 3 \cdot 10^{-22} \sim 10^{-22}, \tag{3.1980}$$

which is basically the same result as in a).

c) For a binary system, assuming circular binary orbits, Kepler's third law provides a direct and accurate estimate for the orbital separation distance between the two binaries, i.e.,

$$2G(M_1 + M_2) = \left(\frac{2\pi}{P}\right)^2 (R_1 + R_2)^3, \tag{3.1981}$$

where $G \simeq 6.674 \cdot 10^{-11}$ N m^2/kg^2 is Newton's gravitational constant, M_1 and M_2 are the masses of the two binaries, respectively, R_1 and R_2 are the respective distances to their common orbital center, and P is the orbital period. Assuming $M_1 = M_2 \equiv M/2$ and $R_1 = R_2 \equiv R$, i.e., the binaries are equally heavy and they are at the same distance compared to their orbital center, we obtain (half of) the orbital separation distance R between the two binaries as

$$R = \frac{1}{2}\sqrt[3]{\frac{GMP^2}{4\pi^2}}. \tag{3.1982}$$

Inserting $M = 2.8M_\odot$ and $P = 0.02$ s, we find that

$$R \simeq 155\,600\ \text{m} \sim 100\ \text{km}. \tag{3.1983}$$

Thus, we can only hope to detect inspirals of compact binary systems (e.g., NS–NS, NS–BH, or BH–BH) with Earth-based interferometers such as LIGO.

d) For a spinning neutron star, the amplitude of the gravitational waves h is approximately given by

$$h \propto \frac{2\delta M R^2 \Omega^2}{r}, \tag{3.1984}$$

where δM is the mass of a nonspherical deformation on the equator of the neutron star, R is the radius of the neutron star, $\Omega \equiv 2\pi/P$ is the angular velocity expressed in the spin period P, and r is the distance to the neutron star. In conventional units, the amplitude of the gravitational waves can be written as [see, e.g., R. Prix, *Gravitational Waves from Spinning Neutron Stars*, in: W. Becker (ed.), *Neutron Stars and Pulsars*, Astrophysics and Space Science Library **357**, 651–685, Springer (Berlin, 2009)]

$$h \propto 10^2 \frac{G}{c^4} \frac{\delta M R^2 f^2}{r} \sim 3 \cdot 10^{-25} \left(\frac{\delta M R^2}{10^{32}\ \text{kg m}^2} \right) \left(\frac{f}{100\ \text{Hz}} \right)^2 \left(\frac{100\ \text{pc}}{r} \right), \tag{3.1985}$$

where G is again Newton's gravitational constant, c is the speed of light in vacuum, and f is the spin frequency. Inserting $\delta M = 10^{-6} M_\odot$, $R = 10$ km, $f = 50$ Hz, and $r = 1$ kpc, we find that

$$h \sim 3 \cdot 10^{-25} \cdot 2 \cdot \left(\frac{1}{2} \right)^2 \cdot \frac{1}{10} = 1.5 \cdot 10^{-26} \sim 10^{-26}. \tag{3.1986}$$

Thus, for the spinning neutron star, the order of magnitude of the amplitude of the gravitational waves at Earth is 10^{-26}.

2.136

a) The energy flux (or energy density) in units of power/area of gravitational waves can be estimated by the relation

$$F \propto \frac{c^3}{32\pi^2 G} |\dot{h}|^2 \sim \frac{c^3}{8G} h^2 f^2, \tag{3.1987}$$

where h and f are the amplitude and the frequency, respectively, of the gravitational waves, which are assumed to be monochromatic. Note that the components of the metric tensor play the role of the gravitational potential, so the derivatives of the components of the metric tensor act as the field, and therefore, $F \propto |\dot{h}|^2$. A useful formula for an estimate of the energy flux on Earth due to gravitational waves is given by [see, e.g., I. Ciufolini (ed.) *et al.*, *Gravitational Waves*, IoP (Bristol, 2001)]

$$F \sim 3 \cdot 10^{-3} \left(\frac{h}{10^{-22}} \right)^2 \left(\frac{f}{1\ \text{kHz}} \right)^2 \ \text{W/m}^2, \tag{3.1988}$$

where h and f are now the amplitude and the frequency, respectively, of the gravitational waves as measured on Earth. Inserting $h = 10^{-21}$ and $f = 200$ Hz for GW150914, we find that

$$F \sim 3 \cdot 10^{-3} \cdot 10^2 \cdot \left(\frac{1}{5}\right)^2 \text{ W/m}^2 = 12 \cdot 10^{-3} \text{ W/m}^2 \sim 10^{-2} \text{ W/m}^2, \quad (3.1989)$$

which is an enormous amount of energy flux compared to the observed energy flux in electromagnetic waves.

b) To estimate the energy flux in electromagnetic waves that is received at Earth from a full moon, we use the following assumptions: (i) The solar irradiance is about 1387 W/m^2 at the surface of the Moon and the Earth; (ii) The reflectivity of the Moon is about 12 %; and (iii) The average solid angle subtended by the Moon in the sky at Earth is about $6.4 \cdot 10^{-5}$ steradians, which is approximately the same as that subtended by the Sun, i.e., about $6.8 \cdot 10^{-5}$ steradians. Therefore, the energy flux from a full moon is given by

$$F_{\text{full moon}} = 1\,387 \text{ W/m}^2 \cdot 0.12 \cdot 6.4 \cdot 10^{-5} = 11 \cdot 10^{-3} \text{ W/m}^2. \quad (3.1990)$$

The assumptions made are not totally accurate, so reducing the solar irradiance to 1 000 W/m^2 and the reflectivity of the Moon to 10 % lead to $F_{\text{full moon}} = 6.4 \cdot 10^{-3}$ W/m^2. Thus, the gravitational wave energy flux of GW150914 in a), i.e., $F_{\text{GW}} \sim 12 \cdot 10^{-3}$ W/m^2, is between once or twice the electromagnetic wave energy flux of a full moon [$F_{\text{GW}} \sim (1,2) F_{\text{full moon}}$], although the estimated distance of GW150914 is about 400 Mpc and the distance from the Earth to the Moon is just $1.25 \cdot 10^{-8}$ pc.

2.137

The metric for a linearly expanding spacetime is given by

$$ds^2 = dt^2 - H^2 t^2 dx^2. \quad (3.1991)$$

We want to start at $t = t_0$, $x = 0$ and arrive at $t = t_1$, $x = L$ without accelerating. This implies that we should move along a geodesic. Thus, we want to determine $x(t)$. The Lagrangian is given by

$$\dot{t}^2 - H^2 t^2 \dot{x}^2 = \beta = \text{const.} \quad (3.1992)$$

The Euler–Lagrange equations are

$$\frac{d}{d\tau}(H^2 t^2 \dot{x}) = 0, \quad (3.1993)$$

$$\ddot{t} - H^2 t \dot{x}^2 = 0. \quad (3.1994)$$

From the first equation, we obtain

$$\dot{x} = \frac{A}{H^2 t^2}, \tag{3.1995}$$

where A is a constant. Inserting this in the Lagrangian, we get

$$\dot{t}^2 - H^2 t^2 \left(\frac{A}{H^2 t^2}\right)^2 = \dot{t}^2 - \frac{A^2}{H^2 t^2} = \beta, \tag{3.1996}$$

which can be solved for

$$\dot{t} = \pm\sqrt{\beta + \frac{A^2}{H^2 t^2}} = \pm\frac{1}{Ht}\sqrt{\beta H^2 t^2 + A^2}. \tag{3.1997}$$

Now, we can determine $x(t)$

$$\frac{dx}{dt} = \frac{dx}{d\tau}\frac{d\tau}{dt} = \frac{A}{H^2 t^2}\frac{Ht}{\sqrt{\beta H^2 t^2 + A^2}} = \frac{A}{Ht\sqrt{\beta H^2 t^2 + A^2}}, \tag{3.1998}$$

and integrating on both sides, we obtain

$$\int_0^{x(t)} dx' = \int_{t_0}^{t_1} \frac{A}{Ht\sqrt{\beta H^2 t^2 + A^2}} dt, \tag{3.1999}$$

and thus, we have

$$x(t) = \int_{t_0}^{t} \frac{A}{Ht'\sqrt{\beta H^2 t'^2 + A^2}} dt'. \tag{3.2000}$$

The farthest one could go, $x(t_1) = L$ would be following a lightlike geodesic, i.e., for $\beta = 0$. Then, we obtain

$$L = \int_{t_0}^{t_1} \frac{A}{Ht'\sqrt{A^2}} dt' = \int_{t_0}^{t_1} \frac{1}{Ht'} dt' = \frac{1}{H}\ln\left(\frac{t_1}{t_0}\right). \tag{3.2001}$$

2.138

The cosmological redshift is due to the fact that signals interchanged between two observers will have different travel times if the size of the universe is altered. In the case of the 2-dimensional mini-universe, we assume that $\chi_0 < \chi_1$. (To treat the case of $\chi_1 > \chi_0$, simply substitute $d\chi$ with $-d\chi$ and $\chi_1 - \chi_0$ with $\chi_0 - \chi_1$.) For a lightlike geodesic, we have $ds = 0$, which gives

$$c dt = S(t) d\chi \quad \Rightarrow \quad \int_{t_0}^{t_1} \frac{c\,dt}{S(t)} = \int_{\chi_0}^{\chi_1} d\chi = \chi_1 - \chi_0. \tag{3.2002}$$

We now assume that a light signal is sent from χ_0 at time t and another at time $t + \varepsilon$, they are received at χ_1 at times t' and $t' + \varepsilon'$. Assuming that ε is small, the redshift is given by

$$z = \frac{\lambda_1 - \lambda_0}{\lambda_0} = \frac{\varepsilon'}{\varepsilon} - 1. \tag{3.2003}$$

For the two different signals, we obtain

$$\int_t^{t'} \frac{c\,dt}{S(t)} = \chi_1 - \chi_0 = \int_{t+\varepsilon}^{t'+\varepsilon'} \frac{c\,dt}{S(t)}. \tag{3.2004}$$

Using the fact that ε is small, this can be simplified to

$$\frac{\varepsilon}{S(t)} = \frac{\varepsilon'}{S(t')} \quad \Rightarrow \quad \frac{\varepsilon'}{\varepsilon} = \frac{S(t')}{S(t)} \quad \Rightarrow \quad z = \frac{S(t')}{S(t)} - 1. \tag{3.2005}$$

2.139

The metric components are independent of the coordinate x and thus ∂_x is a Killing vector field, implying that

$$k = -g(N, \partial_x) = -g_{xx} N^x, \tag{3.2006}$$

where N is the 4-frequency of the light signal, is a constant. Since the 4-frequency is a null vector, we also obtain

$$g(N,N) = g_{tt}(N^t)^2 + g_{xx}(N^x)^2 = 0 \quad \Longrightarrow \quad N^t = \sqrt{-\frac{g_{xx}(N^x)^2}{g_{tt}}} = \frac{k}{\sqrt{-g_{tt}g_{xx}}}. \tag{3.2007}$$

The 4-velocity of a comoving observer is given by $V = \gamma \partial_t$ and normalization results in

$$g(V,V) = g_{tt}\gamma^2 = 1 \quad \Longrightarrow \quad \gamma = \frac{1}{\sqrt{g_{tt}}}. \tag{3.2008}$$

The frequency observed by such a comoving observer is therefore

$$\omega = g(N,V) = g_{tt}\gamma N^t = g_{tt}\frac{1}{\sqrt{g_{tt}}}\frac{k}{\sqrt{-g_{tt}g_{xx}}} = \frac{k}{\sqrt{-g_{xx}}}. \tag{3.2009}$$

The ratio between the observed frequency ω_1 and the emitted frequency ω_0 is therefore

$$\frac{\omega_1}{\omega_0} = \sqrt{\frac{g_{xx}(t_0)}{g_{xx}(t_1)}} = e^{(t_0 - t_1)/R}. \tag{3.2010}$$

Since the light signal travels along a null geodesic, its worldline satisfies

$$ds^2 = dt^2 - e^{2t/R}dx^2 = 0 \quad \Longrightarrow \quad \frac{dx}{dt} = e^{-t/R}. \tag{3.2011}$$

Integrating this expression with the initial condition $x(t_0) = x_0$ results in

$$t_1 = -R \ln\left(e^{-t_0/R} - \frac{x_1 - x_0}{R}\right), \tag{3.2012}$$

and inserting this into the frequency ratio leads to

$$z = \frac{\omega_0}{\omega_1} - 1 = \frac{Re^{-t_0/R}}{Re^{-t_0/R} - x_1 + x_0} - 1 = \frac{x_1 - x_0}{Re^{-t_0/R} - x_1 + x_0}. \tag{3.2013}$$

2.140

a) We compute the geodesics for the given metric from the variational principle

$$\delta \int \frac{1}{2}\left(\dot{t}^2 - e^{2t/R}\dot{x}^2\right) d\tau = 0, \tag{3.2014}$$

where we use the same notation as in Problem 2.83. The Euler–Lagrange equations are

$$\ddot{t} + \frac{1}{R}e^{2t/R}\dot{x}^2 = 0, \quad -e^{2t/R}\ddot{x} - \frac{2}{R}e^{2t/R}\dot{x}\dot{t} = 0, \tag{3.2015}$$

and by comparing this with the general form of the geodesics equation

$$\ddot{x}^\mu + \Gamma^\mu_{\alpha\beta}\dot{x}^\alpha\dot{x}^\beta = 0, \tag{3.2016}$$

we can identify the nonzero Christoffel symbols

$$\Gamma^t_{xx} = \frac{1}{R}e^{2t/R}, \quad \Gamma^x_{tx} = \Gamma^x_{xt} = \frac{1}{R}. \tag{3.2017}$$

b) The metric tensor is

$$(g_{\mu\nu}) = \begin{pmatrix} 1 & 0 \\ 0 & -e^{2t/R} \end{pmatrix}, \tag{3.2018}$$

and thus, we have the inverse metric tensor

$$(g^{\mu\nu}) = \begin{pmatrix} 1 & 0 \\ 0 & -e^{-2t/R} \end{pmatrix}. \tag{3.2019}$$

We find that

$$\begin{aligned} g_{\mu\nu}\nabla^\mu\nabla^\nu\Phi &= \nabla^\nu\partial_\nu\Phi = \partial^\nu\partial_\nu\Phi + \Gamma^\nu_{\nu\alpha}\partial^\alpha\Phi \\ &= g^{\mu\nu}\partial_\mu\partial_\nu\Phi + \Gamma^\nu_{\nu\alpha}g^{\alpha\beta}\partial_\beta\Phi = \Phi_{tt} - e^{-2t/R}\Phi_{xx} + \underbrace{\frac{1}{R}\Phi_t}_{=\Gamma^x_{xt}g^{tt}\Phi_t} = 0, \end{aligned} \tag{3.2020}$$

where $\Phi_t = \partial\Phi/\partial t$, etc.

2.141

a) The geodesic equation can be computed from $\dot{t}^2 - \exp(2t/t_H)\dot{r}^2 = 0$, where $\dot{t} = dt/d\tau$, i.e., $dr/dt = \pm\exp(-t/t_H)$ with $r(t_0) = 0$. Using $dr/dt = +\exp(-t/t_H) > 0$, this implies that

$$r(t) = \int_0^r dr' = \int_{t_0}^t e^{-t'/t_H}\, dt' = \left[-t_H e^{-t'/t_H}\right]_{t_0}^t = -t_H\left(e^{-t/t_H} - e^{-t_0/t_H}\right). \tag{3.2021}$$

b) For the line $\theta = \pi/2$ and $\varphi = 0$ at fixed universal time $t = t_0 =$ const., we have $dt = 0$ and $d\Omega = 0$, which imply that

$$ds^2 = -e^{2t/t_H} dr^2 \quad \Rightarrow \quad g_{rr}(t) = -e^{2t/t_H}. \tag{3.2022}$$

Thus, the proper distance between the origin $r = 0$ and a point $r > 0$ is given by

$$d_p(t,r) = \int_{r'=0}^{r'=r} \sqrt{-g_{rr}(t)}\, dr' = \int_0^r \sqrt{e^{2t/t_H}}\, dr' = e^{t/t_H} \int_0^r dr' = re^{t/t_H}. \tag{3.2023}$$

c) The cosmological redshift relates the emitted and received wavelengths as (see solutions to Problems 2.138 and 2.139)

$$\frac{\lambda_{\text{rec}}}{\lambda_{\text{em}}} = \frac{S(t_{\text{rec}})}{S(t_{\text{em}})} = \frac{e^{t_{\text{rec}}/t_H}}{e^{t_{\text{em}}/t_H}} = e^{(t_{\text{rec}}-t_{\text{em}})/t_H}. \tag{3.2024}$$

Thus, using $t_{\text{em}} = t_0$ and $t_{\text{rec}} = t$, the spectral shift of the light ray is given by

$$z \equiv \frac{\lambda_{\text{rec}}}{\lambda_{\text{em}}} - 1 = e^{(t-t_0)/t_H} - 1. \tag{3.2025}$$

2.142

a) At cosmological time t we find that

$$g_{tt} = 1, \quad g_{ij} = -a(t)^2 G_{ij}, \tag{3.2026}$$

where G_{ij} are the components of a Riemannian metric. The proper length of a curve on the cosmological simultaneity is given by

$$\ell = \int \sqrt{-g_{ij}\dot{x}^i\dot{x}^j}\, ds = a(t)\int \sqrt{G_{ij}\dot{x}^i\dot{x}^j}\, ds \equiv a(t)\ell_c, \tag{3.2027}$$

where s is the curve parameter and ℓ_c the comoving length of the curve. Fixing the end points of the curve and minimizing this length gives the proper distance d_p between the points as $d_p = a(t)d_c$, where the comoving distance d_c is the minimal value of the integral

$$d_c = \min\left(\int \sqrt{G_{ij}\dot{x}^i\dot{x}^j}\, ds\right), \tag{3.2028}$$

which is independent of the cosmological time t. It follows that

$$v_p = \frac{d(d_p)}{dt} = \dot{a}(t)d_c = \frac{\dot{a}}{a}d_p. \tag{3.2029}$$

b) Hubble's law describes the observation that objects (read: galaxies) are moving away from the Earth at velocities proportional to their distances. The further the galaxies are, the faster they are moving away. The velocities of the galaxies are measured by their redshifts. The velocities arise from the expansion of the universe. Indeed, the proper distance between galaxies increases even though they are comoving.

2.143

a) If we use the given approximation, i.e., k is small and $\rho \simeq \rho_\Lambda$, we can approximate the first Friedmann equation as

$$\frac{\dot{a}^2}{a^2} = \frac{8\pi G}{3}\rho_\Lambda. \tag{3.2030}$$

This equation has the solution

$$a(t) = a_0 e^{(t-t_0)/\Delta\tau}, \quad \Delta\tau = \sqrt{\frac{3}{8\pi G\rho_\Lambda}}. \tag{3.2031}$$

Inserting this solution into the expression for Ω, we obtain

$$1 - \Omega \propto e^{-2(t-t_0)/\Delta\tau}. \tag{3.2032}$$

We observe that the larger t becomes, the smaller $1-\Omega$ becomes, and thus, $1-\Omega$ gets closer and closer to zero, which means that Ω gets closer and closer to one.

b) Today, we have a universe with a curvature that is very close to zero, which without inflation would need a lot of fine-tuning to maintain. After inflation, a spacetime is obtained, which is very close to being flat as per the solution to a), even if the initial spacetime had a lot of curvature.

2.144

a) The scaling of the energy density for an ideal fluid with equation-of-state parameter w is given by $\rho = \rho_0 a^{-3(1+w)}$. The density of component i today is given by

$$\rho_{0,i} = \rho_c\Omega_i, \tag{3.2033}$$

where ρ_c is the critical density today. This means that

$$\rho_{0,\Lambda} = \rho_c\Omega_\Lambda \simeq 0.7\rho_c \quad \rightarrow \quad \rho_\Lambda = 0.7\rho_c a^{-3(1-1)} = 0.7\rho_c,$$
$$\rho_{0,m} = \rho_c\Omega_m \simeq 0.3\rho_c \quad \rightarrow \quad \rho_m = 0.3\rho_c a^{-3},$$

$$\rho_{0,r} = \rho_c \Omega_r \simeq 10^{-4} \rho_c \quad \rightarrow \quad \rho_r = 10^{-4} \rho_c a^{-3\left(1+\frac{1}{3}\right)} = 10^{-4} \rho_c a^{-4}, \tag{3.2034}$$

since $w = -1$ for a cosmological constant, $w = 0$ for a matter gas, and $w = \frac{1}{3}$ for a radiation gas. The ratio between the energy densities of matter and the cosmological constant is therefore given by

$$\frac{\rho_\Lambda}{\rho_m} = \frac{\Omega_\Lambda}{\Omega_m} \frac{1}{a^{-3}}. \tag{3.2035}$$

The energy densities were therefore the same when this ratio was equal to one, i.e., when

$$a^3 = \frac{\Omega_m}{\Omega_\Lambda} \quad \Rightarrow \quad a = \left(\frac{\Omega_m}{\Omega_\Lambda}\right)^{1/3} \simeq \left(\frac{0.3}{0.7}\right)^{1/3} \simeq 0.75. \tag{3.2036}$$

Note that this is not *very* long ago. A naïve estimate with $H_0 \simeq 70$ km/s/Mpc $\simeq 2 \cdot 10^{-18}$ s^{-1} gives a couple of billion years ago (similar to the birth of the solar system).

b) Similarly, the ratio between matter and radiation energy densities is given by

$$\frac{\rho_m}{\rho_r} = \frac{\Omega_m}{\Omega_r} \frac{a^{-3}}{a^{-4}} = \frac{\Omega_m}{\Omega_r} a. \tag{3.2037}$$

This ratio is equal to one when

$$a = \frac{\Omega_r}{\Omega_m} \simeq \frac{10^{-4}}{0.3} = \frac{1}{3\,000}. \tag{3.2038}$$

As the cosmological redshift is given by $1 + z = \frac{a_0}{a}$, we find that

$$z = \frac{1}{a} - 1 \simeq 3\,000, \tag{3.2039}$$

at matter-radiation equality.

2.145

From the relation

$$\dot{a} = \sqrt{\frac{8\pi G}{3}} \sqrt{\sum_i \rho_{0i} a^{-1-3w_i}}, \tag{3.2040}$$

with the definitions

$$\rho_c = \frac{3H^2}{8\pi G} \quad \text{and} \quad \rho_{0i} = \rho_{0c} \Omega_i, \tag{3.2041}$$

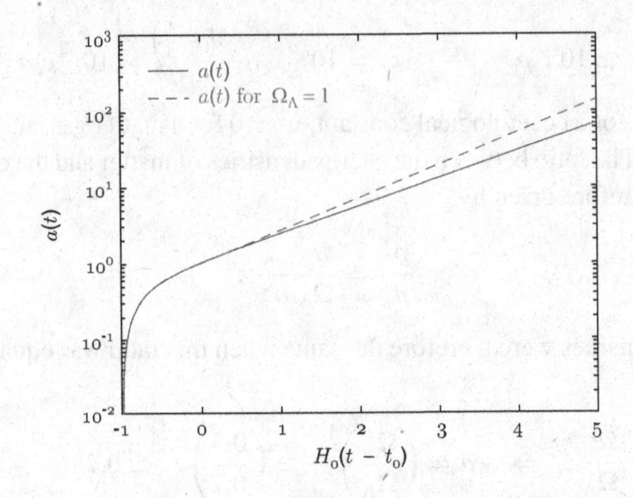

Figure 3.18 The scale factor $a(t)$ for the solution to Problem 2.145. The solution for $\Omega_\Lambda = 1$ and $\Omega_m = 0$ is also shown for times $t > t_0$ for comparison.

we find that

$$H_0^2 = \frac{8\pi G\rho_{0c}}{3} \implies \dot{a} = H_0\sqrt{\sum_i \Omega_i a^{-1-3w_i}}. \tag{3.2042}$$

In our given scenario, there are two components, $i = m =$ matter and $i = \Lambda =$ dark energy with corresponding parameters

$$\Omega_m = 0.3, \quad \Omega_\Lambda = 0.7, \quad w_m = 0, \quad w_\Lambda = -1. \tag{3.2043}$$

The differential equation governing the scale factor is therefore

$$H_0 = \frac{\dot{a}}{\sqrt{0.3a^{-1} + 0.7a^2}}. \tag{3.2044}$$

Integrating this from t_0 to t, we find that

$$H_0(t - t_0) = \int_1^{a(t)} \frac{da}{\sqrt{0.3a^{-1} + 0.7a^2}}. \tag{3.2045}$$

The result of numerically evaluating this expression is shown in Figure 3.18. Note that the solution quickly becomes dominated by the cosmological constant after $t = t_0$ as the matter component dilutes. However, as the cosmological constant is smaller by a factor of 0.7 compared to the case where $\Omega_\Lambda = 1$, the exponential factor in the large time limit (and therefore, $H(t \to \infty)$) is also smaller by a factor 0.7.

Unlike the $\Omega_\Lambda = 1$ solution, our solution has a time t for which $a(t) = 0$. Defining this time as $t = 0$, we find that (from looking at where the graph intersects zero)

$$-H_0 t_0 \simeq -0.96 \quad \Longrightarrow \quad t_0 \simeq \frac{0.96}{H_0}. \tag{3.2046}$$

Current experimental data restricts H_0 to be around 70 km/s/Mpc or, equivalently, $H_0 \simeq 7 \cdot 10^{-11}$ yr^{-1}, leading to

$$t_0 \simeq 1.3 \cdot 10^{10} \text{ years}, \tag{3.2047}$$

estimating the universe to be around 13 billion years old.

2.146

a) The universe has a Killing vector field ∂_φ. Therefore, the quantity $g(\partial_\varphi, \dot\gamma) = -a^2\dot\varphi = \text{constant} = -k$. The 4-velocity of a comoving observer is given by $U = \partial_t$, and therefore, at $t = t_0$

$$g(U, \dot\gamma) = \frac{1}{\sqrt{1-v^2}} = \dot t. \tag{3.2048}$$

At $t = t_0$, we also have

$$g(\dot\gamma, \dot\gamma) = 1 = \dot t^2 - a_0^2\dot\varphi^2 = \frac{1}{1-v^2} - a_0^2\dot\varphi^2 = \frac{1}{1-v^2} - \frac{k}{a_0^2}$$
$$\Rightarrow \quad \frac{k^2}{a_0^2} = \frac{1}{1-v^2} - 1 = \frac{v^2}{1-v^2} \quad \Rightarrow \quad k = \frac{v a_0}{\sqrt{1-v^2}}. \tag{3.2049}$$

From the 4-velocity having constant magnitude equals to 1, we deduce

$$1 = \dot t^2 - a^2\dot\varphi^2 = \dot t^2 - \frac{k^2}{a^2} \quad \Rightarrow \quad \dot t = \sqrt{1 + \frac{k^2}{a^2}}. \tag{3.2050}$$

Dividing this by $\dot\varphi = k/a^2$ leads to the separable differential equation

$$\frac{dt}{d\varphi} = \frac{\dot t}{\dot\varphi} = \frac{a^2\sqrt{1+k^2/a^2}}{k} = \frac{a}{k}\sqrt{a^2+k^2}. \tag{3.2051}$$

In order for the object to complete an entire lap around the universe in finite time t_1, we therefore require that

$$2\pi = \int_0^{2\pi} d\varphi = \int_{t_0}^{t_1} \frac{k}{a\sqrt{k^2+a^2}}\, dt, \tag{3.2052}$$

for some finite t_1. In other words, we have

$$\int_{t_0}^{\infty} \frac{k}{a\sqrt{k^2+a^2}}\,dt > 2\pi \quad \Rightarrow \quad \int_{t_0}^{\infty} \frac{v a_0}{a(t)} \frac{dt}{\sqrt{(1-v^2)a(t)^2 + a_0^2 v^2}} > 2\pi. \tag{3.2053}$$

b) We know that $g(\dot{\gamma}, \partial_\varphi) = -a^2\dot{\varphi} = -k$ is constant and that $\dot{t} = \sqrt{1 + k^2/a^2}$. The relative velocity $v(t)$ at time t is therefore given by

$$\gamma = g(\partial_t, \dot{\gamma}) = \frac{1}{\sqrt{1 - v(t)^2}}. \tag{3.2054}$$

We find that

$$\gamma = g(\partial_t, \dot{\gamma}) = \dot{t} = \sqrt{1 + \frac{k^2}{a(t)^2}} = \sqrt{1 + \frac{a_0^2}{a(t)^2}\frac{v^2}{1-v^2}}. \tag{3.2055}$$

Solving for $v(t)$, we obtain

$$v(t) = \frac{a_0 v}{\sqrt{a(t)^2 + v^2[a_0^2 - a(t)^2]}}. \tag{3.2056}$$

Note that $v(t_0) = v$ and that $v(t) \to 0$ as $a(t)/a_0 \to \infty$.

2.147

For a flat universe containing matter and radiation only, the following must be satisfied

$$1 = \Omega_m + \Omega_r = \Omega_m + x \quad \Rightarrow \quad \Omega_m = 1 - x. \tag{3.2057}$$

The time evolution of the scale factor $a(t)$ can be found through the relationship

$$\dot{a} = \sqrt{\frac{8\pi G}{3}}\sqrt{\rho_{0,m}a^{-1} + \rho_{0,r}a^{-2}} = H_0\sqrt{\Omega_m a^{-1} + \Omega_r a^{-2}}, \tag{3.2058}$$

where it has been assumed that $a_0 = 1$ and used that $\sqrt{\frac{8\pi G}{3}} = \frac{H_0^2}{\rho_{0,c}}$. This is a separable differential equation and we find that

$$t_0 - t_1 = \int_{t_1}^{t_0} dt = \frac{1}{H_0}\int_{a_1}^{a_0} \frac{da}{\sqrt{\Omega_m a^{-1} + \Omega_r a^{-2}}} = \frac{1}{H_0}\int_{a_1}^{1} \frac{a\,da}{\sqrt{(1-x)a + x}}, \tag{3.2059}$$

since we work under the assumption that $a_0 = 1$. Performing this integral, e.g., by partial integration, leads to

$$t_0 - t_1 = \frac{1}{H_0}\frac{2}{3}\left[\frac{\sqrt{(1-x)a + x}}{(1-x)^2}[(1-x)a - 2x]\right]_{a_1}^{1}. \tag{3.2060}$$

The energy density ratio between matter and radiation scales as $\rho_m/\rho_r \sim a(t)$ due to $\rho_m \propto a^{-3}$ and $\rho_r \propto a^{-4}$. Since

$$\frac{\rho_{0,m}}{\rho_{0,r}} = \frac{1-x}{x} \simeq \frac{1}{x}, \tag{3.2061}$$

we find that $\rho_m \simeq \rho_r$ when

$$\frac{\rho_m}{\rho_r} = \frac{\rho_{0,m}}{\rho_{0,r}}\frac{a(t_1)}{a_0} = \frac{a(t_1)}{x} \quad \Rightarrow \quad a(t_1) \simeq x. \tag{3.2062}$$

It follows that

$$\begin{aligned} t_0 - t_1 &= \frac{2}{3H_0}\left\{\frac{\sqrt{(1-x)+x}}{(1-x)^2}[(1-x)-2x] - \frac{\sqrt{(1-x)x+x}}{(1-x)^2}[(1-x)x-2x]\right\} \\ &\simeq \frac{2}{3H_0}[(1+2x)(1-3x)] \simeq \frac{2}{3H_0}(1-x). \end{aligned} \tag{3.2063}$$

The time since matter-radiation equality is therefore just slightly less than the age $\frac{2}{3H_0}$ of a fully matter-dominated universe.

2.148

The curvature parameter is given by

$$|\Omega_K| = \left|-\frac{1}{H^2a^2}\right| = \left|-\frac{1}{\left(\frac{\dot{a}}{a}\right)^2 a^2}\right| = \frac{1}{\dot{a}^2}. \tag{3.2064}$$

The time derivative of $|\Omega_K|$ is therefore

$$\frac{d|\Omega_K|}{dt} = \frac{d(1/\dot{a}^2)}{dt} = -\frac{2}{\dot{a}^3}\ddot{a}. \tag{3.2065}$$

For an expanding universe, $\dot{a} > 0$, which implies that $\dot{a}^3 > 0$, and thus, the requirement that $|\Omega_K|$ decreases with time is given by

$$\frac{d|\Omega_K|}{dt} = -\frac{2}{\dot{a}^3}\ddot{a} < 0 \quad \Rightarrow \quad \ddot{a} > 0. \tag{3.2066}$$

That $|\Omega_K|$ decreases with time is therefore equivalent to the scale factor growth accelerating. The Friedmann acceleration equation is

$$\ddot{a} = -\frac{4\pi G}{3}a(\rho + 3p) > 0 \quad \Rightarrow \quad \rho + 3p < 0 \quad \Rightarrow \quad \frac{p}{\rho} < -\frac{1}{3}. \tag{3.2067}$$

The condition on w for $|\Omega_K|$ to decrease with time is therefore $w < -\frac{1}{3}$. In particular, note that this is the case for a cosmological constant, where $w = -1$.

2.149

a) From the evolution equations, we find that

$$\dot{a} = \sqrt{\frac{8\pi G\rho_{0,c}}{3}}a^{-\frac{1+3w}{2}} = H_0 a^{-k}, \tag{3.2068}$$

where $k = \frac{1}{2} + \frac{3w}{2}$. It follows that $\dot{a}a^k = H_0$, and therefore,

$$H_0(t - t_0) = \int_{t_0}^{t} H_0\, dt = \int_{a_0}^{a(t)} a^k\, da = \int_1^{a(t)} a^k\, da$$
$$= \begin{cases} \frac{a(t)^{k+1}-1}{k+1}, & k \neq -1 \leftrightarrow w \neq -1 \\ \ln a(t), & k = -1 \leftrightarrow w = -1 \end{cases}. \tag{3.2069}$$

Solving for $a(t)$ leads to

$$a(t) = \begin{cases} [(k+1)H_0(t-t_0)+1]^{\frac{2}{3(1+w)}}, & w \neq -1 \\ \exp H_0(t-t_0), & w = 1 \end{cases}. \tag{3.2070}$$

b) For the Hubble parameter, we find that

$$H(t) = \frac{\dot{a}}{a} = H_0 a^{k+1} = \frac{H_0}{1 + \frac{3}{2}(1+w)H_0(t-t_0)}, \quad w \neq -1. \tag{3.2071}$$

For $w = -1$, we find that $H(t) = H_0$, which is also equal to the above expression with $w = -1$ inserted. Thus, generally

$$H(t) = \frac{H_0}{1 + \frac{3}{2}(1+w)H_0(t-t_0)}. \tag{3.2072}$$

Note that $H(t)$ decreases with time whenever $w > -1$.

2.150

a) The matter part of the action is given by

$$S_m = \int \mathcal{L}\sqrt{|\bar{g}|}d^4x = \int \left[g^{\mu\nu}(\partial_\mu\phi)(\partial_\nu\phi) - V(\phi)\right]\sqrt{|\bar{g}|}d^4x. \tag{3.2073}$$

Taking the variation of this leads to

$$\delta S_m = \int \left[\frac{1}{2}(\partial_\mu\phi)(\partial_\nu\phi)\delta g^{\mu\nu} + g^{\mu\nu}(\partial_\mu\phi)(\partial_\nu\delta\phi) - V'(\phi)\delta\phi\right]\sqrt{|\bar{g}|}d^4x$$
$$+ \int \mathcal{L}\delta\sqrt{|\bar{g}|}d^4x. \tag{3.2074}$$

The equation of motion for the scalar field ϕ itself is found by varying the action with respect to ϕ only. From the variation of the action derived in a), we then find

$$\delta S_m = \int \left[g^{\mu\nu}(\partial_\mu \phi)(\partial_\nu \delta\phi) - V'(\phi)\delta\phi\right]\sqrt{|g|}d^4x = 0. \tag{3.2075}$$

Partial integration of the first term leads to

$$\delta S_m = -\int \delta\phi\left[\partial_\nu\left\{g^{\mu\nu}\sqrt{|g|}\partial_\mu\phi\right\} + V'(\phi)\sqrt{|g|}\right]d^4x = 0. \tag{3.2076}$$

Since this should hold regardless of the variation $\delta\phi$, we conclude that

$$\frac{1}{\sqrt{|g|}}\partial_\nu\left(g^{\mu\nu}\sqrt{|g|}\partial_\mu\phi\right) = g^{\mu\nu}\nabla_\mu\nabla_\nu\phi = -V'(\phi). \tag{3.2077}$$

This is the equation of motion for ϕ.

b) With $\partial_i\phi = 0$ and $\partial_t\phi = \dot{\phi}$ and using the result of Problem 2.60, we find that

$$T_{\mu\nu} = U_\mu U_\nu \dot{\phi}^2 - g_{\mu\nu}\left[\frac{1}{2}\dot{\phi}^2 - V(\phi)\right], \tag{3.2078}$$

where $U_\mu = \delta^0_\mu$. We note that $g^{\mu\nu}U_\mu U_\nu = 1$ and U is therefore a 4-velocity field. Comparing this to the stress–energy tensor of an ideal fluid $T_{\mu\nu} = U_\mu U_\nu(\rho_0 + p) - g_{\mu\nu}p$, we obtain

$$p = \frac{1}{2}\dot{\phi}^2 - V(\phi) \quad \text{and} \quad \rho_0 = \frac{1}{2}\dot{\phi}^2 + V(\phi). \tag{3.2079}$$

The equation-of-state parameter is therefore given by

$$w = \frac{p}{\rho_0} = \frac{\dot{\phi}^2 - 2V(\phi)}{\dot{\phi}^2 + 2V(\phi)}. \tag{3.2080}$$

Note that when $\dot{\phi}^2 \ll V(\phi)$, we find $w \simeq -1$. This is the equation-of-state parameter for a cosmological constant.

Bibliography

Textbooks

Bambi, Cosimo (2018) *Introduction to General Relativity: A Course for Undergraduate Student of Physics* (Springer)

Baumgarte, Thomas W. and Stuart L. Shapiro (2021) *Numerical Relativity: Starting from Scratch* (Cambridge University Press)

Baumgarte, Thomas W. and Stuart L. Shapiro (2010) *Numerical Relativity: Solving Einstein's Equations on the Computer* (Cambridge University Press)

Carroll, Sean M. (2019) *Spacetime and Geometry: An Introduction to General Relativity* (Cambridge University Press)

Cheng, Ta-Pei (2015) *A College Course on Relativity and Cosmology* (Oxford University Press)

Cheng, Ta-Pei (2013) *Einstein's Physics: Atoms, Quanta, and Relativity Derived, Explained, and Appraised* (Oxford University Press)

Cheng, Ta-Pei (2009) *Relativity, Gravitation and Cosmology: A Basic Introduction*, 2nd ed. (Oxford University Press)

Christodoulides, Costas (2016) *The Special Theory of Relativity: Foundations, Theory, Verification, Applications* (Springer)

D'Inverno, Ray (1992) *Introducing Einstein's Relativity* (Oxford University Press)

Dirac, Paul A. M. (1996) *General Theory of Relativity* (Princeton University Press)

Dunningham, Jacob and Vlatko Vedral (2018) *Introductory Quantum Physics and Relativity*, 2nd ed. (World Scientific)

Faraoni, Valerio (2013) *Special Relativity* (Springer)

Fischer, Kurt (2013) *Relativity for Everyone: How Space-Time Bends* (Springer)

Freund, Jürgen (2008) *Special Relativity for Beginners: A Textbook for Undergraduates* (World Scientific)

Gray, Norman (2019) *A Student's Guide to General Relativity* (Cambridge University Press)

Greiner, Walter (2004) *Classical Mechanics: Point Particles and Relativity* (Springer)

Gourgoulhon, Éric (2012) *3+1 Formalism in General Relativity: Bases of Numerical Relativity* (Springer)

Guidry, Mike (2019) *Modern General Relativity: Black Holes, Gravitational Waves, and Cosmology* (Cambridge University Press)

Günther, Helmut and Volker Müller (2019) *The Special Theory of Relativity: Einstein's World in New Axiomatics* (Springer)

Hartle, James B. (2003) *Gravity: An Introduction to Einstein's General Relativity* (Addison-Wesley)
Hobson, M. P., G. P. Efstathiou, and A. N. Lasenby (2006) *General Relativity: An Introduction for Physicists* (Cambridge University Press)
Khriplovich, I. B. (2005) *General Relativity* (Springer)
Lawden, Derek F. (2003) *Introduction to Tensor Calculus, Relativity and Cosmology* (Dover)
Misner, Charles W., Kip S. Thorne, and John Archibald Wheeler (2018) *Gravitation* (Princeton University Press)
Narlikar, Jayant V. (2010) *An Introduction to Relativity* (Cambridge University Press)
Nolting, Wolfgang (2017) *Theoretical Physics 4: Special Theory of Relativity* (Springer)
Padmanabhan, T. (2010) *Gravitation: Foundations and Frontiers* (Cambridge University Press)
Plebanski, Jerzy and Andrzej Krasinski (2006) *An Introduction to General Relativity and Cosmology* (Cambridge University Press)
Poisson, Eric (2004) *A Relativist's Toolkit: The Mathematics of Black-Hole Mechanics* (Cambridge University Press)
Rindler, Wolfgang (2006) *Relativity: Special, General, and Cosmological*, 2nd ed. (Oxford University Press)
Rindler, Wolfgang (1991) *Introduction to Special Relativity*, 2nd ed. (Oxford University Press)
Ryder, Lewis (2009) *Introduction to General Relativity* (Cambridge University Press)
Schutz, Bernard (2009) *A First Course in General Relativity*, 2nd ed. (Cambridge University Press)
Schutz, Bernard (2003) *Gravity from the Ground Up: An Introductory Guide to Gravity and General Relativity* (Cambridge University Press)
Schwarz, Patricia M. and John H. Schwarz (2004) *Special Relativity: From Einstein to Strings* (Cambridge University Press)
Stephani, Hans (2004) *Relativity: An Introduction to Special and General Relativity*, 3rd ed. (Cambridge University Press)
Straumann, Norbert (2013) *General Relativity*, 2nd ed. (Springer)
Takeuchi, Tatsu (2010) *An Illustrated Guide to Relativity* (Cambridge University Press)
Wald, Robert M. (1984) *General Relativity* (University of Chicago Press)
Walecka, John D. (2007) *Introduction to General Relativity* (World Scientific)
Weinberg, Steven (2013) *Gravitation and Cosmology: Principles and Applications of the General Theory of Relativity* (Wiley)
Wiltshire, David L., Matt Visser, and Susan M. Scott (eds.) (2009) *The Kerr Spacetime: Rotating Black Holes in General Relativity* (Cambridge University Press)
Woodhouse, Nicholas M. J. (2003) *Special Relativity* (Springer)

Problem Books

Lightman, Alan P., William H. Press, Richard H. Price, and Saul A. Teukolsky (2018) *Problem Book in Relativity and Gravitation* (Princeton University Press)
Moore, Thomas A. (2012) *A General Relativity Workbook* (University Science Books)
Scott, Robert B. (2016) *A Student's Manual for a First Course in General Relativity* (Cambridge University Press)
Tsamparlis, Michael (2019) *Special Relativity: An Introduction with 200 Problems and Solutions* (Springer)

Index to the Problems and Solutions

Printed in the United States
by Baker & Taylor Publisher Services